Male Infertility: Understanding, Causes and Treatment

Rajender Singh • Kiran Singh
Editors

Male Infertility: Understanding, Causes and Treatment

Springer

Editors
Rajender Singh
Endocrinology Division
Central Drug Research Institute
Lucknow, Uttar Pradesh
India

Kiran Singh
Department of Molecular and Human Genetics
Institute of Science, Banaras Hindu University
Varanasi, Uttar Pradesh
India

ISBN 978-981-10-4016-0 ISBN 978-981-10-4017-7 (eBook)
DOI 10.1007/978-981-10-4017-7

Library of Congress Control Number: 2017943463

Printed on acid-free paper

This Springer imprint is published by Springer Nature
The registered company is Springer Nature Singapore Pte Ltd.
The registered company address is: 152 Beach Road, #21-01/04 Gateway East, Singapore 189721, Singapore

Preface

Male infertility presents a spectrum of phenotypes often with a complex etiology. About 15% of couples worldwide suffer from infertility, and male factors contribute to roughly one third, female factors contribute to another one third, and the remaining are due to combined male and female factors. In-depth physical and physiological examinations reveal exact etiology in only a few cases, resulting in the classification of the remaining as idiopathic. Due to complex and partially understood etiology, the treatment of male infertility is not straightforward. In the cases with a defined etiology, specific treatments offer hope, but in the cases with idiopathic infertility, empirical and generalized treatments are advised often. This puts impetus on a thorough understanding of the process of spermatogenesis and fertility to craft new avenues for infertility treatment.

This book aims at providing a comprehensive coverage of male infertility. It is divided in three sections; the first section introduces the reader to spermatogenesis and male fertility to provide a reasonable understanding of spermatogenic failure and male infertility. This covers the overview of the male reproductive system, its genesis, sperm production, maturation, and post-ejaculation changes that are necessary for male fertility. The second section deals with a thorough coverage of the known and plausible causes of male infertility. The foremost among these are genetic causes, such as Y deletions and other gene mutations, cytogenetic defects, and congenital syndromic forms of male infertility. Among environmental and lifestyle factors, obesity, oxidative stress, and sexually transmitted infections are discussed. The latest developments in the genetic, epigenetic, and proteome-related causes of male infertility have been covered toward the end of this section.

The third section of the book is dedicated to the management of male infertility. Since the etiology of infertility is complex, a number of different therapeutic or prophylactic measures are advised depending upon the severity of the disorder and the depth of investigation. This section entails nutritional, lifestyle, and other prophylactic measures that can be adopted to avoid loss of fertility. Other chapters in this section emphasize on specific and empirical treatments of male infertility. Toward the end, fertility preservation options for cancer patients are detailed. Upon failure of most of the treatments, ARTs are suggested, which have revolutionized the field of infertility treatment. A detailed description of ARTs is beyond the scope of this book; however, an overview of these techniques with opportunities and challenges has been discussed.

The book has been composed and designed to serve a broad reader base from basic scientists and postgraduate students to doctors in reproductive medicine. Most of the material is composed in a way that even the patients can read and understand it to benefit their fight against infertility. A thorough understanding of the disease is the key to successful treatment; therefore, the first section is highly relevant for clinicians and patients. The second section is largely related to the causes and would be apt for the researchers and postgraduate students who aim a career in the field of reproductive medicine. Since the third section largely deals with the therapeutic and prophylactic measures, it would cater to the patients and layman for adopting measures (nutritional, lifestyle, and prophylactic) to delay or avoid the development of infertility and to clinicians for offering counseling to infertility patients.

Uttar Pradesh, India Rajender Singh
Uttar Pradesh, India Kiran Singh

About the Editors

Rajender Singh is a senior scientist at the Central Drug Research Institute (CDRI), Lucknow, India. He obtained his doctoral degree in clinical reproductive genetics from the Centre for Cellular and Molecular Biology (CCMB), Hyderabad, India. Dr. Singh is working in the area of male reproductive biology for the last more than 10 years. He is a recipient of Young Scientist Awards from the Council of Scientific and Industrial Research (CSIR), the Indian National Science Academy (INSA), the Indian Science Congress Association (ISCA), and the Indian Society of Human Genetic (ISHG). He has also received the Innovative Young Biotechnologist Award (IYBA) from the Department of Biotechnology, Government of India. Dr. Singh is working on understanding and treatment of male infertility. His research interests include genetic and epigenetic basis of male infertility. He has published more than 100 research articles and has guided doctoral thesis students.

Kiran Singh is working as an assistant professor in the Department of Molecular and Human Genetics, Institute of Science, Banaras Hindu University, Varanasi, India. She has made significant contributions both to the teaching and research field. She teaches various courses such as clinical genetics, reproductive genetics, population genetics, cytogenetics, etc. to postgraduate students. Her research interests include reproductive health, particularly genetics of human male infertility and miscarriage. She has received honors and recognition from diverse scientific reproductive societies/institutes as invited speakers in national and international conferences, seminars, and meetings. Dr. Singh has supervised many Ph.D., M.D., and M.S. students for their thesis.

Contents

Contributors

Neeraj Kumar Agrawal Department of Endocrinology and Metabolism, Institute of Medical Sciences, Banaras Hindu University, Varanasi, Uttar Pradesh, India

Akhtar Ali Centre for Genetic Disorders, Institute of Science, Banaras Hindu University, Varanasi, Uttar Pradesh, India

N.H. Balasinor Neuroendocrinology, National Institute for Research in Reproductive Health (ICMR), Parel, Mumbai, India

Arnab Banerjee Department of Biological Sciences, BITS Pilani KK Birla Goa Campus, Sancoale, Goa, India

Sandeep Kumar Bansal Male Reproductive Biology Laboratory, CSIR-Central Drug Research Institute, Lucknow, Uttar Pradesh, India

Sudha Bhagwati Male Reproductive Biology Laboratory, CSIR-Central Drug Research Institute, Lucknow, Uttar Pradesh, India

Shilpa Bisht Laboratory for Molecular Reproduction and Genetics, Department of Anatomy, AIIMS, New Delhi, India

Rima Dada Laboratory for Molecular Reproduction and Genetics, Department of Anatomy, AIIMS, New Delhi, India

Rohit Kumar Deshmukh CSIR-Centre for Cellular and Molecular Biology, Hyderabad, Telangana, India

Sharvari Deshpande Neuroendocrinology, National Institute for Research in Reproductive Health (ICMR), Parel, Mumbai, India

Vidhu Dhawan Laboratory for Molecular Reproduction and Genetics, Department of Anatomy, AIIMS, New Delhi, India

Pawan K. Dubey Centre for Genetic Disorders, Institute of Science, Banaras Hindu University, Varanasi, Uttar Pradesh, India

Joseph R.D. Fernandes Department of Biological Sciences, BITS Pilani KK Birla Goa Campus, Sancoale, Goa, India

Avinash Gaikwad, M.Sc. Department of Anatomy and Developmental Biology, Monash University, Clayton, VIC, Australia

Rahul Gajbhiye, M.B.B.S., Ph.D. Andrology Clinic, Department of Clinical Research, National Institute for Research in Reproductive Health, Indian Council of Medical Research, Parel, Mumbai, India

Gopal Gupta Male Reproductive Biology Laboratory, CSIR-Central Drug Research Institute, Lucknow, Uttar Pradesh, India

Bhavana Kushwaha Male Reproductive Biology Laboratory, CSIR-Central Drug Research Institute, Lucknow, Uttar Pradesh, India

Priyanka Mishra Male Reproductive Biology Laboratory, CSIR-Central Drug Research Institute, Lucknow, Uttar Pradesh, India

Sweta Mohan Neuroendocrinology, National Institute for Research in Reproductive Health (ICMR), Parel, Mumbai, India

Sujit Kumar Mohanty Male Reproductive Biology Laboratory, CSIR-Central Drug Research Institute, Lucknow, Uttar Pradesh, India

Priyanka Parte Department of Gamete Immunobiology, National Institute for Research in Reproductive Health, Mumbai, Maharashtra, India

Aniket Patankar Department of Gamete Immunobiology, National Institute for Research in Reproductive Health, Mumbai, Maharashtra, India

Pranav Patni Male Reproductive Biology Laboratory, CSIR-Central Drug Research Institute, Lucknow, Uttar Pradesh, India

Vertika Singh Department of Molecular and Human Genetics, Institute of Science, Banaras Hindu University, Varanasi, Uttar Pradesh, India

Kiran Singh Department of Molecular and Human Genetics, Institute of Science, Banaras Hindu University, Varanasi, Uttar Pradesh, India

Rajender Singh Male Reproductive Biology Laboratory, CSIR-Central Drug Research Institute, Lucknow, Uttar Pradesh, India

Ashima Sinha Department of Biophysics, All India Institute of Medical Sciences, New Delhi, India

Archana Bharadwaj Siva CSIR-Centre for Cellular and Molecular Biology, Hyderabad, Telangana, India

Anima Tripathi Department of Pharmaceutics, IIT-BHU, Varanasi, Uttar Pradesh, India

Sameer Trivedi Department of Urology and Division of Andrology, Institute of Medical Sciences, Banaras Hindu University, Varanasi, Utter Pradesh, India

Rajesh Verma Department of Endocrinology and Metabolism, Institute of Medical Sciences, Banaras Hindu University, Varanasi, Uttar Pradesh, India

Savita Yadav Department of Biophysics, All India Institute of Medical Sciences, New Delhi, India

Abbreviations

a-CGH	Array-comparative genomic hybridization
ACTH	Adrenocorticotrophic hormone
AIDS	Acquired immune deficiency syndrome
AIS	Androgen insensitivity syndrome
AMH	Anti-mullerian hormone
AR	Androgen receptor
ART	Assisted reproductive technology
ASAs	Anti-sperm antibodies
AZF	Azoospermic factor
BMI	Basal metabolic index
BPA	Bisphenol A
CBAVD	Congenital bilateral absence of vas deferens
cKO	Conditional knockout
CMV	Cytomegalovirus
CNS	Central nervous system
CNVs	Copy number variations
COH	Controlled ovarian hyperstimulation
CREB	Cyclic AMP-responsive element binding
CSC	Cigarette smoke condensate
DFI	DNA fragmentation index
DSD	Disorder of sexual development
EDCs	Endocrine-disrupting compounds
EMWs	Electromagnetic waves
ER	Estrogen receptor
ESHRE	European Society of Human Reproduction and Embryology
FA	Fatty acid
FAS	Fetal alcohol syndrome
FISH	Fluorescence in situ hybridization
FSH	Follicle-stimulating hormone
FSHR	Follicle-stimulating hormone receptor
GIFT	Gamete intrafallopian transfer
GnRH	Gonadotropin-releasing hormone
GU	Gonorrhea urethritis
HBV	Hepatitis B virus

hCG	Human chorionic gonadotropin
HCMV	Human cytomegalovirus
HCV	Hepatitis C virus
HFCS	High-fructose corn syrup
HH	Hypogonadotropic Hypogonadism
HIV	Human immunodeficiency virus
hMG	Human menopausal gonadotropin
HPG	Hypothalamic–pituitary–gonadal axis
HPV	Human papillomavirus
HSV	Herpes simplex virus
ICSH	Interstitial cell-stimulating hormone
ICSI	Intracytoplasmic sperm injection
INSL3	Insulin-like factor 3
IUI	Intrauterine insemination
IVF	In vitro fertilization
KS	Klinefelter's syndrome
LH	Luteinizing hormone
MA	Maturation arrest
MACS	Magnetic activated cell sorting
MDA	Malondialdehyde
MESA	Microsurgical epididymal sperm aspiration
MIC	Minimum inhibitory concentration
MSCI	Meiotic sex chromosome inactivation
MSY	Male-specific region of Y chromosome
NAAT	Nucleic acid amplification test
NAC	N-acetylcysteine
NGS	Next-generation sequencing
NGU	Nongonococcal urethritis
NOA	Nonobstructive azoospermia
NOS	Nitrogen species
NRY	Non-recombining portion of Y chromosome
OAT	Oligoasthenoteratozoospermia
PAE	Paternal age effect
PCR	Polymerase chain reaction
PE	Premature ejaculation
PESA	Percutaneous epididymal sperm aspiration
PFOA	Perfluorooctanoic acid
PGC	Primordial germ cell
PGD	Prenatal genetic diagnosis
PTCs	Peritubular cells
PTMs	Posttranslational modifications
PUFA	Polyunsaturated fatty acid
PZD	Partial zona dissection
QTL	Quantitative trait loci
RE	Retrograde ejaculation

RF	Radio frequency
ROS	Reactive oxygen species
RTI	Reproductive tract infection
SAM	S-Adenosylmethionine
SAXI	Sexual antagonism-driven X inactivation hypothesis
SCOS	Sertoli cell-only syndrome
SNP	Single nucleotide polymorphism
SOD	Superoxide dismutase
SRY	Sex-determining region Y
SSCs	Spermatogonial stem cells
STD	Sexually transmitted disease
STIs	Sexually transmitted infections
STSs	Sequence-tagged sites
SUZI	Subzonal insemination
TDS	Testicular dysgenesis syndrome
TESA	Testicular sperm aspiration
TESE	Testicular sperm extraction
WHO	World Health Organization
WT	Wilms' tumor
XCI	X-chromosome inactivation
YCC	Y Chromosome Consortium
ZIFT	Zygote intrafallopian transfer

Part I

Understanding Spermatogenesis and Male Fertility

Overview of the Male Reproductive System

1

Sujit Kumar Mohanty and Rajender Singh

Abstract

The human reproductive system consists of primary and secondary organs to facilitate the process of reproduction. The male reproductive system is specialized for the production of male gametes and their transportation to the female reproductive tract that is mediated by supporting fluids and production of testosterone. The organs of the male reproductive system consist of the paired testis (site of testosterone and sperm production), scrotum (compartment for testis localization), epididymis, vas deferens, seminal vesicles, prostate gland, bulbourethral gland, ejaculatory duct, urethra, and penis. The accessory organs facilitate the process of sperm maturation and transportation. Sperm with the secretions of seminal vesicles, prostate, and bulbourethral glands constitute semen (the ejaculate). The penis and urethra help in delivering the ejaculate to the female reproductive tract. This chapter provides an introduction to the male reproductive system and its functions.

Keywords

Male reproductive system • Male reproductive organs • Spermatogenesis and fertility • Testicular cells • Sperm production

Key Points

- Testes are the primary male sex organs as they are sperm factory.
- The highest numbers of cell divisions in human body take place in the testes, producing about 4.7 million sperm/g testes everyday.

S.K. Mohanty • R. Singh (✉)
Male Reproductive Biology Laboratory, CSIR-Central Drug Research Institute, Lucknow, Uttar Pradesh, India
e-mail: rajender_singh@cdri.res.in

R. Singh, K. Singh (eds.), *Male Infertility: Understanding, Causes and Treatment*, DOI 10.1007/978-981-10-4017-7_1

- A number of accessory glands, such as the prostate, seminal vesicle, and bulbourethral gland, pour their secretions that mix with sperm mass to constitute the ejaculate (semen).
- An average human ejaculate contains about 200 million sperm.

1.1 Origin of the Reproductive Organs

The reproductive system is developed from the embryonic intermediate mesoderm. The development of permanent reproductive organs of the adults starts within the bipotential gonad in the embryonic stage. The bipotential gonad, capable of forming both male and female structures, is made up of embryonic structures containing several ducts and can differentiate into male or female structures under the effect of several developmental sex-specific gene expressions. Some of the ducts disappear just before the end of fetal life. These embryonic structures are the Wolffian and Müllerian ducts, also known as mesonephric and paramesonephric ducts, respectively. The Wolffian ducts (mesonephric duct) develop as male reproductive organs, while the Müllerian ducts give rise to the female reproductive organs. The alternate ducts regress depending upon sex of the fetus.

1.2 Anatomy and Physiology of the Male Reproductive System

1.2.1 Scrotum

Testes are located behind the penis in a pouch of skin-covered, highly pigmented, muscular sac, called the scrotum (Fig. 1.1). The unique muscles (dartos muscle and cremaster muscle) make up the wall of scrotum. These muscles are involved in contracting and relaxing the testicles (also called as testes), moving them closer to the

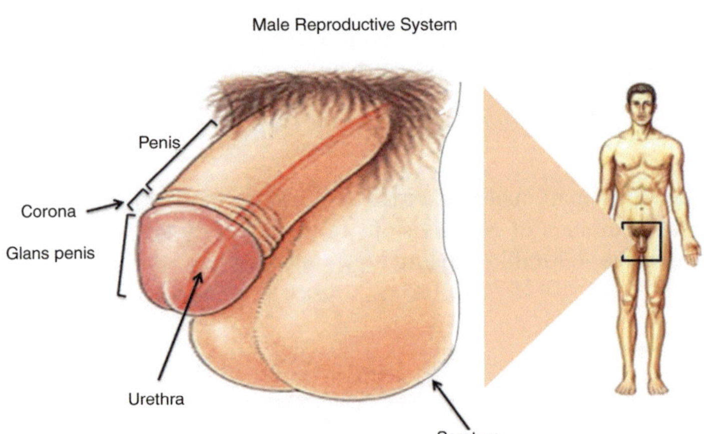

Fig. 1.1 External view of the male reproductive system

body to decrease the surface area to retain heat or far away from the body to increase the scrotal surface area, which promotes heat loss. When the testes are closer to the body, they become warm to support spermatogenesis; alternatively, when the core body temperature increases above the ideal range of spermatogenesis, the wall of the scrotum relaxes to move testes far from the body.

1.2.2 Testes

The testes are the male gonads responsible for the production of both sperm and testosterone and are active throughout the reproductive lifespan of a male. Each testis is found inside its own pouch on one side of the scrotum and is connected to the scrotum by a spermatic cord and cremaster muscles, the muscle layer of scrotum. Each testis is internally divided into 300–400 structures, known as lobules (Fig. 1.2). Each lobule contains a section of seminiferous tubules lined with epithelial cells. Within the seminiferous tubule, the epithelial cells contain stem cells that divide and differentiate into spermatozoa through the process of spermatogenesis (Goldstein and Schlegel 2013). The lumen of the seminiferous tubules helps in facilitating the development of spermatozoa at the hollow center of the tubules, where sperm are produced and pushed into the ductal systems of the testes. Exclusively, from the lumen of the seminiferous tubules, sperm travel toward the straight tubules and from there into a fine meshwork of tubules, called the rete testes (Shupnik and Schrefflofer 1997).

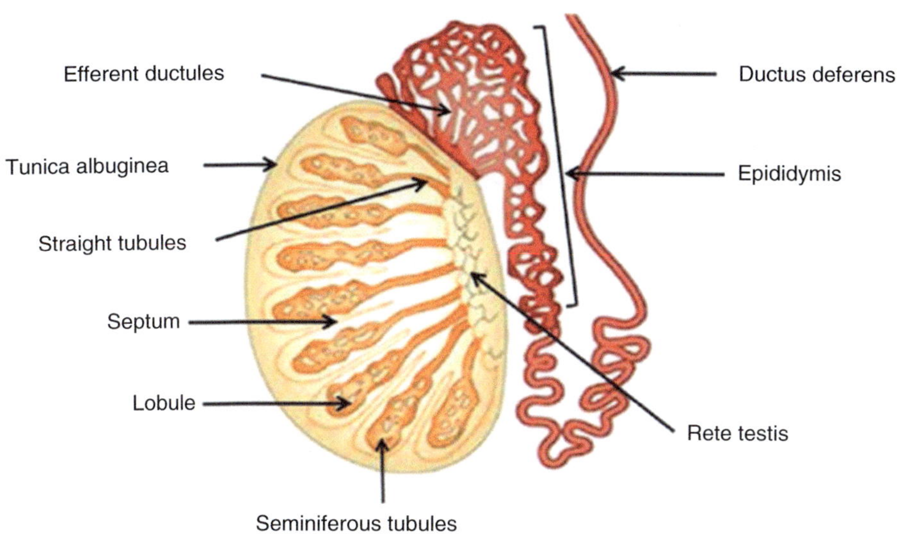

Fig. 1.2 A schematic view of the human testis showing seminiferous tubules (the site of sperm production), epididymis (the site of sperm maturation and storage), and vas deferens (the site of exit)

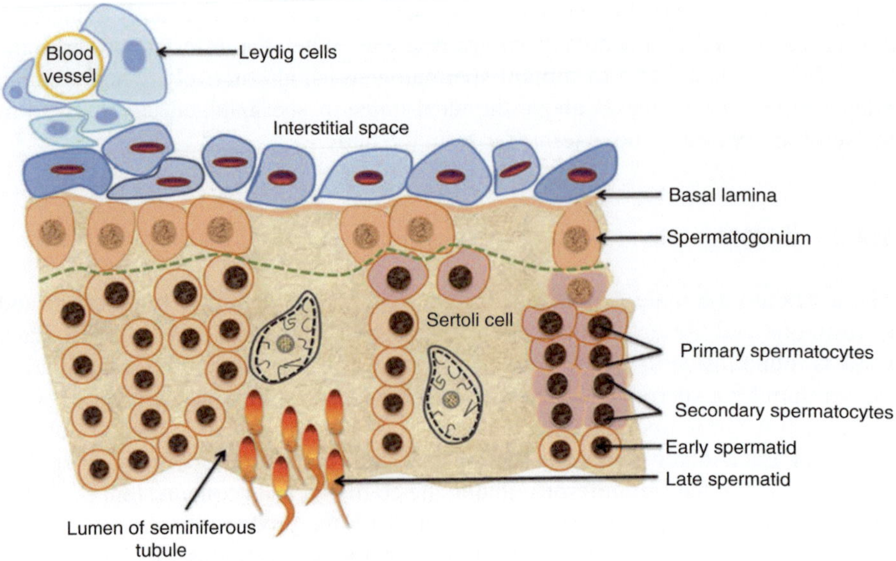

Fig. 1.3 A cross section view of seminiferous tubule showing the arrangement of somatic cells (Sertoli, Leydig, and interstitial) and various stages of germ cell development

1.2.2.1 Leydig Cells

The Leydig cells or interstitial cells are located adjacent to the seminiferous tubules in the space between the neighboring tubules in testis (Fig. 1.3). These cells possibly have their origin in the mesonephros and develop outside the testicular cord in the testes. Several gap junctions permit direct communication between the Leydig cells. Leydig cells can be differentiated from other testicular cells by the presence of round nucleus, prominent nucleolus, and crystals in the cytoplasm (due to the presence of cholesterol lipid droplets). Cholesterol in the cytoplasm of the Leydig cells is used for testosterone production (Haider 2004). The process does not begin until puberty when LH stimulates Leydig cell to produce and secrete testosterone. Testosterone, secreted by the Leydig cells, acts on the Sertoli cells, which in turn regulate the development, maturation, and differentiation of the germ cells.

1.2.2.2 Sertoli Cells

The Sertoli cells are a kind of sustentacular cells or nurse cells that are nondividing somatic cells that rest on the basement membrane and form the wall of the tubules (Fig. 1.3). Sertoli cells are unique as they have an irregular-shaped prominent nucleus, Sertoli germ cell connections, as well as unique tight junctional complexes between the adjacent Sertoli cell membranes. The Sertoli cells synthesize a variety of essential products necessary for germ cell survival, thus making a unique and favorable environment in the basal compartment for the maturation of germ cells. For the development and differentiation of the germ cells, the Sertoli cells perform a number of functions (Goldstein and Schlegel 2013). First, they provide physical support to the germ cells and facilitate their progression toward the lumen of the seminiferous tubules. Second, they provide a suitable microenvironment necessary for germ cell maturation and differentiation. Third, they form tight junctions and provide immune privilege to the postmeiotic germ cells. Fourth, during the maturation of germ cell, the Sertoli cells consume the unneeded portion of spermatozoa by the process of phagocytosis.

1.2.2.3 Blood–Testis Barrier

The blood–testis barrier is ideally termed as Sertoli cell barrier. This barrier splits the seminiferous epithelium into the basal and apical (adluminal) compartments. The adluminal compartment contains postmeiotic germ cells and spermatozoa, but the basal compartment contains spermatogonia (mother stem cells) and immature spermatocytes (Dym and Fawcett 1970). The functional components of the barrier include the tight junctions (TJ), the basal ectoplasmic specialization (basal ES), the basal tubulobulbar complex (basal TBC) (both are testis-specific actin-based adherens junction [AJ] types), and the desmosome-like junctions, which are present adjacent to the seminiferous epithelium. The blood-testis barrier serves a number of functions: (1) sequester the postmeiotic germ cells from the immune system to avoid elicitation of immune response, (2) regulate the entry of nutrients and other molecules from the blood stream to the testes, and (3) maintain a gradient of biochemical composition between the two compartments.

1.2.2.4 Germ Cells

The mother germ cells, the spermatogonia, line the basement membrane inside the tubule (Fig. 1.3). Spermatogonia are the stem cells, also called as spermatogonial stem cells (SSCs), line the periphery of seminiferous tubules. SSCs give rise to other spermatogonia (2n) by the process of mitosis. Spermatogonia give rise to primary spermatocytes, which upon meiosis give rise to secondary spermatocytes and spermatids (1n). Spermatids ultimately differentiate into spermatozoa (1n).

1.2.2.5 Spermatogenesis

Spermatogenesis is the process of production of spermatozoa from the male primordial germ cells (spermatogonia) in the testes (Fig. 1.4). Spermatogonia, the most immature male germ cells, lie along the basement membrane of the tubule in the basal compartment. The formation of highly specialized spermatozoa from the spermatogonia requires approximately 64 days. The first mitotic division takes place in the fetal testis, producing the spermatogonia and primary spermatocytes that are present at birth, and the complete functional sperm are not formed until the onset of puberty. These spermatogonial cells undergo several mitotic divisions to produce large number of cells that will either participate in stem cell renewal or go to produce daughter cells, which will later become spermatocytes. There are two meiotic divisions involving primary and secondary spermatocytes (2n) that ultimately give rise to four haploid spermatids. After

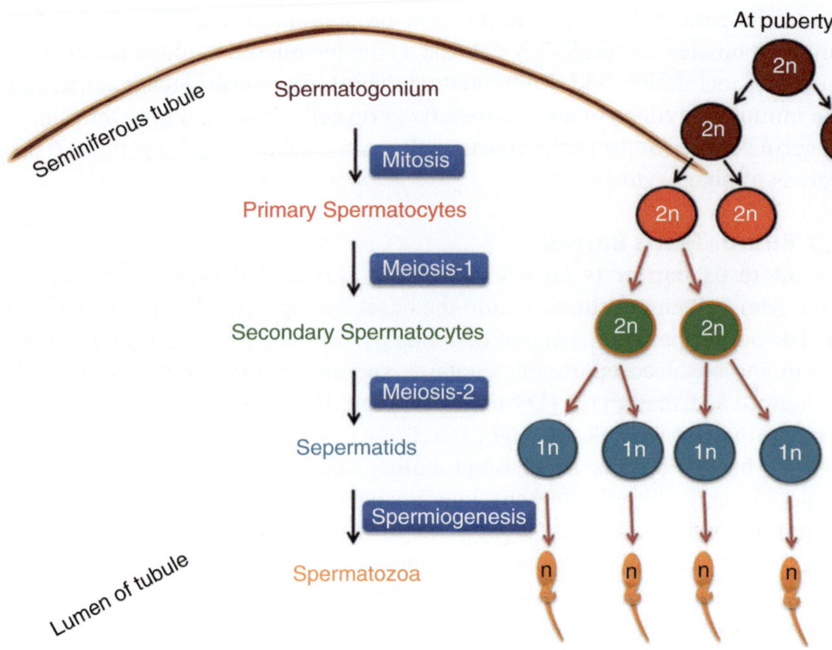

Fig. 1.4 An overview of the process of spermatogenesis

metamorphosis, the round spermatids shed most of the cytoplasm to become elongated mature spermatozoa that are capable of motility, and this process of differentiation is known as spermiogenesis (Clermont 1972). This transformation consists of the development of the acrosome, condensation of chromatin, formation of the flagellum, and migration of cytoplasmic organelles. The spermatozoa, thus formed, are released into the epididymis where they complete their maturation and gain motility.

1.3 Testosterone

The cluster of the Leydig cells, which resides in the interstitial space created by the adjacent seminiferous tubules, produces testosterone, a primary androgen that is required for spermatogenesis and development and maintenance of male secondary sexual characters. The secretion of testosterone by the Leydig cells occurs by the seventh week of development in the male embryos. This initial release of testosterone results in the anatomical differentiation of the male reproductive organs. To ensure the proper functionality of the male reproductive system, a continuous and regulated secretion of testosterone is essential. A sustained release of the normal concentration of testosterone promotes spermatogenesis, whereas low level of testosterone may lead to male infertility. In the testis, high local concentration of testosterone is required to promote spermatogenesis (Shupnik and Schrefflofer 1997). Moreover, testosterone is also released into the blood circulation and plays a significant role in muscle development, bone growth, and the development of secondary sex characteristics. The regulation of testosterone concentration throughout the body is significant for male reproductive functions. In the brain, the hypothalamus and pituitary gland control the production of testosterone. The gonadotropin-releasing hormone (GnRH) secreted from the hypothalamus binds to the GnRH receptors on the anterior pituitary gland to stimulate the release of luteinizing hormone (LH) and follicle-stimulating hormone (FSH). These two hormones play a significant role for reproductive functions in both males and females. In males, FSH promotes spermatogenesis, and LH upon binding to its receptors on the Leydig cells, stimulating the production of testosterone.

1.4 Sperm Transport

1.4.1 Epididymis

The epididymis is made up of coiled tubes that enclose around the superior and posterior edges of the testes where newly formed sperm continue to mature (Fig. 1.2). It takes an average of 12 days for sperm to travel through the coil of epididymis. As they travel along the length of the epididymis, sperm mature further and acquire the ability to move under their own power (Bedford et al. 1994). The length of epididymis delays the release of immature sperm, providing them a proper time to mature. The mature sperm are stored in the distal end of epididymis until ejaculation occurs.

1.4.2 Duct System

During ejaculation, the mature sperm come out of the epididymis and move toward the vas deferens with the help of smooth muscle contraction (Figs. 1.2 and 1.5). The vas deferens is a thick, muscular tube that is packed together inside the scrotum with connective tissue, blood vessels, and nerves into a structure called the spermatic cord. Each vas deferens of the epididymis extends into the abdominal cavity through the inguinal canal, located in the abdominal wall. From here, the vas deferens prolongs to the pelvic cavity and finally reaches behind the bladder, where these end in a region called the ampulla (Goldstein and Schlegel 2013). The mass of semen is produced with the help of three accessory glands of the male reproductive system: the seminal vesicle, the prostate, and the bulbourethral glands.

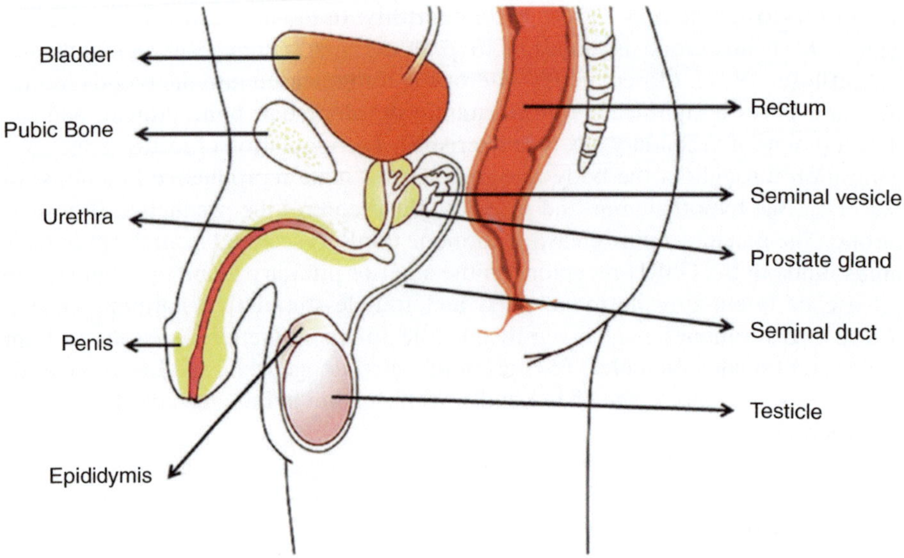

Fig. 1.5 Cross-sectional view of the internal and external organs of the male reproductive system

1.5 Seminal Vesicle

As sperm reach the ampulla from the vas deferens at the time of ejaculation, they mix with fluid from the associated seminal vesicle (Fig. 1.5). The seminal vesicles are basically glands that add approximately 60% of the semen volume. The fluid contains maximum amount of sugars, which are used by sperm to generate ATP to permit movement through the female reproductive tract. Another major component of their secretion is seminogelin proteins that provide thickness to semen, required for coagulation immediately upon ejaculation (Goldstein and Schlegel 2013). The fluid, which now have both sperm and seminal vesicle secretions, moves into the associated ejaculatory duct, which is a small structure formed from the ampulla of the vas deferens and the duct of the seminal vesicle. The paired ejaculatory ducts transport the seminal fluid into the prostate gland.

1.6 Prostate Gland

It is a muscular gland that encloses the first inch of the prostatic urethra as it appears from the bladder (Fig. 1.5). The size of the gland is like a walnut and secretes an alkaline, milky fluid to the passing seminal fluid called as semen. The contraction of the smooth muscles of the prostate gland during ejaculation helps in discharging semen from the urethra (Goldstein and Schlegel 2013). The prostate gland adds a variety of proteases, which help in semen liquefaction to facilitate the release of sperm.

1.7 Bulbourethral Gland

The bulbourethral glands, also called as Cowper's glands, are located underneath the prostate gland that release a thick, salty fluid that lubricates the end of the urethra and the vagina and helps in cleaning the remaining urine from the penile urethra (Fig. 1.5). The fluid from these accessory glands is released in two phases, the first phase after the male becomes sexually stimulated and second phase shortly before the release of semen. The most interesting function of this gland during unfavorable environment in the female reproductive tract is that the alkalinity of seminal fluid helps neutralize the acidic vaginal pH and turns the environment favorable to permit sperm mobility.

1.8 Urethra

The urethra is the last part of the urinary tract that crosses the corpus spongiosum (Figs. 1.1 and 1.5). Urethral opening, known as meatus, lies on the top of the glans penis. It is both, a way for urine and for the ejaculation of semen.

1.9 Penis

The penis is the male external genitalia (Figs. 1.1 and 1.5). It is made up of special tissue that helps in erection and allows the process of insemination. The shaft of the penis surrounds the penile urethra. The shaft is composed of three column-like chambers of erectile tissue that cover the length of the shaft. Each of the two lateral chambers is called a corpus cavernosum (Goldstein and Schlegel 2013). Together, these make the bulk of the penis. The corpus spongiosum surrounds the spongy or penile urethra. The end of the penis, called the glans penis, has a high concentration of nerve endings, ensuing a very sensitive skin that influences the probability of ejaculation. The skin from the shaft extends down over the glans and forms a collar, called the prepuce (foreskin). The prepuce also contains a dense concentration of nerve ending, and both lubricate and protect the sensitive skin of the glans penis (Rowley et al. 1970).

1.10 Structure of Human Mature Sperm

Spermatozoon is a unique cell that is capable of limited independent survival and motility that is dispensed out of human body (Fig. 1.6). Each day, approximately 100–300 million sperm are produced. A sperm cell is classified into distinct regions on the basis of their appearance and function. The head of sperm contains an elongated haploid nucleus with 22 autosomes and an X or Y chromosome with a very small amount of cytoplasm. The anterior portion, covered by a cap-like structure, is called as acrosome, which is filled with lysozymes important for penetrating egg

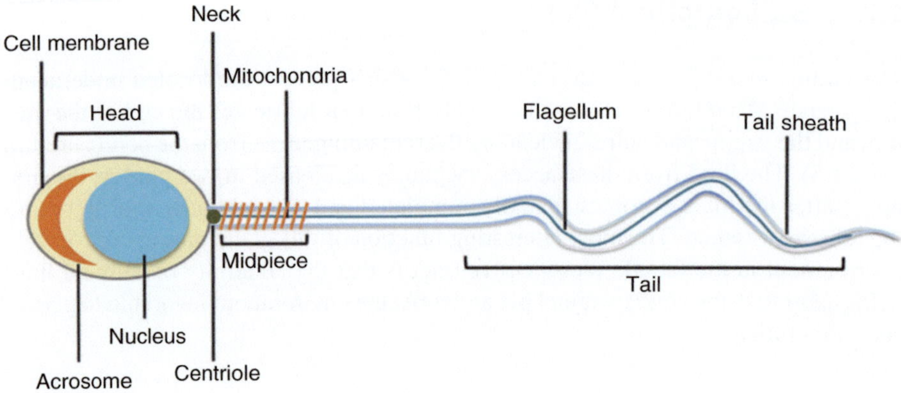

Fig. 1.6 Structure of human spermatozoa showing head (with nucleus and acrosome), mid-piece (with sheet of mitochondria), and tail that provides motility

during fertilization (Clermont 1972). The mid-piece possesses tightly packed mitochondria, which produce energy in the form of ATP for movement of tail that enables sperm motility essential for fertilization. The tail is made up of flagella, which extends from the neck and mid-piece, and consists of typical 9 + 2 microtubule arrangement.

Acknowledgments The authors would like to thank the Council of Scientific and Industrial Research (CSIR), Govt. of India for financial help under the network project, PROGRAM (BSC0101).

References

Bedford JM (1994) The status and the state of the human epididymis. Hum Reprod 9(11):2187–2199

Bedford JM, Calvin H, Cooper GW (1973) The maturation of spermatozoa in the human epididymis. J Reprod Fertil Suppl 18:199–213

Clermont Y (1972) Kinetics of spermatogenesis in mammals: seminiferous epithelium cycle and spermatogonial renewal. Physiol Rev 52(1):198–236

Dym M, Fawcett DW (1970) The blood–testis barrier in the rat and the physiological compartmentation of the seminiferous epithelium. Biol Reprod 3(3):308–326

Goldstein M, Schlegel P (2013) Surgical and medical management of male infertility. Cambridge University Press, Cambridge

Haider SG (2004) Cell biology of Leydig cells in the testis. Int Rev Cytol 233:181–241

Rowley MJ, Teshima F, Heller CG (1970) Duration of transit of spermatozoa through the human male ductular system. Fertil Steril 21(5):390–396

Shupnik MA, Schrefflofer DA (1997) Molecular aspects of steroid hormone action in the male reproductive axis. J Androl 18(4):341–344

Embryonic Development of the Testis

2

Pranav Patni, Sujit Kumar Mohanty, and Rajender Singh

Abstract

Testicular development is a very interesting aspect of male reproduction and fertility. The primordial germ cells migrate from yolk sac to the genital ridge along the wall of the hindgut and the dorsal mesentery to ultimately settle in the genital ridge that would give rise to testes. Sex-determining region of the Y chromosome (SRY) is the principal driver of testes development. Testes development culminates into descent of fully formed testes in the scrotum, which is necessary for facilitating spermatogenesis. Failure of testicular descent results in their retention in the inguinal canal or abdomen, often associated with azoospermia and infertility. This chapter provides a brief overview of the process of testicular development and descent with a glimpse of the consequences of the failure of testicular descent.

Keywords

Testis development • Germ cell migration • Testes descent • Cryptorchidism Azoospermia

Key Points

- The testes differentiate themselves earlier than the ovaries, namely, in the course of the 7th week.
- SRY gene on the Y chromosome is the primary driver for testis development, which in tum drives male sexual differentiation.
- Interestingly, the primordial germ cells migrate from yolk sac to the genital ridge along the wall of the hindgut and the dorsal mesentery.

P. Patni • R. Singh (✉) • S.K. Mohanty
Male Reproductive Biology Laboratory, CSIR-Central Drug Research Institute,
Lucknow, Uttar Pradesh, India
e-mail: rajender_singh@cdri.res.in

R. Singh, K. Singh (eds.), *Male Infertility: Understanding, Causes and Treatment*, DOI 10.1007/978-981-10-4017-7_2

- Migration of testis to the scrotum completes by the end of the 8th month, which is followed by small changes up to shortly after birth.
- Failure of testes descent can result in a number of abnormal positions of testes from inguinal canal to abdomen, called as cryptorchidism, which is often associated with azoospermia.

2.1 Introduction

Functional gonads are essential for sexual reproduction and the survival of higher animal species. The development of gonads is a particularly interesting event that is highly orchestrated in the form of origin, migration, and final settlement of the germ cells in the gonads. All these critical events take place in a tightly regulated temporal fashion. In humans, the sexual development and differentiation takes place at three levels: (1) chromosomal level, which is decided at the time of gamete fusion; (2) gonad level, which is the development of either testes or ovaries; and (3) physiological level, which is the development of secondary sexual organs and characters under the influence of hormones secreted by gonads. The primordial gonads are bipotential and have the capability to differentiate into ovaries or testis depending upon the molecular signals. The identification of the testis-determining gene, SRY, was a major discovery in the signals that set the path for development of either testis or ovaries. SRY is believed to be the master regulator of gonadal development, the absence of which results in ovarian development. Though a number of genes have been identified to be important for ovarian development, no master regulator of ovarian development has been identified.

In humans, the first important step of sexual differentiation takes place during the initial 7 weeks of the embryonic development that consists of several successive events starting with the establishment of genetic sex, development of the gonadal ridge, and immigration of primordial germ cells trailed by a sexually dimorphic differentiation of the gonadal anlagen into either testes or ovaries. Until this point of time, it is denoted as the indifferent stage of gonadal development, and no morphologically distinct sex differences can be observed in the developing human gonads. This developmental phase has a major influence on the later events of male as well as female paths since it establishes the hormonal dimorphism. This chapter details the differentiation of male gonads, covering the events from its first appearance through maturation to ultimate migration in the scrotum.

2.2 Overview of the Development of the Testes

The differentiation of the testes takes place more quickly than ovaries in the course of the 7th week (44 days). SRY is the principal driver gene for testicular differentiation. SRY initiates a series of gene expression, ultimately helping testes differentiation. The primordial germ cells are the bipotential cells that can give rise to both spermatozoa and oocytes. These are diploid like all other somatic cells and can

Table 2.1 Chronology of important early events in human male sex differentiation

Event	Age at start (dpc)	Size CRL (mm)
Genetic sex	0	
PGC migration from yolk sac	28	4
Formation of gonadal ridge	32	5
PGCs reach gonadal ridge	37	10
Male sex determination	42	15
Leydig cells appear	55	30
Androgen, INSL3 detectable	63	40
Testicular descent (first phase)	67	

dpc days post conception, *CRL* crown rump length ("sitting height")

already be found in human embryos in the primary ectoderm (epiblast) in the 2nd week of gestation. The first step in the organogenesis of testes is the differentiation of Sertoli's supporting cells (Karl and Capel 1998). The appearance of these cells in mice comes from the gonadal ridge, precisely the pluripotent coelomic epithelial cells (Table 2.1). The gene expression events triggered by the SRY result in the formation of intercellular membrane connections that would surround the primordial germ cells. With this, the gonadal cords start rising into the medulla. In males, cells of the mesonephric origin start accumulating on the outer side of gonadal cords and form peritubular myocytes. Gonadal cords develop into testicular cords and later into seminiferous tubules. The efferent ductules develop link with rete testis and the mesonephric duct. The influence of testosterone toward the end of the 8th week directs tight coiling of the cranial part of the mesonephric duct to develop into epididymis, while the exterior part of the duct remains as the deferens duct.

Post 8th week, certain mesenchymal cells in the testicular cord develop into Leydig cells, which further drive testosterone production. The mesenchyme between the testicular cords leads to the development of septa that divide each testis into lobules. The exact origin of these cells remains unknown. The mesenchyme at this stage also forms connective tissue layer between the testicular cords and the future tunica albuginea. The coelomic epithelium finally transforms into a mesothelium, just like other cavities. Finally, the testes are migrated to the scrotum. Migration of the testis apparently involves two phases; the initial stage is transabdominal migration, and the second stage is passage through the inguinal canal.

2.3 Formation of the Primitive Gonads

By day 32 post conception (pc), the gonadal anlagen can be recognized as combined bipotential structures in the developing human embryo. No morphological sexual dimorphism can be seen at this stage of development. Primordial germ cells (PGCs), which develop into gonocytes later on, cannot be observed at this time of gonadal development (Shawlot and Behringer 1995; Torres et al. 1995; Park and Jameson 2005).

The mesonephros at this stage also has primordium of the adrenal glands and the urinary system. The development of the urogenital system is regulated mainly by two transcriptional regulators: Wilms' tumor-associated gene 1 (WT1) (tumor suppressor) and steroidogenic factor-1 (SF1) (the orphan nuclear receptor). WT1 is a DNA- and RNA-binding protein with transcriptional and posttranscriptional regulation capacity. It is expressed in gonadal stromal, coelomic epithelial and immature Sertoli cells, interacts with the cAMP-responsive element-binding protein CITED 2, and is regulated by "paired box gene 2" (PAX2). SF1, expressed in the gonadal ridge, is a transcriptional regulator of steroid hydrolases, gonadotropins, and aromatase and is involved in the stabilization of intermediate mesoderm, follicle development, and ovulation. Furthermore, SF1 helps in regulating the anti-Müllerian hormone (AMH), dosage-sensitive sex reversal congenital adrenal hypoplasia critical region on the X chromosome protein 1 (DAX1), and steroidogenic acute regulatory protein (StAR). Normally these genes are expressed in the somatic testicular compartment and are important for normal testicular cord formation, and for the beginning of steroidogenesis they help in the differentiation of the Leydig and Sertoli cells. Sf1 knockout in mice results in the failure of gonadal and adrenal development, whereas the corresponding loss of function mutations in humans has a less prominent gonadal phenotype and adrenal insufficiency (Biason-Lauber and Schoenle 2000; Achermann et al. 2002; Park and Jameson 2005).

2.4 Cell Lineages

2.4.1 Primordial Germ Cells

By the end of the 5th week of conception, three different lineages of somatic cell types with bipotential fate dependent on their future paths constitute the gonadal anlagen. At this stage, immigrating primordial germ cells (PGCs) are colonizing the gonadal structures. After the final localization in the gonad, they are specified as gonocytes. The PGCs differentiate from epiblast-derived stem cells in the yolk sac. Expression of alkaline phosphatase, OCT3/4, and c-kit by the germ cells at this stage can be used to differentiate them from other cells. The PGCs migrate to the genital ridge under guidance of the extracellular matrix proteins expressed along the dorsal mesentery of the hindgut (Fig. 2.1). During the transit, the PGCs undergo active mitotic proliferation and expand in numbers by the time they reach gonadal anlagen. (Bendel-Stenzel et al. 1998; Wylie 1999). Gonocytes continue proliferation in the early testis shortly after determination and later become mitotically quiescent. The entry into meiosis is not allowed until much later in time. This decision is governed by the somatic cells since XY PGCs residing in an ovary follow the female path (McLaren and Southee 1997). Nevertheless, entry into meiosis may also be activated by mechanisms intrinsic to the germ cells (Morelli and Cohen 2005). Differentiation of the somatic cells (Leydig and Sertoli cells) in the male gonad continues even in

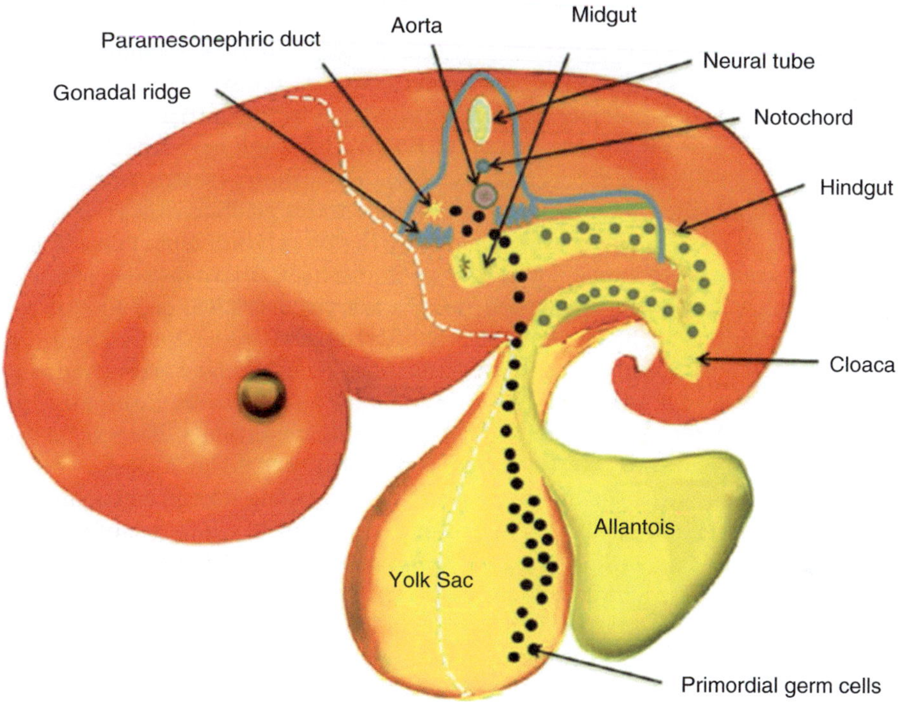

Fig. 2.1 Migration of the primordial germ cells from yolk sac to the genital ridge along the wall of hindgut and the dorsal mesentery

the absence of germ cells. This results in testosterone production that ensures pubertal development, but they are infertile due to Sertoli-cell-only syndrome (SCOS) (Söder 2007).

2.4.2 Somatic Cell Lineages in the Male Testis

Upon completion of the 6th week of embryonic development, four cell lineages consisting of Sertoli cells, Leydig cells, and peritubular cells and gonocytes can be identified in the indifferent gonad. The crucial somatic cell lineages are Sertoli cells, Leydig cells, and peritubular cells. Failure of differentiation and function of any of these lineages would result in severe phenotypes with respect to adult gonadal function and fertility.

2.4.2.1 Sertoli Cells

Sertoli cells are essential for testicular histogenesis and future functions. In adult testis, the Sertoli cells are nurse cells for spermatogenesis, creating niches for differentiation of spermatogonial stem cells and providing structural support, nutrients, and growth factors for the developing germ cells. Due to the fact that sperm output in the adult testis is related to the number of Sertoli cells, the control of Sertoli cell proliferation in the developing testis is very important for future production of male germ cells (Petersen and Söder 2006). Pituitary follicle-stimulating hormone (FSH) and its receptor (FSHR) are important factors for Sertoli cell development. Any functional impairments of the FSHR may result in reduced fertility, a reduction of Sertoli cell numbers and, therefore, a reduction of testicular size combined with a reduction in circulating testosterone levels (Huhtaniemi et al. 1987; Simoni et al. 1997; O'Shaughnessy et al. 2006). Sertoli cell differentiation and proliferation is one of the most important steps in male sex determination. The fetal hypothalamic-pituitary-gonadal (HPG) axis is not yet operative, and FSH is not available during the first phase of Sertoli cell differentiation (Söder 2007). Therefore, Sertoli cell proliferation at this stage is controlled by other regulators (Petersen et al. 2001, 2002; Petersen and Söder 2006). A number of endocrine disruptors and inflammatory factors can disrupt Sertoli cell differentiation and proliferation at this stage (Petersen et al. 2002, 2004; Petersen and Söder 2006; Söder 2007).

Pre-Sertoli cells are first defined as cells of the supporting lineage expressing the sex-determining region on the Y chromosome (SRY). After SRY expression, the SRY-related HMG box 9 (SOX9), a gene with predominantly testis promoting activity, is expressed by the Sertoli cells and that leads to an upregulation of AMH, fibroblast growth factor 9 (FGF9), and prostaglandin D2 (PGD2). These genes affect the differentiation of the reproductive tract and therefore define male sex determination. This procedure is quickly trailed by morphological changes in the primitive gonad, therefore, embracing testicular elements such as the arrangement of testicular lines.

2.4.2.2 Leydig Cells

In the developing male, Leydig cells constitute another crucial testicular cell lineage. Leydig cells originate from steroidogenic precursor cells that migrate from the coelomic epithelium and mesonephric mesenchyme to colonize the indifferent gonad (Merchant-Larios and Moreno-Mendoza 1998; O'Shaughnessy et al. 2006; Söder 2007). These cells start to proliferate and differentiate at week 7 of human embryonic development under the influence of the Sertoli cell signals, such as AMH, desert hedgehog (DHH), and FGF9 (Clark et al. 2000; Colvin et al. 2001). The first generation of the Leydig cells is fetal type, which appears after the testes determination. Other Leydig cell types appear before puberty and after achieving puberty (Ge et al. 2006; Colvin et al. 2001). At the 8th week of human gestation, fetal-type Leydig cells start producing testosterone and other androgens (Svechnikov et al. 2010). Initially, they are regulated by the placental human chorionic gonadotropin (hCG), which shares on Leydig cells signaling receptors with pituitary LH, though the latter appears much later in the development when the HPG axis becomes established in the beginning of the second trimester of human pregnancy. At mid gestation, they constitute 40% of the total testicular cell mass. Leydig cells are situated in the interstitial compartment of the testis and increase their number during the first 2–3 months after birth (Svechnikov et al. 2010).

In addition to testosterone, a crucial hormone for differentiation of male external and internal genitalia, Leydig cells also produce SF1 that is necessary for steroidogenesis (Achermann et al. 2002) and insulin-like factor-3 (INSL3). INSL3 and its receptor RXFP2, together with androgens and AMH, are involved in the process of testicular descent. The first transabdominal phase of testicular descent occurs in human fetuses during weeks 8–16. Apart from its role in testicular descent, INSL3 seems to be an important paracrine mediator in male gonad and serves as a useful marker of Leydig cell differentiation (Ferlin et al. 2006).

In the coelomic epithelium, adrenocortical and gonadal steroidogenic cells share an embryonic origin and exist as one lineage before divergence into the gonadal and adrenocortical paths (Mesiano and Jaffe 1997). Additionally, the expression of adrenocorticotropic hormone (ACTH) receptor on fetal Leydig cells makes ACTH (ACTH) an important regulator of Leydig cell development. A common origin of testicular and adrenocortical tissue is also supported by abnormalities that affect both these organs together (Stikkelbroeck et al. 2001). In a similar way to Sertoli cells, Leydig cells represent an obvious target of disruptive actions of xenobiotics and EDCs (Söder 2007). In adult animals, these cells demonstrate a large regenerative capacity. Several growth factors have been implicated in Leydig cell regeneration and survival (Yan et al. 2000). Yet it is not yet clearly identified if this regeneration is driven by the resident Leydig precursor cells. A second possible hypothesis suggests that peritubular testis cells also represent a reserve pool of steroidogenic cells.

2.4.2.3 Peritubular Cells

Peritubular cells (PTCs) are required for early histogenesis of the seminiferous cords along the basal membrane of the seminiferous tubuli. Together with the basal membrane and the Sertoli cells, they form the blood-testis barrier and provide physical support for the Sertoli cells. In the postpubertal testis, they are supposed to add contractile forces, which are thought to be necessary for pushing tubular fluid and sperm release (Söder 2007). By the chemotactic signals received from the Sertoli cells, early PTCs and the cells contributing to the vasculature of the testis migrate from the adjacent mesonephros (Cupp et al. 2003). This migration process is a crucial step in sex determination and is SRY dependent. Normal SRY expression is related to GATA4, a gene also expressed by the PTCs. GATA4 also activates steroidogenic genes such as StAR, CYP11A, CYP17, CYP19, and HSD3B2, which are mainly expressed in the Leydig cells. Considering this and the fact that they are highly proliferative cells, PTCs demonstrate important features for normal testis development (Capel et al. 1999; Schmahl and Capel 2003), but their precise role in adult testicular function is still not known. Data accumulated lately indicate their possible role as a reserve or stem cell pool (Haider et al. 1995) and that they might be involved in the regeneration of Leydig cells after a disruptive injury.

2.5 Testicular Descent

Function of the postpubertal testes is dependent on their scrotal position. The procedure of testicular descent consists of two phases: the first transabdominal phase of descent followed by the inguino-scrotal phase aiming to transfer the testes to a scrotal position (Fig. 2.2). The first phase begins soon after testis determination and

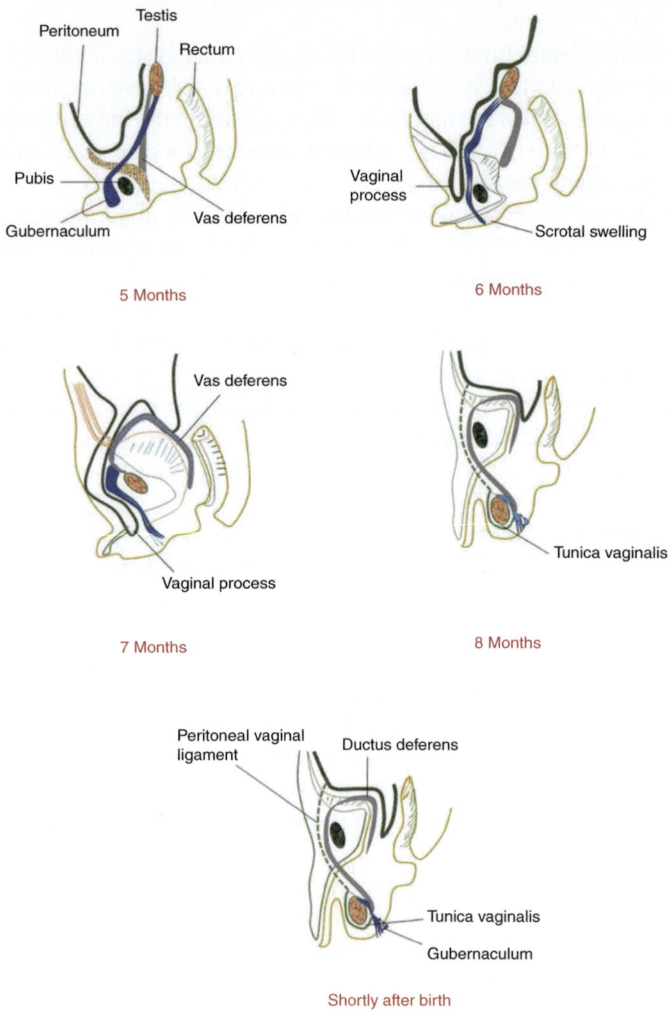

Fig. 2.2 The course of testicular descent: Between the 7th and the 12th week of gestation, the gubernaculum shortens and pulls the testes, the deferent duct, and its vessels onward. By the 6th month, the testes reach the orifice of the inguinal canal and cross it during the 7th month to reach their final position in the scrotum by the 8th month. After this, the inguinal canal contracts around the spermatic cord to complete the process. In the first year of life, the upper part of the vaginal process becomes obliterated, and peritoneo-vaginal ligament remains there. The lower portion persists as the tunica vaginalis testis

differentiation of Leydig cells and guides the testis from a position in the upper abdomen to the inner opening of the inguinal channel in the pelvic part of the abdomen. The testes with the epididymis and the proximal part of vas deferens finally move through the inguinal canal after week 18 of gestation. During the final 2 months of pregnancy, the testes usually take their scrotal position.

2.6 Perinatal Events in Testicular Maturation

During the third trimester of pregnancy, the fetal testes still produce large quantities of androgen but less than the peak activity at mid gestation. Although closer to birth, at term age, the hormonal activity of the testes declines, but it is still clearly measurable. However, soon after birth in both sexes of primates, but best recognized in human males, the first few months are a period of a very high hormonal activity of the testes and the hypothalamic-pituitary axis (Grumbach 2005). This period is often referred to as the mini-puberty and characterized by a hormonal surge of gonadotropins and testosterone. This is associated with proliferation of the Sertoli cells and some extent of germ cell development, i.e., transformation of gonocytes to Ad spermatogonia, at a time when gonadotropin, testosterone, and inhibin B reach high levels. More detailed studies have shown that LH value begins to increase 2 weeks after birth and decline to prepubertal values by 1 year of age in both sexes. FSH value also begins to increase 2 weeks after birth and decline to prepubertal levels by 1 year of age in boys and 2 years of age in girls. In parallel, testosterone level in boys often reach a peak of 10–15 nmol/L during the 2nd month of postnatal life but then decline to prepubertal low levels at 6 months of age (Forest 1975). The biological role of "mini-puberty" for future testicular and male reproductive function is unknown, but it has been speculated that it may play a role in the germ cell maturation and for the development of male gender identity.

2.7 Cryptorchidism: The Failure of Testicular Descent

Cryptorchidism or undescended testicle is a common developmental abnormality. Cryptorchidism is a stage in which testes fail to descend in the scrotal sac (Fig. 2.3). Since spermatogenesis takes place at a temperature 2–3 ° less than the body, failure of testicular descent leads to spermatogenic failure. The prevalence of cryptorchidism in the newborns is approximately 1–3%, but in premature children it increases to approximately 30% (Kolon et al. 2004). Cryptorchidism is of two types, unilateral and bilateral cryptorchidism. The prevalence of azoospermia in unilateral cryptorchidism is 13%, but this incidence increases up to 89% in bilateral cryptorchidism, suggesting that most of the cryptorchid individuals are azoospermic (Hadziselimovic and Herzog 2001). Bilateral cryptorchidism is more common as compared to unilateral cryptorchidism. The etiology of cryptorchidism is complex, involving a wide range of risk factors such as chromosomal, genetic and epigenetic alterations, hormonal imbalances, exposure to environmental toxicants, and the effect of endocrine disruptors.

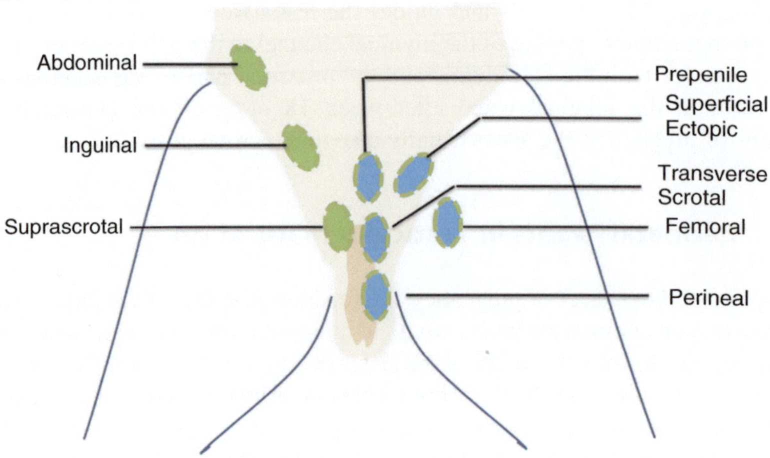

Fig. 2.3 Failure of testicular descent can take place at several levels, resulting in a variety of abnormal testicular positions from inguinal canal to abdominal. On the left side, normal testicular descent is shown (final scrotal position now shown), while the right side shows various positions of maldescent

2.8 Testicular Descent: Associated Disorders

Infertility is the primary disorder in cryptorchidism. Cryptorchidism is associated with spermatogenic alterations, which may range from normozoospermia to subfertility and azoospermia (Zimmermann et al. 1997, 1999). The severity of spermatogenic failure depends on the presence of unilateral or bilateral cryptorchid condition. Excryptorchid individuals may also display defects in spermatogenesis. A retrospective study described arthrogryposis multiplex congenita (AMC), a condition defined by the presence of multiple joint contractures at the time of birth to be associated with cryptorchidism (Fallat et al. 1992). The association of limb deformities was described by an external indirect compression of the inguino-scrotal region during the third trimester (Fallat et al. 1992).

Hypospadias, a congenital midline fusion defect of urethra leading to abnormal location of urethral meatus in males, is associated with increased risk of cryptorchidism (Tasian et al. 2010). Also, the incidence of hypospadias severity increased the risk of acquired cryptorchidism; however, the mechanism is still unexplained (Itesako et al. 2011). Further, the patients having disorder of sexual development (DSD) often have cryptorchid testis/gonads with ambiguous genitalia (Matsumoto et al. 2012). The risk of testicular cancer is 3–8 times high in cryptorchid individuals, and around 5–10% of patients with testicular cancer are excryptorchid (Whitaker 1988; Møller et al. 1996). It is thus an established risk factor for testicular germ cell tumor (TGCT). A recent study described that altered regulation of growth factor expression in the spermatogonial stem cell (SSC) somatic cell niche may impair the fine balance between SSC self-renewal and differentiation, which may drive the stem cells toward neoplastic transformation in cryptorchid individuals (Ferguson and Agoulnik 2015).

Acknowledgments The authors would like to thank the Council of Scientific and Industrial Research (CSIR), Govt. of India, for financial help under the network project, PROGRAM (BSC0101).

References

Achermann JC, Ozisik G, Ito M, Orun UA, Harmanci K, Gurakan B, Jameson JL (2002) Gonadal determination and adrenal development are regulated by the orphan nuclear receptor steroidogenic factor-1, in a dose-dependent manner. J Clin Endocrinol Metabol 87(4):1829–1833

Bendel-Stenzel M, Anderson R, Heasman J, Wylie C (1998) The origin and migration of primordial germ cells in the mouse. Semin Cell Dev Biol 9(4):393–400

Biason-Lauber A, Schoenle EJ (2000) Apparently normal ovarian differentiation in a prepubertal girl with transcriptionally inactive steroidogenic factor 1 (NR5A1/SF-1) and adrenocortical insufficiency. Am J Hum Genet 67(6):1563–1568

Capel B, Albrecht KH, Washburn LL, Eicher EM (1999) Migration of mesonephric cells into the mammalian gonad depends on Sry. Mech Dev 84(1):127–131

Clark AM, Garland KK, Russell LD (2000) Desert hedgehog (Dhh) gene is required in the mouse testis for formation of adult-type Leydig cells and normal development of peritubular cells and seminiferous tubules. Biol Reprod 63(6):1825–1838

Colvin JS, Green RP, Schmahl J, Capel B, Ornitz DM (2001) Male-to-female sex reversal in mice lacking fibroblast growth factor 9. Cell 104(6):875–889

Cupp AS, Uzumcu M, Skinner MK (2003) Chemotactic role of neurotropin 3 in the embryonic testis that facilitates male sex determination. Biol Reprod 68(6):2033–2037

Fallat ME, Hersh JH, Hutson JM (1992) Theories on the relationship between cryptorchidism and arthrogryposis. Pediatr Surg Int 7(4):271–273

Ferguson L, Agoulnik AI (2015) Testicular cancer and cryptorchidism. Front Endocrinol 4:32

Ferlin A, Arredi B, Zuccarello D, Garolla A, Selice R, Foresta C (2006) Paracrine and endocrine roles of insulin-like factor 3. J Endocrinol Invest 29(7):657–664

Forest MG (1975) 4-Differentiation and development of the male. Clin Endocrinol Metab 4(3):569–596

Ge RS, Dong Q, Sottas CM, Papadopoulos V, Zirkin BR, Hardy MP (2006) In search of rat stem Leydig cells: identification, isolation, and lineage-specific development. Proc Natl Acad Sci U S A 103(8):2719–2724

Grumbach M (2005) A window of opportunity: the diagnosis of gonadotropin deficiency in the male infant. J Clin Endocrinol Metab 90:3122–3127

Hadziselimovic F, Herzog B (2001) The importance of both an early orchidopexy and germ cell maturation for fertility. Lancet 358(9288):1156–1157

Haider SG, Laue D, Schwochau G, Hilscher B (1995) Morphological studies on the origin of adult-type Leydig cells in rat testis. Ital J Anat Embryol 100:535–541

Huhtaniemi IT, Yamamoto M, Ranta T, Jalkanen J, Jaffe RB (1987) Follicle-stimulating hormone receptors appear earlier in the primate fetal testis than in the ovary*. J Clin Endocrinol Metab 65(6):1210–1214

Itesako T, Nara K, Matsui F, Matsumoto F, Shimada K (2011) Acquired undescended testes in boys with hypospadias. J Urol 185(6):2440–2443

Karl J, Capel B (1998) Sertoli cells of the mouse testis originate from the coelomic epithelium. Dev Biol 203(2):323–333

Kolon TF, Patel RP, Huff DS (2004) Cryptorchidism: diagnosis, treatment, and long-term prognosis. Urol Clin North Am 31(3):469–480

Matsumoto F, Yamauchi K, Matsui F, Shimada K, Ida S (2012) Acquired cryptorchidism in a boy with disorder of sex development. Clin Pediatr Endocrinol 21(1):1

McLaren A, Southee D (1997) Entry of mouse embryonic germ cells into meiosis. Dev Biol 187(1):107–113

Merchant-Larios H, Moreno-Mendoza N (1998) Mesonephric stromal cells differentiate into Leydig cells in the mouse fetal testis. Exp Cell Res 244(1):230–238
Mesiano S, Jaffe RB (1997) Developmental and functional biology of the primate fetal adrenal cortex 1. Endocr Rev 18(3):378–403
Møller H, Prener A, Skakkebxk NE (1996) Testicular cancer, cryptorchidism, inguinal hernia, testicular atrophy, and genital malformations: case-control studies in Denmark. Cancer Causes Control 7(2):264–274
Morelli MA, Cohen PE (2005) Not all germ cells are created equal: aspects of sexual dimorphism in mammalian meiosis. Reproduction 130(6):761–781
O'Shaughnessy PJ, Baker PJ, Johnston H (2006) The foetal Leydig cell–differentiation, function and regulation. Int J Androl 29(1):90–95
Park SY, Jameson JL (2005) Minireview: transcriptional regulation of gonadal development and differentiation. Endocrinology 146(3):1035–1042
Petersen C, Söder O (2006) The sertoli cell–a hormonal target and 'super'nurse for germ cells that determines testicular size. Horm Res Paediatr 66(4):153–161
Petersen C, Boitani C, Fröysa B, Söder O (2001) Transforming growth factor-α stimulates proliferation of rat Sertoli cells. Mol Cell Endocrinol 181(1):221–227
Petersen C, Boitani C, Fröysa B, Söder O (2002) Interleukin-1 is a potent growth factor for immature rat Sertoli cells. Mol Cell Endocrinol 186(1):37–47
Petersen C, Fröysa B, Söder O (2004) Endotoxin and proinflammatory cytokines modulate Sertoli cell proliferation in vitro. J Reprod Immunol 61(1):13–30
Schmahl J, Capel B (2003) Cell proliferation is necessary for the determination of male fate in the gonad. Dev Biol 258(2):264–276
Shawlot W, Behringer RR (1995) Requirement for Liml in head-organizer function. Nature 374(6521):425–430
Simoni M, Gromoll J, Nieschlag E (1997) The follicle-stimulating hormone receptor: biochemistry, molecular biology, physiology, and pathophysiology 1. Endocr Rev 18(6):739–773
Söder O (2007) Sexual dimorphism of gonadal development. Best Pract Res Clin Endocrinol Metab 21(3):381–391
Stikkelbroeck NM, Otten BJ, Pasic A, Jager GJ, Sweep CF, Noordam K, Hermus AR (2001) High prevalence of testicular adrenal rest tumors, impaired spermatogenesis, and Leydig cell failure in adolescent and adult males with congenital adrenal hyperplasia. J Clin Endocrinol Metab 86(12):5721–5728
Svechnikov K, Landreh L, Weisser J, Izzo G, Colon E, Svechnikova I, Söder O (2010) Origin, development and regulation of human Leydig cells. Horm Res Paediatr 73(2):93–101
Tasian GE, Zaid H, Cabana MD, Baskin LS (2010) Proximal hypospadias and risk of acquired cryptorchidism. J Urol 184(2):715–720
Torres M, Gómez-Pardo E, Dressler GR, Gruss P (1995) Pax-2 controls multiple steps of urogenital development. Development 121(12):4057–4065
Whitaker RH (1988) Neoplasia in cryptorchid men. Semin Urol 6(2):107–109
Wylie C (1999) Germ cells. Cell 96(2):165–174
Yan W, Kero J, Huhtaniemi I, Toppari J (2000) Stem cell factor functions as a survival factor for mature Leydig cells and a growth factor for precursor Leydig cells after ethylene dimethane sulfonate treatment: implication of a role of the stem cell factor/c-Kit system in Leydig cell development. Dev Biol 227(1):169–182
Zimmermann S, Schöttler P, Engel W, Adham IM (1997) Mouse Leydig insulin-like (Ley I-L) gene: structure and expression during testis and ovary development. Mol Reprod Dev 47(1):30–38
Zimmermann S, Steding G, Emmen JM, Brinkmann AO, Nayernia K, Holstein AF, Engel W, Adham IM (1999) Targeted disruption of the Insl3 gene causes bilateral cryptorchidism. Mol Endocrinol 13(5):681–691

HPG Axis: The Central Regulator of Spermatogenesis and Male Fertility

3

Vertika Singh, Neeraj Kumar Agrawal, Rajesh Verma, and Kiran Singh

Abstract

Pituitary gonadotropins have been established as essential components for the differentiation of the male reproductive organs. Human sexual maturation and spermatogenesis are intricately regulated by the hypothalamic-pituitary-gonadal (HPG) axis, which eventually determines the reproductive potential of an organism. Alterations affecting this fine balance can severely impair sexual development and fertility. These defects may result from mutations, small deletions or polymorphic changes within the regulatory genes involved in the biosynthesis of hormones, hormone receptors, growth factors and their associated signal transduction pathways. This present chapter summarizes the functioning and regulation of the HPG axis, its control over spermatogenesis by means of FSH and LH synthesis, and the impact of endocrine disruptors on this central axis regulating fertility.

Keywords

HPG axis • Gonadotropins • Spermatogenesis • Fertility • Endocrine disruptors GnRH

V. Singh • K. Singh (✉)
Department of Molecular and Human Genetics, Institute of Science, Banaras Hindu University, Varanasi, Uttar Pradesh, India
e-mail: singhk4@rediffmail.com

N.K. Agrawal • R. Verma
Department of Endocrinology and Metabolism, Institute of Medical Sciences, Banaras Hindu University, Varanasi, Uttar Pradesh, India

R. Singh, K. Singh (eds.), *Male Infertility: Understanding, Causes and Treatment*, DOI 10.1007/978-981-10-4017-7_3

Key Points

- Human sexual maturation and spermatogenesis are intricately regulated by the hypothalamic-pituitary-gonadal (HPG) axis.
- Gonadotropin-inhibiting hormone (GnIH) and kisspeptin are two hypothalamic neuropeptides regulating the HPG axis.
- Melatonin affects the process of sexual maturation and reproductive functions by stimulating the HPG axis.
- INSL3 is considered as a biomarker of Leydig cell functionality in males.
- LH receptor mutations are associated with Leydig cell hyperplasia and spermatogenic arrest.
- Endocrine-disrupting compounds (EDCs) may affect fertility by interfering with endocrine actions of the HPG axis.

3.1 Introduction

The endocrine regulation of spermatogenesis is largely directed by the neuroendocrine actions along the hypothalamic-pituitary-gonadal (HPG) axis. Proper development and organization of the HPG axis are indispensable for normal reproductive competence. The fundamental molecule that regulates the function of the HPG axis is gonadotropin-releasing hormone (GnRH). GnRH neurons are believed to originate in the olfactory placode, which further migrate to their destinations in the brain. The episodic and timely secretion of GnRH from the hypothalamus along with GnRH receptor (GnRH-R) activation in pituitary gonadotrophs is crucial for optimum gonadotropin synthesis and secretion. Dysregulation of any of these functions may result in delayed or complete absence of puberty, leading to infertility (Catt et al. 1985; Rasmussen 1993; Krsmanovic et al. 1996; Terasawa 1998; Moenter et al. 2003; Krsmanovic et al. 2009).

A robust and pulsatile release of GnRH from the hypothalamus regulates the secretion of two major endocrine signals from the pituitary gland: follicle-stimulating hormone (FSH) and luteinizing hormone (LH). These are heterodimeric glycoprotein hormones that act via GnRH receptors in the testis. Low level of GnRH results in decreased FSH and LH secretion, which gives rise to hypogonadotropic hypogonadism (HH), resulting in low androgen secretion and impaired spermatogenesis (Seminara et al. 2000). Adequate functionality of the LH and FSH receptors plays a crucial role in relaying the functions of HPG axis. Besides, genetic alterations within the regulatory genes involved in the biosynthesis of hormones, growth factors, hormone receptors and their associated signal transduction pathways may lead to the impairment of fertility. Infertile patients with altered secretion of HPG hormones are tested for serum FSH, LH, total free testosterone, oestradiol and prolactin levels. This chapter recites the regulation of spermatogenesis by the HPG axis, the effect of endocrine disruptors and genetic causes on its regulation, which can have implications in understanding and treatment of male infertility.

3.2 The Hypothalamic–Pituitary–Gonadal Axis

The hypothalamus, the pituitary, and the testis form an integrated system for the appropriate secretion of male hormones and maintenance of normal spermatogenic functions. The hypothalamus secretes GnRH, which in turn stimulates the gonadotroph cells of pituitary to secrete FSH and LH (Fig. 3.1). These hormones play a vital role in regulating the gonadal functions. Inhibins, activins and steroid hormones, the secretory products of gonads, affect the secretion of gonadotrophins. Recent evidences suggest that internal and external factors such as stress hormones, leptin and the opioid system also influence the HPG axis by modulating the secretion of GnRH and gonadotrophins.

Recently, findings on the functions of RFamide peptides have emerged into picture. RFamides are small peptides possessing Arg-Phe-NH2 motif at the C-terminus. Two groups of RFamide peptides are known to actively participate in HPG

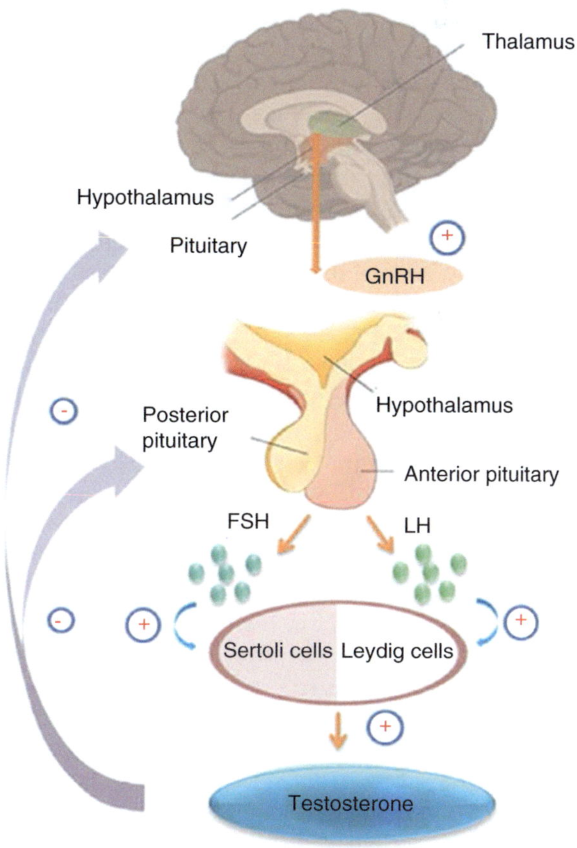

Fig. 3.1 Schematic diagram showing male hypothalamic–pituitary–gonadal (HPG) axis

regulation: gonadotropin-inhibiting hormone (GnIH), their related peptides and the group of kisspeptins. Kisspeptins and GnIH are two hypothalamic neuropeptides. They are critical players in the regulation of the reproductive axis. Kisspeptins act as stimulators of the reproductive axis, while GnIH is the inhibitory antagonist (Ubuka et al. 2005; Pinilla et al. 2012; Ubuka et al. 2013; Wahab et al. 2015). Recently, a group of researchers has found that serum kisspeptin levels were significantly lower in infertile males as compared to fertile, suggesting that kisspeptin might be associated with fertility problems in males (Ramzan et al. 2015). Together, it is likely that an integrated functional interaction of both these hypothalamic neuropeptides plays an important role as a regulator and gatekeeper in sustaining reproductive competence.

3.3 GnRH Neurons: Origin and Development

GnRH is an indispensable peptide, with both endocrine and neuromodulatory roles in vertebrates. An intricate crosstalk between various developmental and neuroendocrine signalling pathways regulates the ontogeny and homoeostasis of GnRH neurons including the production, secretion and action of GnRH (Wierman et al. 2011). The development of the olfactory system and GnRH neurons is intimately connected, modulated by common cell surface receptors. During early embryogenesis, GnRH neurons originate from the neural crest within the olfactory placode; however, the recent evidence suggests that GnRH cells have multiple embryonic origins and transiently associate with the developing olfactory system while migrating to ventral forebrain locations. After penetrating through the cribriform plate, the GnRH neurons reach the hypothalamus, where they disengage from the olfactory axonal guides, lose motility and disperse further into the basal lamina of the brain before undergoing terminal differentiation (Forni and Wray 2015) (Fig. 3.2).

In mice, this migratory development begins at embryonic day E10.5 and is completed by E17.5 (Schwarting et al. 2007). A coordinated interplay of various cell adhesion molecules, axonal guidance cues and extracellular matrix proteins, neurotransmitters and transcription factors and growth factors are involved in the regulation and synchronization of the migratory events of GnRH neurons (Forni and Wray 2015; Kim 2015). GnRH neurons spread their axonal processes across the medial eminence of the hypothalamus through which pulsatile GnRH is secreted into circulation via the hypophyseal portal system. GnRH is temporarily secreted at 3–6 months postnatally, sometimes called as “mini-puberty”. GnRH secretion then remains dormant until the inception of puberty, when it gets reactivated to initiate secondary sexual maturation (Wierman et al. 2011). Therefore, the normal development and harmonized actions of the HPG axis are indispensable for GnRH pulse generation and proper reproductive functions.

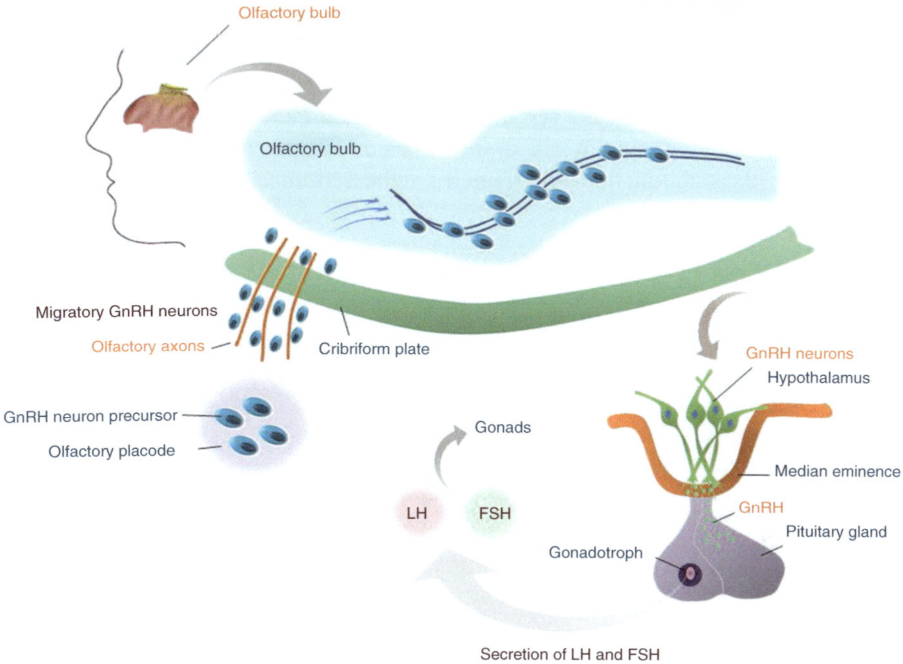

Fig. 3.2 Diagrammatic representation of the migratory events of GnRH neurons. GnRH neurons originate from the neural crest and ectodermal progenitors within the olfactory placode. After penetrating through the cribriform plate along olfactory sensory axons, the GnRH neurons reach the hypothalamus, extending their axons along the median eminence

3.4 Role of FSH and LH in Spermatogenesis

FSH and LH are the primary regulators of spermatogenesis; however, the initiation and maintenance of spermatogenesis are driven by indispensable action of androgens. The functions of FSH and LH on spermatogenesis are mainly regulated through the secretion of Sertoli cell factors, which is mediated directly by FSH and indirectly by LH (via testosterone-androgen receptor).

FSH and LH facilitate their actions by means of specific transmembrane receptors, FSHR and LHR, respectively. FSHR expression is predominantly seen in the Sertoli cells; however, LHR is expressed by the Leydig cells. The gonadotropin response from the pituitary produces two major endocrine signals from the testis: steroidal hormone testosterone production from Leydig cells in response to LH and non-steroidal inhibin production from the Sertoli cells in response to FSH. The secretory actions of testosterone and inhibin occur in a pulsatile and non-pulsatile manner, respectively. Although the classically established endocrine regulation of spermatogenesis occurs via LH and FSH, the paracrine regulation occurs through a coordinated action of inhibins, activins and follistatin hormones. Gonadotropin actions during pubertal stage are essential in synchronizing the behaviour of the primordial germ cell, Leydig cells and Sertoli cells for orchestrating spermatogenesis. Hormonal insufficiency during these stages may affect the scrotal decent and testicular development in adults. Contrarily, in adults, the germ cell dysfunctions as a result of hormonal imbalance occur largely through functional deficiencies in the Sertoli cells.

3.5 FSH and LH in Human Male Infertility

The role of FSH in human male infertility has been extensively analysed using various association studies. Though the knockout studies in animal models have provided an indirect clue towards the functional impairments, the exact mechanism underlying the pathogenesis in humans is still uncovered. FSHβ and FSH receptor knockout male mice are fertile, however, with reduced testicular size and spermatogenic impairment (Kumar et al. 1997). However, upcoming reports have highlighted that loss of function mutation in the FSHβ gene leads to azoospermia in men, with low testosterone and delayed virilization in a few cases (Lindstedt et al. 1998; Phillip et al. 1998; Layman et al. 2002). Nevertheless, a few infertile cases with an idiopathic FSH deficiency, but normal virilization and testosterone levels, have also been identified (Mantovani et al. 2003; Giltay et al. 2004; Murao et al. 2008). Another study reported a homozygous, inactivating *FSHR* mutation in five men with spermatogenic failure of variable phenotype. Although these cases were fertile, elevated FSH level and oligozoospermia phenotype were clearly evident (Tapanainen et al. 1997).

The phenotypic spectrum of *LHR* mutations is more complex and often associated with abnormal development of external genitalia and impaired sexual differentiation. *LHR* mutations widely occur in association with deficiencies of testicular decent and pseudohermaphroditism (Berthezène et al. 1976; Gromoll et al. 2000; Richter-Unruh et al. 2002; Simoni et al. 2008; Richard et al. 2011; Latronico and Arnhold 2012; Kossack et al. 2013). *LHR* mutations in these cases were associated with Leydig cell hyperplasia and spermatogenic arrest. Interestingly, in one of the studies, a patient with a deletion at exon 10 of the *LHR* displayed an azoospermia phenotype with delayed onset of puberty (Gromoll et al. 2000). A study evidenced homozygous missense mutation (Q54R) in the LHβ gene in a male with pubertal delay, low testosterone and spermatogenic arrest (Weiss et al. 1992). This mutation preserved the hormone synthesis and immunoreactivity, but prevented its binding to the LH receptors (Weiss et al. 1992). Two phenotypically milder LH receptor mutations have been described in patients with micropenis (S616Y, I625K), pubertal failure and infertility (S616Y). One of the patients with LH receptor mutation showed the absence of mature Leydig cells and spermatogenesis arrest at spermatid stage (Martens et al. 1998).

3.6 Endocrine Disruptors: Modulators of the HPG Axis

Many natural and synthetic compounds affect endocrine organs and the signalling pathways that impair human health. During the last century, the profusion of synthetic chemicals developed for consumer products and commercial industrial processes has upraised significant apprehensions with respect to their ill effects on health (Diamanti-Kandarakis et al. 2009).

Endocrine-disrupting compounds (EDCs) are synthetic or natural compounds that interfere with endogenous endocrine actions (Zawatski and Lee 2013). The US Environmental Protection Agency (EPA) has defined endocrine disruptors as "an exogenous agent that interferes with synthesis, secretion, transport, metabolism, binding action, or elimination of natural blood-borne hormones that are present in the body and are responsible for homoeostasis, reproduction, and developmental process".

It was believed that EDCs exert their actions predominantly through nuclear hormone receptors, including androgen receptors (ARs), oestrogen receptors (ERs), progesterone receptors (PRs), thyroid receptors (TRs) and retinoid receptors among others. But today, the advancement in basic scientific research has shown a much broader perspective of their action. Now, we know that EDCs act via nuclear receptors, nonnuclear steroid hormone receptors (e.g. membrane ERs), nonsteroid receptors, orphan receptors (e.g. aryl hydrocarbon receptor (AhR)), enzymatic pathways involved in steroid biosynthesis and/or metabolism, and various other mechanisms regulating the endocrine and reproductive functions. Commonly used EDCs include pharmacological compounds such as diethylstilbestrol (DES), a synthetic oestrogen as well as industrial or agricultural chemicals such as plasticizers or insecticides, industrial solvents/lubricants and their byproducts [polychlorinated biphenyls (PCBs) (Pocar et al. 2011), polybrominated biphenyls (PBBs) (Darnerud 2008), dioxins (Shi et al. 2007)], plastics [bisphenol A (BPA)] (Rubin 2011), plasticizers (phthalates) (Hauseur and Cal 2005), pesticides [methoxychlor, chlorpyrifos, dichlorodiphenyltrichloroethane (DDT) (Hayes et al. 2011)] and fungicides (vinclozolin) (Fig. 3.3).

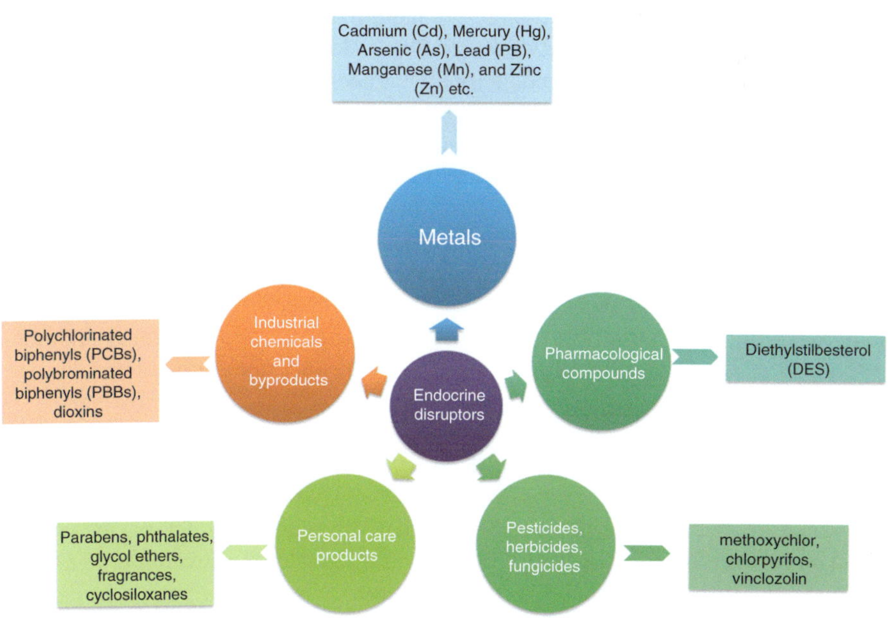

Fig. 3.3 Figure illustrating some of the important endocrine-disrupting molecules and their sources

Contemporaneous exposers can also disrupt the HPG axis. Natural chemicals in edible food products such as phytoestrogens, including genistein and coumestrol, are also shown to act as endocrine disruptors (Diamanti-Kandarakis et al. 2009).

EDCs interfere with the hormonal pathways of the HPG axis through a multitude of mechanisms. They compete for oestrogen receptor binding and activation, post-receptor signalling pathways and modulating the synthesis, bioactivity or degradation of hormones, receptors and cofactors. Studies have demonstrated that pubertal timing can be influenced by either prenatal or postnatal exposure to EDCs. EDCs exposure can disrupt many aspects of the HPG axis during early stages of the CNS development and sexual differentiation. These include modulation of neuroendocrine organization and feedback loops and gonadal sex steroid synthesis (Navarro et al. 2009; Walker and Gore 2011). Upcoming evidences have suggested that exposure to EDCs can adversely affect not only the organism that comes in contact with it but also the future progeny of the exposed individuals (Anway et al. 2005; Anway and Skinner 2006; Guerrero-Bosagna et al. 2010). Exposure to EDCs may disrupt the normal functioning of the hypothalamic circuitry, which may substantially inhibit GnRH, LH and FSH release for regulation of sexual development and gametogenesis (Milardi et al. 2008). In rats, it was demonstrated that exposure to DES (Lassurguere et al. 2003), PCB (Gore et al. 2002) and atrazine (Hayes et al. 2011) causes disruption in the HPG axis, leading to gonadal insufficiency.

Kisspeptins are a group of neuropeptides encoded by the *KISS1* gene, produced mainly by neuronal clusters at hypothalamic nuclei and are broadly recognized as the fundamental activators of the HPG axis at the onset of puberty (Smith et al. 2006). Kisspeptin and its G-protein-coupled receptor act as gatekeepers to control the secretion of GnRH, thereby regulating the anterior pituitary hormones and testicular hormones such as testosterone, activing and inhibin B (Roseweir and Millar 2009; Silveira et al. 2010; Hamlin and Guillette 2011). In rats, the neonatal exposure to oestrogenic EDCs such as BPA and genistein has been shown to inhibit the kisspeptin synthesis (Bateman and Patisaul 2008; Navarro et al. 2009). These studies have provided initial clues regarding the mechanism of action of some of the EDCs.

Hormones play a primary role in coordinating the regulation of mammalian spermatogenesis, which in turn depends on a functional hypothalamic-pituitary-testis axis. Uncovering the mechanisms of action of EDCs on reproductive outcomes is of significant interest nowadays. It will further provide insights into the biological effects of the EDCs on reproduction, embryonic development and fertility. The study of their transgenerational effects would be particularly interesting as they can potentially affect the fertility of the exposed and the upcoming generations.

3.7 Melatonin and HPG Axis in Reproductive Health

Melatonin (N-acetyl-5-methoxy-tryptamine), a principal product of the pineal gland, is produced predominantly during the dark phase of the circadian cycle. This hormone plays an essential role in the regulation of circadian changes in various physiological aspects and neuroendocrine functions. In mammals, melatonin can affect the process of sexual maturation and reproductive functions by stimulating the HPG axis (Shi et al. 2013).

It has been well documented that melatonin exerts its effects on the HPG axis to ultimately regulate the testicular function. The effects of melatonin have been recently shown to be relayed by other factors, which are GnIH (gonadotropin-inhibiting hormone) and kisspeptins. Both these factors belong to a group of RFamide peptides. Mammalian GnIH is also called as RFamide-related peptide (RFRP) (Osugi et al. 2014). Two types of GnIH are characterized in humans to regulate the HPG axis in males, that is, RFRP-1 and RFRP-3 (Tsutsui et al. 2013). GnIH is a hypothalamic factor that inhibits the HPG axis. It was first of all identified in the quail (Tsutsui et al. 2000). Melatonin also seems to act directly on GnIH neurons through its receptor to induce GnIH expression in Aves (Ubuka et al. 2006). Overall, melatonin acts indirectly on the GnRH neuronal activity. Chronic treatment of mature birds with GnIH for 2 weeks resulted in decreased plasma testosterone concentrations and release of gonadotropins (Ubuka et al. 2006; Nargund 2015), suggesting their important effect on the HPG axis. Disturbed sleep and melatonin synthesis have been shown to affect night testosterone production (Wurtman 2014).

Recently, a peptide hormone INSL3 has emerged into picture. In mammals INSL3 is synthesized by the interstitial Leydig cells in males. Several other organs are reported to secrete this hormone, but the circulatory form is exclusively derived from the testis. INSL3 acts in both autocrine and paracrine manner in the testis. The functions of INSL3 are believed to be regulated by the HPG axis. Acting as a downstream effector molecule, it buffers the action of both LH and FSH for proper steroidogenesis and reproductive functions (Ivell et al. 2014). *INSL3* gene expression regulates the process of testicular descent, and its disruption results in bilateral cryptorchidism in males (Nef et al. 1999; Zimmermann et al. 1999). INSL3 is considered as a biomarker for Leydig cell functionality in males (Ivell et al. 2014).

Conclusion

In both, males and females, gametogenesis is controlled by the HPG axis that refers to the GnRH-gonadotropins-steroids axis. This regulates gametogenesis by releasing FSH and LH from the anterior pituitary and the synthesis of steroid hormones in gonads. An understanding of the reproductive axis is essential for the assessment of abnormal development of the genitalia, hypergonadism, hypogonadism, infertility and various other reproductive dysfunctions. This axis normally functions in a tightly regulated manner to produce circulating steroids essential for male sexual development, function and fertility. A number of genetic mutations and endocrine disruptors are known to result in alterations in the HPG regulation and contribute to infertility. Further understanding of the HPG regulation and the mechanism of action of EDCs would help in better understanding and management of male infertility.

References

Anway MD, Skinner MK (2006) Epigenetic transgenerational actions of endocrine disruptors. Endocrinology 147(6):s43–s49

Anway MD, Cupp AS, Uzumcu M, Skinner MK (2005) Epigenetic transgenerational actions of endocrine disruptors and male fertility. Science 308(5727):1466–1469

Bateman HL, Patisaul HB (2008) Disrupted female reproductive physiology following neonatal exposure to phytoestrogens or estrogen specific ligands is associated with decreased GnRH activation and kisspeptin fiber density in the hypothalamus. Neurotoxicology 29(6):988–997

Berthezène F, Forest MG, Grimaud JA, Claustrat B, Mornex R (1976) Leydig-cell agenesis: a cause of male pseudohermaphroditism. N Engl J Med 295(18):969–972

Catt KJ, Loumaye E, Wynn PC, Iwashita M, Hirota K, Morgan RO, Chang JP (1985) GnRH receptors and actions in the control of reproductive function. J Steroid Biochem 23(5):677–689

Darnerud PO (2008) Brominated flame retardants as possible endocrine disrupters. Int J Androl 31(2):152–160

Diamanti-Kandarakis E, Bourguignon JP, Giudice LC, Hauser R, Prins GS, Soto AM, Zoeller RT, Gore AC (2009) Endocrine-disrupting chemicals: an Endocrine Society Scientific Statement. Endocr Rev 30(4):293–342

Forni PE, Wray S (2015) GnRH, anosmia and hypogonadotropic hypogonadism–where are we? Front Neuroendocrinol 36:165–177

Giltay JC, Deege M, Blankenstein RA, Kastrop PM, Wijmenga C, Lock TT (2004) Apparent primary follicle-stimulating hormone deficiency is a rare cause of treatable male infertility. Fertil Steril 81(3):693–696

Gore AC, Wu TJ, Oung T, Lee JB, Woller MJ (2002) A novel mechanism for endocrine-disrupting effects of polychlorinated biphenyls: direct effects on gonadotropin-releasing hormone Neurones. J Neuroendocrinol 14(10):814–823

Gromoll J, Eiholzer U, Nieschlag E, Simoni M (2000) Male hypogonadism caused by homozygous deletion of exon 10 of the luteinizing hormone (LH) receptor: differential action of human chorionic gonadotropin and LH. J Clin Endocrinol Metab 85(6):2281–2286

Guerrero-Bosagna C, Settles M, Lucker B, Skinner MK (2010) Epigenetic transgenerational actions of vinclozolin on promoter regions of the sperm epigenome. PLoS One 5(9):e13100

Hamlin HJ, Guillette LJ (2011) Embryos as targets of endocrine disrupting contaminants in wildlife. Birth Defects Res C 93(1):19–33

Hauser R, Calafat AM (2005) Phthalates and human health. Occup Environ Med 62(11):806–818

Hayes TB, Anderson LL, Beasley VR, de Solla SR, Iguchi T, Ingraham H, Kestemont P, Kniewald J, Kniewald Z, Langlois VS, Luque EH (2011) Demasculinization and feminization of male gonads by atrazine: consistent effects across vertebrate classes. J Steroid Biochem Mol Biol 127(1):64–73

Ivell R, Heng K, Anand-Ivell R (2014) Insulin-like factor 3 and the HPG axis in the male. Frontiers in Endocrinology 5:6

Kim SH (2015) Congenital hypogonadotropic hypogonadism and Kallmann syndrome: past, present, and future. Endocrinol Metab 30(4):456–466

Kossack N, Troppmann B, Richter-Unruh A, Kleinau G, Gromoll J (2013) Aberrant transcription of the LHCGR gene caused by a mutation in exon 6A leads to Leydig cell hypoplasia type II. Mol Cell Endocrinol 366(1):59–67

Krsmanovic LZ, Hu L, Leung PK, Feng H, Catt KJ (2009) The hypothalamic GnRH pulse generator: multiple regulatory mechanisms. Trends Endocrinol Metab 20(8):402–408

Krsmanovic LZ, Stojilkovic SS, Catt KJ (1996) Pulsatile gonadotropin-releasing hormone release and its regulation. Trends Endocrinol Metab 7(2):56–59

Kumar TR, Wang Y, Lu N, Matzuk MM (1997) Follicle stimulating hormone is required for ovarian follicle maturation but not male fertility. Nature genetics 15(2):201–204

Lassurguère J, Livera G, Habert R, Jégou B (2003) Time-and dose-related effects of estradiol and diethylstilbestrol on the morphology and function of the fetal rat testis in culture. Toxicol Sci 73(1):160–169

Latronico AC, Arnhold IJ 2012. Inactivating mutations of the human luteinizing hormone receptor in both sexes. In: Seminars in reproductive medicine, vol 30, no. 5. Thieme Medical, Stuttgart, pp 382–386

Layman LC, Porto AL, Xie J, Da Motta LACR, Da Motta LDC, Weiser W, Sluss PM (2002) FSHβ gene mutations in a female with partial breast development and a male sibling with normal puberty and azoospermia. J Clin Endocrinol Metab 87(8):3702–3707

Lindstedt G, Nyström E, Matthews C, Ernest I, Janson PO, Chatterjee K (1998) Follitropin (FSH) deficiency in an infertile male due to FSHβ gene mutation. A syndrome of normal puberty and virilization but under-developed testicles with azoospermia, low FSH but high lutropin and normal serum testosterone concentrations. Clin Chem Lab Med 36(8):663–665

Mantovani G, Borgato S, Beck-Peccoz P, Romoli R, Borretta G, Persani L (2003) Isolated follicle-stimulating hormone (FSH) deficiency in a young man with normal virilization who did not have mutations in the FSHβ gene. Fertil Steril 79(2):434–436

Martens JW, Verhoef-Post M, Abelin N, Ezabella M, Toledo SP, Brunner HG, Themmen AP (1998) A homozygous mutation in the luteinizing hormone receptor causes partial Leydig cell hypoplasia: correlation between receptor activity and phenotype. Mol Endocrinol 12(6):775–784

Milardi D, Giampietro A, Baldelli R, Pontecorvi A, De Marinis L (2008) Fertility and hypopituitarism. J Endocrinol Invest 31(9 Suppl):71–74

Moenter SM, DeFazio RA, Pitts GR, Nunemaker CS (2003) Mechanisms underlying episodic gonadotropin-releasing hormone secretion. Front Neuroendocrinol 24(2):79–93

Murao K, Imachi H, Muraoka T, Fujiwara M, Kushida Y, Haba R, Ishida T (2008) Isolated follicle-stimulating hormone (FSH) deficiency without mutation of the FSHβ gene and successful treatment with human menopausal gonadotropin. Fertil Steril 90(5):2012–2e17

Nargund VH (2015) Effects of psychological stress on male fertility. Nat Rev Urol 12(7):373–382

Navarro VM, Sanchez-Garrido MA, Castellano JM, Roa J, Garcia-Galiano D, Pineda R, Aguilar E, Pinilla L, Tena-Sempere M (2009) Persistent impairment of hypothalamic KiSS-1 system after exposures to estrogenic compounds at critical periods of brain sex differentiation. Endocrinology 150(5):2359–2367

Nef S, Parada LF (1999) Cryptorchidism in mice mutant for Insl3. Nature genetics 22(3):295–299

Osugi T, Ubuka T, Tsutsui K (2014) Review: evolution of GnIH and related peptides structure and function in the chordates. Front Neurosci 8:255

Phillip M, Arbelle JE, Segev Y, Parvari R (1998) Male hypogonadism due to a mutation in the gene for the β-subunit of follicle-stimulating hormone. N Engl J Med 338(24):1729–1732

Pinilla L, Aguilar E, Dieguez C, Millar RP, Tena-Sempere M (2012) Kisspeptins and reproduction: physiological roles and regulatory mechanisms. Physiol Rev 92(3):1235–1316

Pocar P, Fiandanese N, Secchi C, Berrini A, Fischer B, Schmidt JS, Schaedlich K, Rhind SM, Zhang Z, Borromeo V (2011) Effects of polychlorinated biphenyls in CD-1 mice: reproductive toxicity and intergenerational transmission. Toxicol Sci 126(1):213–226

Ramzan MH, Ramzan M, Ramzan F, Wahab F, Jelani M, Khan MA, Shah M (2015) Insight into the serum kisspeptin levels in infertile males. Arch Iran Med 18(1):12–17

Rasmussen DD (1993) Diurnal modulation of rat hypothalamic gonadotropin-releasing hormone release by melatonin in vitro. J Endocrinol Invest 16(1):1–7

Richard N, Leprince C, Gruchy N, Pigny P, Andrieux J, Mittre H, Manouvrier S, Lahlou N, Weill J, Kottler ML (2011) Identification by array-comparative genomic hybridization (array-CGH) of a large deletion of luteinizing hormone receptor gene combined with a missense mutation in a patient diagnosed with a 46, XY disorder of sex development and application to prenatal diagnosis. Endocr J 58(9):769–776

Richter-Unruh A, Martens JWM, Verhoef-Post M, Wessels HT, Kors WA, Sinnecker GHG, Boehmer A, Drop SLS, Toledo SPA, Brunner HG, Themmen APN (2002) Leydig cell hypoplasia: cases with new mutations, new polymorphisms and cases without mutations in the luteinizing hormone receptor gene. Clin Endocrinol (Oxf) 56(1):103–112

Roseweir AK, Millar RP (2009) The role of kisspeptin in the control of gonadotrophin secretion. Hum Reprod Update 15(2):203–212

Rubin BS (2011) Bisphenol a: an endocrine disruptor with widespread exposure and multiple effects. J Steroid Biochem Mol Biol 127(1):27–34

Schwarting GA, Wierman ME, Tobet SA (2007) Gonadotropin-releasing hormone neuronal migration. Semin Reprod Med 25(5):305–312

Seminara SB, Beranova M, Oliveira LM, Martin KA, Crowley WF Jr, Hall JE (2000) Successful use of pulsatile gonadotropin-releasing hormone (GnRH) for ovulation induction and pregnancy in a patient with GnRH receptor mutations 1. J Clin Endocrinol Metab 85(2):556–562
Shi L, Li N, Bo L, Xu Z (2013) Melatonin and hypothalamic–pituitary–gonadal axis. Curr Med Chem 20(15):2017–2031
Shi Z, Valdez KE, Ting AY, Franczak A, Gum SL, Petroff BK (2007) Ovarian endocrine disruption underlies premature reproductive senescence following environmentally relevant chronic exposure to the aryl hydrocarbon receptor agonist 2, 3, 7, 8-tetrachlorodibenzo-p-dioxin. Biol Reprod 76(2):198–202
Silveira LF, Teles MG, Trarbach EB, Latronico AC, 2010. Role of kisspeptin/GPR54 system in human reproductive axis. In: Kallmann syndrome hypogonadotropic hypogonadism, vol 39. Karger, Basel, pp 13–24
Simoni M, Tüttelmann F, Michel C, Böckenfeld Y, Nieschlag E, Gromoll J (2008) Polymorphisms of the luteinizing hormone/chorionic gonadotropin receptor gene: association with maldescended testes and male infertility. Pharmacogenet Genomics 18(3):193–200
Smith JT, Clifton DK, Steiner RA (2006) Regulation of the neuroendocrine reproductive axis by kisspeptin-GPR54 signaling. Reproduction 131(4):623–630
Tapanainen JS, Aittomäki K, Min J, Vaskivuo T, Huhtaniemi IT (1997) Men homozygous for an inactivating mutation of the follicle-stimulating hormone (FSH) receptor gene present variable suppression of spermatogenesis and fertility. Nat Genet 15(2):205–206
Terasawa E (1998) Cellular mechanism of pulsatile LHRH release 1. Gen Comp Endocrinol 112(3):283–295
Tsutsui K, Saigoh E, Ukena K, Teranishi H, Fujisawa Y, Kikuchi M, Ishii S, Sharp PJ (2000) A novel avian hypothalamic peptide inhibiting gonadotropin release. Biochem Biophys Res Commun 275(2):661–667
Tsutsui K, Ubuka T, Bentley GE, Kriegsfeld L (2013) Review: regulatory mechanisms of gonadotropin-inhibitory hormone (GnIH) synthesis and release in photoperiodic animals. Front Neurosci 7:60
Ubuka T, Bentley GE, Ukena K, Wingfield JC, Tsutsui K (2005) Melatonin induces the expression of gonadotropin-inhibitory hormone in the avian brain. Proc Natl Acad Sci U S A 102(8):3052–3057
Ubuka T, Son YL, Bentley GE, Millar RP, Tsutsui K (2013) Gonadotropin-inhibitory hormone (GnIH), GnIH receptor and cell signaling. Gen Comp Endocrinol 190:10–17
Ubuka T, Ukena K, Sharp PJ, Bentley GE, Tsutsui K (2006) Gonadotropin-inhibitory hormone inhibits gonadal development and maintenance by decreasing gonadotropin synthesis and release in male quail. Endocrinology 147(3):1187–1194
Wahab F, Shahab M, Behr R (2015) The involvement of gonadotropin inhibitory hormone and kisspeptin in the metabolic regulation of reproduction. J Endocrinol 225(2):R49–R66
Walker DM, Gore AC (2011) Transgenerational neuroendocrine disruption of reproduction. Nat Rev Endocrinol 7(4):197–207
Weiss J, Axelrod L, Whitcomb RW, Harris PE, Crowley WF, Jameson JL (1992) Hypogonadism caused by a single amino acid substitution in the β subunit of luteinizing hormone. N Engl J Med 326(3):179–183
Wierman ME, Kiseljak-Vassiliades K, Tobet S (2011) Gonadotropin-releasing hormone (GnRH) neuron migration: initiation, maintenance and cessation as critical steps to ensure normal reproductive function. Front Neuroendocrinol 32(1):43–52
Wurtman RJ (2014) Low doses of melatonin promote sleep onset and maintenance in older people—an update. US Neurol 10:117–119
Zawatski W, Lee MM (2013) Male pubertal development: are endocrine-disrupting compounds shifting the norms? J Endocrinol 218(2):R1–R12
Zimmermann S, Steding G, Emmen JM, Brinkmann AO, Nayernia K, Holstein AF, Engel W, Adham IM (1999) Targeted disruption of the Insl3 gene causes bilateral cryptorchidism. Molecular Endocrinology 13(5):681–691

Sperm Maturation in Epididymis

4

Gopal Gupta

Abstract

In mammals, sperm are produced in the testes to attain (a) the cellular morphology for swimming actively up to the site of fertilization and (b) the haploid nucleus to fuse with that of ova for transferring the paternal genetic information. However, the spermatozoa cannot achieve these functions until they travel through the epididymis, where they spend ~1–2 weeks time to undergo "maturation" that bestows them with motility and fertilizing ability. Epididymis is thus a unique organ that is crucial for male fertility. The tall columnar cells of the epididymal epithelium create a special region-specific milieu (fluid) by active absorption and secretion of components such as proteins, enzymes, hormones, electrolytes, and organic molecules. The proximal segment secretes a variety of proteins and other factors that cause surface modifications on sperm and instigate cell signaling to make them potentially motile and fertile, while the distal end (cauda) creates a special environment for storing mature sperm for long periods of time until ejaculation, in a viable but physically quiescent state (in most mammals like mouse, rat, and man). This chapter discusses in brief the epididymal maturation of mammalian sperm.

Keywords

Sperm maturation • Epididymis • Male fertility • Male infertility

---

G. Gupta
Male Reproductive Biology Laboratory,
CSIR-Central Drug Research Institute, Lucknow, Uttar Pradesh, India
e-mail: g_gupta@cdri.res.in

R. Singh, K. Singh (eds.), *Male Infertility: Understanding, Causes and Treatment*, DOI 10.1007/978-981-10-4017-7_4

Key Points

- Testicular sperm are progressively immotile and infertile, and they gain these properties after maturation in the epididymis.
- Epididymis is a dynamic, tubular organ that receives sperm from the testis and passes them on to the vas deferens.
- A special, progressively changing epididymal fluid environment ensures surface modifications and cell signaling in sperm cells, required for motility and fertility.
- The epididymal function is under strict endocrine and paracrine regulations.

4.1 Introduction

The mammalian epididymis consists of a single, long, coiled, and convoluted tubule that transports, concentrates, matures, and stores testicular sperm before ejaculation (Turner 2008). A few fine efferent ducts carry the testicular sperm, which are immotile and incapable of fertilizing the ova, from the rete testis to the epididymal duct for maturation into competent male gametes that are progressively motile and capable of undergoing capacitation, acrosome reaction, and fertilization. The efferent ducts do not serve merely as a conduit for transporting testicular sperm to the epididymis but also initiate the process of fluid reabsorption, thereby removing 50–96% of luminal fluid component and increasing the concentration of sperm several fold (Hess 2002). The active fluid resorption by the efferent ducts is supported by the presence of Na^+/K^+ ATPase, Na^+/H^+exchanger-3 (NHE-3) and cystic fibrosis transmembrane regulator (CFTR)-Cl^- channels, aquaporins, and histological features supporting endocytosis (Hess 2002; Rodríguez and Hinton 2003). Estrogen signaling through estrogen receptor-α may also play a vital role in fluid resorption in efferent ducts as ERαKO male mice have abnormally dilated efferent ducts due to defects in fluid transport, which prevents sperm transport and causes infertility (Hess et al. 1997). The present chapter provides a brief overview of the important process of sperm maturation in epididymis, which is a prerequisite for natural fertility.

4.2 Epididymal Morphology

The epididymis is found adhered as an "appendix" to the upper and lateral side of the testis, and the word "epididymis" is derived from the word "didymi," which is the ancient nomenclature for the testes. Epididymis can broadly be distinguished into three functionally distinct segments, the proximal head or "caput," the middle body or "corpus," and the distal tail or "cauda." The proximal segments (caput and corpus) have been assigned the function of providing a special milieu to sperm for the development of their motility and fertility, while the distal cauda creates a unique environment to store sperm viable and physically quiescent (in most mammals, including rat, mouse, and human) for long periods of time until ejaculation.

The organ (epididymis) itself measures just a few centimeters (depending on species); however, the uncoiled epididymal tubule may vary in length from about 3.2 m in rat to 6–7 m in man and 80 m in stallion (Turner 2008), and it takes ~1–2 weeks time for sperm to travel through the epididymis. The epididymal epithelium consists of tall columnar principal cells with microvilli whose height decreases from proximal to distal segment and are involved in secretion, absorption, and phagocytosis (Turner 2008).

4.3 Epididymis: The Site of Sperm Maturation

Though the differentiation and maturation of sperm cells in the testis is mostly under genomic regulation, the post-testicular maturation of sperm is controlled by external factors in the epididymis. While possessing the distinct sperm morphology, testicular sperm entering the epididymis are both progressively immotile and incapable of fertilizing the ova (infertile), and they gain both these properties during their transit through the epididymis (Fig. 4.1). Thus sperm present only in the distal segments of epididymis (mostly cauda) are potentially motile and fertile. However, sperm are highly specialized cells with extremely condensed and inactive nucleus, which has been rendered transcriptionally and translationally quiescent (almost

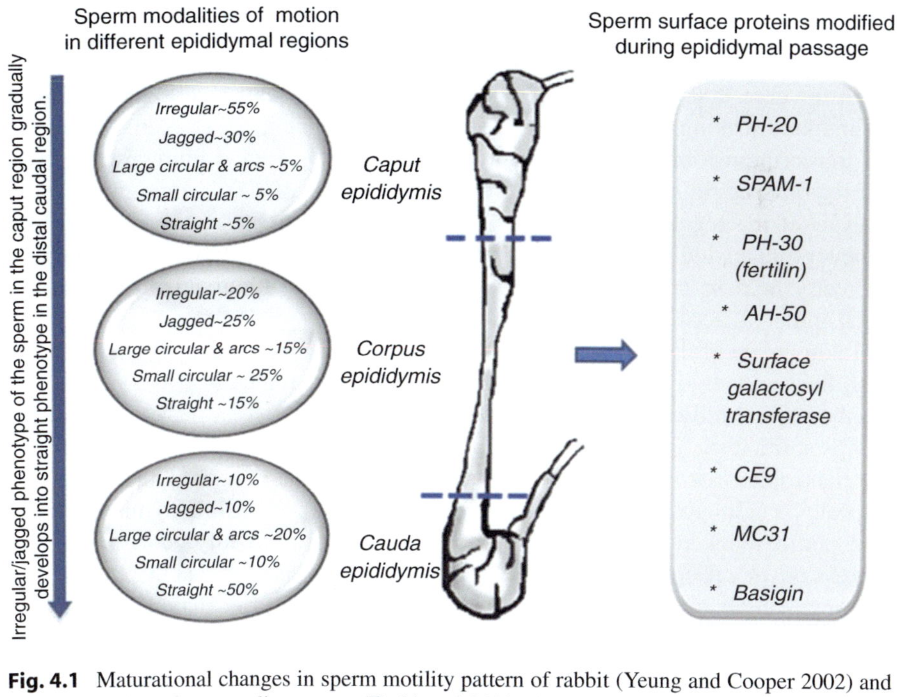

Fig. 4.1 Maturational changes in sperm motility pattern of rabbit (Yeung and Cooper 2002) and surface proteins of mammalian sperm (Toshimori 2003) during epididymal transit. The proteins are not enlisted in a segment-specific order

completely) during spermiogenesis in the testis. Hence, maturational changes that occur in the epididymis are dependent on the epididymal milieu/fluid, the composition of which is governed by active, segment-specific absorption and secretion of proteins, enzymes, hormones, ions/electrolytes, and other small organic molecules by the epididymal epithelial cells. The sperm bathe in a precisely regulated composition of epididymal fluid, which is conducive for sperm maturation and storage (survival) and whose composition changes from caput to cauda in relation to the distinct functions of these regions. It is pertinent to mention that the normal functioning of the epididymis is regulated by androgens (endocrine factors) and the presence of sperm and plasma (paracrine factors) (Robaire et al. 2007; Turner et al. 2007). The segment-specific epididymal physiology and energy metabolism in relation to sperm maturation and storage have long been known to be sensitive to the endocrine and paracrine factors in rats (Brooks 1981) and rhesus monkeys (Gupta et al. 1992, 1993, 1994). However, recent efforts have stressed more upon the segment-specific transcriptome, proteome, and secretome profile of the epididymis.

4.4 Epididymal Transcriptome and Proteome

A detailed microarray analysis of the transcriptome of mouse epididymis identified 2186 genes with segment-specific expression difference of ≥fourfold. Subsequent qRT-PCR indicated the strict segment-specific expression of cystatin-8 and Ros1 proto-oncogene in the proximal caput, glutathione peroxidase-5 and clusterin in the distal caput and corpus, and cysteine-rich secretory protein-1 (Crisp-1) in the cauda (Johnston et al. 2005). Subsequent analysis of rat epididymal proteome by the same group found striking similarities and some differences (Jelinsky et al. 2007). A similar transcriptome analysis of human epididymis discovered epididymis abundant and region-specific expression of several genes (Li et al. 2008). However, further analysis of the regulatory regions of the differentially expressed genes and their protein products indicated that these would either directly be secreted or otherwise indirectly help in creating the special epididymal milieu for sperm maturation (Johnston et al. 2005).

Taking view of the fact that epididymal duct's protein components help in creating the segment-specific physiology of the epididymal duct necessary for creating the special milieu (fluid) that is critical for sperm maturation, other research groups studied the proteome of the epididymal tissue. Using high-resolution 2D gel electrophoresis followed by mass spectrometry, Yuan et al. (2006) identified 28 proteins belonging primarily to basic cellular metabolism, amino acid metabolism, antioxidant system, and smooth muscle tissue showing segment-specific expression in rat epididymis. Interestingly, antioxidant enzymes like inducible carbonyl anhydrase and peroxiredoxin-4 were localized to the distal cauda region, while enzymes catalyzing amino-acid metabolism were abundant in the caput region, which is the primary site for protein synthesis and secretion (Yuan et al. 2006). The strict, region-specific synthesis and secretion of proteins indicated their precise regulatory mechanism in the epididymis. Recently small noncoding

RNAs have emerged as key players in regulation of epididymal function. A genome wide profiling of miRNA signatures in mouse epididymis identified 218 miRNAs expressed specifically in the epithelial cells. While 75% of these had equivalent levels along the entire epididymal tract, a small cohort had region-specific expression, like the miR-204-5p and miR-196b-5p, which were down- and upregulated by ~39- and 45-fold in caput and cauda, respectively. Besides, 79 miRNAs displayed conserved expression in mouse, rat, and human tissues, which included the let-7 family of miRNAs that have been implicated in regulation of androgen signaling (Nixon et al. 2015).

4.5 Epididymal Secretome

In the proximal caput, almost all of the proteins in the rete testis fluid are absorbed and subsequently replaced by epididymis-specific proteins/components secreted mainly by the proximal segments (Fig. 4.1). For example, testicular clusterin and transferrin secreted by the Sertoli cells are completely removed from the epididymal plasma and replaced by more than 100 proteins with high polymorphism in their molecular weights and isoelectric points (Gatti et al. 2004). In a preliminary study, the rat epididymal tubules were found to synthesize and secrete proteins in a region-specific manner with the most prominent secreted bands in caput epididymis consistent with the heavy and light chains of epididymal clusterin and the most secreted protein in cauda epididymis being a 25 kDa protein consistent with protein D (Turner et al. 1994). Subsequently, more detailed investigation of the boar epididymal lumen revealed that the protein secretion was highly regionalized with the maximum number of proteins being secreted by the distal caput and minimum number by the proximal caput and cauda (Syntin et al. 1996). Some of the major proteins identified included epididymal clusterin, glutathione peroxidase, retinol-binding protein, lactoferrin, EP4, p-*N*-acetyl-hexosaminidase, α-mannosidase, and procathepsin L (Syntin et al. 1996). Likewise, in the stallion epididymal epithelium, about 117 proteins were secreted, out of which, just 18 proteins made up 92.5% of total secretory activity, comprising mainly of lactoferrin (41.2%) and epididymal clusterin (24.6%). Other major proteins secreted were albumin, prostaglandin D2 synthase (PGDS), glutathione peroxidase (GPX), cholesterol transfer protein (HE1/CTP), and hexosaminidase. The caput epididymis was characterized by the secretion of clusterin (53%), PGDS (44%), GPX (6%) and the corpus segment by the secretion of lactoferrin (60%), clusterin (29%), hexosaminidase (10%), and procathepsin-D (6.9%), while the cauda segment was marked mainly by the secretion of lactoferrin (2–4 mg/mL). The corpus region was characterized by the secretion of highest number of proteins, possessing the highest concentration of proteins (60–80 mg/mL) and spermatozoa (85%) in the luminal fluid (Fouchecourt et al. 2000). The epididymal fluid is also shown to contain both soluble and particulate matter. The presence of small membrane vesicles named "epididymosomes" has been described in this fluid, which are small vesicles of 25–50 nm in diameter, having a different composition from the surrounding fluid (Gatti et al. 2004; Guyonnet et al.

2011). These vesicles transfer specific proteins and signaling molecules from the epididymal epithelium to sperm for aiding maturation, e.g., the Wnt ligand for LRP receptors to instigate Wnt signaling (Koch et al. 2015).

4.6 Maturational Changes in Sperm During Epididymal Transit

The most important event during sperm maturation in epididymis is the modification of sperm surface proteins. Thus, the sperm membrane undergoes constant remodeling during epididymal transit. This includes proteolytic removal of surface proteins, their redistribution to different loci on sperm, changes in their molecular weights and antigenicity, modification of their side chain glycosyl units like D-galactose and *N*-acetyl-D-galactosamine residues by glycosylation/deglycosylation. The enzymes required for glycosyl modification like the glycosidases and the glycosyltransferases like fucosyltransferase, galactosyltransferase, and sialyltransferase are present in the epididymal fluid. Some of these modified components have been shown to participate in acrosome reaction, interaction, binding, and penetration of the zona pellucida (Toshimori 2003). Sperm proteins that undergo modifications through protolysis include angiotensin 1-converting enzyme, disintegrin and metalloprotease (ADAM) gene family, sperm adhesion molecule-1 (Spam1 or PH-20 hyaluronidase), basigin, and α-D-mannosidase (Sipila et al. 2009).

In an elegant study, Suryawanshi et al. (2011) compared the proteome of caudal sperm with that of testicular sperm and identified 140 extra proteins on the caudal sperm out of which nine were novel and primarily involved in metabolic processes (Suryawanshi et al. 2011). In yet another study, Baker et al. (2011) studied the changes in sperm phosphoproteins and total proteins using titanium dioxide. A total of 53 phosphoproteins were significantly modified during epididymal transit, and this was confirmed for ornithine-decarboxylase antizyme 3, heat-shock protein90R, and testis lipid-binding protein (by immunoblotting), which underwent a major loss during epididymal passage. Recently, transcription-independent Wnt signaling has been shown in mouse sperm through Wnt ligands released from the epididymal epithelium in epididymosomes, and male mice mutant for the Wnt regulator cyclin Y-like 1 are sterile due to immotile and malformed spermatozoa. The signaling is mediated through GSK3 by reducing of global protein poly-ubiquitination to maintain protein homeostasis, inhibiting the septin 4 phosphorylation to establish a membrane diffusion barrier in the sperm tail and inhibiting the protein phosphatase 1 to initiate sperm motility (Koch et al. 2015).

4.7 Development of Motility Potential in Sperm During Epididymal Transit

The testicular sperm entering the epididymis are immotile, in most mammals. The development of sperm motility during epididymal transit and the role of epididymis have been studied by a number of simple and ingenious *in vivo* and *in vitro* experiments by different investigators. Motility development involves the initiation of flagellar movement followed by coordination of this movement into a waveform whiplash motion causing the propulsion of sperm into a progressively moving cell. The amplitude of lateral displacement of head (ALH) is more in the proximal region of epididymis with low progressive motility. However, as sperm move to distal segment of epididymis, the ALH decreases and progressive motility increases. In mouse, sperm are immotile at the start of epididymal tubule but rapidly develop motility in the proximal caput region, though the flagellar beat is erratic with negligible progression. The motion becomes circular when sperm reach the proximal corpus region of epididymis. Most heterogeneous sperm motility is observed in the mid-corpus region and most homogeneous motility in the mid-cauda region of epididymis (Soler et al. 1994).

In rabbits, testicular sperm are motile but nonprogressive. Forward progression develops in distal caput, but sperm in the proximal cauda display maximum motility percentage (Sanchez et al. 1996). On the other hand in monkeys, the motion of sperm in the initial segment is sluggish and irregular, which becomes more erratic as sperm move to more distal region of epididymis due to an increase in flagellation. The most drastic change in sperm motility pattern becomes evident between distal caput and proximal corpus, though maximum values for motility parameters are achieved only in distal corpus with full kinematic development in proximal cauda (Yeung et al. 1996). Ligating the efferent ducts to retain sperm in the rete testis increases their rate of flagellar beating, but heads remain static, and no forward motility is developed, while the motility of rat sperm trapped in the caput epididymis by ligation for a few days were comparable to caput sperm of non-ligated controls, with sperm displaying circular path (Burgos and Tovar 1974). In rabbits and guinea pigs, sperm aging in corpus region developed forward motility with fertility, while in hamster, increased motility was not associated with fertility (Cooper 2012). However, *in vitro* incubation of sperm in rete testis and epididymal fluid was ineffective in promoting forward motility in rams and bulls (Cooper 2012).

Concluding Remarks

Once released into the epididymis by the efferent ducts, the immotile testicular sperm bathe in a progressively and constantly changing epididymal fluid environment. The composition of the fluid is continuously adjusted by the secretory and absorptive activities of the epididymal epithelium to ensure exposure of sperm to required biological stimulus causing well-programmed cell signaling and surface modifications in a region-specific manner. The recent discovery of epididymosomes released by the epididymal epithelium into the lumen as minute vesicles containing special proteins and signaling molecules for sperm maturation further indicates that the process is far from being simple. These changes bestow the sperm with progressive motility and the ability to undergo post-ejaculatory events like capacitation, acrosome reaction, and fertilization. A large number of sperm proteins have been shown to undergo glycosylation/deglycosylation and other posttranslational modifications during epididymal maturation, yet the exact role of these proteins in the entire process is far from being understood completely.

Acknowledgment I would like to thank the Council of Scientific and Industrial Research (CSIR), Govt. of India, for financial help, particularly for the network project on competent gamete generation that supported my work on sperm motility (PROGRAM, BSC0101). This is CDRI communication number 9455.

References

Baker MA, Smith ND, Hetherington L, Pelzing M, Condina MR, Aitken RJ (2011) Use of titanium dioxide to find phosphopeptide and total protein changes during epididymal sperm maturation. J Proteome Res 10(3):1004–1017

Brooks DE (1981) Metabolic activity in the epididymis and its regulation by androgens. Physiol Rev 61(3):515–555

Burgos MH, Tovar ES (1974) Sperm motility in the rat epididymis. Fertil Steril 25(11):985–991

Cooper TG (2012) The epididymis, sperm maturation and fertilisation. Springer Science & Business Media, New York

Fouchécourt S, Métayer S, Locatelli A, Dacheux F, Dacheux JL (2000) Stallion epididymal fluid proteome: qualitative and quantitative characterization; secretion and dynamic changes of major proteins. Biol Reprod 62(6):1790–1803

Gatti JL, Castella S, Dacheux F, Ecroyd H, Metayer S, Thimon V, Dacheux JL (2004) Post-testicular sperm environment and fertility. Anim Reprod Sci 82:321–339

Gupta G, Srivastava A, Setty BS (1992) Effect of efferentiectomy on enzymes of glycolytic pathway, HMP pathway and TCA cycle in epididymis and vas deferens of rhesus monkey. Indian J Exp Biol 30(11):1062–1065

Gupta G, Srivastava A, Setty BS (1993) Androgenic regulation of glycolytic and HMP pathway in epididymis and vas deferens of rhesus monkey. Indian J Exp Biol 31(4):305–311

Gupta G, Srivastava A, Setty BS (1994) Activities and androgenic regulation of kreb cycle enzymes in the epididymis and vas deferens of rhesus monkey. Endocr Res 20(3):275–290

Guyonnet B, Dacheux F, Dacheux JL, Gatti JL (2011) The epididymal transcriptome and proteome provide some insights into new epididymal regulations. J Androl 32(6):651–664

Hess RA (2002) The efferent ductules: structure and functions. In: Robaire B, Hinton BT (eds) The epididymis: from molecules to clinical practice. Springer, New York, pp 49–80

Hess RA, Bunick D, Lee KH, Bahr J, Taylor JA, Korach KS, Lubahn DB (1997) A role for oestrogens in the male reproductive system. Nature 390(6659):509–512

Jelinsky SA, Turner TT, Bang HJ, Finger JN, Solarz MK, Wilson E, Brown EL, Kopf GS, Johnston DS (2007) The rat epididymal transcriptome: comparison of segmental gene expression in the rat and mouse epididymides. Biol Reprod 76(4):561–570

Johnston DS, Jelinsky SA, Bang HJ, DiCandeloro P, Wilson E, Kopf GS, Turner TT (2005) The mouse epididymal transcriptome: transcriptional profiling of segmental gene expression in the epididymis. Biol Reprod 73(3):404–413

Koch S, Acebron SP, Herbst J, Hatiboglu G, Niehrs C (2015) Post-transcriptional Wnt signaling governs epididymal sperm maturation. Cell 163(5):1225–1236

Li JY, Wang HY, Liu J, Liu Q, Zhang JS, Wan FC, Liu FJ, Jin SH, Zhang YL (2008) Transcriptome analysis of a cDNA library from adult human epididymis. DNA Res 15(3):115–122

Nixon B, Stanger SJ, Mihalas BP, Reilly JN, Anderson AL, Dun MD, Tyagi S, Holt JE, McLaughlin EA (2015) Next generation sequencing analysis reveals segmental patterns of micro RNA expression in mouse epididymal epithelial cells. PLoS One 10(8):e0135605

Pérez-Sánchez F, Tablado L, Yeung CH, Cooper TG, Soler C (1996) Changes in the motility patterns of spermatozoa from the rabbit epididymis as assessed by computer-aided sperm motion analysis. Mol Reprod Dev 45(3):364–371

Robaire B, Seenundun S, Hamzeh M, Lamour S (2007) Androgenic regulation of novel genes in the epididymis. Asian J Androl 9(4):545–553

Rodríguez CM, Hinton BT (2003) The testicular and epididymal luminal fluid microenvironment. In: Tulsiani D (ed) Introduction to mammalian reproduction. Springer, New York, pp 61–77

Sipilä P, Jalkanen J, Huhtaniemi IT, Poutanen M (2009) Novel epididymal proteins as targets for the development of post-testicular male contraception. Reproduction 137(3):379–389

Soler C, Yeung CH, Cooper TG (1994) Development of sperm motility patterns in the murine epididymis. Int J Androl 17(5):271–278

Suryawanshi AR, Khan SA, Gajbhiye RK, Gurav MY, Khole VV (2011) Differential proteomics leads to identification of domain-specific epididymal sperm proteins. J Androl 32(3):240–259

Syntin P, Dacheux F, Druart X, Gatti JL, Okamura N, Dacheux JL (1996) Characterization and identification of proteins secreted in the various regions of the adult boar epididymis. Biol Reprod 55(5):956–974

Toshimori K (2003) Testicular and epididymal maturation of mammalian spermatozoa. In: Tulsiani D (ed) Introduction to mammalian reproduction. Springer, New York, pp 93–111

Turner TT (2008) De Graaf's thread: the human epididymis. J Androl 29(3):237–250

Turner TT, Avery EA, Sawchuk TJ (1994) Assessment of protein synthesis and secretion by rat seminiferous and epididymal tubules in vivo. Int J Androl 17(4):205–213

Turner TT, Johnston DS, Finger JN, Jelinsky SA (2007) Differential gene expression among the proximal segments of the rat epididymis is lost after efferent duct ligation. Biol Reprod 77(1):165–171

Yeung CH, Morrell JM, Cooper TG, Weinbauer GF, Hodges JK, Nieschlag E (1996) Maturation of sperm motility in the epididymis of the common marmoset (*Callithrix jacchus*) and the cynomolgus monkey (*Macaca fascicularis*). Int J Androl 19(2):113–121

Yuan H, Liu A, Zhang L, Zhou H, Wang Y, Zhang H, Wang G, Zeng R, Zhang Y, Chen Z (2006) Proteomic profiling of regionalized proteins in rat epididymis indicates consistency between specialized distribution and protein functions. J Proteome Res 5(2):299–307

Sperm Capacitation: The Obligate Requirement for Male Fertility

5

Rohit Kumar Deshmukh and Archana Bharadwaj Siva

Abstract

Capacitation is defined as an ensemble of several physiological, molecular and cellular changes in the spermatozoa, making them fertilization competent. It is considered as an obligate requirement for sperm fertility, since failures in sperm capacitation affect the fertilization potential. This chapter discusses the hallmarks of capacitation, including molecular changes involved in this phenomenon. Laboratory-based studies on human spermatozoa (molecular studies and sperm function tests based on capacitation and its associated events: hyperactivation, acrosome reaction and tyrosine phosphorylation) have been discussed with a view to highlight the pressing need for translating this information into the clinical practice. Additionally, a requirement to develop molecular markers/sperm function tests based on protein tyrosine phosphorylation has been emphasized. The latter have come to the fore with increasing incidence of infertility and frequent use (and need) of assisted reproductive technologies like IVF and ICSI.

Keywords

Sperm capacitation • Male infertility • Molecular marker • Sperm function tests • Hyperactivation • Acrosome reaction • Tyrosine phosphorylation • ARTs

Key Points

- Sperm capacitation, discovered in 1951, independently by CR Austin and MC Chang, is considered as an obligate requirement for sperm fertility.

R.K. Deshmukh • A.B. Siva (✉)
CSIR-Centre for Cellular and Molecular Biology, Hyderabad, Telangana, India
e-mail: abs@ccmb.res.in

R. Singh, K. Singh (eds.), *Male Infertility: Understanding, Causes and Treatment*, DOI 10.1007/978-981-10-4017-7_5

- The last six decades have seen a considerable rise in laboratory-based studies on human sperm capacitation and its associated phenomena: hyperactivation, acrosome reaction and protein tyrosine phosphorylation.
- Clinical tests based on the identified molecular markers are rather scarce, with one test, viz. Androvia Cap-Score™ showing promising results in being able to discriminate fertile from infertile men.
- In the present era of assisted reproductive techniques (ARTs), especially ICSI, it is mandatory to develop reliable sperm function tests based on capacitation and other related phenomena to ensure the selection of the "healthiest" spermatozoa.

5.1 Introduction

In mammals, after having gone through the journey of formation in testis and maturation in epididymis; spermatozoa, the male gamete, isn't quite ready yet to marry the female gamete, the oocyte. It still has to undergo a whole battery of changes—this time—in the female reproductive tract, to fertilize the oocyte (Fig. 5.1). This

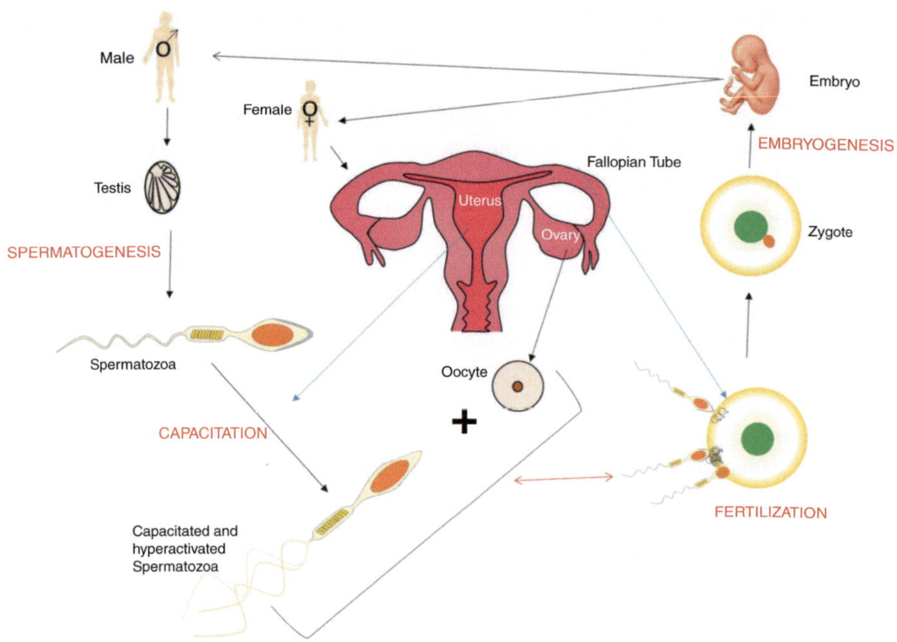

Fig. 5.1 Life cycle of sperm: After production in testis in sufficient numbers with normal shape, the spermatozoa undergo maturation in epididymis, gain motility and undergo capacitation in the female reproductive tract, then acrosome react after oocyte binding and penetrate and activate the egg, resulting in successful fertilization. The *blue arrows* indicate the site of event

ensemble of post-ejaculation changes in the spermatozoa has been collectively called as sperm "capacitation". Capacitation renders the spermatozoa functionally mature.

Origin of spermatozoa in the testis is followed by its capacitation (after ejaculation) in the female reproductive tract and ultimately fertilization with oocyte in the fallopian tube. Sperm contribution to fertilization to assess the "male factors" is usually estimated through evaluation of semen parameters, namely, sperm count, morphology and motility (World Health Organization 2010). Quite often, in spite of these parameters being normal and these males being termed as normozoospermic (normal count, motility and morphology); the infertility still exists in the male partner. Such cases of idiopathic (unknown etiology) infertility have been attributed substantially to the problems in sperm capacitation (Tucker et al. 1987; Matzuk and Lamb 2002; Esposito et al. 2004; Hildebrand et al. 2010; Nandi and Homburg 2016).

5.2 What Is Sperm Capacitation?

Sperm capacitation has been defined as the "ensemble of all the physiological, molecular and cellular changes in the spermatozoa, which are necessary to make it fertilization competent". It was independently discovered by Austin and Chang in 1951 (Austin 1951; Chang 1951). Although discovered more than half a century ago, capacitation is still regarded as a "poorly understood" phenomenon, owing to the fact that each mammalian species has its unique features at the physical (time of capacitation) and molecular level (Chang 1984) that are difficult to monitor, since it takes place in the female reproductive tract (either in the oviduct or in the vicinity of the egg).

Sperm capacitation is a prerequisite for successful fertilization as evidenced from the observations that a block in capacitation causes male infertility (Tucker et al. 1987; Matzuk and Lamb 2002; Esposito et al. 2004; Hildebrand et al. 2010). Therefore, there has been a pressing need to understand sperm capacitation in all the individual species making it a focus of investigations of many gamete biologists worldwide. Most progress in understanding the phenomenon of capacitation has been because of *in vitro* methods for capacitation (Yanagimachi 1969). In the procedure, freshly ejaculated or epididymal spermatozoa are washed and incubated at physiological conditions in a defined medium that mimics the female oviductal fluid (Dow and Bavister 1989). The medium normally has the following composition: electrolytes, metabolic energy source and a macromolecule to allow for cholesterol efflux like serum albumin (Yanagimachi 1969, 1994). Several *in vitro* studies have revealed that during capacitation, spermatozoa undergo a number of biochemical and biophysical changes (Fig. 5.2), such as increase in membrane fluidity (Davis et al. 1980; Cross 1998; Buffone et al. 2009; Salvolini et al. 2013), activation of trans-bilayer signalling events (Go and Wolf 1985; Visconti et al. 1998; Gadella and Harrison 2000; Flesch et al. 2001; Sheriff and Ali 2010; Ickowicz et al. 2012), changes in redox status of spermatozoa leading to generation of reactive oxygen species (ROS) (de Lamirande and Gagnon 1992; Aitken 1995; O'Flaherty et al.

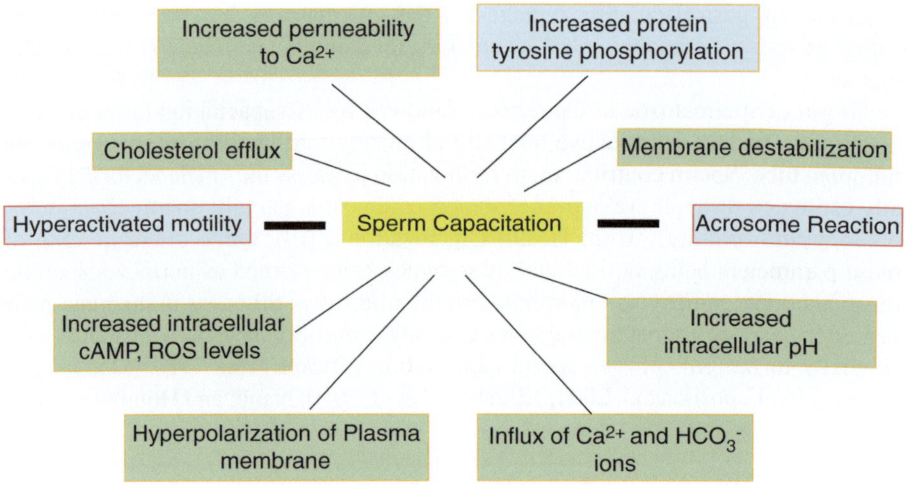

Fig. 5.2 A schematic representation of capacitation and its associated hallmarks (in *blue*) and biochemical and biophysical changes

2006; Musset et al. 2012), removal of stabilizing proteins (Shivaji et al. 1990; Villemure et al. 2003; Leahy and Gadella 2011) and phosphorylation of proteins (Leyton and Saling 1989; Visconti et al. 1995; Mitra and Shivaji 2004; Arcelay et al. 2008; Mitchell et al. 2008; Kota et al. 2009; Katoh et al. 2014).

5.3 Hallmarks of Capacitation

Capacitation is generally monitored by recording protein tyrosine phosphorylation (pY), hyperactivation (Yanagimachi 1994; Kulanand and Shivaji 2001; Baker et al. 2006) and acrosome reaction (Ward and Storey 1984; Meizel and Turner 1991; Aitken 1995; Curry and Watson 1995; Mitra and Shivaji 2004; Varano et al. 2008; Bragado et al. 2012; Jaldety and Breitbart 2015), which are also considered as the "hallmarks of capacitation" (Fig. 5.2). Capacitation changes lead to the transformation in the motility pattern of spermatozoa from a progressively motile cell to a more vigorous, but less progressive, motile cell (Yanagimachi 1969; Suarez and Dai 1992; Mortimer and Swan 1995; Ho and Suarez 2001). This type of motility is termed as "hyperactivation", and subsequent to this, capacitation ends with the ability of spermatozoa to undergo "acrosome reaction", during which the spermatozoa releases the hydrolytic enzymes to facilitate its penetration and fusion with the oocyte—finally leading to fertilization. The increase in pY is another distinctive feature of the mammalian spermatozoa associated with capacitation. This molecular change is considered as an important characteristic of mammalian capacitation and has been addressed by various groups worldwide in varied animal models (Visconti and Kopf 1998; Visconti et al. 1999; Kulanand and Shivaji 2001; Lefièvre et al. 2002; Jha et al. 2003; Shivaji et al. 2007, 2009; Arcelay et al. 2008; Mitchell et al. 2008; Kota et al. 2009).

5.3.1 Hyperactivation

Hyperactivation, which is defined as "a distinct change in the sperm motility from a symmetrical to an asymmetrical pattern, is crucial for fertilization" (Yanagimachi 1969; Suarez 2008). The mammalian spermatozoa, while in the epididymis are immotile. But when released in the female reproductive tract/culture media, they quickly begin to swim and get hyperactivated (Morton et al. 1974), which imparts sperm the ability to traverse through the mucus-filled, labyrinthine lumen of the oviduct to reach the female gamete. Hyperactivation also helps the spermatozoa in penetrating the cumulus oophorus and the zona pellucida (Suarez et al. 1991; Suarez 2008). This activated spermatozoon generates a near symmetrical flagellar beat, which is called as a "planar motility" pattern. This planar motility propels the spermatozoa in an almost linear trajectory (Suarez and Dai 1992; Mortimer and Swan 1995; Ho et al. 2002). The amplitude of the flagellar bend is usually increased only on one side of the hyperactivated spermatozoa. This increased uneven amplitude leads to a circular, wriggling and whiplash type of motility pattern of the spermatozoa as shown in Fig. 5.3, and these movements are assessed objectively by using the computer-assisted sperm analysis (CASA) system (Shivaji et al. 1995; Panneerdoss et al. 2012). Hyperactivation is initiated and maintained by the involvement of a number of physiological factors like calcium, bicarbonate, cAMP and metabolic substrates (Visconti et al. 1999).

Fig. 5.3 Change in sperm movement from progressive to hyperactivated during sperm capacitation

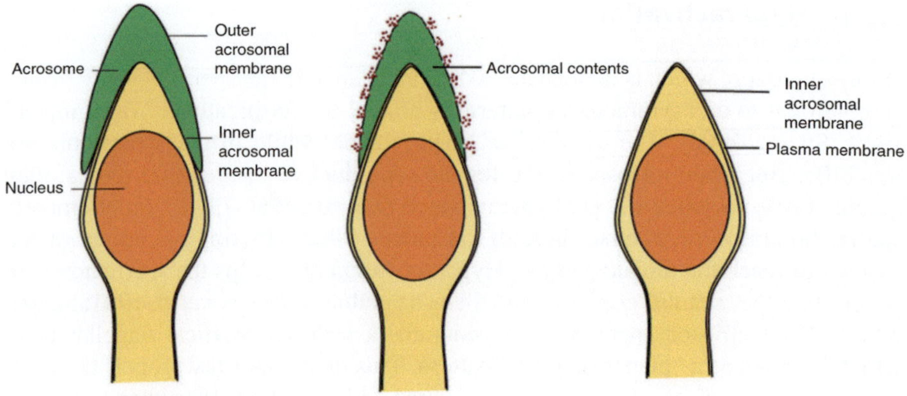

Fig. 5.4 Schematic representation of various stages in the progression of the sperm acrosome reaction (adapted from Curry and Watson 1995)

5.3.2 Acrosome Reaction

Acrosome reaction is an absolute crucial step for successful fertilization, as it is due to acrosomal secretions alone that the sperm makes its progress through the investments surrounding the egg. In fact, males with spermatozoa lacking the acrosome are infertile (Baccetti et al. 1991). During the acrosome reaction, multiple fusions occur between the plasma membrane and the outer acrosomal membrane in the anterior region of the head. These multiple fusions lead to the formation of extensive hybrid membrane vesicles and subsequent exposure of the inner acrosomal membrane and acrosomal contents (Cardullo and Florman 1993). These stages of acrosome reaction have been depicted in Fig. 5.4.

5.3.3 Protein Tyrosine Phosphorylation

Protein tyrosine phosphorylation (pY), a post-translational event, is also considered as hallmark of capacitation. pY is a regulatory mechanism which controls many processes, such as cell cycle control, cytoskeleton assembly, cellular growth, receptor regulation and ionic current modulation (Hunter 2000; Pawson 2004; Vizel et al. 2015). The first evidence of protein tyrosine phosphorylation in spermatozoa was provided by Leyton and Saling (1989) in mouse. Later, Visconti et al. (1995) showed a correlation between sperm capacitation and protein tyrosine phosphorylation in mouse spermatozoa, and soon this increase was demonstrated in spermatozoa of various other species during capacitation, including human (Leclerc et al. 1996; Osheroff et al. 1999), hamster (Kulanand and Shivaji 2001), cat (Pukazhenthi et al. 1998), pig (Tardif et al. 2001), boar (Kalab et al. 1998), bovine (Galantino-Homer et al. 1997, 2004), equine (Pommer et al. 2003), cynomolgus monkey (Mahony and Gwathmey 1999), tammar wallaby and brushtail possum (Sidhu et al. 2004), guinea pig (Kong et al. 2008) and ram (Grasa et al. 2006).

Naz and Rajesh (2004) proposed a model for tyrosine phosphorylation pathways during sperm capacitation. The model suggests that sperm capacitation involves three main signalling pathways, namely, a cAMP/PKA-dependent pathway (pathway I) [unique to spermatozoa], a receptor tyrosine kinase pathway (pathway II) and a non-receptor protein tyrosine kinase pathway (pathway III). A crosstalk between tyrosine kinase and cAMP-dependent kinase signalling pathways in human sperm motility regulation is a unique feature in spermatozoa (Bajpai and Doncel 2003). SRC family kinases (SFKs) known to play an important role in this capacitation-associated increase in protein tyrosine phosphorylation (Battistone et al. 2013) are shown to be downstream of PKA. The target proteins for PKA could be protein tyrosine kinase(s) or protein tyrosine phosphatase(s) or both. These kinase(s) and phosphatase(s) then regulate the downstream phosphorylation of their substrate proteins at their tyrosine residues leading to a cascade of signalling events. Till date, a number of kinases have been identified (Table 5.1), which are involved

Table 5.1 List of kinases identified in spermatozoa from several species

Kinase	Species (reference)
Receptor tyrosine kinases	
EGFR	Human (Breitbart and Etkovitz 2011); ram (Luna et al. 2012); bull (Etkovitz et al. 2009); boar (Awda and Buhr 2010)
IGFR1/IGF1	Human (Wang et al. 2015)
Tyrosine kinase-32	Porcine (Tardif et al. 2003)
FGFR1	Mouse (Cotton et al. 2006)
Non-receptor tyrosine kinases	
SRC	Mouse (Krapf et al. 2012); human (Lawson et al. 2008; Mitchell et al. 2008)
LYN	Mouse (Goupil et al. 2011); bovine (Lalancette et al. 2006)
FYN	Human (Kumar and Meizel 2005); mouse (Luo et al. 2012); rat (Kierszenbaum et al. 2009)
YES	Human (Cheng and Mruk 2012); porcine (Bragado et al. 2012)
HCK	Mouse (Goupil et al. 2011); bovine (Bordeleau and Leclerc 2008)
LCK	Hamster (Singh DK et al. 2017)
PYK2	Human (Battistone et al. 2014); mouse (Chieffi et al. 2003; Roa-Espitia et al. 2016); Bovine (González-Fernández et al. 2013); stallion (Rotfeld et al. 2012)
FER	Mouse (Alvau et al. 2016)
Serine threonine kinases	
Protein kinase A	Human (Leclerc et al. 1996); bovine (Galantino-Homer et al. 1997); porcine (Tardif et al. 2001); hamster (Kulanand and Shivaji et al. 2001)
Protein kinase B/AKT	Human (Aquila et al. 2005); mouse (Feng et al. 2005); boar (Aparicio et al. 2007); Stallion (Gallardo Bolaños et al. 2014)
ERK 1/2	Human (Almog et al. 2008); mouse (Nixon et al. 2010); guinea pig (Chen et al. 2005); boar (Awda and Buhr 2010)
Phosphoinositide 3-kinase (PI3K)	Human (Sagare-Patil et al. 2013)

in the process of capacitation, and the list is still expanding (Lawson et al. 2008; Mitchell et al. 2008; Varano et al. 2008; Goupil et al. 2011; Battistone et al. 2013; Wang et al. 2015). Although several kinases have been identified in the spermatozoa (Table 5.1), their functional relevance is seen only *in vitro* and mostly in animal models. The importance of the identified kinases and thus the regulation of tyrosine phosphorylation in male fertility/infertility has not yet been explored much.

5.4 Diagnosis and Prognosis of Male Infertility/Fertility: Importance of Capacitation-Based Sperm Function Tests

In humans, the prognosis and diagnosis of male fertility has been a subject of research worldwide. As mentioned earlier, a good percentage of human pregnancy failures can be attributed to decreased male fertility or male factor infertility (Thonneau et al. 1991; Sharlip et al. 2002; Lee and Foo 2014). To evaluate human sperm fertility, there has always been a consistent effort to get in place sperm function tests, owing to low predictive power of standard seminal parameters (motility, concentration and morphology) (Oehninger 1995; Carrell 2000; Muller 2000; Aitken 2006; Lefièvre et al. 2007; Vasan 2011; De Jonge and Barratt 2013; Esteves et al. 2014; Oehninger et al. 2014). Attempts have been made in laboratories for decades to design sperm function tests based on capacitation and its associated events/parameters for predicting male fertility.

Sperm penetration tests, including the sperm mucus penetration test and sperm penetration assay, are being routinely used in fertility centres. In addition, various biochemical and biophysical changes during capacitation (Zaneveld et al. 1991; Benoff 1993; Martínez and Morros 1996; Cross 1998; Travis and Kopf 2002; Visconti et al. 2002, 2011; Mitra and Shivaji 2005; Signorelli et al. 2012; Aitken and Nixon 2013) also are being utilized for designing sperm-function tests, for instance, determining the cholesterol efflux, examining activation of ion channels, evaluating protein phosphorylation changes, measuring intracellular calcium and pH and reactive oxygen species, monitoring hyperactivation and acrosome reaction, etc. Three of these events/changes are discussed in the following sections.

5.4.1 Monitoring Hyperactivation (HA)

One of the indicators of capacitation is the display of HA by spermatozoa (Burkman 1984). Sperm motility, hyperactivation and related motility kinematic parameters like average path velocity (VAP), curvilinear velocity (VCL), straight line velocity (VSL), linearity (LIN), amplitude of lateral head displacement (ALH), straightness (STR) and beat cross frequency (BCF) are assessed using CASA (Larsen et al. 2000; Freour et al. 2009). Based on the aforesaid kinematic parameters, namely, VCL, LIN and ALH, the non-hyperactivated spermatozoa (exhibiting planar motility pattern) can be differentiated from the hyperactivated spermatozoa (exhibiting either circular or helical motility patterns) using the SORT facility of the CASA (Youn et al. 2011).

Impaired sperm hyperactivation (HA) has been observed in human patients with infertility (Wong et al. 1993; Munier et al. 2004; Wiser et al. 2014). Wiser et al. evaluated spermatozoa from the normal patients who were to undergo IVF. They found that patients with increased hyperactivated motility had significantly higher fertilization rate compared to the group with no increased hyperactivated motility. Several groups have also found a good correlation between sperm hyperactivation, zona-induced acrosome reaction and zona binding (Liu et al. 2007); sperm motility, capacitation and tyrosine phosphorylation (Yunes et al. 2003; Buffone et al. 2005); and oocyte penetration (Wang et al. 1991), thus presenting HA as a good prognostic parameter for sperm fertility.

5.4.2 Monitoring Acrosome Reaction (AR)

Only capacitated spermatozoa are known to undergo acrosome reaction, underscoring its importance in predicting sperm capacitation and fertility potential of spermatozoa (Bielfeld et al. 1994). Acrosomal status in human spermatozoa is monitored with the fluorescent conjugated lectins (PNA, peanut agglutinin, and PSA, *Pisum sativum* agglutinin) (Cross and Meizel 1989). Additionally, several methods of assessing induced AR *in vitro* have been designed, where the ability of spermatozoa to acrosome react in the presence of calcium-mobilizing agents, such as calcium ionophore (A23187) or the physiological inducers like progesterone and zona pellucida proteins, is assessed (Brucker and Lipford 1995**;** Bastiaan et al. 2002). There are other fluorescent tests to evaluate the acrosome, like chlortetracyclin (CTC) staining, in which staining can differentiate three different sperm populations: the uncapacitated and acrosome intact (F pattern), the capacitated and acrosome intact (B pattern) and the capacitated and acrosome reacted (AR pattern) (Kholkute et al. 1992; Dasgupta et al. 1994).

The fact that *in vivo*, acrosome reaction is induced by progesterone and zona proteins, evaluation of induced acrosome reaction is routinely used as a predictor of sperm quality for utilization in clinics for assisted reproductive technologies (ARTs) (Shimizu et al. 1993; Coetzee et al. 1994; Fusi et al. 1994; Yovich et al. 1994; Glazier et al. 2000; Makkar et al. 2003). Quite often, spontaneous acrosome reaction is also evaluated and correlated with sperm fertility (Bielsa et al. 1994; Parinaud et al. 1995; Tavalaee et al. 2014; Wiser et al. 2014).

5.4.3 Monitoring Tyrosine Phosphorylation (pY)

In human spermatozoa, increase in global protein tyrosine phosphorylation occurs during capacitation and is correlated with the fertilizing ability of the spermatozoa (Yunes et al. 2003; Liu et al. 2006; Barbonetti et al. 2008, 2010; Mendeluk et al. 2010; Kwon et al. 2014; Sati et al. 2014).

In spite of the importance of pY in human sperm capacitation, laboratory studies and clinic-based sperm-function tests on pY are very scarce. Such studies have to be in place to determine the predictive capability of pY of sperm fertility. As discussed for the kinases as well earlier, profiling of infertile patients' samples with appropriate controls is essential to develop sperm function tests based on this important molecular event during sperm capacitation.

5.5 From Bench to Clinics: Male Fertility Biomarkers and ARTs

There has been a steady rise in the molecular studies on the role of capacitation and its associated events (hyperactivation, acrosome reaction and tyrosine phosphorylation) in male fertility, *in vitro* (Fig. 5.5a, b). In spite of such extensive work being carried out at the laboratory level, these studies do not seem to have found application in the clinics yet. There are only a handful of clinics globally which seem to offer basic sperm capacitation/acrosome reaction tests as a part of routine sperm analysis, e.g. FIVMadrid; Poma Fertility; Androvia Life Sciences; University of Utah Hospitals and Clinics; the Male Fertility Lab, University of Washington; and Genetics & IVF Institute (references for website information). This data/information presented and discussed here is based on literature survey and searches on the World Wide Web, and real picture regarding the clinical usage of sperm capacitation tests might differ and remains to be determined.

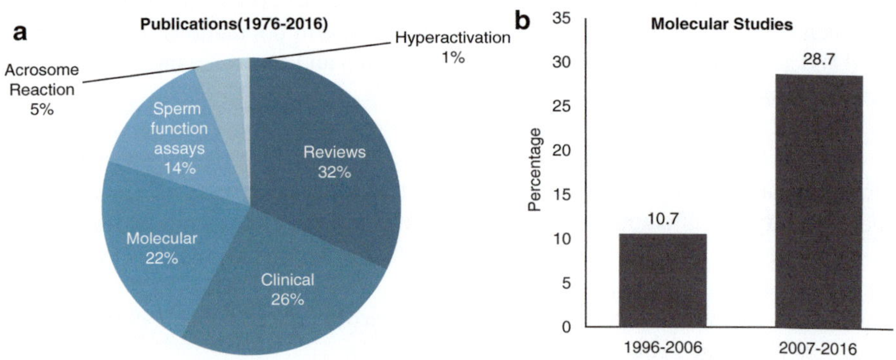

Fig. 5.5 (**a**) Relative percentages of studies in different categories (as shown) conducted over the last four decades worldwide (total number of studies =363). (**b**) Threefold increase in molecular studies has taken place in the last decade (2007–2016) as compared to the previous one (1996–2006). The publications were taken from PubMed (http://www.ncbi.nlm.nih.gov/pubmed/). The search was done using the following terms: [sperm capacitation, fertilization, infertility, human(s), infertility, infertile], [sperm, sperm capacitation, human(s), infertility, infertile, fertilization, hyperactivation] and [sperm, sperm capacitation, human(s), infertility, infertile, fertilization, acrosome reaction]. Based on their content, the publications (related to humans) were assigned to six categories and then percentages calculated, as shown in the pie chart

There is a pressing need to evaluate the potential of the capacitation-associated sperm molecules/events and sperm function tests as biomarkers (or predictors) of sperm fertility/infertility. One promising sperm function/molecular test in this direction has been the "Androvia Cap-Score™ test"—a clinical test based on sperm surface ganglioside, GM1 (http://www.androvialifesciences.com/cap-score-sperm-function-test/). This test is based on the work of Dr. Alex Travis and is based on the localization of GM1 on sperm head (Buttke et al. 2006; Selvaraj et al. 2007). GM1 is a sperm membrane component that regulates the opening and closing of specific calcium ion channels on the surface of sperm head. Androvia uses technology that identifies the ability of sperm to undergo capacitation. Since capacitation, hyperactivation and the acrosome reaction require an influx of calcium ions, by identifying the presence and location of GM1 in the sperm membrane across a number of sperm and identifying how many sperm are undergoing capacitation, a "**Cap-Score™**" can be generated that is predictive of the fertilizing ability of sperm in the ejaculate. The company Androvia claims that their preliminary research has already validated the ability of the test to discriminate between fertile and infertile populations of men, thus gaining clinical significance as a molecular marker/sperm function test.

Sperm capacitation and its associated events are the very basis of intrauterine insemination (IUI) and *in vitro* fertilization (IVF), the first line of ART management for couples with unexplained infertility/subfertility (Muratori et al. 2011; Wiser et al. 2014; Tosti and Ménézo 2016). In the cases of IUI and IVF, where cryopreserved spermatozoa are used, knowledge of sperm capacitation is especially useful for extending the health and life span of the sperm (and thus success of the ART), since it is known that freeze-thawed spermatozoa exhibit a precocious acrosome reaction-like phenotype, suggesting capacitation-like event during the process of cryopreservation (Gomez et al. 1997). Though extensively used in domestic species (such as bovine, pigs and dogs), it is well known and accepted that cryopreservation damages sperm, with a large number of cells losing their fertility potential after freezing/thawing (Cormier and Bailey 2003). Knowledge about mechanisms involved in capacitation/acrosome reaction would help in efforts towards minimizing the cryo-damage to spermatozoa and improve the success rate in ARTs, as being used in the livestock industry (Singh et al. 2014; Layek et al. 2016).

The life cycle of sperm is complex and involves a series of events, which have to be perfect for successful fertilization—viz. production in testis in sufficient numbers with normal shape, maturation in epididymis, gain of motility, successful capacitation, hyperactivation and acrosome reaction, oocyte binding and penetration, activation of the ovum and ultimately successful fertilization. All these parameters ought to be looked at in defining a "healthy" spermatozoon, and defects in any of these complex events can cause male infertility. The use of ICSI (intracytoplasmic sperm injection) bypasses many of these events, increasing the risk of choosing the "compromised spermatozoa". To avoid this, as already emphasized, it is imperative to develop new pre-ART molecular markers/sperm-function tests, for use in the clinics (Muratori et al. 2011; Natali and Turek 2011). Although few efforts to define predictive tests for ICSI success have already begun with limited success (Vural et al. 2005; Setti et al. 2012; Brown et al. 2013; Breznik et al. 2013; Meerschaut et al. 2013), further research in this direction is much needed.

Concluding Remarks

It is well accepted now that conventional semen analysis is unable to precisely predict sperm fertility potential, thus warranting search of biomarkers for fertility/infertility based on newer research (Weber et al. 2005; Lewis 2007; Lamb 2010). Attempts to translate the molecular information about capacitation—from laboratories to clinic—and to develop capacitation-based molecular markers/sperm function assays (besides other tests) is the need of the hour, especially in the era of assisted reproductive technologies like ICSI. The pre-ART tests would permit the clinicians and the infertile couples to make a more informed decision about the treatment/procedure and be assured of its success. It, thus, becomes necessary to continue improving our understanding of sperm capacitation, not only for the basic understanding of sperm physiology but also to understand its functionality, both *in vivo* and *in vitro*, ultimately translating into higher success rate in assisted reproductive technologies.

Acknowledgements The author is in receipt of grant from CSIR (XIIth FYP grant "PROGRAM") for research on molecular basis of sperm capacitation and gratefully acknowledges the support.

References

Aitken RJ (1995) Free radicals, lipid peroxidation and sperm function. Reprod Fertil Dev 7(4):659–668

Aitken RJ (2006) Sperm function tests and fertility. Int J Androl 29(1):69–75

Aitken RJ, Nixon B (2013) Sperm capacitation: a distant landscape glimpsed but unexplored| NOVA. The University of Newcastle's Digital Repository

Almog T, Lazar S, Reiss N, Etkovitz N, Milch E, Rahamim N, Dobkin-Bekman M, Rotem R, Kalina M, Ramon J, Raziel A (2008) Identification of extracellular signal-regulated kinase 1/2 and p38 MAPK as regulators of human sperm motility and acrosome reaction and as predictors of poor spermatozoan quality. J Biol Chem 283(21):14479–14489

Alvau A, Battistone MA, Gervasi MG, Navarrete FA, Xu X, Sánchez-Cárdenas C, De la Vega-Beltran JL, Da Ros VG, Greer P, Darszon A, Krapf D (2016) The tyrosine kinase FER is responsible for the capacitation-associated increase in tyrosine phosphorylation in murine sperm. Development 143(13):2325–2333

Androvia LifeSciences (2016). http://www.androvialifesciences.com/cap-score-sperm-function-test/. Accessed 11 Sept 2016

Aparicio IM, Bragado MJ, Gil MC, Garcia-Herreros M, Gonzalez-Fernandez L, Tapia JA, Garcia-Marin LJ (2007) Porcine sperm motility is regulated by serine phosphorylation of the glycogen synthase kinase-3α. Reproduction 134(3):435–444

Aquila S, Gentile M, Middea E, Catalano S, Andò S (2005) Autocrine regulation of insulin secretion in human ejaculated spermatozoa. Endocrinology 146(2):552–557

Arcelay E, Salicioni AM, Wertheimer E, Visconti PE (2008) Identification of proteins undergoing tyrosine phosphorylation during mouse sperm capacitation. Int J Dev Biol 52(5–6):463–472

Austin CR (1951) Observations on the penetration of the sperm in the mammalian egg. Aust J Sci Res B 4(4):581–596

Awda BJ, Buhr MM (2010) Extracellular signal-regulated kinases (ERKs) pathway and reactive oxygen species regulate tyrosine phosphorylation in capacitating boar spermatozoa. Biol Reprod 83(5):750–758

Baccetti B, Burrini AG, Collodel G, Piomboni P, Renieri T (1991) A "miniacrosome" sperm defect causing infertility in two brothers. J Androl 12(2):104–111
Bajpai M, Doncel GF (2003) Involvement of tyrosine kinase and cAMP-dependent kinase cross-talk in the regulation of human sperm motility. Reproduction 126(2):183–195
Baker MA, Hetherington L, Aitken RJ (2006) Identification of SRC as a key PKA-stimulated tyrosine kinase involved in the capacitation-associated hyperactivation of murine spermatozoa. J Cell Sci 119(15):3182–3192
Barbonetti A, Vassallo MRC, Cinque B, Antonangelo C, Sciarretta F, Santucci R, D'Angeli A, Francavilla S, Francavilla F (2008) Dynamics of the global tyrosine phosphorylation during capacitation and acquisition of the ability to fuse with oocytes in human spermatozoa. Biol Reprod 79(4):649–656
Barbonetti A, Vassallo MRC, Cordeschi G, Venetis D, Carboni A, Sperandio A, Felzani G, Francavilla S, Francavilla F (2010) Protein tyrosine phosphorylation of the human sperm head during capacitation: immunolocalization and relationship with acquisition of sperm-fertilizing ability. Asian J Androl 12(6):853–861
Bastiaan HS, Menkveld R, Oehninger S, Franken DR (2002) Zona pellucida induced acrosome reaction, sperm morphology, and sperm–zona binding assessments among subfertile men. J Assist Reprod Genet 19(7):329–334
Battistone MA, Da Ros VG, Salicioni AM, Navarrete FA, Krapf D, Visconti PE, Cuasnicu PS (2013) Functional human sperm capacitation requires both bicarbonate-dependent PKA activation and down-regulation of Ser/Thr phosphatases by Src family kinases. Mol Hum Reprod 19(9):570–580
Battistone MA, Alvau A, Salicioni AM, Visconti PE, Da Ros VG, Cuasnicú PS (2014) Evidence for the involvement of proline-rich tyrosine kinase 2 in tyrosine phosphorylation downstream of protein kinase A activation during human sperm capacitation. Mol Hum Reprod 20(11):1054–1066
Benoff S (1993) Preliminaries to fertilization: the role of cholesterol during capacitation of human spermatozoa. Hum Reprod 8(12):2001–2006
Bielfeld P, Anderson RA, Mack SR, De Jonge CJ, Zaneveld LJ (1994) Are capacitation or calcium ion influx required for the human sperm acrosome reaction? Fertil Steril 62(6):1255–1261
Bielsa MA, Andolz P, Gris JM, Martinez P, Egozcue J (1994) Andrology: which semen parameters have a predictive value for pregnancy in infertile couples? Hum Reprod 9(10):1887–1890
Bordeleau LJ, Leclerc P (2008) Expression of hck-tr, a truncated form of the src-related tyrosine kinase hck, in bovine spermatozoa and testis. Mol Reprod Dev 75:828–837
Bragado MJ, Gil MC, Martin-Hidalgo D, de Llera AH, Bravo N, Moreno AD, Garcia-Marin LJ (2012) Src family tyrosine kinase regulates acrosome reaction but not motility in porcine spermatozoa. Reproduction 144(1):67–75
Breitbart H, Etkovitz N (2011) Role and regulation of EGFR in actin remodeling in sperm capacitation and the acrosome reaction. Asian J Androl 13(1):106–110
Breznik BP, Kovačič B, Vlaisavljević V (2013) Are sperm DNA fragmentation, hyperactivation, and hyaluronan-binding ability predictive for fertilization and embryo development in in vitro fertilization and intracytoplasmic sperm injection? Fertil Steril 99(5):1233–1241
Brown DB, Merryman DC, Rivnay B, Houserman VL, Long CA, Honea KL (2013) Evaluating a novel panel of sperm function tests for utility in predicting intracytoplasmic sperm injection (ICSI) outcome. J Assist Reprod Genet 30(4):461–477
Brucker C, Lipford GB (1995) The human sperm acrosome reaction: physiology and regulatory mechanisms. An update. Hum Reprod Update 1(1):51–62
Buffone MG, Calamera JC, Verstraeten SV, Doncel GF (2005) Capacitation-associated protein tyrosine phosphorylation and membrane fluidity changes are impaired in the spermatozoa of asthenozoospermic patients. Reproduction 129(6):697–705
Buffone MG, Doncel GF, Calamera JC, Verstraeten SV (2009) Capacitation-associated changes in membrane fluidity in asthenozoospermic human spermatozoa. Int J Androl 32(4):360–375
Burkman LJ (1984) Characterization of hyperactivated motility by human spermatozoa during capacitation: comparison of fertile and oligozoospermic sperm populations. Arch Androl 13(2–3):153–165
Buttke DE, Nelson JL, Schlegel PN, Hunnicutt GR, Travis AJ (2006) Visualization of GM1 with cholera toxin B in live epididymal versus ejaculated bull, mouse, and human spermatozoa. Biol Reprod 74(5):889–895

Cardullo RA, Florman HM (1993) Strategies and methods for evaluating acrosome reaction. Methods Enzymol 225:136
Carrell DT (2000) Semen analysis at the turn of the century: an evaluation of potential uses of new sperm function assays. Arch Androl 44(1):65–75
Chang MC (1951) Fertilizing capacity of spermatozoa deposited into the fallopian tubes. Nature 168:697–698
Chang MC (1984) The meaning of sperm capacitation a historical perspective. J Androl 5(2):45–50
Chen WY, Ni Y, Pan YM, Shi QX, Yuan YY, Chen AJ, Mao LZ, Yu SQ, Roldan ER (2005) GABA, progesterone and zona pellucida activation of PLA2 and regulation by MEK-ERK1/2 during acrosomal exocytosis in Guinea pig spermatozoa. FEBS Lett 579(21):4692–4700
Cheng CY and Mruk DD (2012) The Blood-Testis Barrier and Its Implications for Male Contraception. Pharmacol Rev 64:16–64. doi: 10.1124/pr.110.002790
Chieffi P, Barchi M, Di Agostino S, Rossi P, Tramontano D, Geremia R (2003) Prolin-rich tyrosine kinase 2 (PYK2) expression and localization in mouse testis. Mol Reprod Dev 65(3):330–335
Coetzee K, Olmedo J, Lombard CJ (1994) Induced acrosome reactions as fertility predictor. J Assist Reprod Genet 11(9):470–473
Cormier N, Bailey JL (2003) A differential mechanism is involved during heparin-and cryopreservation-induced capacitation of bovine spermatozoa. Biol Reprod 69(1):177–185
Cotton L, Gibbs GM, Sanchez-Partida LG, Morrison JR, de Kretser DM, O'Bryan MK (2006) FGFR-1 signaling is involved in spermiogenesis and sperm capacitation. J Cell Sci 119(1):75–84
Cross NL (1998) Role of cholesterol in sperm capacitation. Biol Reprod 59(1):7–11
Cross NL, Meizel S (1989) Methods for evaluating the acrosomal status of mammalian sperm. Biol Reprod 41(4):635–641
Curry MR, Watson PF (1995) Sperm structure and function. Gametes—the spermatozoon. Cambridge University Press, Cambridge, pp 45–69
DasGupta S, O'Toole C, Mills CL, Fraser LR (1994) Fertilization and early embryology: effect of pentoxifylline and progesterone on human sperm capacitation and acrosomal exocytosis. Hum Reprod 9(11):2103–2109
Davis BK, Byrne R, Bedigian K (1980) Studies on the mechanism of capacitation: albumin-mediated changes in plasma membrane lipids during in vitro incubation of rat sperm cells. Proc Natl Acad Sci 77(3):1546–1550
De Jonge CJ, Barratt CL (2013) Methods for the assessment of sperm capacitation and acrosome reaction excluding the sperm penetration assay. Spermatogenesis: Methods and Protocols, 113–118.
Dow MP, Bavister BD (1989) Direct contact is required between serum albumin and hamster spermatozoa for capacitation in vitro. Gamete Res 23(2):171–180
Esposito G, Jaiswal BS, Xie F, Krajnc-Franken MA, Robben TJ, Strik AM, Kuil C, Philipsen RL, van Duin M, Conti M, Gossen JA (2004) Mice deficient for soluble adenylyl cyclase are infertile because of a severe sperm motility defect. Proc Natl Acad Sci U S A 101(9):2993–2998
Esteves SC, Sharma RK, Gosálvez J, Agarwal A (2014) A translational medicine appraisal of specialized andrology testing in unexplained male infertility. Int Urol Nephrol 46(6):1037–1052
Etkovitz N, Tirosh Y, Chazan R, Jaldety Y, Daniel L, Rubinstein S, Breitbart H (2009) Bovine sperm acrosome reaction induced by G-protein-coupled receptor agonists is mediated by epidermal growth factor receptor transactivation. Dev Biol 334:447–457
Feng HL, Sandlow JI, Zheng LJ (2005) C-kit receptor and its possible function in human spermatozoa. Mol Reprod Dev 70(1):103–110
FIVMadrid, http://fivmadrid.es/en/treatments-fivmadrid/semen-analysis-sperm-capacitation-test/, accessed on 11th September 2016
Flesch FM, Brouwers JF, Nievelstein PF, Verkleij AJ, van Golde LM, Colenbrander B, Gadella BM (2001) Bicarbonate stimulated phospholipid scrambling induces cholesterol redistribution and enables cholesterol depletion in the sperm plasma membrane. J Cell Sci 114(19):3543–3555
Freour T, Jean M, Mirallie S, Langlois ML, Dubourdieu S, Barriere P (2009) Predictive value of CASA parameters in IUI with frozen donor sperm. Int J Androl 32(5):498–504

Fusi FM, Viganò P, Daverio R, Busacca M, Vignali M (1994) Effects of the coculture with human endometrial cells on the function of spermatozoa from subfertile men. Fertil Steril 61(1):160–167

Gadella BM, Harrison RA (2000) The capacitating agent bicarbonate induces protein kinase A-dependent changes in phospholipid transbilayer behavior in the sperm plasma membrane. Development 127(11):2407–2420

Galantino-Homer HL, Visconti PE, Kopf GS (1997) Regulation of protein tyrosine phosphorylation during bovine sperm capacitation by a cyclic adenosine 3'5'-monophosphate-dependent pathway. Biol Reprod 56(3):707–719

Galantino-Homer HL, Florman HM, Storey BT, Dobrinski I, Kopf GS (2004) Bovine sperm capacitation: assessment of phosphodiesterase activity and intracellular alkalinization on capacitation-associated protein tyrosine phosphorylation. Mol Reprod Dev 67(4):487–500

Gallardo Bolaños JM, Balao da Silva C, Martin Munoz P, Plaza Davila M, Ezquerra J, Aparicio IM, Tapia JA, Ortega Ferrusola C, Peña FJ (2014) Caspase activation, hydrogen peroxide production and Akt dephosphorylation occur during stallion sperm senescence. Reprod Domest Anim 49(4):657–664

Genetics & IVF Institute (2016). http://www.givf.com/fertility/maleinfertility.shtml. Accessed 11 Sept 2016

Glazier DB, Diamond SM, Marmar JL, Gibbs M, Corson SL (2000) A modified acrosome induction test. Arch Androl 44(1):59–64

Go KJ, Wolf DP (1985) Albumin-mediated changes in sperm sterol content during capacitation. Biol Reprod 32(1):145–153

Gomez MC, Catt JW, Gillan L, Evans G, Maxwell WM (1997) Effect of culture, incubation and acrosome reaction of fresh and frozen-thawed ram spermatozoa for in vitro fertilization and intracytoplasmic sperm injection. Reprod Fertil Dev 9(7):665–673

González-Fernández L, Macías-García B, Loux SC, Varner DD, Hinrichs K (2013) Focal adhesion kinases and calcium/calmodulin-dependent protein kinases regulate protein tyrosine phosphorylation in stallion sperm. Biol Reprod 88(6):138

Goupil S, La Salle S, Trasler JM, Bordeleau LJ and Leclerc P. (2011) Developmental expression of SRC-related tyrosine kinases in the mouse testis. J Androl 2011;32:95–110

Grasa P, Cebrián-Pérez JÁ, Muiño-Blanco T (2006) Signal transduction mechanisms involved in in vitro ram sperm capacitation. Reproduction 132(5):721–732

Hildebrand MS, Avenarius MR, Fellous M, Zhang Y, Meyer NC, Auer J, Serres C, Kahrizi K, Najmabadi H, Beckmann JS, Smith RJ (2010) Genetic male infertility and mutation of CATSPER ion channels. Eur J Hum Genet 18(11):1178–1184

Ho HC, Suarez SS (2001) Hyperactivation of mammalian spermatozoa: function and regulation. Reproduction 122:519–526

Ho HC, Granish KA, Suarez SS (2002) Hyperactivated motility of bull sperm is triggered at the axoneme by Ca 2+ and not cAMP. Dev Biol 250(1):208–217

Hunter T (2000) Signaling—2000 and beyond. Cell 100(1):113–127

Ickowicz D, Finkelstein M, Breitbart H (2012) Mechanism of sperm capacitation and the acrosome reaction: role of protein kinases. Asian J Androl 14(6):816–821

Jaldety Y, Breitbart H (2015) ERK1/2 mediates sperm acrosome reaction through elevation of intracellular calcium concentration. Zygote 23(05):652–661

Jha KN, Kameshwari DB, Shivaji S (2003) Role of signaling pathways in regulating the capacitation of mammalian spermatozoa. Cell Mol Biol (Noisy-le-Grand) 49(3):329–340

Kalab P, Pěknicová J, Geussova G, Moos J (1998) Regulation of protein tyrosine phosphorylation in boar sperm through a cAMP-dependent pathway. Mol Reprod Dev 51(3):304–314

Katoh Y, Takebayashi K, Kikuchi A, Iki A, Kikuchi K, Tamba M, Kawashima A, Matsuda M, Okamura N (2014) Porcine sperm capacitation involves tyrosine phosphorylation and activation of aldose reductase. Reproduction 148(4):389–401

Kholkute SD, Meherji P, Puri CP (1992) Capacitation and the acrosome reaction in sperm from men with various semen profiles monitored by a chlortetracycline fluorescence assay. Int J Androl 15(1):43–53

Kierszenbaum AL, Rivkin E, Talmor-Cohen A, Shalgi R, Tres LL (2009) Expression of full-length and truncated Fyn tyrosine kinase transcripts and encoded proteins during spermatogenesis and localization during acrosome biogenesis and fertilization. Mol Reprod Dev 76:832–843. doi:10.1002/mrd.21049

Kong LJ, Shao B, Wang GL, Dai TT, Xu L, Huang JY (2008) Capacitation-associated changes in protein-tyrosine-phosphorylation, hyperactivation and acrosome reaction in Guinea pig sperm. Asian Australas J Anim Sci 21(2):181

Kota V, Dhople VM, Shivaji S (2009) Tyrosine phosphoproteome of hamster spermatozoa: role of glycerol-3-phosphate dehydrogenase 2 in sperm capacitation. Proteomics 9(7):1809–1826

Krapf D, Ruan YC, Wertheimer EV, Battistone MA, Pawlak JB, Sanjay A, Pilder SH, Cuasnicu P, Breton S, Visconti PE (2012). cSrc is necessary for epididymal development and is incorporated into sperm during epididymal transit. Dev Biol 369:43–53. doi: 10.1016/j.ydbio.2012.06.017

Kulanand J, Shivaji S (2001) Capacitation associated changes in protein tyrosine phosphorylation, hyperactivation and acrosome reaction in hamster spermatozoa. Andrologia 33:95–104

Kumar P, Meizel S (2005) Nicotinic acetylcholine receptor subunits and associated proteins in human sperm. J Biol Chem 280:25928–25935

Kwon WS, Rahman MS, Pang MG (2014) Diagnosis and prognosis of male infertility in mammal: the focusing of tyrosine phosphorylation and phosphotyrosine proteins. J Proteome Res 13(11):4505–4517

Lalancette C, Faure RL, Leclerc P (2006) Identification of the proteins present in the bull sperm cytosolic fraction enriched in tyrosine kinase activity: a proteomic approach. Proteomics 6:4523–4540

Lamb DJ (2010) Semen analysis in 21st century medicine: the need for sperm function testing. Asian J Androl 12(1):64–70

Lamirande E, Gagnon C (1992) Reactive oxygen species and human spermatozoa. J Androl 13(5):379–386

Larsen L, Scheike T, Jensen TK, Bonde JP, Ernst E, Hjollund NH, Zhou Y, Skakkebæk NE, Giwercman A, Danish First Pregnancy Planner Study Team (2000) Computer-assisted semen analysis parameters as predictors for fertility of men from the general population. Hum Reprod 15(7):1562–1567

Lawson C, Goupil S, Leclerc P (2008) Increased activity of the human sperm tyrosine kinase SRC by the cAMP-dependent pathway in the presence of calcium. Biol Reprod 79:657–666. doi: 10.1095/biolreprod.108.070367

Layek SS, Mohanty TK, Kumaresan A, Parks JE (2016) Cryopreservation of bull semen: evolution from egg yolk based to soybean based extenders. Anim Reprod Sci 172:1–9

Leahy T, Gadella BM (2011) Sperm surface changes and physiological consequences induced by sperm handling and storage. Reproduction 142(6):759–778

Leclerc P, de Lamirande E, Gagnon C (1996) Cyclic adenosine 3', 5'monophosphate-dependent regulation of protein tyrosine phosphorylation in relation to human sperm capacitation and motility. Biol Reprod 55(3):684–692

Lee LK, Foo KY (2014) Recent insights on the significance of transcriptomic and metabolomic analysis of male factor infertility. Clin Biochem 47(10):973–982

Lefièvre L, Jha KN, Lamirande E, Visconti PE, Gagnon C (2002) Activation of protein kinase a during human sperm capacitation and acrosome reaction. J Androl 23(5):709–716

Lefièvre L, Bedu-Addo K, Conner SJ, Machado-Oliveira GS, Chen Y, Kirkman-Brown JC, Afnan MA, Publicover SJ, Ford WCL, Barratt CL (2007) Counting sperm does not add up any more: time for a new equation? Reproduction 133(4):675–684

Lewis SE (2007) Is sperm evaluation useful in predicting human fertility? Reproduction 134(1):31–40

Leyton L, Saling P (1989) 95 kd sperm proteins bind ZP3 and serve as tyrosine kinase substrates in response to zona binding. Cell 57(7):1123–1130

Liu DY, Clarke GN, Baker HWG (2006) Tyrosine phosphorylation on capacitated human sperm tail detected by immunofluorescence correlates strongly with sperm–zona pellucida (ZP) binding but not with the ZP-induced acrosome reaction. Hum Reprod 21(4):1002–1008

Liu DY, Liu ML, Clarke GN, Baker HWG (2007) Hyperactivation of capacitated human sperm correlates with the zona pellucida-induced acrosome reaction of zona pellucida-bound sperm. Hum Reprod 22(10):2632–2638

Luna C, Colás C, Pérez-Pé R, Cebrián-Pérez JA, Muiño-Blanco T (2012) A novel epidermal growth factor-dependent extracellular signal-regulated MAP kinase cascade involved in sperm functionality in sheep. Biol Reprod 87(4):93

Luo J, Gupta V, Kern B, Tash JS, Sanchez G, Blanco G, Kinsey WH (2012) Role of FYN kinase in spermatogenesis: defects characteristic of Fynnull sperm in mice. Biol Reprod 86:1–8. doi: 10.1095/biolreprod.111.093864

Mahony MC, Gwathmey T (1999) Protein tyrosine phosphorylation during hyperactivated motility of cynomolgus monkey (Macaca fascicularis) spermatozoa. Biol Reprod 60:1239–1243

Makkar G, Ng EHY, Yeung WSB, Ho PC (2003) The significance of the ionophore-challenged acrosome reaction in the prediction of successful outcome of controlled ovarian stimulation and intrauterine insemination. Hum Reprod 18(3):534–539

Martínez P, Morros A (1996) Membrane lipid dynamics during human sperm capacitation. Front Biosci 1:d103–d117

Matzuk MM, Lamb DJ (2002) Genetic dissection of mammalian fertility pathways. Translocations 45:46XY

Meerschaut FV, Leybaert L, Nikiforaki D, Qian C, Heindryckx B, De Sutter P (2013) Diagnostic and prognostic value of calcium oscillatory pattern analysis for patients with ICSI fertilization failure. Hum Reprod 28(1):87–98

Meizel S, Turner KO (1991) Progesterone acts at the plasma membrane of human sperm. Mol Cell Endocrinol 77(1–3):R1–R5

Mendeluk GR, Sardi-Segovia LM, Chenlo PH, Pugliese MN, Repetto H, Curi S, Ariagno J, Prentki Santos E, Paez P, Passanante EG, Palaoro LA (2010) Assessment of human sperm protein tyrosine phosphorylation by immunocytochemistry in a clinical andrology laboratory. Preliminary data. Biotech Histochem 84(6):321–328

Mitchell LA, Nixon B, Baker MA, Aitken RJ (2008) Investigation of the role of SRC in capacitation-associated tyrosine phosphorylation of human spermatozoa. Mol Hum Reprod 14(4):235–243

Mitra K, Shivaji S (2004) Novel tyrosine-phosphorylated post-pyruvate metabolic enzyme, dihydrolipoamide dehydrogenase, involved in capacitation of hamster spermatozoa. Biol Reprod 70(4):887–899

Mitra K, Shivaji S (2005) Proteins implicated in sperm capacitation. Indian J Exp Biol 43(11):1001

Mortimer ST, Swan MA (1995) Variable kinematics of capacitating human spermatozoa. Hum Reprod 10:3178–3182

Morton B, Harrigan-Lum J, Albagli L, Jooss T (1974) The activation of motility in quiescent hamster sperm from the epididymis by calcium and cyclic nucleotides. Biochem Biophys Res Commun 56(2):372–379

Muller CH (2000) Rationale, interpretation, andrology lab corner validation, and uses of sperm function tests. J Androl 21(1):10–30

Munire M, Shimizu Y, Sakata Y, Minaguchi R, Aso T (2004) Impaired hyperactivation of human sperm in patients with infertility. J Med Dent Sci 51(1):99–104

Muratori M, Marchiani S, Tamburrino L, Forti G, Luconi M, Baldi E (2011) Markers of human sperm functions in the ICSI era. Front Biosci 16:1344–1363

Musset B, Clark RA, DeCoursey TE, Petheo GL, Geiszt M, Chen Y, Cornell JE, Eddy CA, Brzyski RG, El Jamali A (2012) NOX5 in human spermatozoa expression, function, and regulation. J Biol Chem 287(12):9376–9388

Nandi A, Homburg R (2016) Unexplained subfertility: diagnosis and management. Obstetrician Gynaecologist 18(2):107–115

Natali A, Turek PJ (2011) An assessment of new sperm tests for male infertility. Urology 77(5):1027–1034

Naz RK, Rajesh PB (2004). Role of tyrosine phosphorylation in sperm capacitation / acrosome reaction. Reprod Biol Endocrinol 2:75

Nixon B, Bielanowicz A, Anderson AL, Walsh A, Hall T, Mccloghry A, Aitken RJ (2010) Elucidation of the signaling pathways that underpin capacitation-associated surface phosphotyrosine expression in mouse spermatozoa. J Cell Physiol 224(1):71–83

O'Flaherty C, de Lamirande E, Gagnon C (2006) Reactive oxygen species modulate independent protein phosphorylation pathways during human sperm capacitation. Free Radic Biol Med 40(6):1045–1055

Oehninger S (1995) An update on the laboratory assessment of male fertility. Hum Reprod 10(suppl 1):38–45

Oehninger S, Franken DR, Ombelet W (2014) Sperm functional tests. Fertil Steril 102(6):1528–1533

Osheroff JE, Visconti PE, Valenzuela JP, Travis AJ, Alvarez J, Kopf GS (1999) Regulation of human sperm capacitation by a cholesterol efflux-stimulated signal transduction pathway leading to protein kinase A-mediated up-regulation of protein tyrosine phosphorylation. Mol Hum Reprod 5(11):1017–1026

Panneerdoss S, Siva AB, Kameshwari DB, Rangaraj N, Shivaji S (2012) Association of lactate, intracellular pH, and intracellular calcium during capacitation and acrosome reaction: contribution of hamster sperm dihydrolipoamide dehydrogenase, the E3 subunit of pyruvate dehydrogenase complex. J Androl 33(4):699–710

Parinaud J, Vieitez G, Moutaffian H, Richoilley G, Labal B (1995) Andrology: variations in spontaneous and induced acrosome reaction: correlations with semen parameters and in-vitro fertilization results. Hum Reprod 10(8):2085–2089

Pawson T (2004) Specificity in signal transduction: from phosphotyrosine-SH2 domain interactions to complex cellular systems. Cell 116(2):191–203

Poma Fertility (2016). www.pomafertility.com/patients/infertility/acrosome-reaction-test-ar/. Accessed 11 Sept 2016

Pommer AC, Rutllant J, Meyers SA (2003) Phosphorylation of protein tyrosine residues in fresh and cryopreserved stallion spermatozoa under capacitating conditions. Biol Reprod 68(4):1208–1214

Pukazhenthi BS, Long JA, Wildt DE, Ottinger MA, Armstrong DL, Howard J. (1998) Regulation of sperm function by protein tyrosine phosphorylation in diverse wild felid species. J Androl 19:675–685

Roa-Espitia AL, Hernández-Rendón ER, Baltiérrez-Hoyos R, Muñoz-Gotera RJ, Cote-Vélez A, Jiménez I, González-Márquez H, Hernández-González EO (2016) Focal adhesion kinase is required for actin polymerization and remodeling of the cytoskeleton during sperm capacitation. Biol Open 5(9):1189–1199

Rotfeld H, Hillman P, Ickowicz D, Breitbart H (2012) PKA and CaMKII mediate PI3K activation in bovine sperm by inhibition of the PKC/PP1 cascade. Reproduction 147(3):347–356

Sagare-Patil V, Vernekar M, Galvankar M, Modi D (2013) Progesterone utilizes the PI3K-AKT pathway in human spermatozoa to regulate motility and hyperactivation but not acrosome reaction. Mol Cell Endocrinol 374:82–91

Salvolini E, Buldreghini E, Lucarini G, Vignini A, Lenzi A, Di Primio R, Balercia G (2013) Involvement of sperm plasma membrane and cytoskeletal proteins in human male infertility. Fertil Steril 99(3):697–704

Sati L, Cayli S, Delpiano E, Sakkas D, Huszar G (2014) The pattern of tyrosine phosphorylation in human sperm in response to binding to zona pellucida or hyaluronic acid. Reprod Sci 21(5):573–581

Selvaraj V, Buttke DE, Asano A, McElwee JL, Wolff CA, Nelson JL, Klaus AV, Hunnicutt GR, Travis AJ (2007) GM1 dynamics as a marker for membrane changes associated with the process of capacitation in murine and bovine spermatozoa. J Androl 28(4):588–599

Setti AS, Braga DPAF, Figueira RCS, Iaconelli A Jr, Borges E Jr (2012) The predictive value of high-magnification sperm morphology examination on ICSI outcomes in the presence of oocyte dysmorphisms. J Assist Reprod Genet 29(11):1241–1247

Sharlip ID, Jarow JP, Belker AM, Lipshultz LI, Sigman M, Thomas AJ, Schlegel PN, Howards SS, Nehra A, Damewood MD, Overstreet JW (2002) Best practice policies for male infertility. Fertil Steril 77(5):873–882

Sheriff DS, Ali EF (2010) Perspective on plasma membrane cholesterol efflux and spermatozoal function. J Hum Reprod Sci 3(2):68

Shimizu Y, Nord EP, Bronson RA (1993) Progesterone-evoked increases in sperm [Ca 2+] i correlate with the egg penetrating ability of sperm from fertile but not infertile men. Fertil Steril 60(3):526–532

Shivaji S, Scheit KH, Bhargava PM (1990) Proteins of seminal plasma. Wiley

Shivaji S, Peedicayil J, Devi LG (1995) Analysis of the motility parameters of in vitro hyperactivated hamster spermatozoa. Mol Reprod Dev 42(2):233–247

Shivaji S, Kumar V, Mitra K, Jha KN (2007) Mammalian sperm capacitation: role of phosphotyrosine proteins. Soc Reprod Fertil Suppl 63:295–312

Shivaji S, Kota V, Siva AB (2009) The role of mitochondrial proteins in sperm capacitation. J Reprod Immunol 83(1):14–18

Sidhu KS, Mate KE, Gunasekera T, Veal D, Hetherington L, Baker MA, Aitken RJ, Rodger JC (2004) A flow cytometric assay for global estimation of tyrosine phosphorylation associated with capacitation of spermatozoa from two marsupial species, the tammar wallaby (*Macropus eugenii*) and the brushtail possum (*Trichosurus vulpecula*). Reproduction 127(1):95–103

Signorelli J, Diaz ES, Morales P (2012) Kinases, phosphatases and proteases during sperm capacitation. Cell Tissue Res 349(3):765–782

Singh DK, Deshmukh RK, Narayanan PK, Shivaji S, Siva AB (2017) SRC family kinases in hamster spermatozoa: Evidence for the presence of LCK. Reproduction. Accepted

Singh VK, Kumar R, Atreja SK (2014) Cryo-survival, cryo-capacitation and oxidative stress assessment of buffalo spermatozoa cryopreserved in new soya milk extender. Livest Sci 160:214–218

Suarez SS (2008) Control of hyperactivation in sperm. Hum Reprod Update 14(6):647–657

Suarez SS, Dai X (1992) Hyperactivation enhances mouse sperm capacity for penetrating viscoelastic media. Biol Reprod 46(4):686–691

Suarez SS, Katz DF, Owen DH, Andrew JB, Powell RL (1991) Evidence for the function of hyperactivated motility in sperm. Biol Reprod 44(2):375–381

Tardif S, Dubé C, Chevalier S, Bailey JL (2001) Capacitation is associated with tyrosine phosphorylation and tyrosine kinase-like activity of pig sperm proteins. Biol Reprod 65(3):784–792

Tardif S, Dubé C, Bailey JL (2003) Porcine sperm capacitation and tyrosine kinase activity are dependent on bicarbonate and calcium but protein tyrosine phosphorylation is only associated with calcium. Biol Reprod 68(1):207–213

Tavalaee M, Deemeh MR, Arbabian M, Kiyani A, Nasr-Esfahani MH (2014) Relationship between fertilization rate and early apoptosis in sperm population of infertile individuals. Andrologia 46(1):36–41

The Male Fertility Lab University of Washington (2016). http://faculty.washington.edu/cmuller/MFL/Sperm_Function_Tests.html. Accessed 11 Sept 2016

Thonneau P, Marchand S, Tallec A, Ferial ML, Ducot B, Lansac J, Lopes P, Tabaste JM, Spira A (1991) Incidence and main causes of infertility in a resident population (1 850 000) of three French regions (1988–1989). Hum Reprod 6(6):811–816

Tosti E, Ménézo Y (2016) Gamete activation: basic knowledge and clinical applications. Hum Reprod Update 22(4):420–439

Travis AJ, Kopf GS (2002) The role of cholesterol efflux in regulating the fertilization potential of mammalian spermatozoa. J Clin Invest 110(6):731–736

Tucker MJ, Leong MKH, Leung CKM, Wong CJY, Chan HHY (1987) Is delayed capacitation a complicating factor in the treatment of idiopathic infertility by intrauterine insemination? J Assist Reprod Genet 4(4):245–247

University of Utah Hospitals & Clinics (2016). http://healthcare.utah.edu/andrology/tests.php. Accessed 11 Sept 2016

Varano G, Lombardi A, Cantini G, Forti G, Baldi E, Luconi M (2008) Src activation triggers capacitation and acrosome reaction but not motility in human spermatozoa. Hum Reprod 23(12):2652–2662

Vasan SS (2011) Semen analysis and sperm function tests: how much to test? Indian J Urol (IJU) J Urol Soc India 27(1):41

Villemure M, Lazure C, Manjunath P (2003) Isolation and characterization of gelatin-binding proteins from goat seminal plasma. Reprod Biol Endocrinol 1(1):1

Visconti PE, Kopf GS (1998) Regulation of protein phosphorylation during sperm capacitation. Biol Reprod 59(1):1–6

Visconti PE, Bailey JL, Moore GD, Pan D, Olds-Clarke P, Kopf GS (1995) Capacitation of mouse spermatozoa. I. Correlation between the capacitation state and protein tyrosine phosphorylation. Development 121(4):1129–1137
Visconti PE, Galantino-Homer HANNAH, Moore GD, Bailey JL, Ning X, Fornes M, Kopf GS (1998) The molecular basis of sperm capacitation. J Androl 19(2):242–248
Visconti PE, Stewart-Savage J, Blasco A, Battaglia L, Miranda P, Kopf GS, Tezón JG (1999) Roles of bicarbonate, cAMP, and protein tyrosine phosphorylation on capacitation and the spontaneous acrosome reaction of hamster sperm. Biol Reprod 61(1):76–84
Visconti PE, Westbrook VA, Chertihin O, Demarco I, Sleight S, Diekman AB (2002) Novel signaling pathways involved in sperm acquisition of fertilizing capacity. J Reprod Immunol 53(1):133–150
Visconti PE, Krapf D, de la Vega-Beltrán JL, Acevedo JJ, Darszon A (2011) Ion channels, phosphorylation and mammalian sperm capacitation. Asian J Androl 13(3):395–405
Vizel R, Hillman P, Ickowicz D, Breitbart H (2015) AKAP3 degradation in sperm capacitation is regulated by its tyrosine phosphorylation. Biochimica et Biophysica Acta (BBA)-General Subjects 1850(9):1912–1920
Vural B, Sofuoglu K, Caliskan E, Delikara N, Aksoy E, Uslu H, Karan A (2005) Predictors of intracytoplasmic sperm injection (ICSI) outcome in couples with and without male factor infertility. Clin Exp Obstet Gynecol 32(3):158–162
Wang C, Leung A, Tsoi WL, Leung J, Ng V, Lee KF, Chan SY (1991) Evaluation of human sperm hyperactivated motility and its relationship with the zona-free hamster oocyte sperm penetration assay. J Androl 12(4):253–257
Wang J, Qi L, Huang S, Zhou T, Guo Y, Wang G, Guo X, Zhou Z, Sha J (2015) Quantitative phosphoproteomics analysis reveals a key role of insulin growth factor 1 receptor (IGF1R) tyrosine kinase in human sperm capacitation. Mol Cell Proteomics 14(4):1104–1112
Ward CR, Storey BT (1984) Determination of the time course of capacitation in mouse spermatozoa using a chlortetracycline fluorescence assay. Dev Biol 104(2):287–296
Weber RF, Dohle GR, Romijn JC (2005) Clinical laboratory evaluation of male subfertility. Adv Clin Chem 40:317–364
Wiser A, Sachar S, Ghetler Y, Shulman A, Breitbart H (2014) Assessment of sperm hyperactivated motility and acrosome reaction can discriminate the use of spermatozoa for conventional in vitro fertilisation or intracytoplasmic sperm injection: preliminary results. Andrologia 46(3):313–315
World Health Organization (2010) WHO laboratory manual for the examination and processing of human semen. World Health Organization, Geneva
Yanagimachi R (1969) In vitro capacitation of hamster spermatozoa by follicular fluid. J Reprod Fertil 18(2):275–286
Yanagimachi R (1994) Mammalian fertilization. Physiol Reprod 1:189–317
Youn JS, Cha SH, Park CW, Yang KM, Kim JY, Koong MK, Kang IS, Song IO, Han SC (2011) Predictive value of sperm motility characteristics assessed by computer-assisted sperm analysis in intrauterine insemination with superovulation in couples with unexplained infertility. Clin Exp Reprod Med 38(1):47–52
Yovich JM, Edirisinghe WR, Yovich JL (1994) Use of the acrosome reaction to ionophore challenge test in managing patients in an assisted reproduction program: a prospective, double-blind, randomized controlled study. Fertil Steril 61(5):902–910
Yunes R, Doncel GF, Acosta AA (2003) Incidence of sperm-tail tyrosine phosphorylation and hyperactivated motility in normozoospermic and asthenozoospermic human sperm samples. Biocell-Mendoza 27(1):29–36
Zaneveld LJD, De Jonge CJ, Anderson RA, Mack SR (1991) Human sperm capacitation and the acrosome reaction. Hum Reprod 6(9):1265–1274

Genomic Landscape of Human Y Chromosome and Male Infertility

6

Vertika Singh and Kiran Singh

Abstract

Initially thought to be functionally inert, the Y chromosome has now been established not only as a regulator organizer of sex determination and a functional hub for spermatogenesis but also as a genetic center involved in mediating autosomal functions and genome-wide expressions. The whole genome and transcriptome analysis of Y chromosome across different species have shed light on the origin, comparative gene content, and long-term providence of this interesting chromosome. Comparative studies further provided insights into the evolutionary and molecular forces driving Y degeneration toward evolutionary destiny. In the due course of evolution, the Y chromosome has undergone dynamic transformations and has evolved autonomously, gaining a lot of distinctive characteristics that no other chromosome possesses. An unusual architecture and dynamic nature has made it the most remarkable chromosome for genetic and molecular studies.

Keywords

Y chromosome • SRY and spermatogenic genes • Azoospermic factor (AZF) Spermatogenesis • Male infertility

Key Points

- Y is the most unstable chromosome characterized by having undergone drastic structural changes with respect to size and content.
- Y chromosome is distinguished by rooted pedigree of 153 Y chromosome haplogroups around the world.

V. Singh • K. Singh (✉)
Department of Molecular and Human Genetics, Institute of Science,
Banaras Hindu University, Varanasi, Uttar Pradesh, India
e-mail: singhk4@rediffmail.com

R. Singh, K. Singh (eds.), *Male Infertility: Understanding, Causes and Treatment*, DOI 10.1007/978-981-10-4017-7_6

- Tiepolo and Zuffardi in 1976 were the first to propose the presence of spermatogenic genes on the Y chromosome.
- Y chromosome has lost nearly 640 genes it once shared with the X chromosome.

6.1 Introduction

Y chromosome, which has been previously thought as a "biological wasteland," a "nonrecombining desert," or a "gene-poor chromosome," is now largely known for its functional significance during sex determination and spermatogenesis (Quintana-Murci and Fellous 2001). Y chromosome has always been in a state of an evolutionary drive, which created multiple Y chromosomes distinguished now by a rooted pedigree of at least 153 Y chromosome haplogroups around the world. This chromosome spans around 60 Mb length, out of which 3 Mb belongs to the pseudoautosomal region (PAR) involved in pairing with the X chromosome during meiosis and the rest 57 Mb to the nonrecombining region of Y chromosome (NRY), which harbors the heterochromatic and euchromatic regions. Skaletsky renamed this region as MSY (male-specific region of Y chromosome) as the designation NRY fails to justify the dynamic evolutionary events occurring on Y chromosome (Skaletsky et al. 2003). Most of the genes on the Y chromosome are present in the euchromatic region. The MSY comprised of 16 coding genes spanning around 10.2 Mb that are by and large single copy and ampliconic multi-copy class genes (Skaletsky et al. 2003). The absence of recombination across a large portion of the Y chromosome has hindered the construction of a Y chromosome linkage map, thus Y chromosome mapping has been based on naturally occurring deletions (Foresta et al. 2001). Verngaud and colleagues in 1986 mapped the Y chromosome into seven deletion intervals (1–7), out of which the short arm and centromere contained intervals 1–4, distal to proximal, and the euchromatic part represented by intervals 5 and 6, proximal to distal, and the heterochromatic region defined by interval 7. In 1992, Vollarath and colleagues further divided this seven interval map into 43 subintervals, which is the most largely accepted map (Foresta et al. 2001). This 43 interval deletion map of human Y chromosome relies exclusively on PCR technology containing specific monomorphic molecular markers across the whole Y chromosome, occurring only once in human genome called as STS (sequence-tagged sites). The euchromatic region of MSY encompasses a total of 23 Mb that contains 156 transcription units and 78 protein-coding genes that encode 27 distinct proteins (Skaletsky et al. 2003; Simoni et al. 2004). Vollrath et al. initially tried to map the Y chromosome and subdivided the Yq11 region; known to be the most frequently deleted region in infertile patients into 223 intervals termed 5A to 5Q and 6A to 6F. On the basis of sequence-tagged site deletion map, Vogt further established 25 intervals of D1 to D25 at Yq11 region.

DNA sequencing of the Y chromosome identified a strikingly unique feature, eight palindromes within the Yq region, designated as 1 to 8 from distal to proximal. These palindromic regions comprised of an array of ampliconic segments containing very long, near identical, direct, and indirect repeats, (Skaletsky et al. 2003) harboring most of the multi-copy genes. Intrachromosomal recombination events between these amplicons are believed to cause a high rate of de novo Y chromosome microdeletions, which are known to be a major cause of male infertility.

6.2 Y Chromosome and the Azoospermia Factor Region (AZF)

The role of Y chromosome as a functional niche for genes involved in the regulation of spermatogenesis has been appreciated since the mid-1970s. Tiepolo and Zuffardi in 1976 were the first to propose a correlation between Y chromosome deletions and human male infertility. By analyzing the karyotype of 1170 men, they observed large deletions in six infertile azoospermic males, which spanned the entire heterochromatic region (Yq12) and some adjacent euchromatic regions (Yq11). Two of these cases had their father carrying a normal Y chromosome indicating a de novo origin of these mutations. This study suggested the significance of these deletions as a cause of azoospermia, which further suggested that a genetic factor at Yq11 is important for male germ cell development. This was then called as "azoospermia factor" (AZF) region (Tiepolo and Zuffardi 1976). However, the genetic complexity of the AZF region remained unanswered until the STS- and YAC-based mapping on patients with microdeletions revealed that the determinants of these deletions display a tripartite organization (Vogt et al. 1996; Foresta et al. 2001). These particular regions regulating spermatogenesis were termed as AZFa, AZF, and AZFc from proximal to distal Yq. Furthermore, a fourth region, AZFd has been proposed, whose existence is still controversial.

Kuroda-Kawaguchi et al. (2001) showed that 47 out of 48 men carried a common proximal and distal breakpoint in 229 Kb amplicons flanking the AZFc region. This finding was further supported by the study of Repping et al. (2002) who showed that the majority of AZFb and AZFc deletions can be explained on the basis of recombination between the palindromic regions containing the ampliconic sequences. Intrachromosomal nonallelic homologous recombination (NAHR) events are more pronounced in the MSY region, mostly at the AZFc locus, which explains the high frequency of deletions at AZFc locus, among other AZF loci (Skaletsky et al. 2003; Machev et al. 2004; Vogt 2005).

6.2.1 AZFa

AZFa region is located in the proximal Yq11.21 region (D3-D6) at deletion interval 5 (subinterval 5C). The region spans around 1.1 Mb (Wimmer et al. 2003). It harbors single copy genes located in the X-degenerate region of the Y chromosome (Qureshi et al. 1996; Pryor et al. 1998). The genes of this region have been shown to be essential for normal spermatogenic functions (Vogt et al. 1992; Reijo et al. 1995). Deletions in the AZFa region have been shown to be associated with azoospermia (Georgiou et al. 2006; Krausz et al. 2006). The candidate genes found in this region are *USP9Y*, *DBY*, and *UTY* and about 11 pseudogenes. The very first gene identified and shown to be absent in infertile patients was *DFFRY* (*Drosophila* fat facets related Y), currently known as *USP9Y* (ubiquitin-specific protease 9, Y chromosome). This gene spans around 16 Kb with 17 exons and functions as a C-terminal ubiquitin-specific protease 9Y, involved in the regulation of protein metabolism (Sun et al. 1999; Kleiman et al. 2007). This gene functions as a "fine-tuner," increasing the efficiency of spermatogenesis. *USP9Y* encodes a nine-residue peptide, which has been shown to represent a new minor histocompatibility antigen (H-Y antigen) involved in graft rejection. Majority of the infertile males carrying AZFa deletion show the complete absence of this interval. Recently, an additional anonymous expressed sequence tag (AZFaT1) was mapped proximal to *USP9Y*, and the absence of *USP9Y* and/or AZFaT1 has been shown to be associated with an oligozoospermia phenotype, while a more severe phenotype (Sertoli cell-only syndrome) reflected the supplementary loss of *DBY* (Sargent et al. 1999). *DBY* is another functional single copy gene belonging to the AZFa region, which codes for an ATP-dependent RNA helicase in humans, thus playing a significant role during premeiotic spermatogonial stages. It consists of 17 exons spanning a length of 16 kb with a testis-specific expression pattern (Ditton et al. 2004). AZFa deletions occur as a consequence of homologous intrachromosomal recombination between two human endogenous retroviral sequences, HERV15uq1 and HERV15yq2, located at the proximal Yq11 region. Complete AZFa deletion removes around 792 kb of the sequence including both the functional genes of this region. Partial deletion of AZFa region has been shown to be associated with hypospermatogenesis; however, the complete deletion results incomplete loss of germ cell production and maturation leading to a Sertoli cell-only phenotype (Kamp et al. 2000; Suganthi et al. 2014). These studies have brought to the conclusion that the genes harbored in this region might play a role in fine-tuning rather than an indispensable role in regulating spermatogenesis.

6.2.2 AZFb

The AZFb region is located between subintervals 5 M to 6B (Vogt 1997). The region makes a 1.5 Mb overlap with the AZFc region and spans about 3.2 Mb, comprising several genes essential for normal spermatogenesis (Ferlin et al. 2003). After the identification of *RBMY*, several other single copy and multi-copy genes were discovered, some of which belonged solely to the AZFb region, while others were shared between the overlap between the AZFb and AZFc regions. *EIF1AY*, *RPS4Y2*, and *SMCY* (present on X-degenerate region) and *HSFY*, *XKRY*, *PRY*, and *RBMY* (located on the ampliconic region) are some of the important spermatogenic candidate genes located in the AZFb region. *RBMY* was the first evidence among the AZF candidate genes to be identified in the AZFb region. It is present in multiple copies in all eutherian placental mammals, encoding a testis-specific RNA-binding protein (Skaletsky et al. 2003). The *RBMY* gene is exclusively expressed in germ line in the testis (spermatogonia, spermatocyte, and round spermatid). Homologous recombination between the palindrome P1 of the Yq results in complete AZFb deletions. Complete AZFb deletion removes 6.23 Mb spanning 32 genes including all the members of testis-specific gene families located in the AZFb region. AZFb deletion has been shown to be associated with severe spermatogenesis failure with the loss of genes like *SMCY*, *EIFIAY*, *RPS4Y2*, and *HSFY* (Ferlin et al. 2003). An altered expression of *HSFY* was shown to be associated with severe infertile phenotype such as SCOS (Sertoli cell-only syndrome) and maturation arrest (Sato et al. 2006). The Y-linked gene (*HSFY*) encodes a protein similar to those regulated by the heat shock factor family, which plays an important role in sperm function. The *HSFY* expression is also seen in Sertoli cells and spermatogenic cells (Shinka et al. 2004). Recent reports also suggest the role of *PRY* and *EIF1AY* genes in spermatogenic impairment (Foresta et al. 2001; Sato et al. 2006). Another significant class of deletion includes the deletions extending from P5 to the distal arm of P1 (P5/distal-P1 deletions) and from P4 to the distal arm of P1 (P4/distal-P1 deletions) called the AZFbc deletions. The P5/distal-P1 deletion spans around 7.66 Mb region including 42 genes, whereas the P4/distal-P1 deletion removes around 7.03 Mb region taking along 38 gene copies. These deletions occur due to nonhomologous recombination between the P5/distal-P1 and P4/distal-P1. However, some reports have provided contradictory results suggesting homologous recombination to be the mechanism behind such kind of deletions. AZFbc deletion results in impaired spermatogenesis with variable phenotypes.

6.2.3 AZFc

AZFc is the most commonly deleted and best studied region of Y chromosome (MSY) mapped to the distal part of Yq or the deletion subintervals 6C–6E, hosting a number of spermatogenesis-related gene families (Yang et al. 2015; Silber 2000). Multiple studies have confirmed that complete AZFc deletion leads to azoospermia or severe oligozoospermia in different ethnic and geographical populations (Yang et al. 2015). However, whether or not different Y chromosome haplotypes/haplogroups show an association with infertility is questionable (McElreavey et al. 2000; Carvalho et al. 2003; Singh and Raman 2009). The region is composed of massive areas of nearly identical, repeated amplicons which are arranged in direct repeats, indirect repeats, or palindromes (Fig. 6.1).

The ampliconic regions undergo a frequent nonallelic homologous recombination (NAHR) giving rise to a variety of structural mutations. The AZFc region spans around 3.5 Mb, containing seven gene families with a total of 19 transcription units exclusively expressed in the testis. Among all the three major AZF regions, AZFc is considered to be the most divisive due to its variable nature (Navarro-Costa et al. 2010). The AZFc amplicons are organized in sequence families, with five different families color coded as blue, green, gray, and yellow each displaying a particular genetic signature harboring a total of 13 different ampliconic units (Kuroda-Kawaguchi et al. 2001; Navarro-Costa et al. 2010). The significance of these

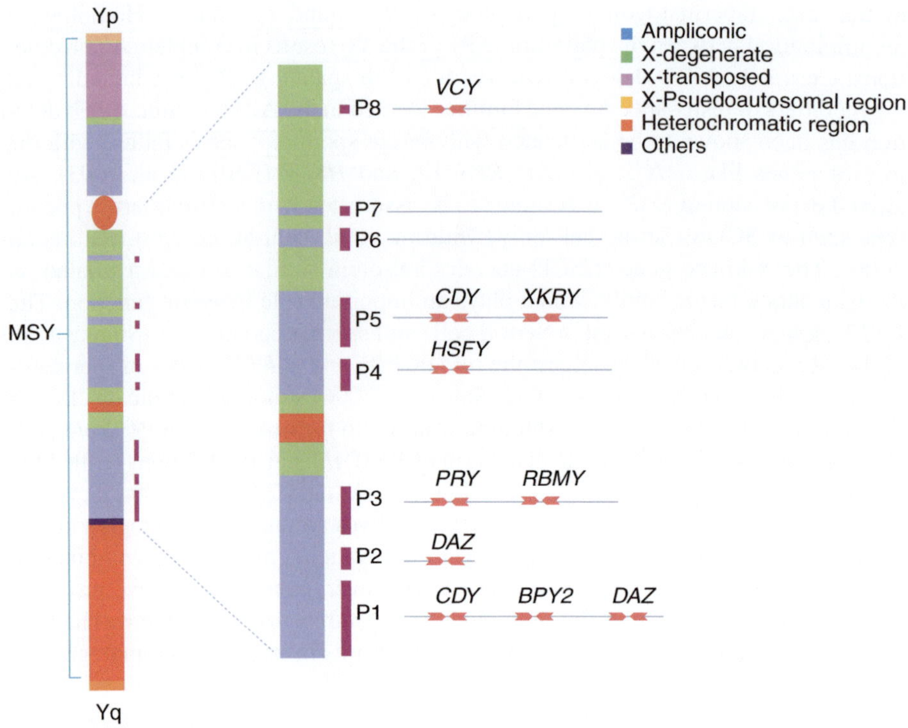

Fig. 6.1 Diagrammatic illustration of the organization of the euchromatic part of human Y chromosome. A substantial fraction of Y chromosome is made up of large ampliconic repeat units, organized as tandem arrays or as inverted repeats (palindromes). The eight palindromes in human cover up to 54% (5.5 Mb) of the ampliconic sequence (P1–P8)

ampliconic regions lies in harboring the genes responsible for spermatogenesis whose variable gene dosage due to change in ampliconic copy number might result in phenotypic alteration at the level of spermatogenesis (Navarro-Costa et al. 2010). Studies from Repping and colleagues in 2006 have demonstrated that AZFc rearrangements are the hotspots for large-scale Y chromosomal structural variations which occur due to high mutation rate in the interval as demonstrated by detection of 11 different AZFc architectures in four Y chromosomes representing the major evolutionary lineages in Y genealogy. It has been thought that the sequencing of AZFc across different Y evolutionary lineage may show new light on the genetic regulation of spermatogenesis (Navarro-Costa et al. 2010). Lack of recombination with a chromosome partner has been suggested to drive the use of amplicons for intrachromosomal recombination in the AZFc region to ensure genetic variability (Yen 2001; Repping et al. 2006; Lange et al. 2009; Navarro-Costa 2010). AZFc region has also evidenced non-homology-based recombination method through activation of nonhomologous DNA end joining (NHEJ) (Costa et al. 2008; Yang et al. 2008; Navarro-Costa et al. 2010). NHEJ is more frequent in nonduplicated region as it does not require DNA pairing for successful ligation (Navarro-Costa et al. 2010; Lieber 2010).

Recently, several types of AZFc partial deletions have been recognized including the gr/gr, b2/b3, and b1/b3 subdeletions. The gr/gr deletions are now newly defined as "gr/gr deletion rearrangements" which is further divided into five rearrangement types, that is, simple gr/gr deletion, gr/gr deletion-b2/b4 duplication, gr/gr deletion-b2/b4 multiple duplication, gr/gr deletion-CDY1, and DAZ amplification (Krausz et al. 2008; Shahid et al. 2011; Choi et al. 2012). These deletions occur as a result of recombination between the sub-amplicons located within the AZFc locus. Among all the gr/gr deletions are the most commonly detected deletions which excise around 1.6 Mb of the AZFc region covering two copies of *DAZ* gene, one copy of *CDY1* gene and one of the three copies of *BPY2* gene. The frequency of gr/gr deletions varies from 2.1 to 12.5% among all cases. gr/gr are the group of deletions resulting from recombination between the amplicons g1/g2, r1/r3, and r2/r4. As

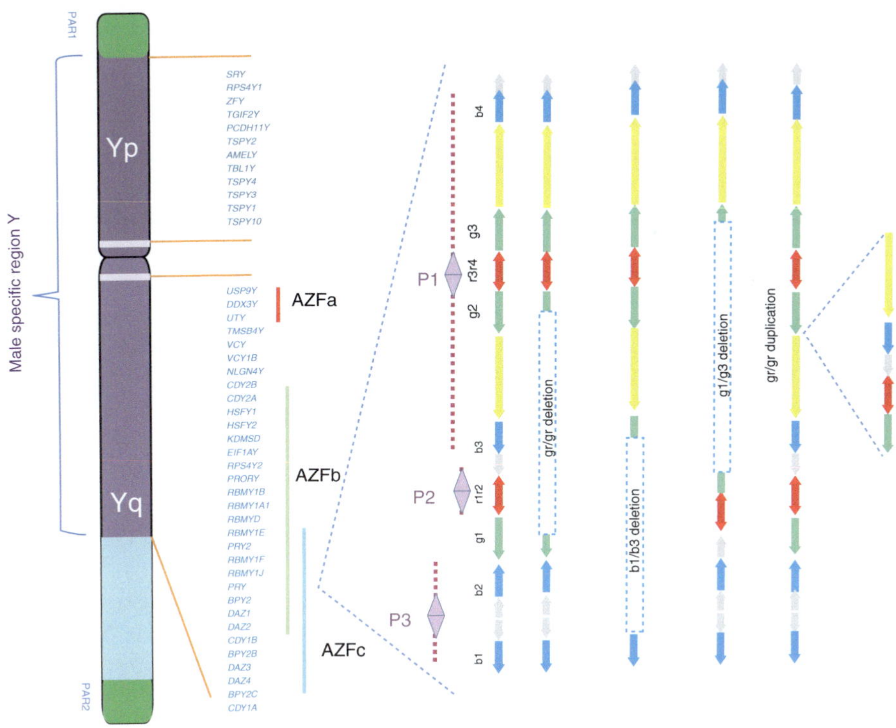

Fig. 6.2 Schematic diagram showing three AZF regions (AZFa, AZFb, and AZFc), their gene content, and various AZFc rearrangements that result in gr/gr, b1/b3, and g1/g3 deletions and gr/gr duplication. The organization of the symmetrical array of ampliconic repeats mapping to the AZFc loci (palindrome P1 to P3) is depicted by inverted triangles

these deletions are vertically transmitted, they are likely to reduce the sperm count subsequently reducing the fertility potential of the offspring (Fig. 6.2).

6.3 Sex-Determining Region Y (*SRY*)

The Sry locus (sex-determining region of the Y chromosome) is an evolutionary-conserved locus on mammalian Y chromosome responsible for testis determination in males. It functions as a developmental switch in the embryonic genital ridge of males, driving a bipotential gonad toward testicular differentiation (Gubbay et al. 1990; Sinclair Griffiths et al. 1990; Koopman et al. 1991; Tanaka and Nishinakamura 2014). Its primary function includes the differentiation of pre-Sertoli cells, an essential event in the testis differentiation of a bipotential gonad. *SRY* initiates the development of the testis by binding to a testis-specific enhancer of *SOX9*, a highly conserved gene that plays a central role in testis developmental program. The mutational analysis of the C-terminal domain of *SRY* suggests its function in regulating the conformation of *SRY*, a change of which may influence the *SRY* function. The N-terminal domain contains the nuclear localization signal (NLS), a mutation in which results in a decline in nuclear importation which somewhat explicate some of the cases of human sex reversal (Tanaka and Nishinakamura 2014). The *Sry* locus contains a highly conserved high mobility group (HMG) box region responsible for DNA binding. HMG domain of *Sry* elicits crucial events in the developmental process. Mutations in HMG box have been shown to be associated with sex reversal, while the mutations outside this region which show little conservation have normal sex determination (Schafer and Goodfellow 1996; Ely et al. 2010). These studies emphasize the functionality of HMG box as the sole functional unit of *Sry* locus. Present studies however show that *Sry* may also be involved in other functions evidenced by its expression in the brain, kidney, and adrenal gland of adult males (Milsted et al. 2004; Turner 2007; Turner et al. 2009; Turner et al. 2011). Recently its overexpression has been reported in some hepatocellular carcinoma cell lines such as K2 cells. The study also suggested a hypothetical model of Sry and *SGF29* pathway in male-specific malignancy of hepatocellular carcinoma by explaining that the altered expression of sry causes the augmentation of *SGF29* which is integrated into STAGA complex to enhance the C-myc target gene expression (Kurabe et al. 2015).

The HMG box of *SRY* and Sox proteins binds to the minor groove of DNA by recognizing a consensus DNA sequence AACAAT. Sry belongs to SOX B family of SOX loci, which includes Sox1, Sox2, Sox3, and Sry. In most of the mammals, Sox1 and Sox2 are autosomal while Sox3 is X-linked. It is proposed that Sry has originated from a mutation in Sox3 on the primordial mammalian X chromosome leading to the advent of Y chromosome. Multiple Sry copies on a single Y chromosome have been reported in some rodent species (Bullejos et al. 1997). These multiple copies occur probably as a result of repetitive organization of mammalian Y chromosome leading to many species-specific duplications and deletions. In human males exposed to high back radiations, multiple copies of Sry have been identified which were absent in normal control males (Premi et al. 2006).

6.4 Y Chromosome Has Significance Beyond Sex Determination and Spermatogenesis

The role of Y chromosome in sex determination has been already established. The recent advances in this area have shown that it is undergoing a rapid evolutionary deterioration. A large number of studies have speculated that Y chromosome could completely decay within the next 10 million years. Recent advances in this field however have provided some unexpected insights by analyzing the Y chromosome evolution independently in two separate sets of mammals covering more than 15 different species including human, chimpanzee, rhesus monkeys, bulls, marmosets, mice, rats, dogs, and opossums. They strikingly found a small but stable group of

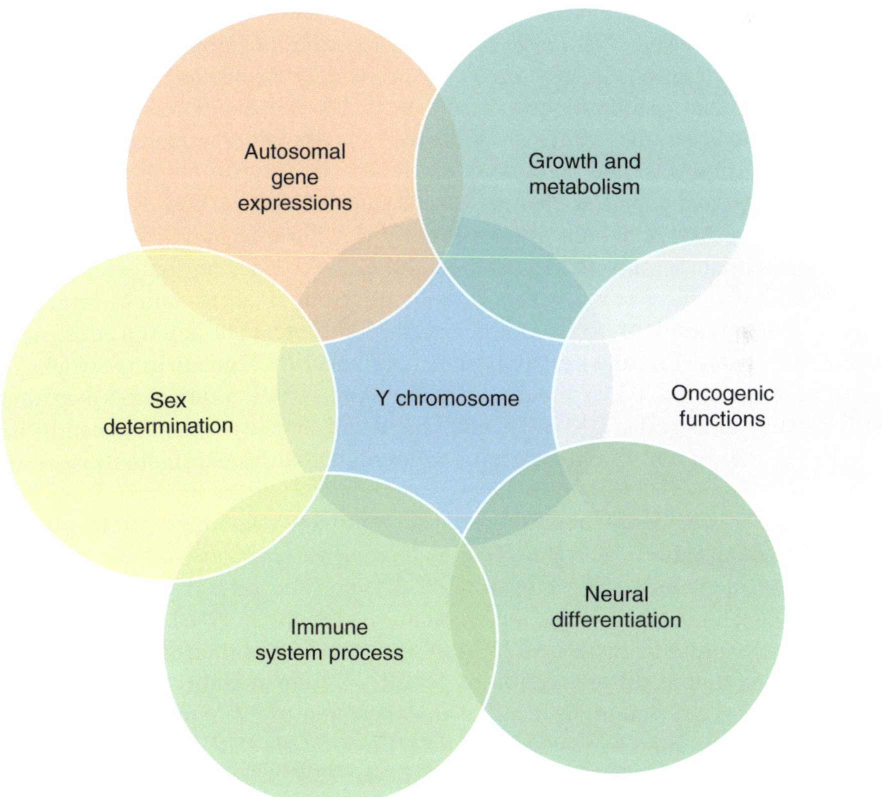

Fig. 6.3 Diagrammatic illustration of Y-linked regulatory functions beyond sex determination and reproduction

essential regulatory genes on Y chromosome that have endured over a long evolutionary period of time, while the surrounding genes were decaying. These genes have been studied to play a critically important role by regulating the expression of other genes throughout the genome (Fig. 6.3).

The reason for the continued endurance of these Y regulatory genes is their dose-dependent nature. These studies suggested the role of Y chromosome beyond sex determination and fertility. A few studies have observed that large deletions in the long arm of Y chromosome alter spermatogenesis and sex ratio distortion to a very little extent, showing its non-detrimental effect.

Sex-specific differences are observed in many diseases including cardiovascular diseases, which have higher evidence in males. A study using rat model has identified Y chromosome as a contributor to hypertension in males. Recently a group has established that the hormones and sex chromosomes have opposite contribution to hypertension generating effects that minimizes the overall sex differences.

In an experiment using FGF mice model to study hypertension, it has been shown that the males with intact gonad have increased blood pressure as compared to XY female, whereas gonadectomized (GDX) XX mice show increased mean arterial pressure in comparison with GDX XY mice, regardless of gonadal sex (Ji et al. 2010). Genetic variations in Y chromosome are widely associated with increase in male diastolic and systolic blood pressure in Polish and Scottish men (Charchar et al. 2002); however, the studies on cardiovascular risk factors such as blood pressure, cholesterol levels, and body mass index lack an association with Y chromosome in Polish and Japanese men. A recent study showed a 50% higher risk of coronary artery disease in men with inherited Y chromosome haplogroup I as compared to other haplogroups (Charchar et al. 2012).

A few studies have recently reported the differential expression of immunological and inflammatory pathway genes in Y chromosome haplogroups, with haplogroup I showing downregulation of *UTY* and *PRKY* genes in macrophage cells (Bloomer et al. 2013). Y chromosome linked variations are also associated with paucity in B cells, NK cells, and iNKT cell development, suggesting its role in regulation and maintenance of immunological homeostasis (Sun et al. 2013).

Although it is evident that chromosome compliment mediates sexually dimorphic expression pattern of some proteins leading to functional differences, the contribution of male-specific region of Y chromosome gene expression during neuronal development remains insufficient. A group of researchers recently reported the expression profile of 23 MSY genes and 15 of their X-linked homologues during neural differentiation of NTERA-2 human embryonal carcinoma cell line NT2. They found an increase in expression of 12 Y-linked genes over neuronal differentiation which included *RBMY1*, *EIF1AY*, *DDX3Y*, *HSFY1*, *BPY2*, *PCDH11Y*, *UTY*, *RPSY1*, *USP9Y*, *SRY*, *PRY*, and *ZFY*. They further showed that SiRNA-mediated knockdown of *DDX3Y*, a DEAD box helicase enzyme in neural progenitor cells, impairs cell cycle progression and apoptotic process consequently interrupting differentiation. They suggested MSY genes to play an important role in neural differentiation and established that *DDX3Y* could function in regulating neural cell development process in a sexually dimorphic manner (Vakilian et al. 2015).

The significance of Y chromosome has also been documented for its functional association with susceptibility toward numerous infectious diseases. It has been demonstrated that Y chromosome linked natural genetic variation affects the survival rate of B6 Y chromosome consomic mice infected with coxsaekievirus B3 (CVB3) (Case et al. 2012). Y chromosome haplogroups has been shown to be associated with progression of AIDS in European Americans (Sezgin et al. 2009). Additionally an age-related loss of Y chromosome (LOY) has been frequently reported in normal hematopoietic cells (Jacobs et al. 1963; Pierre and Hoagland 1972). A recent study has analyzed 1153 elderly men and reported that LOY in peripheral blood was associated with risks of all-cause mortality and non-hematological cancer mortality. The study has suggested the LOY in blood to be a predictive biomarker of male carcinogenesis (Forsberg et al. 2014).

6.5 Oncogenic Role of Y Chromosome

Nowadays a considerable attention has been focused on understanding the Y chromosomal aberrations in modulating susceptibility toward cancer progression. A higher incidence of the same has been found in males as compared to the females. Large Y chromosomal deletions and altered transcriptional events have been reported in numerous cancers (Mitelman et al. 2014). The genetic mapping studies have identified a distinctive locus on human MSY, the gonadoblastoma locus (GBY) located on the short arm of Y chromosome expanding over a region of 1–2 Mb proximal to the centromere. This locus harbors a gene predisposing dysgenetic gonads to develop gonadoblastoma (Page 1987). Further studies identified *TSPY* as a putative gene for this locus. This gene is present in multiple copies in the BGY region. The expression of gene is confined preferentially to the testicular germ cells; however, various studies have demonstrated its expression in somatic cells under disease condition such as cancer. A recent investigation based on data mining study on hepatocellular carcinoma (HCC) using RNA-Seq gene expression data of 27 pairs of male tumor and nontumor-paired samples at the Cancer Genome Atlas (TCGA) project has shown a consistent upregulation of genes such as *TGIF2LY* and *VCY* in 30% cases of liver cancer with frequent downregulation of *ZFY* and *DAZ1* in 70% of cases. These observations strongly suggest that Y-linked genes may predispose germ cells toward oncogenic progression and multistep process of tumorigenesis (Fig. 6.4).

Loss of Y chromosome from peripheral blood mononuclear cells (PBMC) in human is associated with increased risk of cancer (Forsberg et al. 2014). Recently a common deletion in Yp11.2 region has been observed in prostate tumor using a bacterial artificial chromosome microarray containing clones derived from Y

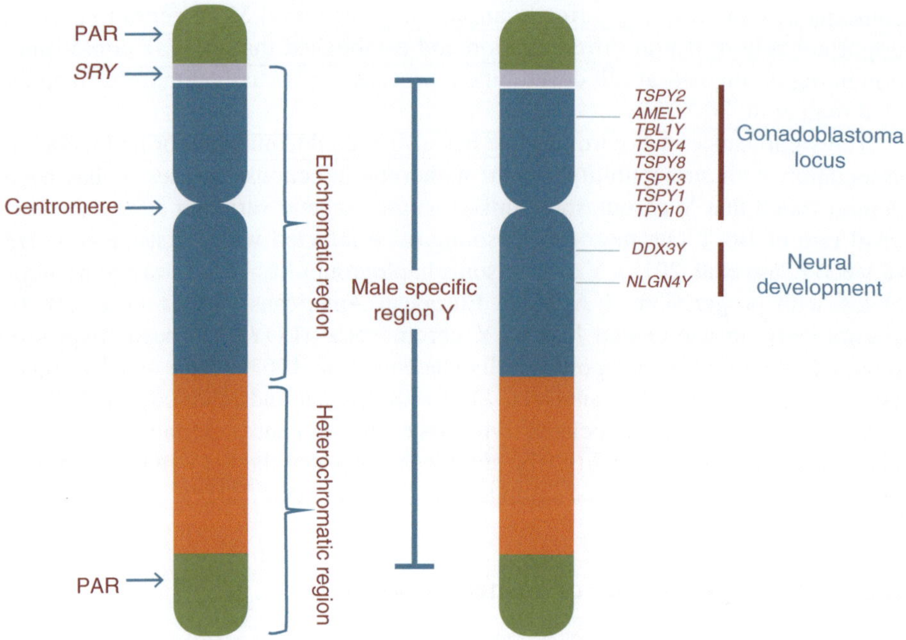

Fig. 6.4 Schematic diagram of Y chromosome showing (**a**) the euchromatic and the heterochromatic sequences on the male-specific region and the location of pseudoautosomal regions involved in pairing with X chromosome during meiosis and (**b**) indication of loci involved in the development of gonadoblastoma and some of the candidate Y genes involved in regulating neural differentiation and development

chromosome. Copy number variations in *TSPY* have also found to be associated with increased risk of prostate cancer (Bianchi 2009). However the role of Y chromosome in human oncogenesis is still controversial and needs further deep research insight.

6.6 Gene Conversions

The Y chromosome palindromes are particularly prone to such events reflecting a conversion rate of 2.8×10^{-4} per duplicated base per 25 years generation (Rozen et al. 2003). The present studies have reported that most of these events are selectively neutral in terms of reproductive fitness (de Vries et al. 2002; Repping et al. 2003; Machev et al. 2004; Zhang et al. 2006; Navarro-Costa et al. 2007; Giachini et al. 2008; Stouffs et al. 2008; Navarro-Costa et al. 2010).

However, the conversion pattern reflects variation between Y chromosome evolutionary lineages (Navarro-Costa et al. 2007). The mechanism of gene conversion pathway serves as a genetic correction mechanism for ampliconic genes (Skaletsky et al. 2003; Lange et al. 2009). This model explains a directional bias in the gene conversion event following the replacement of defective coding sequence with unaffected template.

The gene conversions act as a significant driver for AZFc variability playing an unpredictable role in functional regulation of interval (Navarro-Costa et al. 2010) protecting these genes from degeneration (Charlesworth 2003; Hawley 2003; Rozen et al. 2003). Thus it is expected that gene conversions might slow down the degeneration of Y-linked genes; nevertheless, the nonrecombining regions are thought to accumulate gene duplications because of an ineffective selection and elimination of duplications due to small population size (Lynch and Conery 2003; Lynch and Walsh 2007) though the growing belief that gene conversions might oppose Y degeneration has been criticized (Graves 2004). The upcoming reports suggest that the Y-Y gene conversions are much more frequent as compared to the X-Y gene conversion events.

6.7 Y Chromosome: Evolution and Degeneration

Although the mammalian X and Y chromosomes have evolved from the same autosomal ancestors around 200–300 million years ago, it has been established as highly divergent structures in the present day form. Restriction of recombination followed by a subsequent genetic loss has resulted in the morphological differentiation of sex chromosome (Bachtrog et al. 2014). Lack of recombination over most of the Y chromosome indicates less effective mechanism of nature in preventing the accumulation of deleterious mutations and driving the fixation of beneficial ones, subsequently leading to genetic erosion of Y chromosome (Bachtrog and Charlesworth 2001). Considering the gene content, Y chromosome has lost nearly all of the approximately 640 genes it once shared with the X chromosome (Hughes et al. 2015). Recent theories suggest that Y chromosome has evolved gradually by gene loss during a time scale, the pace of which slows down ultimately precipitating as paucity of genes and stasis (Hughes et al. 2010). The first step in the evolution of Y chromosome underlies the acquisition of a male-determining gene on one member of a pair of autosomes that ultimately will become the sex chromosome. The origin of a heteromorphic sex chromosome eventually after the acquisition of a male-determining gene requires suppression of recombination between the homomorphic proto-sex chromosomes, which in turn allows the Y chromosome to evolve independently of its X homologue (Bachtrog 2013). Numerous evolutionary models have been proposed for the Y chromosome degeneration over an evolutionary time scale, and the common feature among all of those is that the natural selection is not favored on a nonrecombining chromosome (Fig. 6.5).

An individual in a population is subjected to recurrent mutations; the beneficial ones get accumulated over time whereas the deleterious are selected against. On a recombining chromosome, the selection acts independently; however, an absence of recombination poses selection to operate on whole chromosome, which will be fixed if the acquired mutation is beneficial or whole of the chromosome is eliminated if it carries a deleterious mutation (Bachtrog 2013). It has been estimated that

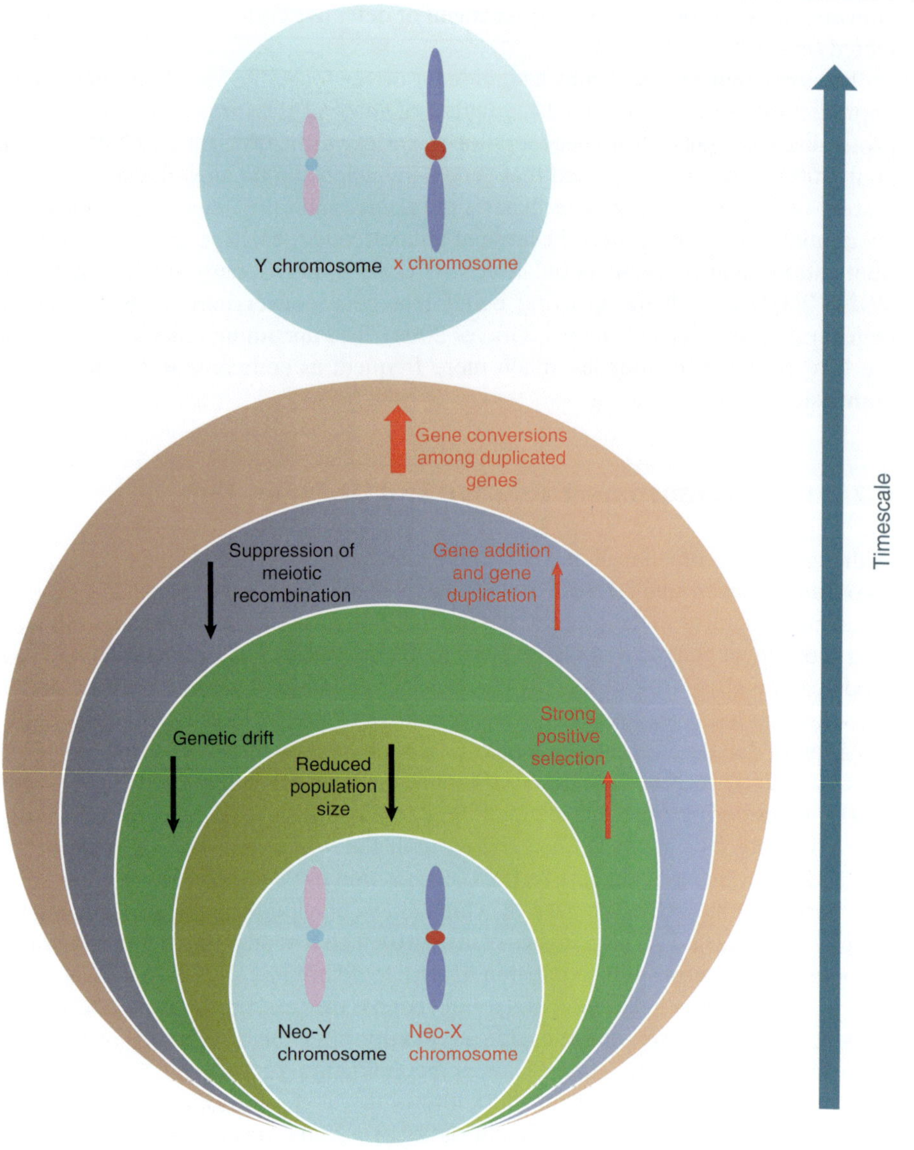

Fig. 6.5 Diagrammatic illustration of the steps in the evolution of nonrecombining sex chromosomes. In an evolutionary trajectory, the Y chromosome degenerated significantly over a period of time. The reduced population size subjected the Y to genetic drift, subsequently reducing the efficiency of natural selection. The absence of recombination over the majority of Y chromosome further enforced the suppression of natural selection; however, strong purifying selection and gene conversion events are some of the factors that counteracted the degenerative forces, maintaining the gene content of the Y chromosome

the human sex chromosomes originated around 150 million years ago from a pair of autosomes, initiated by the emergence of a male-determining gene sry (Lahn and Page 1999; Graves 1995; Vidal et al. 2001; Repping et al. 2002; Marais et al. 2010). The Y chromosome evolved as a recombinationally inert unit possibly by a mechanism underlying several inversion events which further leads to inefficient selection and reduced polymorphism, the so-called Hill-Robertson effect with the preferable accumulation and establishment of sex-antagonistic genes on Y chromosome (Charlesworth and Charlesworth 2000; Gordo and Charlesworth 2000). The crossing-over events are restricted to the two small regions called pseudoautosomal regions (PAR) that is the sole player for all meiotic crossing-over events in males (Lahn and Page 1999). The recent evidences have revealed that the nonrecombining MSY region of Y chromosome has eventually lost around 97% of genes it initially possessed. While the pseudoautosomal regions had perfectly normal characteristics, the MSY region accumulated a large amount of repeats (~ 80% of its current DNA) which eventually turned into heterochromatin (Skaletsky et al. 2003; Ross et al. 2005). The sequencing of Y chromosome provided a deeper insight to the functionality of MSY region of Y chromosome by proposing that the genes in the MSY region belong to nine gene families called as ampliconic genes. These genes undergo frequent gene conversions within each gene family contradictory to the established view of being a recombinationally inert region. The comparison of interparalogue divergence human chimp divergence estimated the level of gene conversions to be 1000-fold the genome average (Rozen et al. 2003; Marais et al. 2010). Interestingly many studies revealed that the ampliconic genes are testis specific and are involved in the regulation of spermatogenesis (Rozen et al. 2003; Skaletsky et al. 2003; Marais et al. 2010). This suggested a highly intricate mechanism of protecting these genes from degeneration by way of gene conversion and gene duplication events. It is strongly believed that the gene conversion events could eventually slow down the process of Y chromosome degeneration, which is supported by a recent investigation by Marias et al. who suggested that a high gene conversion is essential for an effective gene conversion to be observed. They also observed that the ampliconic regions on Y chromosome evolved gradually for an increased dosage being selected for large-scale duplication gene events. A high level of Y-Y gene conversions seen in human might have been selected to oppose the Y chromosome linked gene degeneration.

6.8 Y Chromosome: Regulation of Autosomal Gene Expression

There is emerging evidence from the studies performed on *Drosophila* that Y chromosome is a member of regulatory genome in males and may directly influence the gene expression, playing an imperative role in regulating male physiological states.

In an experiment using a mouse model of atherosclerosis with male-biased sexual dimorphism, a combined analysis of quantitative trait loci (QTL) mapping along with gene expression profiling (eQTL) of bone marrow-derived macrophages exhibited a remarkable differential expression of genes with a male or female expression bias. Further, Y chromosome in males represented a hotspot for trans-eQTL with around 334 characterized Y chromosome eQTLs. All these studies lead to the establishment of Y chromosome as a global regulator controlling genome-wide expression of various genes in mice (Bhasin et al. 2008). In experiments performed using B-6 Y chromosome consomic strains of mice, it has been demonstrated that Y chromosome regulates the transcriptome epigenetically in $CD4^+$ T cells and exhibits cell-type-specific effects based on autosomal background of mice. A copy number variation analysis in Y chromosome identified an inverse correlation between increase in copy number and upregulated genome-wide expression demonstrating Y chromosome trans-eQTL regulatory property (Case et al. 2013). Y chromosome linked variation affects the differential distribution of androgen receptors on the heart, along with differential chromatin modeling, suggesting its involvement in regulating chromatin dynamics. It has also been shown to regulate genome-wide expression profiles of pathogenic immune cells in males.

6.9 Future Prospects

The Y chromosome possesses unique characteristics in many aspects. Its role in sex determination and reproductive functions has been widely accepted. Its dynamic nature and distinctive properties have been demonstrated to provide extremely important information unraveling the evolutionary lineage of human populations. The upcoming studies have revealed its essential biological roles apart from reproductive functions making this chromosome a very important element of human genome. The theoretical and technological advances will provide a deeper insight into its role in diverse biological phenomena.

References

Bachtrog D (2013) Y-chromosome evolution: emerging insights into processes of Y-chromosome degeneration. Nat Rev Genet 14(2):113–124

Bachtrog D, Charlesworth B (2001) Towards a complete sequence of the human Y chromosome. Genome Biol 2(5):1

Bachtrog D, Mank JE, Peichel CL, Kirkpatrick M, Otto SP, Ashman TL, Hahn MW, Kitano J, Mayrose I, Ming R, Perrin N (2014) Sex determination: why so many ways of doing it? PLoS Biol 12(7):e1001899

Bhasin JM, Chakrabarti E, Peng DQ, Kulkarni A, Chen X, Smith JD (2008) Sex specific gene regulation and expression QTLs in mouse macrophages from a strain intercross. PLoS One 3(1):e1435

Bianchi NO (2009) Y chromosome structural and functional changes in human malignant diseases. Mutat Res/Rev Mutat Res 682(1):21–27

Bloomer LD, Nelson CP, Eales J, Denniff M, Christofidou P, Debiec R, Moore J, Zukowska-Szczechowska E, Goodall AH, Thompson J, Samani NJ (2013) Male-specific region of the Y chromosome and cardiovascular risk phylogenetic analysis and gene expression studies. Arterioscler Thromb Vasc Biol 33(7):1722–1727

Bullejos M, Sánchez A, Burgos M, Hera C, Jiménez R, Díaz de La Guardia R (1997) Multiple, polymorphic copies of SRY in both males and females of the vole *Microtus cabrerae*. Cytogenet Genome Res 79(3–4):167–171

Carvalho C, Fujisawa M, Shirakawa T, Gotoh A, Kamidono S, Freitas Paulo T, Santos SE, Rocha J, Pena SD, Santos FR (2003) Lack of association between Y chromosome haplogroups and male infertility in Japanese men. Am J Med Genet A 116(2):152–158

Case LK, Toussaint L, Moussawi M, Roberts B, Saligrama N, Brossay L, Huber SA, Teuscher C (2012) Y chromosome regulates survival following murine coxsackievirus b3 infection. G3: Genes|Genomes|Genetics 2(1):115–121

Case LK, Wall EH, Dragon JA, Saligrama N, Krementsov DN, Moussawi M, Zachary JF, Huber SA, Blankenhorn EP, Teuscher C (2013) The Y chromosome as a regulatory element shaping immune cell transcriptomes and susceptibility to autoimmune disease. Genome Res 23(9):1474–1485

Charchar FJ, Tomaszewski M, Padmanabhan S, Lacka B, Upton MN, Inglis GC, Anderson NH, McConnachie A, Zukowska-Szczechowska E, Grzeszczak W, Connell JM (2002) The Y chromosome effect on blood pressure in two European populations. Hypertension 39(2):353–356

Charchar FJ, Bloomer LD, Barnes TA, Cowley MJ, Nelson CP, Wang Y, Denniff M, Debiec R, Christofidou P, Nankervis S, Dominiczak AF (2012) Inheritance of coronary artery disease in men: an analysis of the role of the Y chromosome. Lancet 379(9819):915–922

Charlesworth B (2003) The organization and evolution of the human Y chromosome. Genome Biol 4(9):1

Charlesworth B, Charlesworth D (2000) The degeneration of Y chromosomes. Philos Trans R Soc Lond Ser B Biol Sci 355(1403):1563–1572

Choi J, Song SH, Bak CW, Sung SR, Yoon TK, Lee DR, Shim SH (2012) Impaired spermatogenesis and gr/gr deletions related to Y chromosome haplogroups in Korean men. PLoS One 7(8):e43550

Costa P, Gonçalves R, Ferrás C, Fernandes S, Fernandes AT, Sousa M, Barros A (2008) Identification of new breakpoints in AZFb and AZFc. Mol Hum Reprod 14(4):251–258

de Vries JW, Hoffer MJ, Repping S, Hoovers JM, Leschot NJ, van der Veen F (2002) Reduced copy number of DAZ genes in subfertile and infertile men. Fertil Steril 77(1):68–75

Ditton HJ, Zimmer J, Kamp C, Rajpert-De Meyts E, Vogt PH (2004) The AZFa gene DBY (DDX3Y) is widely transcribed but the protein is limited to the male germ cells by translation control. Hum Mol Genet 13(19):2333–2341

Ely D, Underwood A, Dunphy G, Boehme S, Turner M, Milsted A (2010) Review of the Y chromosome, sry and hypertension. Steroids 75(11):747–753

Ferlin A, Moro E, Rossi A, Dallapiccola B, Foresta C (2003) The human Y chromosome's azoospermia factor b (AZFb) region: sequence, structure, and deletion analysis in infertile men. J Med Genet 40(1):18–24

Foresta C, Moro E, Ferlin A (2001) Y chromosome microdeletions and alterations of spermatogenesis 1. Endocr Rev 22(2):226–239

Forsberg LA, Rasi C, Malmqvist N, Davies H, Pasupulati S, Pakalapati G, Sandgren J, de Ståhl TD, Zaghlool A, Giedraitis V, Lannfelt L (2014) Mosaic loss of Y chromosome in peripheral blood is associated with shorter survival and higher risk of cancer. Nat Genet 46(6):624–628

Georgiou I, Syrrou M, Pardalidis N, Karakitsios K, Mantzavinos T, Giotitsas N, Loutradis D, Dimitriadis F, Saito M, Miyagawa I, Tzoumis P (2006) Genetic and epigenetic risks of intracytoplasmic sperm injection method. Asian J Androl 8(6):643–673

Giachini C, Laface I, Guarducci E, Balercia G, Forti G, Krausz C (2008) Partial AZFc deletions and duplications: clinical correlates in the Italian population. Hum Genet 124(4):399–410

Gordo I, Charlesworth B (2000) The degeneration of asexual haploid populations and the speed of Muller's ratchet. Genetics 154(3):1379–1387

Graves JAM (1995) The origin and function of the mammalian Y chromosome and Y-borne genes—an evolving understanding. BioEssays 17(4):311–320

Graves JAM (2004) The degenerate Y chromosome–can conversion save it? Reprod Fertil Dev 16(5):527–534

Gubbay J, Collignon J, Koopman P, Capel B, Economou A, Münsterberg A, Vivian N, Goodfellow P, Lovell-Badge R (1990) A gene mapping to the sex-determining region of the mouse Y chromosome is a member of a novel family of embryonically expressed genes. Nature 346(6281):245–250

Hawley RS (2003) The human Y chromosome: rumors of its death have been greatly exaggerated. Cell 113(7):825–828

Hughes JF, Skaletsky H, Pyntikova T, Graves TA, van Daalen SK, Minx PJ, Fulton RS, McGrath SD, Locke DP, Friedman C, Trask BJ (2010) Chimpanzee and human Y chromosomes are remarkably divergent in structure and gene content. Nature 463(7280):536–539

Hughes JF, Skaletsky H, Koutseva N, Pyntikova T, Page DC (2015) Sex chromosome-to-autosome transposition events counter Y-chromosome gene loss in mammals. Genome Biol 16(1):1

Jacobs PA, Brunton M, Brown WC, Doll R, Goldstein H (1963) Change of human chromosome count distributions with age: evidence for a sex difference. Nature 197:1080

Ji H, Zheng W, Wu X, Liu J, Ecelbarger CM, Watkins R, Arnold AP, Sandberg K (2010) Sex chromosome effects unmasked in angiotensin II–induced hypertension. Hypertension 55(5):1275–1282

Kamp C, Hirschmann P, Voss H, Huellen K, Vogt PH (2000) Two long homologous retroviral sequence blocks in proximal Yq11 cause AZFa microdeletions as a result of intrachromosomal recombination events. Hum Mol Genet 9(17):2563–2572

Kleiman SE, Yogev L, Hauser R, Botchan A, Maymon BS, Paz G, Yavetz H (2007) Expression profile of AZF genes in testicular biopsies of azoospermic men. Hum Reprod 22(1):151–158

Koopman P, Gubbay J, Vivian N, Goodfellow P, Lovell-Badge R (1991) Male development of chromosomally female mice transgenic for sry. Nature 351(6322):117–121

Krausz C, Degl'Innocenti S, Nuti F, Morelli A, Felici F, Sansone M, Varriale G, Forti G (2006) Natural transmission of USP9Y gene mutations: a new perspective on the role of AZFa genes in male fertility. Hum Mol Genet 15(18):2673–2681

Krausz C, Giachini C, Xue Y, O'Bryan MK, Gromoll J, Rajpert-de Meyts E, Oliva R, Aknin-Seifer I, Erdei E, Jorgensen N, Simoni M (2008) Phenotypic variation within European carriers of the Y-chromosomal gr/gr deletion is independent of Y-chromosomal background. J Med Genet 46(1):21–31

Kurabe N, Murakami S, Tashiro F (2015) SGF29 and sry pathway in hepatocarcinogenesis. World J Biol Chem 6(3):139

Kuroda-Kawaguchi T, Skaletsky H, Brown LG, Minx PJ, Cordum HS, Waterston RH, Wilson RK, Silber S, Oates R, Rozen S, Page DC (2001) The AZFc region of the Y chromosome features massive palindromes and uniform recurrent deletions in infertile men. Nat Genet 29(3):279–286

Lahn BT, Page DC (1999) Four evolutionary strata on the human X chromosome. Science 286(5441):964–967

Lange J, Skaletsky H, van Daalen SK, Embry SL, Korver CM, Brown LG, Oates RD, Silber S, Repping S, Page DC (2009) Isodicentric Y chromosomes and sex disorders as byproducts of homologous recombination that maintains palindromes. Cell 138(5):855–869

Lieber MR (2010) The mechanism of double-strand DNA break repair by the nonhomologous DNA end joining pathway. Annu Rev Biochem 79:181

Lynch M, Conery JS (2003) The origins of genome complexity. Science 302(5649):1401–1404

Lynch M, Walsh B (2007) The origins of genome architecture, vol 98. Sinauer Associates, Sunderland

Machev N, Saut N, Longepied G, Terriou P, Navarro A, Levy N, Guichaoua M, Metzler-Guillemain C, Collignon P, Frances AM, Belougne J (2004) Sequence family variant loss from the AZFc interval of the human Y chromosome, but not gene copy loss, is strongly associated with male infertility. J Med Genet 41(11):814–825

Marais GA, Campos PR, Gordo I (2010) Can intra-Y gene conversion oppose the degeneration of the human Y chromosome? A simulation study. Genome Biol Evol 2:347–357
McElreavey K, Krausz C, Bishop CE (2000) The human Y chromosome and male infertility. In: The genetic basis of male infertility. Springer, Berlin Heidelberg, pp 211–232
Milsted A, Serova L, Sabban EL, Dunphy G, Turner ME, Ely DL (2004) Regulation of tyrosine hydroxylase gene transcription by sry. Neurosci Lett 369(3):203–207
Mitelman F, Johansson B, Mertens F (2014) Mitelman database of chromosome aberrations and gene fusions in cancer [Internet]. Center for Bioinformatics: National Cancer Institute
Navarro-Costa P, Pereira L, Alves C, Gusmão L, Proença C, Marques-Vidal P, Rocha T, Correia SC, Jorge S, Neves A, Soares AP (2007) Characterizing partial AZFc deletions of the Y chromosome with amplicon-specific sequence markers. BMC Genomics 8(1):342
Navarro-Costa P, Gonçalves J, Plancha CE (2010) The AZFc region of the Y chromosome: at the crossroads between genetic diversity and male infertility. Hum Reprod Update 16(5): 525–542
Page DC (1987) Hypothesis: a Y-chromosomal gene causes gonadoblastoma in dysgenetic gonads. Development 101(Supplement):151–155
Pierre RV, Hoagland HC (1972) Age-associated aneuploidy: loss of Y chromosome from human bone marrow cells with aging. Cancer 30(4):889–894
Premi S, Srivastava J, Chandy SP, Ahmad J, Ali S (2006) Tandem duplication and copy number polymorphism of the SRY gene in patients with sex chromosome anomalies and males exposed to natural background radiation. Mol Hum Reprod 12(2):113–121
Pryor JL, Kent-First M, Muallem A, Van Bergen AH, Nolten WE, Meisner L, Roberts KP (1998) Microdeletions in the Y chromosome of infertile men. J Urol 159(2):608–609
Quintana-Murci L, Fellous M (2001) The human Y chromosome: the biological role of a "functional wasteland". Biomed Res Int 1(1):18–24
Qureshi SJ, Ross AR, Ma K, Cooke HJ, Chandley AC, Hargreave TB (1996) Polymerase chain reaction screening for Y chromosome microdeletions: a first step towards the diagnosis of genetically-determined spermatogenic failure in men. Mol Hum Reprod 2(10):775–779
Reijo R, Lee TY, Salo P, Alagappan R, Brown LG, Rosenberg M, Rozen S, Jaffe T, Straus D, Hovatta O, Chapelle ADL (1995) Diverse spermatogenic defects in humans caused by Y chromosome deletions encompassing a novel RNA-binding protein gene. Nat Genet 10(4):383–393
Repping S, Skaletsky H, Lange J, Silber S, van der Veen F, Oates RD, Page DC, Rozen S (2002) Recombination between palindromes P5 and P1 on the human Y chromosome causes massive deletions and spermatogenic failure. Am J Hum Genet 71(4):906–922
Repping S, Skaletsky H, Brown L, van Daalen SK, Korver CM, Pyntikova T, Kuroda-Kawaguchi T, de Vries JW, Oates RD, Silber S, van der Veen F (2003) Polymorphism for a 1.6-Mb deletion of the human Y chromosome persists through balance between recurrent mutation and haploid selection. Nat Genet 35(3):247–251
Repping S, van Daalen SK, Brown LG, Korver CM, Lange J, Marszalek JD, Pyntikova T, van der Veen F, Skaletsky H, Page DC, Rozen S (2006) High mutation rates have driven extensive structural polymorphism among human Y chromosomes. Nat Genet 38(4):463–467
Ross MT, Grafham DV, Coffey AJ, Scherer S, McLay K, Muzny D, Platzer M, Howell GR, Burrows C, Bird CP, Frankish A (2005) The DNA sequence of the human X chromosome. Nature 434(7031):325–337
Rozen S, Skaletsky H, Marszalek JD, Minx PJ, Cordum HS, Waterston RH, Wilson RK, Page DC (2003) Abundant gene conversion between arms of palindromes in human and ape Y chromosomes. Nature 423(6942):873–876
Sargent CA, Boucher CA, Kirsch S, Brown G, Weiss B, Trundley A, Burgoyne P, Saut N, Durand C, Levy N, Terriou P (1999) The critical region of overlap defining the AZFa male infertility interval of proximal Yq contains three transcribed sequences. J Med Genet 36(9):670–677
Sato Y, Yoshida K, Shinka T, Nozawa S, Nakahori Y, Iwamoto T (2006) Altered expression pattern of heat shock transcription factor, Y chromosome (HSFY) may be related to altered differentia-

tion of spermatogenic cells in testes with deteriorated spermatogenesis. Fertil Steril 86(3):612–618
Schafer AJ, Goodfellow PN (1996) Sex determination in humans. BioEssays 18(12):955–963
Sezgin E, Lind JM, Shrestha S, Hendrickson S, Goedert JJ, Donfield S, Kirk GD, Phair JP, Troyer JL, O'Brien SJ, Smith MW (2009) Association of Y chromosome haplogroup I with HIV progression, and HAART outcome. Hum Genet 125(3):281–294
Shahid M, Dhillon VS, Khalil HS, Sexana A, Husain SA (2011) Associations of Y-chromosome subdeletion gr/gr with the prevalence of Y-chromosome haplogroups in infertile patients. Eur J Hum Genet 19(1):23–29
Shinka T, Sato Y, Chen G, Naroda T, Kinoshita K, Unemi Y, Tsuji K, Toida K, Iwamoto T, Nakahori Y (2004) Molecular characterization of heat shock-like factor encoded on the human Y chromosome, and implications for male infertility. Biol Reprod 71(1):297–306
Silber SJ (2000) Microsurgical TESE and the distribution of spermatogenesis in non-obstructive azoospermia. Hum Reprod 15(11):2278–2284
Simoni M, Bakker E, Krausz C (2004) EAA/EMQN best practice guidelines for molecular diagnosis of Y-chromosomal microdeletions. State of the art 2004. Int J Androl 27(4):240–249
Sinclair Griffiths BL, Smith MJ, Foster JW, Frischauf AM, Lovell-Badge R, Goodfellow PN (1990) A gene from the human sex-determining region encodes a protein with homology to a conserved DNA-binding motif. Nature 346:240–244
Singh K, Raman R (2009) A386G polymorphism of the DAZL gene is not associated with idiopathic male infertility in North India. J Hum Reprod Sci 2(2):54
Skaletsky H, Kuroda-Kawaguchi T, Minx PJ, Cordum HS, Hillier L, Brown LG, Repping S, Pyntikova T, Ali J, Bieri T, Chinwalla A (2003) The male-specific region of the human Y chromosome is a mosaic of discrete sequence classes. Nature 423(6942):825–837
Stouffs K, Tournaye H, Van der Elst J, Haentjens P, Liebaers I, Lissens W (2008) Do we need to search for gr/gr deletions in infertile men in a clinical setting? Hum Reprod 23(5):1193–1199
Suganthi R, Vijesh VV, Vandana N, Benazir JFA (2014) Y choromosomal microdeletion screening in the workup of male infertility and its current status in India. Int J Fertil Steril 7(4):253
Sun C, Skaletsky H, Birren B, Devon K, Tang Z, Silber S, Oates R, Page DC (1999) An azoospermic man with a de novo point mutation in the Y-chromosomal gene USP9Y. Nat Genet 23(4):429–432
Sun SL, Horino S, Itoh-Nakadai A, Kawabe T, Asao A, Takahashi T, So T, Funayama R, Kondo M, Saitsu H, Matsumoto N (2013) Y chromosome–linked B and NK cell deficiency in mice. J Immunol 190(12):6209–6220
Tanaka SS, Nishinakamura R (2014) Regulation of male sex determination: genital ridge formation and sry activation in mice. Cell Mol Life Sci 71(24):4781–4802
Tiepolo L, Zuffardi O (1976) Localization of factors controlling spermatogenesis in the nonfluorescent portion of the human Y chromosome long arm. Hum Genet 34(2):119–124
Turner JM (2007) Meiotic sex chromosome inactivation. Development 134(10):1823–1831
Turner ME, Farkas J, Dunmire J, Ely D, Milsted A (2009) Which sry locus is the hypertensive Y chromosome locus? Hypertension 53(2):430–435
Turner ME, Ely D, Prokop J, Milsted A (2011) Sry, more than testis determination? Am J Phys Regul Integr Comp Phys 301(3):R561–R571
Vakilian H, Mirzaei M, Sharifi Tabar M, Pooyan P, Habibi Rezaee L, Parker L, Haynes PA, Gourabi H, Baharvand H, Salekdeh GH (2015) DDX3Y, a male-specific region of Y chromosome gene, may modulate neuronal differentiation. J Proteome Res 14(9):3474–3483
Vidal VP, Chaboissier MC, de Rooij DG, Schedl A (2001) Sox9 induces testis development in XX transgenic mice. Nat Genet 28(3):216–217
Vogt PH (1997) Report of the third international workshop on human Y chromosome mapping 1997. Cytogenet Genome Res 79(1–2):1–20
Vogt PH (2005) AZF deletions and Y chromosomal haplogroups: history and update based on sequence. Hum Reprod Update 11(4):319–336

Vogt P, Chandley AC, Hargreave TB, Keil R, Ma K, Sharkey A (1992) Microdeletions in interval 6 of the Y chromosome of males with idiopathic sterility point to disruption of AZF, a human spermatogenesis gene. Hum Genet 89(5):491–496

Vogt PH, Edelmann A, Kirsch S, Henegariu O, Hirschmann P, Kiesewetter F, Köhn FM, Schill WB, Farah S, Ramos C, Hartmann M (1996) Human Y chromosome azoospermia factors (AZF) mapped to different subregions in Yq11. Hum Mol Genet 5(7):933–943

Wimmer R, Kirsch S, Weber A, Rappold GA, Schempp W (2003) The azoospermia region AZFa: an evolutionary view. Cytogenet Genome Res 99(1–4):146–150

Yang Y, Ma MY, Xiao CY, Li L, Li SW, Zhang SZ (2008) Massive deletion in AZFb/b + c and azoospermia with Sertoli cell only and/or maturation arrest. Int J Androl 31(6):573–578

Yang B, Ma YY, Liu YQ, Li L, Yang D, Tu WL, Shen Y, Dong Q, Yang Y (2015) Common AZFc structure may possess the optimal spermatogenesis efficiency relative to the rearranged structures mediated by non-allele homologous recombination. Sci Rep 5:10551

Yen P (2001) The fragility of fertility. Nat Genet 29(3):243–244

Zhang F, Li Z, Wen B, Jiang J, Shao M, Zhao Y, He Y, Song X, Qian J, Lu D, Jin L (2006) A frequent partial AZFc deletion does not render an increased risk of spermatogenic impairment in East Asians. Ann Hum Genet 70(3):304–313

Seminal Decline in Semen Quality in Humans Over the Last 80 years

7

Priyanka Mishra and Rajender Singh

Abstract

During the later half of the twentieth century, there have been scientific debates over a decline in semen quality in past decades. Several studies from all over the world have reported a deterioration in semen quality. Interestingly, the reports of decline are contrasted by only a few studies. At the same time, there are reports of an increase in the incidence of other reproductive system malfunctions, such as testicular cancer, hypospadias, and cryptorchidism. Exposure to estrogenic, antiandrogenic, and maternal lifestyle factors during the fetal period are supposed to contribute to the temporal decline. In this chapter, we have presented a comprehensive review by highlighting the top time course studies reporting instability of semen quality. We have also discussed the putative factors responsible for this decline with an eye on transgenerational impact on human fertility.

Keywords

Sperm count • Semen quality • Sperm motility • Sperm morphology • Infertility

Key Points

- Donald Macomber and Morris B. Sanders undertook a quantitative assessment of spermatozoa in semen samples for the first time in 1929.
- The World Health Organization (WHO) released its first manual for semen analysis and classification of fertile and infertile men in 1980, and five manuals have since been released.

P. Mishra • R. Singh (✉)
Male Reproductive Biology Laboratory, CSIR-Central Drug Research Institute, Lucknow, Uttar Pradesh, India
e-mail: rajender_singh@cdri.res.in

R. Singh, K. Singh (eds.), *Male Infertility: Understanding, Causes and Treatment*, DOI 10.1007/978-981-10-4017-7_7

- Nelson and Bunge for the first time inferred a decline in semen parameters in the year 1974.
- In 1992, Carlsen et al. reported that semen quality has declined by 50% at the global level.
- Decline in semen quality is concomitant with increased incidence of testicular abnormalities such as cancer, hypospadias, and cryptorchidism.
- A number of factors that are responsible for decline in semen quality may be transgenerational in nature, exposing future generations to further decline.

7.1 Introduction

The foundation of modern semen analysis was laid back in 1929, when physician Donald Macomber and Morris. B. Sanders for the first time undertook a quantitative assessment of spermatozoa in semen samples of 294 males. Blood cell counting chamber was used to count sperm and semen evaluation was suggested as a measure of male fertility. It was concluded that sperm count of 60 million per mL in a man could establish successful pregnancy in his female counterpart (Macomber and Sanders 1929). McLeod wrote "I think that if we are to select a count level to represent the demarcation line between "poor and fair" fertility, 60 million per mL would be a wise choice" (MacLeod and Heim 1945). This value was set as a standard limiting value to gauge male fertility. In 1951, normal reference value for sperm count was dropped down to 20 million per mL after a comparative study done by John MacLeod and Gold on 1000 fertile men and 800 infertile men (MacLeod and Gold 1951).

Realizing the diagnostic importance of semen analysis to standardize and bring uniformity in its evaluation across all the labs worldwide, the World Health Organization (WHO) was set forth to provide guidelines and reference values for the analysis of semen parameters. The first manual of WHO was published in 1980, and subsequently other updated versions were released in the years 1987, 1992, 1999, and 2010 (Belsey et al. 1980, Barratt et al. 1995, Cooper et al. 2009, World Health Organization 1987, 1999). Currently, the fifth edition is in circulation, which

Table 7.1 Comparison of reference values provided by the successive WHO manuals

Semen parameters	1980	1987	1992	1999	2010
Volume (mL)	ND	≥2	≥2	≥2	≥1.5
Sperm count ($\times 10^6$) per mL	20–200	≥20	≥20	≥20	≥15
Total sperm count	ND	≥40	≥40	≥40	≥39
Total motility (%)	≥60	≥50	≥50	≥50	≥40
Progressive motility (%)	≥23	≥25%[a]	≥25%[a]	≥25%[a]	≥32%[a,b]
Vitality (%)	ND	≥50	≥75	≥75	≥58
Morphology (%)	80.5	≥50	≥30	≥14	≥4
Leucocyte count (10^6/mL)	<4.7	<1.0	<1.0	<1.0	<1.0

[a]Rapid progressive motility
[b]Slow progressive motility

was released in 2010. Although the current version of WHO manual provides reference values on evidence-based data collected from recent fathers with known time to pregnancy, these new clinical reference values are remarkably low as compared to the previous one (Table 7.1). Interestingly, the gradual lowering of normal reference values ascribed by WHO for evaluating semen parameters relate to the fact that the quality of semen has declined in past decades and are consistent with the reports that came from many regions of the world over the secular changes that happened in semen parameters in past decades.

7.2 Hallmark Studies Describing a Decline Over the Past 80 years

It was in 1974 that Nelson and Bunge first inferred a decline in semen parameters while analyzing semen samples of 386 men presenting themselves for vasectomy in Iowa, USA. Title of the report was "Semen analysis: evidence for changing parameters of male fertility potential." They noticed that mean sperm concentration in these groups of men was 48×10^6 per mL and only 7% had sperm concentration above 100×10^6 per mL, which was far low value than those set by MacLeod and Gold as standard normal range for fertile males (Nelson and Bunge 1974). Following this, other reports also found a decline in sperm count while analyzing semen of fertile male subjects with intermediate values of mean sperm concentration between 70×10^6 to 81×10^6 per mL (Rehan et al. 1975; Sobrero and Rehan 1975; Smith et al. 1977; Zukerman et al. 1977). Leto and Frensilli (1981) observed that several parameters had declined in potential sperm donors in the past 8 years. Downward trends in sperm count, rate of forward progression, viability, and normal morphology were observed.

In 1983, Bostofte et al. published a report on decline of fertility in Danish men. Semen quality of 1077 men observed in 1952 was compared with 1000 men examined in 1972 (Bostofte et al. 1982). They found a fall in sperm count (median values 73.4 million per mL in 1952 and 54.5 million per mL in 1972), deterioration in spermatozoa motility, and an increase in the number of abnormal spermatozoa. Osser et al. (1984) reported similar findings while analyzing 185 men for infertility in Sweden. Median sperm concentration was declined from 109×10^6 per mL in 1960 to 65×10^6 per mL in 1980. Amidst, there were also reports of no change in semen quality. Macleod and Wang (1979) acclaimed on consistent trend in sperm counts found in US men seeking advice for infertility-related problems in their hospital clinic. Similarly, David et al. (1979) reported the presence of higher sperm count values in pre-vasectomy males (since 1973) than those ascribed by Macleod in 1951. Observing all these studies, W. H. James (1980), in his review article, concluded that at least in some places, it seems likely that a secular decline has occurred. These evocative studies on changing semen quality were small and limited to a particular region and thus could not arouse/instigate fervent debate among scientific community.

The issue got revived again in 1992 with the publication of an influential report of Elisabeth Carlsen, Aleksander Giwercman, Niels Keiding, and Niels E Skakkeblek from the Department of Growth and Reproduction, the National University Hospital (Rigshospitalet), and the Panum Institute, Copenhagen, Denmark, which provided an evidence for global decline of semen quality during the past 50 years (1938–1990) (Carlsen et al.1992). In this paper, meta-analysis was done by taking data from 61 published studies from multiple nations, which included males with no history of infertility. A significant decrease in mean sperm count and volume was reported by using linear regression. Sperm count had declined at a rate of 0.94×10^6 per mL per year (1% per year) from 113 million per mL in 1940 to 66 million per mL in 1990. Also, the volume had declined from 3.40 to 2.75 mL, which advocates more pronounced decrease in sperm count. Although this study got many international critics, namely, Brake and Krause 1992, Bromwich et al. 1994, Fisch et al. 1996, Olsen and Rachootin 2003, and Fisch 2008, who put question marks about the retrospective design and the mathematical analysis used in the study, nevertheless, it evoked researchers all round the world to investigate the trend of semen quality in their respective countries.

Since 1992, the scientific literature became crowded with peer-reviewed publications over this concern, and heated debate among scientists and clinicians started. Swan et al. 1997 undertook a reanalysis of Carlsen study and tried to subset its shortcomings by regulating confounding factors (age, abstinence time, semen collection method, and fertility status) and implemented different statistics methods to reach a more comprehensive interpretation. They extracted data for only 56 studies from the United States, Europe, and non-Western countries and found that decline in sperm density in the United States (1.5% per year), Europe, and Australia (3% per year) were greater than the average decline reported by Carlsen et al. (1% per year). Data for non-Western countries was too low to draw a definite conclusion. Swan et al. 2000 extended the Carlsen study by adding 47 new studies to include in analysis a total of 101 studies for the period 1934–1996. Decline in average sperm count was observed during 1934–1996 as seen earlier in the data over 1938–1990. This made the issue even more interesting and invited others to undertake meta-analysis to further explore the decline.

From year 1992 onwards, more attention was paid to design unbiased comprehensive studies with less methodological flaws to extrapolate the outcomes of studies to the general population. To conduct prospective studies, andrologists and clinicians started picking up data from the records of sperm banks or from sperm donors or by selecting male subjects from infertile or subfertile couples (Fisch 2008). Auger et al. 1995 collected semen data for 20 years (1973–1996) from a sperm bank in Paris where the mode of semen collection and the method of semen analysis had remained the same during the whole study period. Analysis reported no change in semen volume during the study period, whereas the mean concentration of sperm decreased by 2.1% per year, from 89–106 million per mL in 1973 to 60–106 million per mL in 1992. During the same period, the percentages of motile and normal spermatozoa decreased by 0.6 and 0.5% per year, respectively.

Deteriorating semen quality in the United Kingdom was reported in a retrospective birth cohort study in 577 men from Scotland analyzed over 11 years (1951 and 1973). Of these subjects, 171 were born before 1959, 120 were born in 1960–1964, 171 in 1965–1969, and 115 in 1970–1974. When comparison of all the four cohorts was undertaken, a later year of birth associated with a lower sperm concentration, a lower total number of sperm, and a lower number of motile sperm in the ejaculate. The median sperm concentration fell from 98 million per mL among donors born before 1959–1978 million per mL among donors born after 1970. The total number of sperm in the ejaculate fell from 301 × 10^6 to 214 × 10^6 per mL, and total number of motile sperm fell from 169.7 × 10^6 to 129 × 10^6 (Irvine et al. 1996). Similarly, Bonde et al. (1998) reported that sperm count is related to the year of birth while analyzing sperm count of 1196 men from three regions of Denmark from 1986 to 1995. Sperm concentration was higher in men born in 1937–1949 whereas lower in male folks born after 1970. Although, results of Bonde et al. (1998) could have been affected by different age and fertility status of the subjects.

Seminal volume and total sperm number trends in men attending subfertility clinics in Athens were examined during the period 1977–1993. Seminal data of 2385 subjects were collected from three andrological laboratories. Analysis indicated a significant decrease in total sperm number over the years with an average value of 154.3 × 10^6 at the beginning in 1977 dropping to 130.1 × 10^6 per mL in 1993 (Adamopoulos et al. 1996). Decline in seminal fluid in Italian population during the past 15 years (1981–1995) was reported by Bilotta et al. (1999) by analyzing the data of fertile semen donors. Mean concentration of spermatozoa was reported to shrink from 88 × 10^6 per mL in 1981–1961 × 10^6 per mL in 1995, mean motility declined from 74 to 66%, and typical morphology fell from 76 to 63%. Similarly, during the study period of 1977–1992, Slama et al. (2004) estimated a 21% decrease (1977–1992) and 47% decrease (1947–1992) in sperm count for French sperm donors.

In the new millennium, large well-controlled studies brought better insight over the concern of instability of semen and indicated its prevalence as a threat across the whole globe. Examination of the urban population of 7780 men (excluding azoospermic) attending an andrology clinic for infertility-related problems in Austria during 1986–2003 revealed that the median sperm concentration dropped down from 27.75 million per mL in 1986 to 4.60 million per mL in 2003 and semen pH increased from 7.4 in 1986 to 7.9 in 2003 (Lackner et al. 2005). In Finland, 858 men from general population were investigated for the semen quality during 1998–2006. It was found that young Finns showed lower sperm counts in the most recent birth cohort compared with few years older cohort. Sperm concentration in most of the young Finns was below the new WHO reference level of 15 million per mL (Jørgensen et al. 2011), whereas during a study period of 1967–1994, no change in sperm count was reported in 234 normal and 5481 infertile Finnish males (Vierula et al. 1996).

While analyzing 10,932 normozoospermic males from infertile couples from Marseille (France), Geoffroy-Siraudin et al. (2012) found declining trends in sperm concentration (1.5% per year), total sperm count (1.6% per year), total motility (0.4% per year), rapid motility (5.5% per year), and normal morphology (2.2% per year). Similarly, Splingart et al. (2012) reported a significant decrease in total sperm count, motility, viability, and normal morphology in Tours, France from 1976 to 2009. Further, a decrease in sperm concentration (1.9% per year) and normal morphology over a 17-year period (1989–2005) was reported among 26,609 males partners of infertile women from all parts of France seeking advice for assisted reproductive technology (ART) procedures (Rolland et al. 2012). A retrospective study of New Zealand conducted across 1987–2007 on 975 sperm donors from Auckland and Wellington showed a decline in sperm concentration from 110×10^6 per mL in 1987 to 50×10^6 per mL in 2007. A drop in semen volume from 3.7 to 3.3 mL in this 20-year duration was also reported (Shine et al. 2008).

Semen parameters were analyzed in a prospective study conducted on German population including 234 young men from Leipzig (East Germany) and 457 men from Hamburg (West Germany). No significant interregional differences were found in sperm concentration, but morphology and motility varied significantly. The median value of sperm concentration in combined population was 44 million per mL, which indicates that young German males have poor semen quality of subfertile range (Paasch et al. 2008). In Denmark, semen quality of annual cohorts of 4867 young men had not declined during 15 years (1996–2010), but there was a significant fall in sperm concentration and total sperm counts as compared to fertile men examined few years back and male partners from historical cohort of infertile couples. Further, only one in four males was found to have optimal sperm concentration (Jørgensen et al. 2012). These findings are consistent with the report of declining trends in conception rates and deteriorating male reproductive health in Danish populations (Jensen et al. 2008).

Trend in semen parameters was also investigated in Sfax city of Tunisia between 1996 and 2007 in a sample of 2940 men in infertile relationships. A decreasing trend in sperm count and percentage of normal morphology was found over the last 12 years (Feki et al. 2009). A recently published report of Haimov-Kochman et al. (2012) provided evidence of adverse trend in semen parameters during the period of 1995–2009. This study was conducted on 2182 samples provided on a weekly basis by 58 young healthy fertile sperm donors. Sperm concentration dropped from 106 million per mL with 79% motility to 68 million per mL with 66% motile sperm. The total motile sperm count per ejaculate also decreased, from 66.4 million to 48.7 million. Decline in sperm count (5.2×10^6 million per mL per year) in Jerusalem across 10 years (1990–1999) was also supported by Almagor et al. (2003). The report also stated that about 38% of sperm donations are being rejected because of unsatisfactory sperm counts.

A US study on 489 semen donors from urban Boston area analyzed semen data using linear regression after adjustment for age, days of abstinence, as well as by the Cochran–Armitage trend test and reported decline in sperm count, total count, and total motile count (Centola et al. 2016). Nonetheless, no change in semen quality was reported in the United States by earlier studies (Fisch et al. 1996; Paulsen et al. 1996; Saidi et al. 1999). Exploration of 29 studies from 1938–1996 of 9612 subfertile men reported that sperm count had not changed during the study period (Saidi et al. 1999). Interestingly, Fisch et al. (1996), analyzed data from sperm banks of New York, California and Minnesota for 1283 men who preserved their sperm prior to vasectomy and reported an increase in sperm concentration over a period of 25 years (1970–1995). In Brazil, semen samples of 764 infertile males examined during 2000–2002 were compared with 1536 infertile men in 2010–2012. The mean sperm concentration per mL decreased remarkably from 61.7

million in 2000–2002 to 26.7 million in 2010–2012 (Borges et al. 2015). Similar unambiguous downfall in sperm count was seen in Sweden (Bendvolt et al. 1991), Belgium (Van Waeleghem et al.1996), Itly (Menchinni et al.1996), Canada (Younglai et al.1998), Norway (Ulstein et al. 1999), Scotland (Sripada et al. 2007), Italy (Vicari et al. 2003), India (Adiga et al. 2008), Argentina (Molina et al. 2010), Spain (Mendiola et al. 2013), and China (Wang et al. 2016) (Table 7.2).

The declining trend in semen quality parameters detailed above is very interesting. In the meantime, increased prevalence of testicular cancers, hypospadias, and cryptorchidism was observed worldwide (Giwercman et al. 1993; Purdue et al. 2005). Emergence of testis cancer increased majorly in the developed countries between 1980 and 2002. The general statistics of increment per year are 2.4% in Sweden, 5.0% in Spain, 2.9% in the UK, 3.0% in Australia, 3.5% in China, while India is at lower risk of only 1.7% (Chia et al. 2010). In Finland, a prospective study was conducted to analyze current trends of semen quality and appearances of testicular cancers. Semen data of 858 men from general population (age 18–19 years) was analyzed from 1998–2006, and the registries of incidence of testicular cancer of 5974 men were observed for 1954–2008 period. This study confirmed an increased incidence of testis neoplasia in the past 15 years in parallel with a decline in semen quality between 1998 and 2006 (Jørgensen et al. 2011).

International Clearing House for Birth Defect Monitoring System (ICBDMS), an organization of WHO, reported that increasing trends of incidence of hypospadias were found during 1960s in Sweden, Norway, Denmark, England, and Hungary during 1970s, while in the United States, the incidence increased from 1970s to 1990s (Toppari et al. 2001). Metropolitan Atlanta Congenital Defects Program (MACDP) and Birth Defects Monitoring Program (BDMP) are the two birth defect surveillance systems of the United States. Analysis of data taken from these registries reported doubling in the rate of occurrence of hypospadias during 1970–1993 from 20.2 to 39.7 per 10,000 births (Paulozzi et al.1997). A longitudinal study conducted during 1977–2005 in Danish boys showed occurrence of hypospadias in 3940 boys among 921,745 live births of male child with increased prevalence from 0.24% in 1977 to 0.52% in 2005 with an annual rate of 2.40% (Lund et al. 2009). Prevalence of cryptorchidism among boys since birth has increased in the UK from 2.7 to 4.1% between the 1950s and 1980s and in Denmark from 1.8 to 8.4% between the 1950s and 1990s. These figures vary from 2.1 to 8.4% in different countries during the last two decades (Paulozzi et al.1999; Virtanen and Toppari 2008).

Interestingly, a decline in fertility rate has occurred in the developed nations in the last decades (Kaufmann et al. 1998; Pearce et al. 1999; Lutz et al. 2003; Yew 2012). There is an average of less than two children per couple in Europe and in Japan (Central Intelligence Agency 2015), whereas in Spain and Italy, this figure is below 1.5 (Bosch 2000). International Committee for Monitoring Assisted Reproductive Technology (ICMART) in 2002 has reported an increase in the use of intracytoplasmic sperm injection (ICSI) worldwide (from 54.4% in 2000 to 60.8% in 2002 in North America, from 45.7 to 53.9% in Europe) and reached 76.1% in Latin America and 92.5% in the Middle East in 2002. According to the European IVF-Monitoring

Table 7.2 List of studies indicating well-defined decline in semen quality

Year	Authors	Sample size	Study period	Location	Fertility status	Major findings
1991	Bendvold et al.	467	1956–1986	Sweden	Infertile	• ↓ Sperm count from 467 million per mL to 305 million per mL • ↓ Percentage normal morphology from 53 to 37%
1992	Carlsen et al.	14,947	1938–1990	Multiple countries	Proven fertile	• ↓ Semen volume from 3.40 to 2.75 mL • ↓ Mean sperm concentration from 113 million/mL to 66 million
1995	Auger et al.	1351	1973–1992	France	Sperm donors	• No change in mean volume • ↓ Mean sperm concentration, 2.1% per year from 89 million per mL to 60 million per mL • ↓ Percentage sperm motility by 0.6% • ↓ Sperm morphology by 0.5%
1996	Irvine et al.	577	1984–1995	Scotland	Sperm donors	• ↓ Sperm concentration from 98 million per mL in older cohort to 78 million per mL among donors born after 1970 • ↓ Total motile sperm count fell from 169 million to 129 million, 2.1% per year • ↑ Motility, 0.18% per year

1996	Van Waelghem et al.	416	1956–1986	Belgium	Sperm donors	• ↑ Semen volume • ↓ Mean sperm concentration • ↓ Sperm morphology from 39.2% in period of 1977–1990 to 26.6% in 1990–1995 • ↓ Rapid progressive motility
1996	Adampolous et al.	2385	1977–1993	Spain	Subfertile	• ↓ Sperm count from 154.3 million per mL to 130.1 million per mL
1996	Menchini et al.	4500	1975–1994	Italy	Normozoospermic male from infertile couples	• ↓ Semen volume from 3.2 to 2.9 mL • ↓ Sperm count from 71.8 to 65.32 million per mL • ↓ Motility from 50 to 32%
1997	Swan et al.	14498	1938–1990	Meta analysis including studies from multiple countries	Proven fertile	• ↓ Sperm density in the United States 1.5% per year • ↓ Sperm density in Australia and Europe 3% per year • No decline in non-Western countries
1998	Bonde et al.	1196	1986–1995	Denmark	Fertility not known	• ↓ Median concentration from 63 million per mL in 1937–1949 to 52 million per mL 1970 onwards • ↓ Median sperm count from 206 million in 1937–1949 to 117 million 1970–1995
1998	Younglai et al.	48,968	1984–1996	Canada	Infertile male	• ↓ Sperm concentration
1999	Ulstein et al.	1108	1974–1994	Norway	Fertile men	• ↓ Sperm variables in consecutive birth cohorts

(continued)

Table 7.2 (continued)

Year	Authors	Sample size	Study period	Location	Fertility status	Major findings
1999	Bilotta et al.	1068	1981–1995	Italy	Semen donors	• ↓ Sperm concentration 31% over study period • ↓ Sperm motility, 8% • ↓ Sperm with typical morphology, 9%
2000	Swan et al.	20162	1934–1996	Meta analysis including studies from multiple countries	Proven fertile	• ↓ Mean sperm concentration from 113 million/mL to 66 million
2003	Almagor et al.	2638	1990–2000	Jerusalem	Healthy males from infertile couples	• ↓ Sperm count by 5.2 million each year • ↓ Motility by 0.5% per year
2003	Vicari et al.	716	1982–1999	Sicily	467 proven fertile, 109 normospermic	• ↓ Sperm density and morphology
2005	Lackner et al.	7780	1986–2003	Austria	Infertile male	• ↓ Median sperm concentration from 27.5 million/mL to 4.60 million per mL
2007	Sripada et al.	4832	1994–2005	Scotland	Normospermic male from subfertile couples	• ↓ Sperm density over a period of 11 years • No similar trend was observed in motility
2008	Shine et al.	975	1987–2007	New Zealand	Unknown fertility status	• ↓ Mean sperm concentration 110 million per mL to 50 million per mL with 2.5% decline per year • ↓ Semen volume, 3.7 to 3.3 mL
2008	Adiga et al.	7770	1993–2005	India	Infertile male	• ↓ Sperm density from 38.18 to 26.61 million per mL • ↓ Sperm motility from 61.6% to 47.1% • ↓ Normal morphology from 40.55% to 19.75%

2009	Feki et al.	1835	1996–2007	Tunisia	Infertile male	• ↓ Mean sperm count from 327.8 million per mL to 260 million per mL • ↓ Normal sperm morphology percentage by 2.6 per year from 42.7% to 16.5% • Motility showed no change
2011	Splingart et al.	1114	1976–2009	France	Male from subfertile couples	• No decline in semen volume • ↓ Total sperm count from 443 million to 300 million • ↓ Motility from 64% to 49% • ↓ Percent of normal from 67% to 26%
2012	Geoffroy–Siraudin et al.	10,932	1988–2007	France	Male partners of infertile couples	• ↓ Sperm concentration, 1.5% per year • ↓ Total sperm count, 1.6% per year • ↓ Rapid motility, 5.5% per year • ↓ Normal morphology, 2.2% per year
2012	Kochman et al.	2182	1995–2009	Jerusalem	Sperm donors	• ↓ Average sperm concentration from 106 million per mL to 68 million per mL • ↓ Motility from 79% to 68% • ↓ Total sperm count from 66.4 to 48.7

(continued)

Table 7.2 (continued)

Year	Authors	Sample size	Study period	Location	Fertility status	Major findings
2015	Borges et al.	2300	2000–2 to 2010–12	Brazil	Subfertile	• ↓ Mean sperm concentration from 61.7 million per mL to 26.7 million per mL • ↓ Total sperm concentration, 183.0 million to 82.8 million • ↓ Normal morphology from 4.6% to 2.7%
2016	Centola et al.	489	2003–2013	Boston, USA	Sperm bank donors	• ↓ Sperm concentration and total sperm count
2016	Wang et al.	5210	2008–2014	China	Sperm bank donors	• ↓ Semen volume, 2.73% per year • ↓ Sperm concentration, 6.89% per year • ↓ Sperm forward motility, 1.37% per year • ↓ Total sperm count, 9.84% per year

Consortium, the number of reported cycles of IVF and ICSI has increased by 4.9% in comparison to 2011 (Kupka et al. 2016). Though the fertility rate is a complex phenomenon that is influenced by a number of variables, a decline is semen quality is one of the important contributors. Similarly, a trend in secular drowning of reproductive hormones, testosterone, and sex hormone-binding globulin levels in serum of men has also been reported (Andersson et al. 2007). Thus, high incidence of the abovementioned reproductive system deformities and decline in reproductive hormone parameters are followed up as evidences of instability of semen parameters over time.

7.3 Factors Alleged for Deteriorating Semen Quality

Semen parameters are highly sensitive biomarkers of male reproductive function. Semen analysis provides information about functioning of various reproductive organs of males, viz., testis, epididymis, accessory gland, seminal vesicle, and prostate and are thus used as a diagnostic tool to access male infertility-related information. Decline in sperm count has occurred in the past 60–70 years, and this fact indicates that the involvement of environmental and lifestyle factors rather than genetic is more likely. A rapid expansion of industries has outpoured the pandora of hazardous chemicals in the immediate environment of human beings. Most of these chemicals are estrogenic and endocrine disruptors. The epithelium of seminiferous tubule is highly sensitive to stress and chemicals. Occupational or domestic exposure to estrogenic chemicals led to impairment of spermatogenesis, which in turn caused low sperm counts and production of defective sperm. Estrogens like compounds bind to estrogen receptors and stimulate pituitary gland for the production of gonadotropins. Gonadotropins influence the secretion of male reproductive hormones and can thus modulate spermatogenesis.

Environmental pollutants cause a generation of reactive oxygen spices (ROS), which if unchecked, can cause deterioration of sperm quality (Hammoud et al. 2010). A study conducted by De Rosa et al. (2003) found that men exposed to traffic pollutants had increase level of methaemoglobin in serum and more amount of lead in semen as compared to those who were not exposed. Prenatal and postnatal exposures have been suggested to affect sperm concentration and reproductive organ development of male child (Jouannet et al. 2001; Pastuszak and Lamb 2013). A follow-up study of two decades by Ramlau-Hansen et al. (2010) reported that alcohol intake at the time of pregnancy associated with lower sperm counts in sons, and most of the pregnant women in Denmark have reported alcohol intake throughout pregnancy. Maternal smoking during pregnancy has also been associated with lower sperm counts in the upcoming male child (Storgaard et al. 2003). Similarly, paternal and maternal smoking has been reported to affect semen quality in the upcoming generations (Axelsson et al. 2013). More interestingly, both alcohol and smoking have been postulated to have transgenerational effects, which may last up to several generations (Lee et al. 2013; Joya et al. 2014; Taki et al. 2014).

Radiations cause detrimental effect on male reproductive health (Aitken et al. 2005). Exposure to cell phone radiations may be a common factor related to a decrease in sperm count, motility, viability, and normal morphology (Agarwal et al.

2008). Radio frequency electromagnetic radiations of mobile phones cause sperm DNA fragmentation and decline in sperm motility (Gorpinchenko et al. 2014). It was also shown that microwave radiation can affect sperm count (Kim et al. 2007) and that 2.45 GHz microwave radiation decreased the diameters of seminiferous tubules (Saygin et al. 2011), thus affecting the Sertoli cell numbers, which in turn impair spermatogenesis. Antiandrogen compounds adversely affect male reproductive health (Kristensen et al. 2011; Nordkap et al. 2012). Inhalation, consumption, or absorption of phthalate chemicals during pregnancy causes birth of male child with low sperm count, and this is due to inhibition of testosterone production (Parks et al. 2000; Sharpe 2005; Swan 2008). About 8% of chemicals display antiestrogenic effects (Vinggaard et al. 2008). It has been found that occurrence of childhood cancer is linked to paternal exposure to hydrocarbons in various forms like benzene, paint, methyl ethyl ketone, plastic and resin fumes, and different types of solvents. A cross-sectional Denmark study on 701 of young men who underwent medical checkup before joining military showed that males with high dietary intake of saturated fat had 38% lesser sperm concentration and 41% lower total sperm count than those who consumed less saturated fat diet (Jensen et al. 2013). Thus, a host of factors, which may vary significantly from one population to the other, have resulted in a significant decline in semen quality over the last several decades. Some of these factors have been discussed in detail in Chap. 23.

7.4 Discussion and Future Directions

Infertility affects about 8–12% of couples, and in 50% cases, male factor is found to cause infertility-related problems. According to the WHO, 60–80 million couples are currently suffering from infertility. In the United States alone, 10% couples are estimated as infertile. The National Centre for Health Statistics has estimated that absolute number of impaired fecundity has increased by 2.7 million women from 4.56 million in 1982 to 7.26 million in 2002. Fertility rate in men less than 30 years has also decreased by about 15% (Kumar and Singh 2015). Simultaneously, demand for assisted reproductions has also grown abruptly in the recent decades (Andersen et al. 2008). Thus, it is tempting to speculate the role of declined semen quality in the hike in prevalence of male infertility in parallel with increasing demand for the assisted reproductive technologies.

In the last few decades, the increased incidence of testis cancer and other male congenital abnormalities like hypospadias and cryptorchidism is in synchronization with a fall in semen quality. A hypothesis underlies that the incidence of testicular neoplasia, undescended testis, hypospadias, and poor semen quality all is part of a single etiology, broadly called as testicular dysgenesis syndrome (TDS), commonly originated by the exposure to estrogenic or antiandrogenic toxins in prenatal period that adversely affect reproductive development of fetus. It is very likely that the rise in infertility goes hand in hand with increase in the frequency of appearance of reproductive organ deformities.

The exposure to endocrine disruptors, chemical toxins like pesticides, fungicides, medications or inadequate nutrition, high fat diet also shows inheritable transgenerational effects, which are epigenetic in nature (Jirtle and Skinner 2007). These factors can modulate methylation pattern of DNA or histone in germ line cells, which can alter gene expression and thus may affect semen parameters up to several generations. Studies in mice have shown that vinclozolin, phthalate, dioxin, tetracycline, high fat diet, and folate are some of the few known compounds exposure to which in ancestral generation can cause lowering of sperm count and reproductive functions in the coming generations (Anway et al. 2006; Dunn and Bale 2011; Manikkam et al. 2012, Nilsson et al. 2012; Zeh et al. 2012; Doyle et al. 2013; Padmanabhan and Watson 2013).

Conclusion

Sufficient scientific and demographical data supports a significant decline in semen quality in most of the populations, with very few exceptions. Diminishing semen quality may impose pandemic threat of various male reproductive system disorders and decline in fertility rates in the coming generations. Adverse trend in semen standard has hit the human reproduction process and thus may bring the subsistence of human species under question. Declining of sperm count would not only bring social and psychological distress at individual level; it would be disappointing for infertile couples as poor semen may lead to cessation of intrauterine insemination-based treatments. Although exposure to antiestrogenic and antiandrogenic compounds during embryonic stage are known causes of drop in sperm count, but a host of other factors await discovery. In addition to decline in sperm count, it is also important to explore what sort of transmissible programming at the epigenetic level in the germ cells could be brought in by the environmental exposure and maternal lifestyle in prenatal period of ensuing fetus. No doubt, immediate long-term follow-up with well-designed studies involving several countries are needed to observe the severity and outcome of changing trend in semen quality. More research studies are needed to identify other potent chemicals that can act as xenoestrogens and endocrine disruptors and also the ways for their rapid detection. It is the time to take the call seriously; otherwise, natural conceptions may be a rare phenomenon in future.

Acknowledgements The authors would like to thank the Council of Scientific and Industrial Research (CSIR) and the Government of India for financial help under the network project, PROGRAM (BSC0101).

References

Adamopoulos DA, Pappa A, Nicopoulou S, Andreou E, Karamertzanis M, Michopoulos J, Deligianni V, Simou M (1996) Andrology: seminal volume and total sperm number trends in men attending subfertllity clinics in the greater Athens area during the period 1977–1993. Hum Reprod 11(9):1936–1941

Adiga SK, Jayaraman V, Kalthur G, Upadhya D, Kumar P (2008) Declining semen quality among south Indian infertile men: a retrospective study. J Hum Reprod Sci 1(1):15

Agarwal A, Deepinder F, Sharma RK, Ranga G, Li J (2008) Effect of cell phone usage on semen analysis in men attending infertility clinic: an observational study. Fertil Steril 89(1):124–128

Aitken RJ, Bennetts LE, Sawyer D, Wiklendt AM, King BV (2005) Impact of radio frequency electromagnetic radiation on DNA integrity in the male germline. Int J Androl 28(3):171–179

Almagor M, Ivnitzki I, Yaffe H, Baras M (2003) Changes in semen quality in Jerusalem between 1990 and 2000: a cross-sectional and longitudinal study. Arch Androl 49(2):139–144

Andersen AN, Goossens V, Ferraretti AP, Bhattacharya S, Felberbaum R, De Mouzon J, Nygren KG (2008) Assisted reproductive technology in Europe, 2004: results generated from European registers by ESHRE. Hum Reprod 23(4):756–771

Andersson AM, Jensen TK, Juul A, Petersen JH, Jørgensen T, Skakkebæk NE (2007) Secular decline in male testosterone and sex hormone binding globulin serum levels in Danish population surveys. J Clin Endocrinol Metab 92(12):4696–4705

Anway MD, Leathers C, Skinner MK (2006) Endocrine disruptor vinclozolin induced epigenetic transgenerational adult-onset disease. Endocrinology 147(12):5515–5523

Auger J, Kunstmann JM, Czyglik F, Jouannet P (1995) Decline in semen quality among fertile men in Paris during the past 20 years. N Engl J Med 332(5):281–285

Axelsson J, Rylander L, Rignell-Hydbom A, Silfver KÅ, Stenqvist A, Giwercman A (2013) The impact of paternal and maternal smoking on semen quality of adolescent men. PLoS One 8(6):e66766

Barratt CLR, Naeeni M, Clements S, Cooke ID (1995) Andrology: clinical value of sperm morphology for in-vivo fertility: comparison between World Health Organization criteria of 1987 and 1992. Hum Reprod 10(3):587–593

Belsey MA, Moghissi KS, Eliasson R, Paulsen CA, Gallegos AJ, Prasad MR (1980) WHO Laboratory manual for the examination of human semen and semen-cervical mucus interaction

Bendvold E, Gottlieb C, Bygdeman M, Eneroth P (1991) Depressed semen quality in Swedish men from barren couples: a study over three decades. Arch Androl 26(3):189–194

Bilotta P, Guglielmo R, Steffe M (1999) Analysis of decline in seminal fluid in the Italian population during the past 15 years. Minerva Ginecol 51(6):223–231

Bonde JPE, Jensen TK, Larsen SB, Abell A, Scheike T, Hjollund NHI, Kolstad HA, Ernst E, Giwercman A, Skakkebæk NE, Keiding N (1998) Year of birth and sperm count in 10 Danish occupational studies. Scand J Work Environ Health:407–413

Borges E Jr, Setti AS, Braga DPDAF, Figueira RDCS, Iaconelli A Jr (2015) Decline in semen quality among infertile men in Brazil during the past 10 years. Int Braz J Urol 41(4):757–763

Bosch X (2000) Spain faces massive decline in population. BMJ 320(7239):891

Bostofte E, Serup J, Rebbe H (1982) Has the fertility of Danish men declined through the years in terms of semen quality? A comparison of semen qualities between 1952 and 1972. Int J Fertil 28(2):91–95

Brake A, Krause W (1992) Decreasing quality of semen. BMJ 305(6867):1498

Bromwich P, Cohen J, Steward I, Walker A (1994) Decline in sperm counts: an artefact of changed reference range of "normal"? BMJ 309(6946):19–22

Carlsen E, Giwercman A, Keiding N, Skakkebæk NE (1992) Evidence for decreasing quality of semen during past 50 years. BMJ 305(6854):609–613

Centola GM, Blanchard A, Demick J, Li S, Eisenberg ML (2016) Decline in sperm count and motility in young adult men from 2003 to 2013: observations from a U.S. sperm Bank. Andrology 4(2):270–276

Central Intelligence Agency (2015) The world Factbook 2014–15. Government Printing Office, Fairfax, VA

Chia VM, Quraishi SM, Devesa SS, Purdue MP, Cook MB, McGlynn KA (2010) International trends in the incidence of testicular cancer, 1973-2002. Cancer Epidemiol Biomark Prev 19(5):1151–1159

Cooper TG, Noonan E, Von Eckardstein S, Auger J, Baker HG, Behre HM, Haugen TB, Kruger T, Wang C, Mbizvo MT, Vogelsong KM (2009) World Health Organization reference values for human semen characteristics. Hum Reprod Update 16(3):231–245

David G, Jouannet P, Martin-Boyce A, Spira A, Schwartz D (1979) Sperm counts in fertile and infertile men. Fertility and Sterility, 31(4):453–455

De Rosa M, Zarrilli S, Paesano L, Carbone U, Boggia B, Petretta M, Maisto A, Cimmino F, Puca G, Colao A, Lombardi G (2003) Traffic pollutants affect fertility in men. Hum Reprod 18(5):1055–1061

Doyle TJ, Bowman JL, Windell VL, McLean DJ, Kim KH (2013) Transgenerational effects of di-(2-ethylhexyl) phthalate on testicular germ cell associations and spermatogonial stem cells in mice. Biol Reprod 88(5):112

Dunn GA, Bale TL (2011) Maternal high-fat diet effects on third-generation female body size via the paternal lineage. Endocrinology 152(6):2228–2236

Feki NC, Abid N, Rebai A, Sellami A, Ayed BB, Guermazi M, Bahloul A, Rebai T, Ammar LK (2009) Semen quality decline among men in infertile relationships: experience over 12 years in the south of Tunisia. J Androl 30(5):541–547

Fisch H (2008) Declining worldwide sperm counts: disproving a myth. Urol Clin N Am 35(2):137–146

Fisch H, Goluboff ET, Olson JH, Feldshuh J, Broder SJ, Barad DH (1996) Semen analyses in 1,283 men from the United States over a 25-year period: no decline in quality. Fertil Steril 65(5):1009–1014

Geoffroy-Siraudin C, Loundou AD, Romain F, Achard V, Courbiere B, Perrard MH, Durand P, Guichaoua MR (2012) Decline of semen quality among 10 932 males consulting for couple infertility over a 20-year period in Marseille, France. Asian journal of andrology, 14(4):584

Giwercman A, Carlsen E, Keiding N, Skakkebaek NE (1993) Evidence for increasing incidence of abnormalities of the human testis: a review. Environ Health Perspect 101(Suppl 2):65

Gorpinchenko I, Nikitin O, Banyra O, Shulyak A (2014) The influence of direct mobile phone radiation on sperm quality. Cent Eur J Urol 67(1):65–71

Haimov-Kochman R, Har-Nir R, Ein-Mor E, Ben-Shoshan V, Greenfield C, Eldar I, Bdolah Y, Hurwitz A (2012) Is the quality of donated semen deteriorating? Findings from a 15 year longitudinal analysis of weekly sperm samples. IMAJ-Israel Med Assoc J 14(6):372

Hammoud A, Carrell DT, Gibson M, Sanderson M, Parker-Jones K, Peterson CM (2010) Decreased sperm motility is associated with air pollution in salt Lake City. Fertil Steril 93(6):1875–1879

Irvine S, Cawood E, Richardson D, MacDonald E, Aitken J (1996) Evidence of deteriorating semen quality in the United Kingdom: birth cohort study in 577 men in Scotland over 11 years. BMJ 312(7029):467–471

James WT (1980) Secular trend in reported sperm counts. Andrologia 12(4):381–388

Jensen TK, Sobotka T, Hansen MA, Pedersen AT, Lutz W, Skakkebæk NE (2008) Declining trends in conception rates in recent birth cohorts of native Danish women: a possible role of deteriorating male reproductive health. Int J Androl 31(2):81–92

Jensen TK, Heitmann BL, Jensen MB, Halldorsson TI, Andersson AM, Skakkebæk NE, Joensen UN, Lauritsen MP, Christiansen P, Dalgård C, Lassen TH (2013) High dietary intake of saturated fat is associated with reduced semen quality among 701 young Danish men from the general population. Am J Clin Nutr 97(2):411–418

Jirtle RL, Skinner MK (2007) Environmental epigenomics and disease susceptibility. Nat Rev Genet 8(4):253–262

Jørgensen N, Vierula M, Jacobsen R, Pukkala E, Perheentupa A, Virtanen HE, Skakkebaek NE, Toppari J (2011) Recent adverse trends in semen quality and testis cancer incidence among Finnish men. Int J Androl 34(4 pt2):e37–e48

Jørgensen N, Joensen UN, Jensen TK, Jensen MB, Almstrup K, Olesen IA, Juul A, Andersson AM, Carlsen E, Petersen JH, Toppari J (2012) Human semen quality in the new millennium: a prospective cross-sectional population-based study of 4867 men. BMJ Open 2(4):e000990

Jouannet P, Wang C, Eustache F, Kold-Jensen T, Auger J (2001) Semen quality and male reproductive health: the controversy about human sperm concentration decline. Apmis, 109(S103), pp.S48–S61
Joya X, Manzano C, Álvarez AT, Mercadal M, Torres F, Salat-Batlle J, Garcia-Algar O (2014) Transgenerational exposure to environmental tobacco smoke. Int J Environ Res Public Health 11(7):7261–7274
Kaufmann RB, Spitz AM, Strauss LT, Morris L, Santelli JS, Koonin LM, Marks JS (1998) The decline in US teen pregnancy rates, 1990–1995. Pediatrics, 102(5):1141–1147
Kim JY, Kim HT, Moon KH, Shin HJ (2007) Long-term exposure of rats to a 2.45 GHz electromagnetic field: effects on reproductive function. Korean J Urol 48(12):1308–1314
Kristensen DM, Skalkam ML, Audouze K, Lesné L, Desdoits-Lethimonier C, Frederiksen H, Brunak S, Skakkebæk NE, Jégou B, Hansen JB, Junker S (2011) Many putative endocrine disruptors inhibit prostaglandin synthesis. Environ Health Perspect 119(4):534
Kumar N, Singh AK (2015) Trends of male factor infertility, an important cause of infertility: a review of literature. J Hum Reprod Sci 8(4):191
Kupka MS, D'Hooghe T, Ferraretti AP, de Mouzon J, Erb K, Castilla JA, Calhaz-Jorge C, De Geyter C, Goossens V, European IVF-Monitoring Consortium (2016) Assisted reproductive technology in Europe, 2011: results generated from European registers by ESHRE. Hum Reprod 31(2):233–248
Lackner J, Schatzl G, Waldhör T, Resch K, Kratzik C, Marberger M (2005) Constant decline in sperm concentration in infertile males in an urban population: experience over 18 years. Fertil Steril 84(6):1657–1661
Lee HJ, Ryu JS, Choi NY, Park YS, Kim YI, Han DW, Ko K, Shin CY, Hwang HS, Kang KS, Ko K (2013) Transgenerational effects of paternal alcohol exposure in mouse offspring. Anim Cells Syst 17(6):429–434
Leto S, Frensilli FJ (1981) Changing parameters of donor semen. Fertil Steril 36(6):766–770
Lund L, Engebjerg MC, Pedersen L, Ehrenstein V, Nørgaard M, Sørensen HT (2009) Prevalence of hypospadias in Danish boys: a longitudinal study, 1977–2005. Eur Urol 55(5):1022–1026
Lutz W, O'Neill BC, Scherbov S (2003) Europe's population at a turning point. Science 299(5615):1991–1992
Macleod J, Gold RZ (1951) The male factor in fertility and infertility: 2. Spermatozoon counts in 1000 men of known fertility and in 1000 cases of infertile marriage. J Urol 66(3):436–449
MacLeod J, Heim LM (1945) Characteristics and variations in semen specimens in 100 normal young men. J Urol 54:474
Macleod J, Wang Y (1979) Male fertility potential in terms of semen quality: a review of the past a study of the present. Fertil Steril 31(2):103–116
Macomber D, Sanders MB (1929) The spermatozoa count. N Engl J Med 200(19):981–984
Manikkam M, Tracey R, Guerrero-Bosagna C, Skinner MK (2012) Dioxin (TCDD) induces epigenetic transgenerational inheritance of adult onset disease and sperm epimutations. PLoS One 7(9):e46249
Menchini-Fabris F, Rossi P, Palego P, Simi S, Turchi P (1996) Declining sperm counts in Italy during the past 20 years. Andrologia 28(6):304–304
Mendiola J, Jørgensen N, Mínguez-Alarcón L, Sarabia-Cos L, López-Espín JJ, Vivero-Salmerón G, Ruiz-Ruiz KJ, Fernández MF, Olea N, Swan SH, Torres-Cantero AM (2013) Sperm counts may have declined in young university students in southern Spain. Andrology 1(3):408–413
Molina RI, Martini AC, Tissera A, Olmedo J, Senestrari D, de Cuneo MF, Ruiz RD (2010) Semen quality and aging: analysis of 9.168 samples in Cordoba. Argentina. Arch Esp Urol 63(3):214–222
Nelson CM, Bunge RG (1974) Semen analysis: evidence for changing parameters of male fertility potential. Fertil Steril 25(6):503–507
Nilsson E, Larsen G, Manikkam M, Guerrero-Bosagna C, Savenkova MI, Skinner MK (2012) Environmentally induced epigenetic transgenerational inheritance of ovarian disease. PLoS One 7(5):e36129

Nordkap L, Joensen UN, Jensen MB, Jørgensen N (2012) Regional differences and temporal trends in male reproductive health disorders: semen quality may be a sensitive marker of environmental exposures. Mol Cell Endocrinol 355(2):221–230
Olsen J, Rachootin P (2003) Invited commentary: monitoring fecundity over time—if we do it, then let's do it right. Am J Epidemiol 157(2):94–97
Osser S, Liedholm P, Ranstam J (1984) Depressed semen quality: a study over two decades. Archives of andrology, 12(1):113–116
Paasch U, Salzbrunn A, Glander HJ, Plambeck K, Salzbrunn H, Grunewald S, Stucke J, Skakkebæk NE, Jørgensen N (2008) Semen quality in sub-fertile range for a significant proportion of young men from the general German population: a co-ordinated, controlled study of 791 men from Hamburg and Leipzig. Int J Androl 31(2):93–102
Padmanabhan N, Watson ED (2013) Lessons from the one-carbon metabolism: passing it along to the next generation. Reprod Biomed Online 27(6):637–643
Parks LG, Ostby JS, Lambright CR, Abbott BD, Klinefelter GR, Barlow NJ, Gray LE (2000) The plasticizer diethylhexyl phthalate induces malformations by decreasing fetal testosterone synthesis during sexual differentiation in the male rat. Toxicol Sci 58(2):339–349
Pastuszak AW, Lamb DJ (2013) Counting your sperm before they fertilize: are sperm counts really declining. Asian J Androl 15(2):179–183
Paulozzi LJ (1999) International trends in rates of hypospadias and cryptorchidism. Environ Health Perspect 107(4):297
Paulozzi LJ, Erickson JD, Jackson RJ (1997) Hypospadias trends in two US surveillance systems. Pediatrics 100(5):831–834
Paulsen CA, Berman NG, Wang C (1996) Data from men in greater Seattle area reveals no downward trend in semen quality: further evidence that deterioration of semen quality is not geographically uniform. Fertil Steril 65(5):1015–1020
Pearce D, Cantisani G, Laihonen A (1999) Changes in fertility and family sizes in Europe. Population Trends, 95:33–40
Purdue MP, Devesa SS, Sigurdson AJ, McGlynn KA (2005) International patterns and trends in testis cancer incidence. Int J Cancer 115(5):822–827
Ramlau-Hansen CH, Toft G, Jensen MS, Strandberg-Larsen K, Hansen ML, Olsen J (2010) Maternal alcohol consumption during pregnancy and semen quality in the male offspring: two decades of follow-up. Hum Reprod 25(9):2340–2345
Rehan NE, Sobrero AJ, Fertig JW (1975) The semen of fertile men: statistical analysis of 1300 men. Fertil Steril 26(6):492–502
Rolland M, Le Moal J, Wagner V, Royère D, De Mouzon J (2012) Decline in semen concentration and morphology in a sample of 26 609 men close to general population between 1989 and 2005 in France. Human Reproduction, p.des415
Saidi JA, Chang DT, Goluboff ET, Bagiella E, Olsen G, Fisch H (1999) Declining sperm counts in the United States? A critical review. J Urol 161(2):460–462
Saygin M, Caliskan S, Karahan N, Koyu A, Gumral N, Uguz AC (2011) Testicular apoptosis and histopathological changes induced by a 2.45 GHz electromagnetic field. Toxicol Ind Health 27(5):455–463
Sharpe RM (2005) Guest editorial: phthalate exposure during pregnancy and lower anogenital index in boys: wider implications for the general population? Environ Health Perspect 113(8):A504
Shine R, Peek J, Birdsall M (2008) Declining sperm quality in New Zealand over 20 years. N Z Med J (Online) 121(1287):50–56
Slama R, Kold-Jensen T, Scheike T, Ducot B, Spira A, Keiding N (2004) How would a decline in sperm concentration over time influence the probability of pregnancy? Epidemiology 15(4):458–465
Smith KD, Rodriguez-Rigau LJ, Steinberger E (1977) Relation between indices of semen analysis and pregnancy rate in infertile couples. Fertil Steril 28(12):1314–1319

Sobrero AJ, Rehan NE (1975) The semen of fertile men. II. Semen characteristics of 100 fertile men. Fertil Steril 26(11):1048–1056

Splingart C, Frapsauce C, Veau S, Barthelemy C, Royère D, Guérif F (2012) Semen variation in a population of fertile donors: evaluation in a French Centre over a 34-year period. Int J Androl 35(3):467–474

Sripada S, Fonseca S, Lee A, Harrild K, Giannaris D, Mathers E, Bhattacharya S (2007) Trends in semen parameters in the northeast of Scotland. J Androl 28(2):313–319

Storgaard L, Bonde JP, Ernst E, Spanô M, Andersen CY, Frydenberg M, Olsen J (2003) Does smoking during pregnancy affect sons' sperm counts? Epidemiology 14(3):278–286

Swan SH (2008) Environmental phthalate exposure in relation to reproductive outcomes and other health endpoints in humans. Environ Res 108(2):177–184

Swan SH, Elkin EP, Fenster L (1997) Have sperm densities declined? A reanalysis of global trend data. Environ Health Perspect 105(11):1228

Swan SH, Elkin EP, Fenster L (2000) The question of declining sperm density revisited: an analysis of 101 studies published 1934-1996. Environ Health Perspect 108(10):961

Taki FA, Pan X, Lee MH, Zhang B (2014) Nicotine exposure and transgenerational impact: a prospective study on small regulatory micro RNAs. Sci Rep 4:7513

Toppari J, Kaleva M, Virtanen HE (2001) Trends in the incidence of cryptorchidism and hypospadias, and methodological limitations of registry-based data. APMIS 109(S103):S37–S42

Ulstein M, Irgens Å, Irgens L (1999) Secular trends in sperm variables for groups of men in fertile and infertile couples. Acta Obstet Gynecol Scand 78(4):332–335

Van Waeleghem K, De Clercq N, Vermeulen L, Schoonjans FRANK, Comhaire F (1996) Deterioration of sperm quality in young healthy Belgian men. Hum Reprod 11(2):325–329

Vicari E, Conticello A, Battiato C, La Vignera S (2003) Sperm characteristics in fertile men and healthy men of the south-east Sicily from year 1982 to 1999. Archivio Italiano di Urologia, Andrologia: Organo Ufficiale [di] SocietÃ Italiana di Ecografia Urologica e Nefrologica/ Associazione Ricerche in Urologia 75(1):28–34

Vierula M, Niemi M, Keiski A, Saaranen M, Saarikoski S, Suominen J (1996) High and unchanged sperm counts of Finnish men. Int J Androl 19(1):11–17

Vinggaard AM, Niemelä J, Wedebye EB, Jensen GE (2008) Screening of 397 chemicals and development of a quantitative structure—activity relationship model for androgen receptor antagonism. Chem Res Toxicol 21(4):813–823

Virtanen HE, Toppari J (2008) Epidemiology and pathogenesis of cryptorchidism. Hum Reprod Update 14(1):49–58

Wang L, Zhang L, Song XH, Zhang HB, Xu CY, Chen ZJ (2016) Decline of semen quality among Chinese sperm bank donors within 7 years (2008–2014). Asian J Androl

World Health Organisation (1999) WHO laboratory manual for the examination of human semen and sperm-cervical mucus interaction. Cambridge University Press, Cambridge

World Health Organization (1987) WHO laboratory manual for the examination of human semen and semen-cervical mucus interaction. World Health Organization, Geneva

Yew LK (2012) Warning bell for developed countries: declining birth rates. Forbes 189(8):30

Younglai EV, Collins JA, Foster WG (1998) Canadian semen quality: an analysis of sperm density among eleven academic fertility centers. Fertil Steril 70(1):76–80

Zeh JA, Bonilla MM, Adrian AJ, Mesfin S, Zeh DW (2012) From father to son: transgenerational effect of tetracycline on sperm viability. Sci Rep 2

Zukerman Z, Rodriguez-Rigau LJ, Smith KD, Steinberger E (1977) Frequency distribution of sperm counts in fertile and infertile males. Fertil Steril 28(12):1310–1313

Part II

Causes of Male Infertility

Syndromic Forms of Male Infertility

8

Vertika Singh, Rajender Singh, and Kiran Singh

Abstract

Syndromes represent abnormalities of more than one organ, and the complex malignancy may be easily identifiable by external appearance or physical examination. However, the exact identification of a syndrome and the complexity of the organs affected may be difficult to identify and often require assistance from cytogenetic and molecular investigations. The molecular basis of various disorders and syndromes has been worked out, and in some cases, molecular diagnosis has become a standard. Interestingly, a number of human syndromes often cosegregate with infertility to little or large degree, and more than 70 such syndromes have been identified. In some syndromes, infertility becomes the primary problem requiring attention; however, other features may be notable well before the onset of puberty. This chapter presents a collection of the syndromic forms of male infertility to illustrate their importance and clinical investigations, with an emphasis on the quantitative loss of fertility associated with them.

Keywords

Syndromes and male infertility • Klinefelter's syndrome • Kallmann syndrome Noonan syndrome • Jacob's syndrome • Cystic fibrosis • Myotonic dystrophy 1 Androgen insensitivity syndrome • Deafness infertility syndrome

V. Singh • K. Singh (✉)
Department of Molecular and Human Genetics, Institute of Science, Banaras Hindu University, Varanasi, Uttar Pradesh, India
e-mail: singhk4@rediffmail.com

R. Singh
Male Reproductive Biology Laboratory, CSIR-Central Drug Research Institute, Lucknow, Uttar Pradesh, India
e-mail: rajender_singh@cdri.res.in

R. Singh, K. Singh (eds.), *Male Infertility: Understanding, Causes and Treatment*, DOI 10.1007/978-981-10-4017-7_8

Key Points

- A number of genetic syndromes cosegregate with male infertility.
- Klinefelter's syndrome is the most frequent disorder of sex chromosomes in humans.
- CBAVD is present in around 1–2% infertile men and 6% obstructive azoospermia cases.
- Around 800 AR mutations are registered in the McGill University database of AR gene.
- Some rare syndromes of infertility include myotonic dystrophy 1, primary ciliary dyskinesia, Kearns–Sayre syndrome, Aarskog–Scott syndrome, persistent Müllerian duct syndrome and Prader–Willi syndrome.

8.1 Introduction

A syndrome is characterised as a disorder that has more than one identifying feature or symptom. Human male infertility is infrequently associated with genetic syndromes. The molecular basis of some syndromes is now well known; however, the basis of other syndromes remains unidentified or idiopathic. A major reason could be the lack of emphasis of infertility workup on the detection of rare syndromes. More than 70 syndromes have been identified to be associated with infertility so far (Hempel and Buchholz 2009). Some of the syndromic forms manifest with infertility as one of the most obvious clinical features, whereas in majority, infertility is coupled with mental retardation and severe malformations. As these individuals are often not concerned with the reproductive health and family planning, they are oblivious of their infertility.

The identification of 46 human chromosomes in 1956 gave birth to the area of genetics, which has diversified tremendously since then. With the expansions of genetic research and its clinical offshoots, numerous genetic factors have been identified to associate with male infertility. Some of the syndromes are associated with chromosomal aberrations, such as Klinefelter's syndrome, Noonan syndrome and 47,XYY syndrome, while others are linked to various autosomal or sex-linked genetic mutations. Due to overlapping features among these syndromes, identification of the molecular aetiology has become a standard method of diagnosis in association with physical examination. This article brings together the molecular defects and phenotypic/anatomical features of the syndromes that have infertility as one of the identifying features.

8.2 Syndromes with Chromosomal Aneuploidy

8.2.1 Klinefelter's Syndrome (47,XXY)

Dr. Harry Klinefelter was the first to identify the clinical presentation of Klinefelter's syndrome in nine men with an array of features: testicular dysgenesis, small and firm testes, elevated serum FSH levels, functional Leydig cells, azoospermia, microorchidism, eunuchoidism and gynecomastia (Klinefelter et al. 1942). For some years till then, it was thought to be an endocrine disorder of unidentified aetiology. Later, in 1959, when the cytogenetic arena was flourishing, Jacobs et al. reported a strong evidence of an extra X chromosome, that is, 47,XXY karyotype, in males with Klinefelter's syndrome (Jacobs 1959). Around 80–90% of Klinefelter's syndrome (KS) cases represent the 'original' karyotype of 47,XXY, while the remaining show either a varying degree of mosaicism (e.g. 47,XXY/46,XY), an additional sex chromosome (48,XXXY; 48,XXYY; 49,XXXXY) or a structurally abnormal X chromosomes (Bojesen et al. 2003; Lanfranco et al. 2004). KS is represented as the most frequent disorder of sex chromosomes in humans, with a global prevalence of around one in 500 males (Nielsen and Wohlert 1991). Because of unusual chromosomal and gonadal features, Klinefelter's syndrome condition has been recognised as a disorder of sexual development (Hughes et al. 2006).

47,XXY males show clinically variable signs that are often age related. In infancy, the KS males may display hypospadias, cryptorchidism or small phallus (Caldwell and Smith 1972). In the toddler years, they may start presenting symptoms of developmental delay and language deficits especially with delayed expressions (Walzer et al. 1978). The school-aged child may display various learning disabilities and behavioural/social problems (Walzer et al. 1978), while the adolescent may display a delayed or incomplete pubertal development with eunuchoid body habitus, small testes and gynecomastia (Robinson et al. 1990). Adults may also develop complications associated with malignancy of breast (Okada et al. 1999). KS often presents a wide spectrum of phenotype in the adulthood; however, it is often associated with primary testicular failure, hypergonadotropic hypogonadism, reduced testicular volume and infertility due to azoospermia and severe oligozoospermia in 90 and 10% of non-mosaic KS patients, respectively (Lanfranco et al. 2004; Ferlin et al. 2007; De Sanctis and Ciccone 2010; Foresta et al. 2012).

KS is present in around 3% of all infertile men, while the frequency increases to 13% in infertile azoospermic population (Van Assche et al. 1996; Vincent et al. 2002; Tüttelmann and Gromoll 2010). KS represents, thus far, the most frequent genetic cause of azoospermia in humans. The azoospermia phenotype in KS develops due to progressive germ cell degeneration that starts at the time of mid-puberty and progresses during puberty and adolescence. The accelerating severity eventually leads to Sertoli cell dysfunction, extensive fibrosis and hyalinisation of seminiferous tubules and Leydig cell hyperplasia (Aksglaede et al. 2006, 2013). The mechanism underlying the global degeneration is still unclear; however, one of the leading hypotheses moves around, an altered dosage of X-linked genes that escapes the process of inactivation (Aksglaede et al. 2006). It has been suggested that the supernumerary X itself prevents the completion of meiosis and disturbs the testicular homeostasis by affecting Sertoli and Leydig cell functions (Aksglaede et al. 2006). Nevertheless, some of the studies have debatably mentioned the elimination of supernumerary X chromosome during meiosis, which was supported by some indirect clues (Foresta et al. 1999; Sciurano et al. 2009). Besides, majority of the reports are in favour of 47,XXY spermatogonia, being able to complete meiosis, which is evident by an increase in the incidence of KS boys to KS fathers (Hall et al. 2002; Staessen et al. 2003; Martin 2008).

Interestingly, a few studies highlighted that a skewed X chromosome inactivation (XCI) and X-linked imprinting may have differential effects on the autistic and schizotypal features observed in KS patients. These reports highlighted the significance of epigenetic processes in the development of KS. Recently, a group of researchers performed a global transcriptome analysis on testicular biopsies obtained from six non-mosaic KS patients with azoospermia. The analysis revealed that a large number of deregulated transcripts belonged to the regulatory pathways of various Sertoli cell and Leydig cell functions (D'Aurora et al. 2015). Similarly, a study demonstrated genome-wide alterations in DNA methylation and gene expression patterns in two regions of the brain from a patient with a 47,XXY karyotype. These genes belonged to the loci which normally escape the XCI in females, thus supporting the hypothesis of X-linked dosage imbalance. So far, no fertility treatment is available for the affected patients.

8.2.2 Jacob's Syndrome (47,XYY)

Avery Sandberg et al. were the first to identify 47,XYY syndrome in a normal 44-year-old man who fathered a Down syndrome child (Sandberg et al. 1961; Hauschka et al. 1962). Few years later, a British geneticist, Patricia Jacobs, described it in detail, and thereafter the presence of an extra Y chromosome was termed as Jacob's syndrome (Jacobs 1974, 1975). 47,XYY syndrome has a prevalence of 1 in 1000 male individuals (Bojesen et al. 2003). 47,XYY genotype in these patients arises due to non-disjunction at the time of the second meiotic division (MII) during spermatogenesis or post-zygotic mitosis (PZM). It is well demonstrated by the presence of an additional Y chromosome in spermatogonia and/or spermatocytes, which gets selective advantage during gametogenesis (Jacobs 1974, 1975; Jacobs and Hassold 1995; Robinson and Jacobs 1999).

The phenotypic features of XYY men are similar to those observed in Klinefelter syndrome, including tall stature, learning disabilities, cognitive impairment and lack of attention (Maclean et al. 1961; Robinson et al. 1989; Ratcliffe 1982; Welch 1985; Ratcliffe et al. 1992; Rovet et al. 1995; Bojesen et al. 2003; Aksglaede et al. 2008); however, they show normal pubertal development and testosterone levels. From the last few years, there has been increasing evidences on the association of 47,XYY complement in the somatic cells and the presence of chromosomally abnormal sperm in the semen (Speed et al. 1991; Blanco et al. 1997; Chevret et al. 1997; Lim et al. 1999; Gonzalez-Merino et al. 2007; Wong et al. 2008). Numerous studies have demonstrated an increase in the presence of sperm mosaicism, aneuploidy or hyperhaploidy in 47,XYY men ranging from 0.57 to 77.8% (Speed et al. 1991; Lim et al. 1999; Morel et al. 1999; Shi and Martin 2000; Wong et al. 2008; Gonzalez-Merino et al. 2007). These men frequently display an increase in the risk of transmission of extra Y chromosome in the offspring (Lim et al. 1999).

It has been proposed that the YY bivalent pairs during meiosis I leave the free X univalent within the vesicle, which subsequently gets eliminated during anaphase, resulting in disomic YY sperm (Lim et al. 1999). An increase in the frequency of

XY disomy can also be explained by this assumption (Lim et al. 1999). Studies by Guttenbach et al. and Kruse et al. have also reported an increasing prevalence of XY and XX disomic sperm, providing strong indication of 47,XXY cells to undergo complete meiotic divisions (Guttenbach et al. 1997; Kruse et al. 1998). However, the presence of extra Y chromosome may impede normal spermatogenesis process, affecting various sperm parameters that define fertility (Milazzo et al. 2006). Nevertheless 47,XYY men are reported to have sperm count between normozoospermia to azoospermia (Faed et al. 1976; Lim et al. 1999; Egozcue et al. 2000; Blanco et al. 2001; Moretti et al. 2007). 47,XYY men with normal sperm counts have potential to achieve a normal pregnancy. However, for men with poor fertility, screening for sperm sex chromosome constitution is strongly recommended.

8.3 Syndromes with Gene Mutations

8.3.1 Kallmann Syndrome

Maestre de San Juan in 1856 was the first to report an association of small testis with the absence of olfactory structures in the brain (de San Juan 1856). Later, the syndrome was clinically recognised and identified in 1944 by an American medical geneticist, Kallmann, who reported an anosmia-associated occurrence of hypogonadism in three affected families and suggested the syndrome to be hereditary (Kallmann 1944). Later in the 1950s, a Swiss anatomist named de Morsier further described that the patients with Kallmann syndrome have underdeveloped or absent olfactory bulbs in association with hypogonadism (de Morsier and Gauthier 1963). A few years thereafter, it was found that the hypogonadism in affected patients developed due to gonadotropin-releasing hormone (GnRH) deficiency (Naftolin et al. 1971). The incidence of Kallmann syndrome has been estimated to be 1 in 8000 boys; however, the prevalence is five times lower in girls. Perhaps in most of the cases, primary amenorrhoea in females remains undervalued and often unexplored (Jones and Kemmann 1976).

Clinical diagnosis of Kallmann syndrome is made by the presence of anosmia together with a diminished libido, erectile dysfunction and lack/delay/stop in pubertal sexual maturation with the absence of secondary sex characters. Though, the status of pituitary and hypothalamus appears normal in Kallmann syndrome, they often present a low serum testosterone level (<100 ng/mL). The adult males show a eunuchoid body habitus, which results due to delayed skeletal maturation (Pallais et al. 1993). The testicular morphology in Kallmann syndrome may show heterogeneous grades of spermatogenic impairments (Nishio et al. 2012); however, spermatogenesis can be easily rescued by hormonal stimulation (Jungwirth et al. 2012). Some nonreproductive characters associated with gene mutations in Kallmann syndrome men include unilateral renal agenesis, congenital ptosis, dyskinesia and/or skeletal abnormalities, involuntary upper limb mirror, cleft lip/palate, ear/hearing defects, coloboma (eye defect), agenesis of one or several teeth (hypodontia), obesity and hyperlaxity of the joints. Nevertheless, it typically cosegregates severe hypogonadotropic hypogonadism with a complete absence of the sense of smell (anosmia). The degree of the hypogonadism may vary significantly between unrelated patients and even between monozygotic twins.

Various histopathological studies have demonstrated that the poor reproductive features of Kallmann syndrome men result due to disrupted embryonic migration of neuroendocrine GnRH from the nose to the brain (Fig. 8.1) (Schwanzel-Fukuda and Pfaff 1989; Teixeira et al. 2010). Kallmann syndrome presents various modes of transmission, which can be X-linked, autosomal recessive, autosomal dominant or

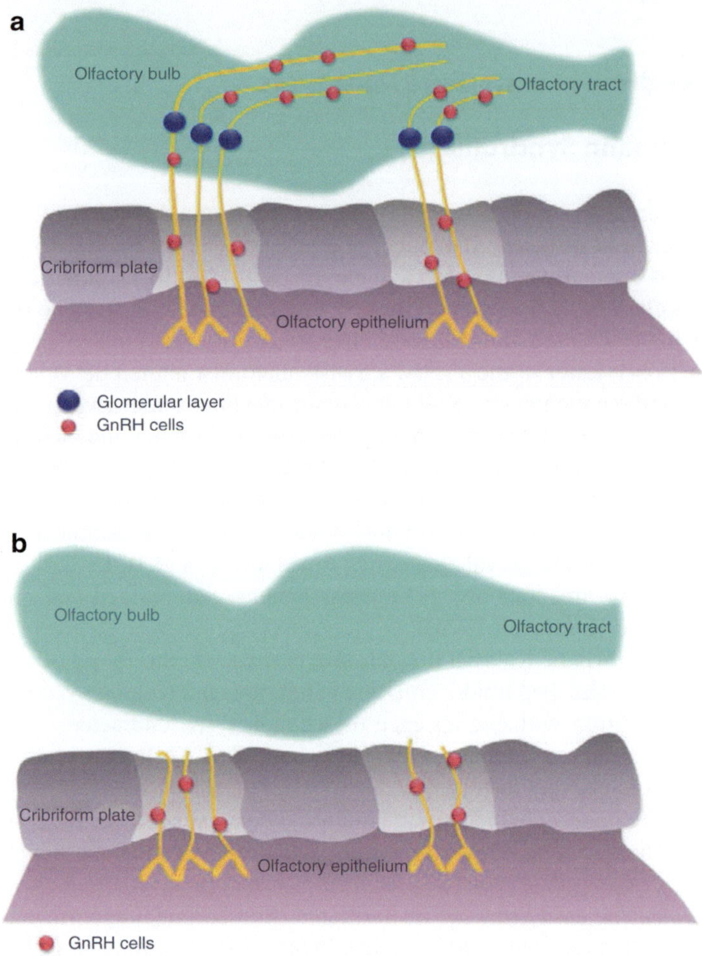

Fig. 8.1 Diagram showing the development and migration of GnRH neurons. In normal condition the GnRH neurons originate in the olfactory placode outside the brain and migrate along the olfactory axons through the cribriform plate to reach the hypothalamus (**a**), while in the KS patients the GnRH neurons fail to migrate and get arrested in the cribriform plate (**b**)

digenic/oligogenic (Dodé and Hardelin 2009; Sykiotis et al. 2010). As this syndrome shows an incomplete penetrance, the genetic linkage analysis in these patients becomes difficult. However, through various candidate gene approaches, including screening of gene mutations associated with the disease phenotype, the researchers have identified various genes, most of which function at the level of GnRH neuron migration. Some of these genes include *KAL-1*, *FGFR1*, *PROKR2* and *PROK2*, *FGF8*, *CHD7*, *WDR11*, *HS6ST1* and *SEMA3A* (Valdes-Socin et al. 2015). The *KAL-1* gene mutation and the *FGFR1/FGF8* gene mutation account for around 8 and 10% of all KS cases, respectively. *KAL-1* gene is mapped to Xp22.32 locus and contains 14 exons. It encodes for 840 amino acid long protein, anosmin-1. It is an extracellular adhesion protein that helps in orchestrating GnRH neuron adhesion and axonal migration. It also acts as a co-receptor for FGF-mediated signalling processes. *KAL-1* gene mutation leads to an abnormal GnRH migration and olfactory neuron disorder (Viswanathan and Eugster 2011). Recently, mutations in *PROKR2* and *PROK2* genes have been identified to occur in approximately 9% of the KS patients (Dodé and Hardelin 2009; Sarfati et al. 2010; Dodé and Ronard 2014). *PROKR2* and *PROK2* genes encode for G protein-coupled, prokineticin receptor-2 and prokineticin-2, respectively. Gene knockout studies have identified loss of *PROKR2* and *PROK2* genes to be associated with hypogonadotropic hypogonadism in conjunction with abnormal GnRH neuron migration.

8.3.2 Androgen Insensitivity Syndrome (AIS)

The end-organ resistance to androgen actions is termed as the androgen insensitivity syndrome (AIS). AIS is a disorder of hormone resistance characterised by a development of female phenotype in XY individual. Pathogenesis of AIS is a result of mutation in X-linked androgen receptor gene (*AR*), which belongs to a class of nuclear receptor family. Knockout studies on animal models have revealed that the presence of *AR* gene expression in Sertoli cell and Leydig cell is crucial for spermatogenesis (Wang et al. 2009). Numerous *AR* mutations have been identified, and around 800 *AR* mutations are registered in the McGill University database of *AR* gene. Around 30% of *AR* mutations are sporadic de novo in nature (Hughes and Deeb 2006). The overall occurrence of AIS varies between 1 in 20,000 and 1 in 99,000 genetic males (Grumbach and Conte 2003).

Hormone resistance syndrome was first characterised by John Morris in 1953, while analysing the clinical features of 82 patients, with female phenotype and bilateral testis (Morris 1953). It was then called as testicular feminisation syndrome; however, AIS is by far the most accepted terminology (Quigley et al. 1995).

AIS can occur due to a varying degree of androgen insensitivity in an XY person (Fig. 8.2). The phenotypes in AIS patients vary on a seven-point scale and are broadly categorised into three categories: partial androgen insensitivity syndrome (PAIS), mild androgen insensitivity syndrome (MAIS) or complete androgen insensitivity syndrome (CAIS) depending on the degree and loss of androgen actions (Quigley et al. 1995).

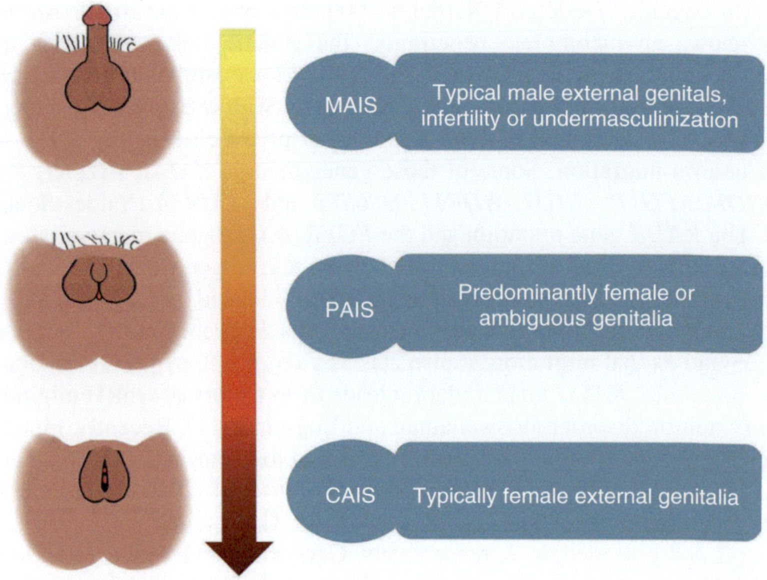

Fig. 8.2 Three major forms of androgen insensitivity syndrome in 46, XY individuals are MAIS, PAIS and CAIS in increasing order of severity

AR gene consists of eight exons. CAIS occurs due to missense mutations occurring throughout the *AR* gene; however, it is most frequent in regions encoding DNA binding and ligand binding domains (Matias et al. 2000; Hughes et al. 2012). Female infants with CAIS are characterised by labial swelling and inguinal hernia; however, the prepubertal girls show a short, blind-ending vagina; a complete absence of Wolffian duct-derived structures such as the epididymis, vas deferens and seminal vesicles; and an absence of the prostate gland. Some authors claim a rare presence of Müllerian duct-derived structures (Dodge et al. 1985; Ulloa-Aguirre et al. 1990; Swanson and Coronel 1993).

PAIS is also referred as incomplete androgen insensitivity syndrome. It exhibits a wide spectrum of phenotypes in comparison to the normal male phenotype, which are not severe enough to be classified as CAIS. All uncharacterised cases of androgen insensitivity syndrome are kept into this class (Aiman et al. 1979; Aiman and Griffin 1982; Morrow et al. 1987). The differentiation of PAIS from CAIS is achieved on the basis of the extent of masculine growth impairment of the external male genitals along with an absence of female characteristic features such as breast which is normally present in all CAIS patients (Quigley et al. 1995). Even though the affected PAIS are infertile or impotent, they usually present normal erectile functions.

In MAIS, the degree of the androgen receptor sensitivity is quite mild in nature. Thus, the effected males of this type are phenotypically normal, containing fully developed male genitalia, though, they may have mild defects in secondary sexual characteristics. Infertility is the most frequently associated symptom of

MAIS. Nevertheless, the presence of infertility is not associated with genital anomalies (Zuccarello et al. 2008). Fertility can however be restored in these patients by high dose of androgen supplementation (Yong et al. 1994). Reduced fertility in MAIS is often manifested as spinal and bulbar muscular atrophy (SBMA). SBMA is caused by an X-linked polymorphic CAG repeat expansion (>35 CAGs) in exon 1 of the AR gene. SBMA is characterised by muscle atrophy and weakness, gynaecomastia, testicular atrophy and infertility.

Management of AIS is performed by optimal dosage of androgens at the time of puberty and beyond.

8.3.3 Noonan Syndrome

Noonan syndrome (NS), also known as pterygium colli syndrome or male Turner syndrome, is a congenital disorder characterised by the presence of short stature, typical dysmorphic facial features and congenital heart defects. The disease was first defined by Noonan and Lezington in 1968 (Noonan and Lexington 1968). As the disease shows strikingly similar features to Turner's syndrome (short stature, webbed neck, low set ears, cubitus valgus, pulmonary stenosis and cardiovascular disorders), it is sometimes referred as male Turner syndrome. The prevalence of NS varies between 1:1000 and 1:2500 live births. NS males often present with infertility associated with testicular atrophy and cryptorchidism (Witt et al. 1988; Nisbct et al. 1999). Most of the NS cases are sporadic; however, it displays a clear autosomal dominant mode of inheritance pattern in some families. The associated locus for this dominant form is mapped to 12q22-qter (Jamieson et al. 1994). Genotype-phenotype studies have reported an association of *PTEN11* gene with the pathogenesis of NS. *PTEN11* gene mutations are present in as high as 50% of the NS cases. Mutations in other genes such as *SOS1*, *RAF1* and *RIT1* are associated with the remaining 20% of the cases. These genes are involved in the regulation of RAS/MAPK cell signalling pathway. However, further research is required to find the exact mechanism associated with the development of this phenotype.

8.3.4 Cystic Fibrosis

Cystic fibrosis (CF) is an autosomal recessive disorder, caused due to mutations in cystic fibrosis transmembrane regulator (*CFTR*) gene. It often manifests as congenital bilateral absence of the vas deferens (CBAVD) in 99% of CF cases. Abnormal development of the mesonephric duct results in bilateral absence of the vas deferens in CF patients (Patrizio and Salameh 1997).

CBAVD is present in around 1–2% infertile men and in 6% obstructive azoospermia cases. The mutational spectrum of *CFTR* gene is significantly heterogeneous, and more than 800 mutations, 70 sequence variants, have been reported (Lewis-Jones et al. 2000). Δ508F is the most common mutation in CF patients; however, the frequency varies across different geographical and ethnic populations. The most frequently found seminal abnormalities in CF patients include azoospermia, reduced semen volume, high acidic pH and low fructose level. Azoospermia in CBAVD occurs due to complete blockage in sperm transportation from testis or epididymis to the outer genital tract. Microsurgical epididymal sperm aspiration and intracytoplasmic sperm injection (ICSI) are the best options for infertile CF patients. More details of this syndrome can be found in Chap. 9.

8.4 Rare Syndromes of Male Infertility

8.4.1 Myotonic Dystrophy 1

Myotonic dystrophy 1, also known as Morbus Curschmann-Steinert or dystrophia myotonica 1, DM1, is caused by an unstable expansion of CTG repeats in the *DMPK* gene (dystrophia myotonica protein kinase), which encodes a serine/threonine protein kinase. A spectrum of phenotypes such as defects of skeletal and smooth muscle, frontal balding, cardiac arrhythmias and abnormalities of the eye, heart, endocrine system and central nervous system are seen in these patients. Males with DM1 may also present symptoms of infertility, which include testicular atrophy, decrease in sperm count, hyalinisation and fibrosis of seminiferous tubules, loss of libido and potency and hypogonadism (Sarkar et al. 2004). Testicular atrophy is the most frequent anomaly observed in around 80% of the cases (Kim et al. 2012a). *DMPK* gene has been mapped to cytogenetic locus, 19q13.3. The 3′UTR region of the gene contains a triplet repeat motif of around 5–35 CTGs (Meschede and Horst 1997), which increases from 50 to several hundred in diseased condition. The age of onset of this disease may decrease from generation to generation in the affected family, and this phenomenon is termed as anticipation (Tsilfidis et al. 1992; Redman et al. 1993; McInnis 1996). Around 73% of the DM1 patients present with oligozoospermia or azoospermia (Klesert et al. 1997).

8.4.2 Primary Ciliary Dyskinesia

In 1933, Kartagener reported a ‘clinical triad’ of ‘sinusitis-bronchiectasis-situs inversus syndrome’ in four patients with familial and hereditary characteristics. ‘This clinical triad’ presenting all three symptoms was termed as complete Kartagener’s syndrome (KS). Cases which showed an absence of situs inversus were referred as incomplete Kartagener’s syndrome (Bent and Smith 1997; Berdon and Willi 2004; Ortega et al. 2007). A more appropriate nomenclature was provided to this syndrome in 1988, as primary ciliary dyskinesia (PCD). Kartagener’s syndrome follows an autosomal recessive inheritance pattern and is seen in about 50% of the PCD cases. It causes simultaneous ciliary dysfunctions in several parts of the body, making it one of the most severe PCD conditions (Cox and Talamo 1979; Afzelius and Eliasson 1982; Bartoloni et al. 2002). The occurrence of Kartagener’s syndrome is around 1 in 30,000 live births. The association of Kartagener’s syndrome with male infertility was first recognised by Afzelius while observing an absence of dynein arms in the spermatozoa and cilia of four patients (Afzelius 1976). The males with this syndrome usually present an infertile phenotype due to loss of sperm motility, which arises due to various ultrastructural defects in sperm tail (Samuel 1987).

Though the genetic research is still going on, mutations in *DNAI1* (the axonemal dynein intermediate chain gene), located on chromosome 9p12–21, and the *DNAH5* gene (axonemal dynein heavy chain gene), located on chromosome 5p15–14, are two significant and widely studied genes in the pathology of Kartagener's syndrome. Mutations in these genes result in the absence of the outer dynein arm of the cilia, leading to abnormal ciliary structure and motor function (Pennarun et al. 1999; Omran et al. 2000; Gravesande and Omran 2005; Zariwala et al. 2006). Afflicted men with immotile spermatozoa have almost no chance of accomplishing pregnancy through natural procedures. Thus, an aid of assisted reproductive technologies is the only option for Kartagener's syndrome patients to initiate a pregnancy.

8.4.3 Kearns–Sayre Syndrome

Kearns–Sayre syndrome (KSS) is one of the multisystem syndromes which occur due to single or large-scale deletions in the mitochondrial genome. The syndrome predominantly affects the neuromuscular and endocrine systems. The diagnosis of KSS is based on the clinical presentation of a classic triad of symptoms: onset of the symptoms before 20 years of age, progressive ophthalmoparesis and pigmentosa retinitis (PR) (Berenbaum et al. 1990). However, it often manifests with other systemic anomalies, such as cardiac conduction defects, cerebellar syndrome, different neurological abnormalities, increased cerebrospinal fluid (CSF) protein concentration and several endocrine disorders (Harvey and Barnett 1992). Reproductive anomalies are reported in around 20–30% of the cases in which the syndrome cosegregates with distinctive presence of cryptorchidism, pubertal delay, low testicular volume and insufficient gonadotropin levels. Mitochondria are strictly derived from mothers; therefore, the syndrome is inherited exclusively through the maternal line. Nevertheless, no genetic locus or mutations have been identified till date, which affects fertility in Kearns–Sayre syndrome males.

8.4.4 Aarskog–Scott Syndrome

Aarskog–Scott syndrome or ASS is a genetically heterogeneous, X-linked recessive disorder (Altıncık et al. 2013). This syndrome is often called as 'faciogenital dysplasia' due to the distinctive presence of facial, genital and skeletal anomalies in these patients. The syndrome was first described by Aarskog and Scott in 1970 (Aarskog 1970; Scott 1971). ASS patients exhibit a wide range of phenotypic heterogeneity, and the symptoms may vary from mild to severe. Minor features typically include the abnormalities of midline and the urogenital system. The diagnosis is predominantly made on the basis of genital anomalies such as shawl scrotum, cryptorchidism, hypertelorism and brachydactyly. The dysplastic changes, however, involve the skeletal abnormalities (Schwartz et al. 2000; Al-Semari et al. 2013). Mutations in the *FGD1* gene (FYVE, RhoGEF and PH domain containing 1) have been identified in around 20% of the ASS cases. This gene is located at Xp11.21 region and encodes for a guanine nucleotide exchange factor (GEF). GEFs are involved in regulating signalling pathways of skeletal development, cytoskeletal reorganisation and morphogenesis. A growth hormone (GH) therapy is the most acceptable and widely used treatment for these patients; however, discrepancies exist in concern with the GH dose, optimal age for the GH therapy and potential adverse effects.

8.4.5 Persistent Müllerian Duct Syndrome

Persistent Müllerian duct syndrome (PMDS) is a rare form of sexual development disorder, first characterised by Nilson in 1939 (Nilson 1939). It is a genetic anomaly characterised by a distinctive presence of Müllerian duct derivatives (i.e. the uterus, cervix, fallopian tubes and vagina) in males, giving it an internal male pseudohermaphrodite phenotype (Yuksel et al. 2006). The development of external genitalia and secondary sexual characteristics, however, occurs normally. The syndrome is caused either by a deficiency of anti-Müllerian hormone (AMH) or its type II receptor (AMHR-II) or due to an insensitivity of the target organ towards Müllerian-inhibiting factor (MIF) (Gutte et al. 2014). The syndrome follows an autosomal recessive mode of inheritance. The genes for *AMH* and *AMHR-II* have been mapped to cytogenetic loci 19p13 and 12q13, respectively. Two types of anatomical variants, male and female types, have been identified in PMDS. The male form is the most frequent, occurring in about 80–90% of PMDS cases. The male form is characterised by the presence of inguinal hernia and unilateral cryptorchidism. The female form contributes to the remaining 10–20% of PMDS cases and is characterised by bilateral cryptorchidism with abdominal testis attached to the fallopian tube (Wu et al. 2000; Dekker et al. 2003). Compromised testicular function and infertility are largely attributed to cryptorchidism in PMDS males. High degree awareness is desirable to diagnose this condition. Early treatment is mandatory to restore fertility and to prevent the development of malignancy in the remnants of Müllerian structures.

8.4.6 Prader–Willi Syndrome

Prader–Willi syndrome (PWS) is a gene imprinting disorder, predominantly characterised by psychomotor retardation in the affected individuals (Holm et al. 1993). It was first described by the Swiss physicists Prader, Labhart and Willi in 1956 (Prader 1956). It occurs at a frequency of 1/10,000–1/30,000 individuals. Deletions and epigenetic alterations in the 15q11.2-q13 region of paternal chromosome 15 lead to syndrome development The syndromic features arise due to the lack of gene expression from the paternally derived chromosome 15q11.2-q13. About 70–75% of the affected individuals with PWS encompass a de novo deletion at paternally inherited chromosome 15q11-q13 region. 20–25% of the patients exhibit maternal uniparental disomy (UPD) at the same locus. The remaining 3% of cases harbour imprinting defects (Wharton and Loechner 1996; Cassidy 1997; Nicholls et al. 1999). A few studies have also reported cytogenetic anomalies, such as chromosomal translocations or rearrangements at 15q11-q13 region (Bittel and Butler 2005; Kim et al. 2012b). Hypogonadism is the most consistent feature of PWS in both males and females. The presence of hypogonadism often cosegregates with clinical presentation of genital hypoplasia, delayed or incomplete puberty and infertility in majority of the cases. Males frequently display features of cryptorchidism, a poor rugated and hypoplastic scrotum and small penis in 80–90% of the PWS cases (Cassidy et al. 2011). Recent reports have demonstrated that primary gonadal failure is the primary contributor to male hypogonadism in PWS cases (Vogels et al. 2008; Siemensma et al. 2011; Radicioni et al. 2012). Infertility, in both the sexes, is almost universal in PWS cases.

8.4.7 Deafness Infertility Syndrome

Deafness infertility syndrome (DIS) was first of all characterised by Avidan et al. in 2003, in a consanguineous family that had three male siblings with deafness and infertility (asthenoteratozoospermia). The syndromic phenotype was attributed to a 70 kb deletion at chromosome 15q15.3 containing genes such as *KIAA0377*, *CKMT1B*, *STRC* and *CATSPER2* (Avidan et al. 2003). It was again identified by Zhang et al. in 2007 in three families presenting large contiguous gene deletion in the locus, but the deleted length was around 100 kb. The syndrome is inherited in an autosomal recessive fashion with a prevalence of <1/1,000,000 individuals. Abundant large inverted repeats in 15q15.3 region make it prone to secondary structure formation and chromosomal rearrangements, resulting in large deletions (Zhang et al. 2007). Affected males are homozygous for the deletion with parents as asymptomatic carriers. However, females with homozygous deletions are deaf, but fertile.

Conclusion

The field of biological research and medical genetics has prospered a lot in the last few decades, enabling the identification and characterisation of various diseases and syndromes. However, the accurate identification of genes and mechanism leading to the appearance of a constellation of features seen in the syndromic cases is yet to be elucidated. Some syndromes may present one or more strikingly specific features unique to that syndrome. However, in some cases the diagnosis becomes really difficult due to overlapping features and the lack of a careful medical examination. In most of the cases of infertility-associated syndromes, lack of awareness and fertility concerns becomes the primary cause of identification failures. Clinical investigations coupled with genetic analysis are needed for proper diagnosis of these syndromes. Males with variable phenotype need thorough clinical and genetic investigations and counselling. High-throughput platform utilisation in genetic mutation detection in these patients will certainly lead to the characterisation of the new syndromic forms of male infertility.

References

Aarskog D (1970) A familial syndrome of short stature associated with facial dysplasia and genital anomalies. J Pediatr 77(5):856–861

Afzelius BA (1976) A human syndrome caused by immotile cilia. Science 193(4250):317–319

Afzelius BA, Eliasson R (1982) Male and female infertility problems in the immotile-cilia syndrome. Eur J Respir Dis Suppl 127:144–147

Aiman J, Griffin JE (1982) The frequency of androgen receptor deficiency in infertile men. J Clin Endocrinol Metabol 54(4):725–732

Aiman J, Griffin JE, Gazak JM, Wilson JD, MacDonald PC (1979) Androgen insensitivity as a cause of infertility in otherwise normal men. N Engl J Med 300(5):223–227

Aksglæde L, Wikström AM, Rajpert-De Meyts E, Dunkel L, Skakkebæk NE, Juul A (2006) Natural history of seminiferous tubule degeneration in Klinefelter syndrome. Hum Reprod Update 12(1):39–48

Aksglaede L, Skakkebaek NE, Juul A (2008) Abnormal sex chromosome constitution and longitudinal growth: serum levels of insulin-like growth factor (IGF)-I, IGF binding protein-3, luteinizing hormone, and testosterone in 109 males with 47, XXY, 47, XYY, or sex-determining region of the Y chromosome (SRY)-positive 46, XX karyotypes. J Clin Endocrinol Metabol 93(1):169–176

Aksglaede L, Link K, Giwercman A, Jørgensen N, Skakkebaek NE, Juul A (2013) 47, XXY Klinefelter syndrome: clinical characteristics and age-specific recommendations for medical management. Am J Med Genet C Semin Med Genet 163(1):55–63

Al-Semari A, Wakil SM, Al-Muhaizea MA, Dababo M, Al-Amr R, Alkuraya F, Meyer BF (2013) Novel FGD1 mutation underlying Aarskog–Scott syndrome with myopathy and distal arthropathy. Clin Dysmorphol 22(1):13–17

Altıncık A, Kaname T, Demir K, Böber E (2013) A novel mutation in a mother and a son with Aarskog–Scott syndrome. J Pediatr Endocrinol Metab 26(3–4):385–388

Avidan N, Tamary H, Dgany O, Cattan D, Pariente A, Thulliez M, Borot N, Moati L, Barthelme A, Shalmon L, Krasnov T (2003) CATSPER2, a human autosomal nonsyndromic male infertility gene. Eur J Hum Genet 11(7):497–502

Bartoloni L, Blouin JL, Pan Y, Gehrig C, Maiti AK, Scamuffa N, Rossier C, Jorissen M, Armengot M, Meeks M, Mitchison HM (2002) Mutations in the DNAH11 (axonemal heavy chain dynein type 11) gene cause one form of situs inversus totalis and most likely primary ciliary dyskinesia. Proc Natl Acad Sci U S A 99(16):10282–10286

Bent JP, Smith RJ (1997) Intraoperative diagnosis of primary ciliary dyskinesia. Otolaryngol Head Neck Surg 116(1):64–67

Berdon WE, Willi U (2004) Situs inversus, bronchiectasis, and sinusitis and its relation to immotile cilia: history of the diseases and their discoverers—manes Kartagener and Bjorn Afzelius. Pediatr Radiol 34(1):38–42

Berenbaum F, Cote D, Pradat P, Rancurel G (1990) Kearns–Sayre syndrome. Neurology 40(1):193–193

Bittel DC, Butler MG (2005) Prader–Willi syndrome: clinical genetics, cytogenetics and molecular biology. Expert Rev Mol Med 7(14):1–20

Blanco J, Rubio C, Simon C, Egozcue J, Vidal F (1997) Increased incidence of disomic sperm nuclei in a 47, XYY male assessed by fluorescent in situ hybridization (FISH). Hum Genet 99(3):413–416

Blanco J, Egozcue J, Vidal F (2001) Meiotic behaviour of the sex chromosomes in three patients with sex chromosome anomalies (47, XXY, mosaic 46, XY/47, XXY and 47, XYY) assessed by fluorescence in-situ hybridization. Hum Reprod 16(5):887–892

Bojesen A, Juul S, Gravholt CH (2003) Prenatal and postnatal prevalence of Klinefelter syndrome: a national registry study. J Clin Endocrinol Metabol 88(2):622–626

Caldwell PD, Smith DW (1972) The XXY (Klinefelter's) syndrome in childhood: detection and treatment. J Pediatr 80(2):250–258

Cassidy SB (1997) Prader–Willi syndrome. J Med Genet 34(11):917–923

Cassidy SB, Schwartz S, Miller JL, Driscoll DJ (2011) Prader–Willi syndrome. Genet Med 14(1):10–26

Chevret E, Rousseaux S, Monteil M, Usson Y, Cozzi J, Pelletier R, Sele B (1997) Meiotic behaviour of sex chromosomes investigated by three-colour FISH on 35 142 sperm nuclei from two 47, XYY males. Hum Genet 99(3):407–412

Cox DW, Talamo RC (1979) Genetic aspects of pediatric lung disease. Pediatr Clin North Am 26(3):467–480

D'Aurora M, Ferlin A, Di Nicola M, Garolla A, De Toni L, Franchi S, Palka G, Foresta C, Stuppia L, Gatta V (2015) Deregulation of Sertoli and leydig cells function in patients with Klinefelter syndrome as evidenced by testis transcriptome analysis. BMC Genomics 16(1):156

de Morsier G, Gauthier G (1963) La dysplasie olfacto-ge'nitale. Pathol Biol 11:1267–1272

de San Juan AM (1856) Falta total de los nerviosolfactorios con anosmia en un individuo en quien existia una atrofia congenita de los testiculos y miembro viril. Siglo Med 131:211

De Sanctis V, Ciccone S (2010) Fertility preservation in adolescents with Klinefelter's syndrome. Pediatr Endocrinol Rev 8:178–181

Dekker HM, de Jong IJ, Sanders J, Wolf RF (2003) Persistent Müllerian duct syndrome. Radiographics 23(2):09–13

Dodé C, Hardelin JP (2009) Kallmann syndrome. Eur J Hum Genet 17(2):139–146

Dodé C, Rondard P (2013) PROK2/PROKR2 signaling and Kallmann syndrome. Front Endocrinol 4:19. doi:10.3389/fendo.2013.00019

Dodge ST, Finkelston MS, Miyazawa K (1985) Testicular feminization with incomplete Müllerian regression. Fertil Steril 43(6):937–938

Egozcue S, Blanco J, Vendrell JM, Garcia F, Veiga A, Aran B, Barri PN, Vidal F, Egozcue J (2000) Human male infertility: chromosome anomalies, meiotic disorders, abnormal spermatozoa and recurrent abortion. Hum Reprod Update 6(1):93–105

Faed M, Robertson J, MacIntosh WG, Grieve J (1976) Spermatogenesis in an infertile XYY man. Hum Genet 33(3):341–347

Ferlin A, Raicu F, Gatta V, Zuccarello D, Palka G, Foresta C (2007) Male infertility: role of genetic background. Reprod Biomed Online 14(6):734–745

Foresta C, Galeazzi C, Bettella A, Marin P, Rossato M, Garolla A, Ferlin A (1999) Analysis of meiosis in intratesticular germ cells from subjects affected by classic Klinefelter's syndrome. J Clin Endocrinol Metabol 84(10):3807–3810

Foresta C, Caretta N, Palego P, Ferlin A, Zuccarello D, Lenzi A, Selice R (2012) Reduced artery diameters in Klinefelter syndrome. Int J Androl 35(5):720–725

Gonzalez-Merino E, Hans C, Abramowicz M, Englert Y, Emiliani S (2007) Aneuploidy study in sperm and preimplantation embryos from nonmosaic 47, XYY men. Fertil Steril 88(3):600–606

Gravesande KSVS, Omran H (2005) Primary ciliary dyskinesia: clinical presentation, diagnosis and genetics. Ann Med 37(6):439–449

Grumbach MM, Conte FA (2003) Disorders of sex differentiation. In: Larsen PR (ed) Williams textbook of endocrinology. Saunders, Philadelphia, pp 842–1002

Gutte AA, Pendharkar PS, Sorte SZ (2014) Transverse testicular ectopia associated with persistent Müllerian duct syndrome—the role of imaging. Br J Radiol 81(967):e176–e178

Guttenbach M, Michelmann HW, Hinney B, Engel W, Schmid M (1997) Segregation of sex chromosomes into sperm nuclei in a man with 47, XXY Klinefelter's karyotype: a FISH analysis. Hum Genet 99(4):474–477

Hall S, Marteau TM, Limbert C, Reid M, Feijóo M, Soares M, Nippert I, Bobrow M, Cameron A, van Diem M, Verschuuren-Bemelmans C (2002) Counselling following the prenatal diagnosis of Klinefelter syndrome: comparisons between geneticists and obstetricians in five European countries. Public Health Genomics 4(4):233–238

Harvey JN, Barnett D (1992) Endocrine dysfunction in Kearns–Sayre syndrome. Clin Endocrinol (Oxf) 37(1):97–104

Hauschka TS, Hasson JE, Goldstein MN, Koepf GF, Sandberg AA (1962) An XYY man with progeny indicating familial tendency to non-disjunction. Am J Hum Genet 14(1):22–30

Hempel M, Buchholz T (2009) Rare syndromes associated with infertility. Reproduktionsmed Endokrinol 6(1):24–26

Holm VA, Cassidy SB, Butler MG, Hanchett JM, Greenswag LR, Whitman BY, Greenberg F (1993) Prader–Willi syndrome: consensus diagnostic criteria. Pediatrics 91(2):398–402

Hughes IA, Deeb A (2006) Androgen resistance. Best Pract Res Clin Endocrinol Metab 20(4):577–598

Hughes IA, Houk C, Ahmed SF, Lee PA, Society LWPE (2006) Consensus statement on management of intersex disorders. J Pediatr Urol 2(3):148–162

Hughes IA, Werner R, Bunch T, Hiort O (2012) Androgen insensitivity syndrome. Semin Reprod Med 30(5):432–442

Jacobs PA (1959) A case of human intersexuality having a possible XXY sex-determining mechanism. Nature 183:302–303

Jacobs PA (1974) A cytogenetic survey of 11,680 newborn infants. Ann Hum Genet 37:359–376

Jacobs PA (1975) XYY genotype. Science 189(4208):1040–1041

Jacobs PA, Hassold TJ (1995) The origin of numerical chromosome abnormalities. Adv Genet 33:101–133

Jamieson CR, van der Burgt I, Brady AF, van Reen M, Elsawi MM, Hol F, Jeffery S, Patton MA, Mariman E (1994) Mapping a gene for Noonan syndrome to the long arm of chromosome 12. Nat Genet 8(4):357–360

Jones JR, Kemmann E (1976) Sella turcica abnormalities in an anovulatory population. Obstet Gynecol 48(1):76–78

Jungwirth A, Giwercman A, Tournaye H, Diemer T, Kopa Z, Dohle G, Krausz C, EAU Working Group on Male Infertility (2012) European Association of Urology guidelines on male infertility: the 2012 update. Eur Urol 62(2):324–332

Kallmann FJ (1944) The genetic aspects of primary eunuchoidism. Am J Ment Defic 48:203–236

Kim WB, Jeong JY, Doo SW, Yang WJ, Song YS, Lee SR, Park JW, Kim DW (2012a) Myotonic dystrophy type 1 presenting as male infertility. Korean J Urol 53(2):134–136

Kim SJ, Miller JL, Kuipers PJ, German JR, Beaudet AL, Sahoo T, Driscoll DJ (2012b) Unique and atypical deletions in Prader–Willi syndrome reveal distinct phenotypes. Eur J Hum Genet 20(3):283–290

Klesert TR, Otten AD, Bird TD, Tapscott SJ (1997) Trinucleotide repeat expansion at the myotonic dystrophy locus reduces expression of DMAHP. Nat Genet 16(4):402–406

Klinefelter HF Jr, Reifenstein EC Jr, Albright F Jr (1942) Syndrome characterized by gynecomastia, aspermatogenesis without A-Leydigism, and increased excretion of follicle-stimulating hormone 1. J Clin Endocrinol Metabol 2(11):615–627

Kruse R, Guttenbach M, Schartmann B, Schubert R, van der Ven H, Schmid M, Propping P (1998) Genetic counseling in a patient with XXY/XXXY/XY mosaic Klinefelter's syndrome: estimate of sex chromosome aberrations in sperm before intracytoplasmic sperm injection. Fertil Steril 69(3):482–485

Lanfranco F, Kamischke A, Zitzmann M, Nieschlag E (2004) Klinefelter's syndrome. Lancet 364(9430):273–283

Lewis-Jones DI, Gazvani MR, Mountford R (2000) Cystic fibrosis in infertility: screening before assisted reproduction: opinion. Hum Reprod 15(11):2415–2417

Lim AST, Fong Y, Yu SL (1999) Analysis of the sex chromosome constitution of sperm in men with a 47, XYY mosaic karyotype by fluorescence in situ hybridization. Fertil Steril 72(1):121–123

Maclean N, Harnden DG, Brown WC (1961) Abnormalities of sex chromosome constitution in newborn babies. Lancet 278(7199):406–408

Martin RH (2008) Meiotic errors in human oogenesis and spermatogenesis. Reprod Biomed Online 16(4):523–531

Matias PM, Donner P, Coelho R, Thomaz M, Peixoto C, Macedo S, Otto N, Joschko S, Scholz P, Wegg A, Bäsler S (2000) Structural evidence for ligand specificity in the binding domain of the human androgen receptor implications for pathogenic gene mutations. J Biol Chem 275(34):26164–26171

McInnis MG (1996) Anticipation: an old idea in new genes. Am J Hum Genet 59(5):973

Meschede D, Horst J (1997) The molecular genetics of male infertility. Mol Hum Reprod 3(5):419–430

Milazzo JP, Rives N, Mousset-Simeon N, Mace B (2006) Chromosome constitution and apoptosis of immature germ cells present in sperm of two 47, XYY infertile males. Hum Reprod 21(7):1749–1758

Morel F, Roux C, Bresson JL (1999) Sex chromosome aneuploidies in sperm of 47, XYY men. Arch Androl 43(1):27–36

Moretti E, Anichini C, Sartini B, Collodel G (2007) Sperm ultrastructure and meiotic segregation in an infertile 47, XYY man. Andrologia 39(6):229–234

Morris JM (1953) The syndrome of testicular feminization in male pseudohermaphrodites. Am J Obstet Gynecol 65(6):1192–1211

Morrow AF, Gyorki S, Warne GL, Burger HG, Bangah ML, Outch KH, Mirovics A, Baker HWG (1987) Variable androgen receptor levels in infertile men. J Clin Endocrinol Metabol 64(6):1115–1121

Naftolin F, Harris GW, Bobrow M (1971) Effect of purified luteinizing hormone releasing factor on normal and hypogonadotrophic anosmic men. Nature 232(5311):496–497

Nicholls RD, Ohta T, Gray TA (1999) Genetic abnormalities in Prader–Willi syndrome and lessons from mouse models. Acta Paediatr 88(s433):99–104

Nielsen J, Wohlert M (1991) Chromosome abnormalities found among 34910 newborn children: results from a 13-year incidence study in Århus, Denmark. Hum Genet 87(1):81–83

Nilson O (1939) Hernia uteri inguinalis beim Manne. Acta Chir Scand 83:231

Nisbct DL, Griffin R, Chitty LS (1999) Prenatal features of Noonan syndrome. Prenat Diagn 19(7):642–647

Nishio H, Mizuno K, Moritoki Y, Kamisawa H, Kojima Y, Mizuno H, Kohri K, Hayashi Y (2012) Clinical features and testicular morphology in patients with Kallmann syndrome. Urology 79(3):684–686

Noonan JAY, Lexington K (1968) Hypertelorism with turner phenotype. A new syndrome with associated congenital heart disease. Am J Dis Child 116:373–380

Okada H, Fujioka H, Tatsumi N, Kanzaki M, Okuda Y, Fujisawa M, Hazama M, Matsumoto O, Gohji K, Arakawa S, Kamidono S (1999) Klinefelter's syndrome in the male infertility clinic. Hum Reprod 14(4):946–952

Omran H, Haffner K, Volkel A, Kuehr J, Ketelsen UP, Ross UH, Konietzko N, Wienker T, Brandis M, Hildebrandt F (2000) Homozygosity mapping of a gene locus for primary ciliary dyskinesia on chromosome 5p and identification of the heavy dynein chain DNAH5 as a candidate gene. Am J Respir Cell Mol Biol 23(5):696–702

Ortega HAV, Vega NDA, Santos BQD, Maia GTDS (2007) Primary ciliary dyskinesia: considerations regarding six cases of Kartagener syndrome. J Bras Pneumol 33(5):602–608

Pallais JC, Au M, Pitteloud N, Seminar AS, Crowley WF (1993) Kallmann syndrome. In: Pagon RA, Bird TD, Dolan CR, Stephens K, Adam MP (eds) GeneReviews™. University of Washington, Seattle

Patrizio P, Salameh WA (1997) Expression of the cystic fibrosis transmembrane conductance regulator (CFTR) mRNA in normal and pathological adult human epididymis. J Reprod Fertil Suppl 53:261–270

Pennarun G, Escudier E, Chapelin C, Bridoux AM, Cacheux V, Roger G, Clément A, Goossens M, Amselem S, Duriez B (1999) Loss-of-function mutations in a human gene related to *Chlamydomonas reinhardtii* dynein IC78 result in primary ciliary dyskinesia. Am J Hum Genet 65(6):1508–1519

Prader A (1956) Ein syndrome von adipositas, kleinwuchs, kryptorchismus und oligophreniee nach myatonieartigerm zustand im Neugeborenenalter. Schweiz Med Wochenschr 86:1260–1261

Quigley CA, Bellis AD, Marschke KB, El-Awady MK, Wilson EM, French FS (1995) Androgen receptor defects: historical, clinical, and molecular perspectives. Endocr Rev 16(3):271–321

Radicioni AF, Di Giorgio G, Grugni G, Cuttini M, Losacco V, Anzuini A, Spera S, Marzano C, Lenzi A, Cappa M, Crino A (2012) Multiple forms of hypogonadism of central, peripheral or combined origin in males with Prader–Willi syndrome. Clin Endocrinol (Oxf) 76(1):72–77

Ratcliffe SG (1982) Speech and learning disorders in children with sex chromosome abnormalities. Dev Med Child Neurol 24(1):80–83

Ratcliffe SG, Pan H, McKie M (1992) Growth during puberty in the XYY boy. Ann Hum Biol 19(6):579–587

Redman JB, Fenwick RG, Fu YH, Pizzuti A, Caskey CT (1993) Relationship between parental trinucleotide GCT repeat length and severity of myotonic dystrophy in offspring. JAMA 269(15):1960–1965

Robinson DO, Jacobs PA (1999) The origin of the extra Y chromosome in males with a 47, XYY karyotype. Hum Mol Genet 8(12):2205–2209

Robinson A, Bender BG, Linden MG, Salbenblatt JA (1989) Sex chromosome aneuploidy: the Denver prospective study. Birth Defects Orig Artic Ser 26(4):59–115

Robinson A, Bender BG, Linden MG (1990) Summary of clinical findings in children and young adults with sex chromosome anomalies. Birth Defects Orig Artic Ser 26(4):225

Rovet J, Netley C, Bailey J, Keenan M, Stewart D (1995) Intelligence and achievement in children with extra X aneuploidy: a longitudinal perspective. Am J Med Genet 60(5):356–363

Samuel I (1987) Kartagener's syndrome with normal spermatozoa. JAMA 258(10):1329–1330

Sandberg AA, Koepf GF, Ishihara T, Hauschka TS (1961) An XYY human male. Lancet 278(7200):488–489

Sarfati J, Dodé C, Young J (2010) Kallmann syndrome caused by mutations in the PROK2 and PROKR2 genes: pathophysiology and genotype-phenotype correlations. In: Quinton R (ed) Kallmann syndrome and hypogonadotropic hypogonadism, vol 39. Karger, Basel, pp 121–132

Sarkar PS, Paul S, Han J, Reddy S (2004) Six5 is required for spermatogenic cell survival and spermiogenesis. Hum Mol Genet 13(14):1421–1431

Schwanzel-Fukuda M, Pfaff DW (1989) Origin of luteinizing hormone-releasing hormone neurons. Nature 338(6211):161–164

Schwartz CE, Gillessen-Kaesbach G, May M, Cappa M, Gorski J, Steindl K, Neri G (2000) Two novel mutations confirm FGD1 is responsible for the Aarskog syndrome. Eur J Hum Genet 8(11):869–874

Sciurano RB, Hisano CL, Rahn MI, Olmedo SB, Valzacchi GR, Coco R, Solari AJ (2009) Focal spermatogenesis originates in euploid germ cells in classical Klinefelter patients. Hum Reprod 24(9):2353–2360

Scott CI (1971) Unusual facies, joint hypermobility, genital anomaly and short stature: a new dysmorphic syndrome. Birth Defects Orig Artic Ser 7(6):240–246

Shi Q, Martin RH (2000) Multicolor fluorescence in situ hybridization analysis of meiotic chromosome segregation in a 47, XYY male and a review of the literature. Am J Med Genet 93(1):40–46

Siemensma EP, de Lind van Wijngaarden RF, Otten BJ, de Jong FH, Hokken-Koelega AC (2011) Testicular failure in boys with Prader–Willi syndrome: longitudinal studies of reproductive hormones. J Clin Endocrinol Metabol 97(3):E452–E459

Speed RM, Faed MJW, Batstone PJ, Baxby K, Barnetson W (1991) Persistence of two Y chromosomes through meiotic prophase and metaphase I in an XYY man. Hum Genet 87(4):416–420

Staessen C, Tournaye H, Van Assche E, Michiels A, Van Landuyt L, Devroey P, Liebaers I, Van Steirteghem A (2003) PGD in 47, XXY Klinefelter's syndrome patients. Hum Reprod Update 9(4):319–330

Swanson ML, Coronel EH (1993) Complete androgen insensitivity with persistent Müllerian structures. A case report. J Reprod Med 38(7):565–568

Sykiotis GP, Plummer L, Hughes VA, Au M, Durrani S, Nayak-Young S, Dwyer AA, Quinton R, Hall JE, Gusella JF, Seminara SB (2010) Oligogenic basis of isolated gonadotropin-releasing hormone deficiency. Proc Natl Acad Sci U S A 107(34):15140–15144

Teixeira L, Guimiot F, Dodé C, Fallet-Bianco C, Millar RP, Delezoide AL, Hardelin JP (2010) Defective migration of neuroendocrine GnRH cells in human arrhinencephalic conditions. J Clin Invest 120(10):3668–3672

Tsilfidis C, MacKenzie AE, Mettler G, Barceló J, Korneluk RG (1992) Correlation between CTG trinucleotide repeat length and frequency of severe congenital myotonic dystrophy. Nat Genet 1(3):192–195

Tüttelmann F, Gromoll J (2010) Novel genetic aspects of Klinefelter's syndrome. Mol Hum Reprod 16(6):386–395

Ulloa-Aguirre A, Mendez JP, Chavez B, Carranza-Lira S, Angeles A, Perez-Palacios G (1990) Incomplete regression of Müllerian ducts in the androgen insensitivity syndrome. Fertil Steril 53(6):1024–1028

Valdes-Socin H, Almanza MR, Fernández-Ladreda MT, Debray FG, Bours V, Beckers A (2015) Reproduction, smell, and neurodevelopmental disorders: genetic defects in different hypogonadotropic hypogonadal syndromes. Front Endocrinol (Lausanne) 5:109

Van Assche E, Bonduelle M, Tournaye H, Joris H, Verheyen G, Devroey P, Van Steirteghem A, Liebaers I (1996) Cytogenetics of infertile men. Hum Reprod 11(Suppl 4):1–26

Vincent MC, Daudin M, Mas P, Massat G, Mieusset R, Pontonnier F, Calvas P, Bujan L, Bourrouillou G (2002) Cytogenetic investigations of infertile men with low sperm counts: a 25-year experience. J Androl 23(1):18–22

Viswanathan V, Eugster EA (2011) Etiology and treatment of hypogonadism in adolescents. Pediatr Clin North Am 58(5):1181–1200

Vogels A, Moerman P, Frijns JP, Bogaert GA (2008) Testicular histology in boys with Prader–Willi syndrome: fertile or infertile? J Urol 180(4):1800–1804

Walzer S, Wolff PH, Bowen D, Silbert AR, Bashir AS, Gerald PS, Richmond JB (1978) A method for the longitudinal study of behavioral development in infants and children: the early development of XXY children. J Child Psychol Psychiatry 19(3):213–229

Wang RS, Yeh S, Tzeng CR, Chang C (2009) Androgen receptor roles in spermatogenesis and fertility: lessons from testicular cell-specific androgen receptor knockout mice. Endocr Rev 30(2):119–132

Welch J (1985) Clinical aspects of the XYY syndrome. Alan R. Liss, New York, pp 323–343

Wharton RH, Loechner KJ (1996) Genetic and clinical advances in Prader–Willi syndrome. Curr Opin Pediatr 8(6):618–624

Witt DR, McGillivray BC, Allanson JE, Hughes HE, Hathaway WE, Zipursky A, Hall JG, Opitz JM, Reynolds JF (1988) Bleeding diathesis in Noonan syndrome: a common association. Am J Med Genet 31(2):305–317

Wong EC, Ferguson KA, Chow V, Ma S (2008) Sperm aneuploidy and meiotic sex chromosome configurations in an infertile XYY male. Hum Reprod 23(2):374–378

Wu HC, Chen JH, Lu HF, Shen WC (2000) Persistent Müllerian duct syndrome with seminoma: CT findings. Am J Roentgenol 174(1):102–104

Yong EL, Ng SC, Roy AC, Yun G, Ratnam SS (1994) Pregnancy after hormonal correction of severe spermatogenic defect due to mutation in androgen receptor gene. Lancet 344(8925):826–827

Yuksel B, Saygun O, Hengirmen S (2006) Persistent Müllerian duct syndrome associated with irreducible inguinal hernia, bilateral cryptorchidism and testicular neoplasia: a case report. Acta Chir Belg 106(1):119–120

Zariwala MA, Leigh MW, Ceppa F, Kennedy MP, Noone PG, Carson JL, Hazucha MJ, Lori A, Horvath J, Olbrich H, Loges NT (2006) Mutations of DNAI1 in primary ciliary dyskinesia: evidence of founder effect in a common mutation. Am J Respir Crit Care Med 174(8):858–866

Zhang Y, Malekpour M, Al-Madani N, Kahrizi K, Zanganeh M, Mohseni M, Mojahedi F, Daneshi A, Najmabadi H, Smith RJ (2007) Sensorineural deafness and male infertility: a contiguous gene deletion syndrome. J Med Genet 44(4):233–240

Zuccarello D, Ferlin A, Vinanzi C, Prana E, Garolla A, Callewaert L, Claessens F, Brinkmann AO, Foresta C (2008) Detailed functional studies on androgen receptor mild mutations demonstrate their association with male infertility. Clin Endocrinol (Oxf) 68(4):580–588

Cystic Fibrosis, *CFTR* Gene, and Male Infertility

9

Rahul Gajbhiye and Avinash Gaikwad

Abstract

An abnormality of cystic fibrosis transmembrane conductance regulator (CFTR) gene is known to be one of the etiologies of male infertility. *CFTR* gene mutations are associated with cystic fibrosis (CF-severe phenotype) to congenital bilateral absence of the vas deferens (CBAVD-mild phenotype). CF is the most common autosomal recessive disorders in the Caucasians, characterized by chronic lung disease, pancreatic insufficiency, rise in sweat chloride, and obstructive azoospermia. The milder phenotype is classified as congenital absence of the vas deferens (bi- or unilateral) (CBAVD or CUAVD) or ejaculatory duct obstruction (EDO). Some of these CAVD cases are associated with unilateral renal anomalies (URA). The role of *CFTR* gene in this subtype of CBAVD-URA is still not understood clearly. The utility of advanced assisted reproductive technologies such as intracytoplasmic sperm injection (ICSI) helps CBAVD males to become biological fathers. If female partner is CF carrier, there is a risk of having a child with CF or CF-related disorders. The currently available CFTR mutation panels cover the most common mutations of Caucasians. Recent studies conducted in South Asian population suggested different spectrum of CFTR mutations than Caucasians. There is a need to develop population-specific *CFTR* gene mutation panels especially for South Asians where CF or CF-related disorders were once considered rare.

R. Gajbhiye, M.B.B.S., Ph.D. (✉)
Andrology Clinic, Department of Clinical Research, National Institute for Research in Reproductive Health, Indian Council of Medical Research,
J M Street, Parel, Mumbai 400 012, India
e-mail: gajbhiyer@nirrh.res.in

A. Gaikwad, M.Sc.
Department of Anatomy and Developmental Biology, Monash University,
19 Innovation walk, Clayton, VIC 3800, Australia

R. Singh, K. Singh (eds.), *Male Infertility: Understanding, Causes and Treatment*, DOI 10.1007/978-981-10-4017-7_9

Keywords
CFTR-RDs • Cystic fibrosis • CBAVD • CUAVD • CBAVD-URA • Male infertility • Genetic counseling

Key Points

- Male infertility is associated with both cystic fibrosis (CF) and congenital bilateral absence of the vas deferens (CBAVD).
- CF and CBAVD are two distinct spectrums of *CFTR* gene abnormalities.
- CBAVD men carry different *CFTR* gene mutations than classic CF.
- Renal anomalies are associated with ~11% men having CBAVD, more common in individuals having congenital unilateral absence of the vas deferens.
- PESE-ICSI is the widely preferred and accepted treatment for men having CBAVD.
- If female partner is CF carrier, there is a risk of having a child with CF or CF-related disorders such as CBAVD.
- *CFTR* gene testing should be offered to both the partners before planning ICSI.
- Population-specific mutation panels are required for accurate diagnosis and calculation of genetic risk in CBAVD men.

9.1 Introduction

Cystic fibrosis (CF) affects multiple organs of the body, and it is associated with mutations in the cystic fibrosis transmembrane conductance regulator (*CFTR*) gene. CF is considered as the most common autosomal recessive disorder in Caucasians with a frequency of 1/2000 (Nielson et al. 1988). Earlier, it was assumed that CF or CF-related disorders (CFTR-RD) are rare in African and Indian populations. However, recently with the advances in the diagnostic techniques as well as due to the increased awareness, CF and CFTR-RDs are increasingly detected in these populations. However, the incidence is still underestimated in Indian and black South African populations. Moreover, studies have reported that the prevalence of CF varies with geographical location (Casals et al. 1992).

CFTR gene is located on chromosome 7q31.2 and contains 27 exons (~250 kb of DNA). More than 1800 CF-causing *CFTR* gene mutations have been reported in the *CFTR* gene mutation database so far. Following are the databases of *CFTR* gene mutations:

- http://www.genet.sickkids.on.ca/
- http://www.umd.be/CFTR/
- http://www.cftr2.org/

There is limited information on exact pathogenicity of the *CFTR* gene mutations reported in different populations. Through CFTR 2 project, functional analysis of identified mutations is being investigated. Six functional classes of CF mutations are described (Fig. 9.1):

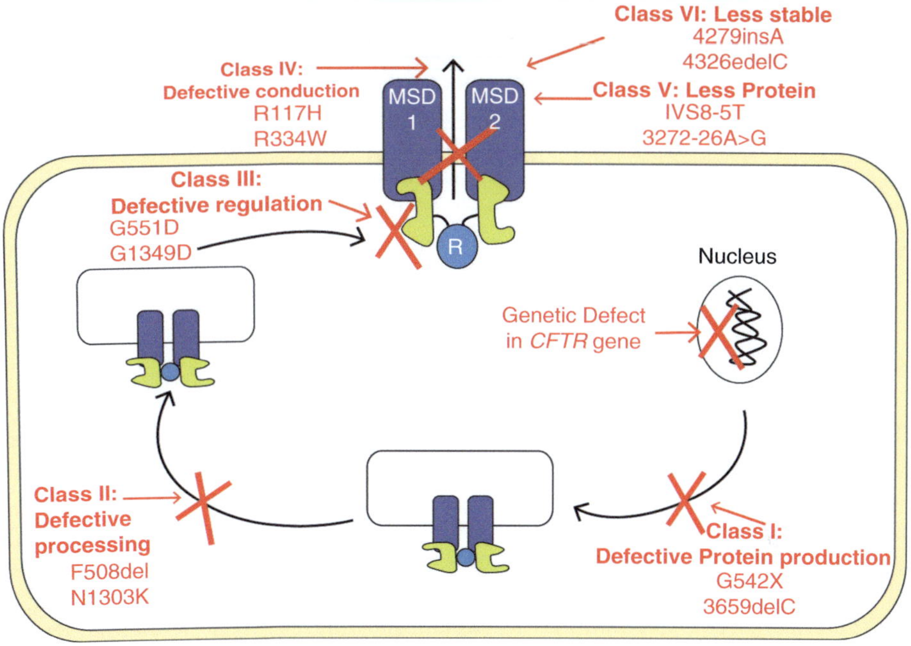

Fig. 9.1 Classes of *CFTR* gene mutations

- *Class I mutations*: CFTR production is stopped early and the protein is defective resulting into nonfunctioning CFTR chloride channels. Accounts for ~10% of *CFTR* gene mutations causing CF worldwide. Mutation leads to premature stop codon, which causes translation of mRNA to stop prematurely.
- *Class II mutations*: No proper processing of CFTR and proteins is destroyed within the cell. F508del (absence of phenylalanine at position 508) is the most commonly reported Class II mutation. F508del occurs in around 88.5% of CF patients worldwide as per the CF registry database.
- *Class III mutations:* CFTR reaches cell surface but it does not open properly to transport chloride. Only a small percentage CF (2–3%) cases have this mutation.
- *Class IV mutations:* Defective conduction of chloride through the channel. These are uncommon mutations and lead to disease ~2% of patients with CF.
- *Class V mutations:* The least common mutations. Splicing defects resulting into improper processing of mRNA are the etiology for Class V mutation.
- *Class VI mutations:* Although function CFTR protein but unstable at cell surface.

The cystic fibrosis transmembrane conductance regulator (CFTR) protein is expressed throughout the epithelial cells in the airways, gastrointestinal tract, and reproductive organs (Quinton 2007). As a result, CF patients manifest symptoms related to multiple organs that include repeated and chronic lung infection, insufficiency of the pancreas, and male infertility. *CFTR* gene mutations are the main etiological factors due to defective electrolyte and fluid transport (Welsh and Fick 1987; Welsh and Smith 1993; Quinton 2007).

In addition to regulating the chloride ion channel in the epithelial cells, CFTR is involved in the following functions: (a) sodium transport through the sodium ion channel, (b) regulation of the chloride flow outside the cell membrane, (c) regulation of the ATP channels, (d) intracellular vesicle transport, (e) acidification of intracellular organelles, (f) inhibition of endogenous calcium-activated chloride channels, and (g) efficient bicarbonate–chloride exchange.

9.2 Pathogenesis

The CFTR protein is an epithelial membrane protein, an ATP-binding cassette (ABC)-transporter-class ion channel. It regulates the chloride ions across epithelial cell membranes. The CFTR protein is made of five domains: two membrane-spanning domains (MSDs) that form the channel pores; two nucleotide-binding domains (NBDs), which control channel gating; and one regulatory domain (R domain), which determines the phosphorylation activity.

There are different hypotheses to explain the role of CFTR abnormalities in developing CF or CFTR-RD. Following are the most relevant hypothesis; it may be possible that the combination of these aspects could contribute to the pathogenesis of the CF or CFTR-RD:

1. *Low-volume hypothesis:* Due to the CFTR dysfunction, there is loss of inhibition of epithelial sodium channels leading to excess sodium and water reabsorption ultimately resulting in dehydration of airway surface materials (Matsui et al. 1998). The low airway surface water volume is not corrected by the epithelium due to the associated loss of chloride. Reduction in periciliary water leads to decrease in the lubricating layer between epithelium and mucus and compresses the cilia by mucus causing inhibition of normal ciliary movement and cough clearance of the mucus. According to this hypothesis, bacteria such as *Pseudomonas aeruginosa* can grow due to the mucus on the epithelium that leads to plaque formation with hypoxic niches (Boucher 2007).
2. *High-salt hypothesis*: Absence of functional CFTR protein leads to retention of excess of sodium and chloride in airway surface liquid. The higher levels of chloride in the periciliary layer then disrupt the function of innate antibiotic molecules such as human β-defensin 1 and thereby allow the growth of bacteria that are normally cleared by normal airways to persist in the lungs (Goldman et al. 1997).
3. *Dysregulation of the host inflammatory response*: Cystic fibrosis cell cultures and uninfected ex vivo tissue samples contain higher concentrations of inflammatory mediators (Freedman et al. 2004). Inflammatory mediators were detected in the lung lavage samples of children as young as 4 weeks of age. The pro-

inflammatory molecules (Interleukin 8, Interleukin 6, TNFα, and arachidonic acid metabolites) were detected CF (Freedman et al. 2004). Studies also reported the activation of NFκB pathway, platelet hyperreactivity, and neutrophil apoptosis abnormalities (Carrabino et al. 2006).

4. *Primary predisposition to infection*: Normally, *P. aeruginosa* binds to functional CFTR, and rapid and self-limiting innate immune response is initiated. In CF, increase in asialo-GM1 in apical cell membranes allows binding of *P aeruginosa* and *Staphylococcus aureus* to the airway epithelium, without CFTR-mediated immune response. The self-limiting response that eliminates *P. aeruginosa* from the airways is lost in CF and at the same time as there is enhanced attachment of bacteria to the epithelial surface.

9.3 Epidemiology

Incidence of CF is reported to be 1 in 2000–3000 in Caucasians with a carrier frequency of 1 in 22–28. High prevalence of CF is reported in North America, Europe, and Australia. Recent studies generated evidence of increased number of CF and CF-related disorders in other ethnic populations residing in Africa, South America, Middle East, and Asia (WHO 2004 and Cystic Fibrosis Foundation Patient Registry 2012 Annual Data Report). It has been reported that there is a variation in birth prevalence due to CF worldwide with different ethnic backgrounds. Prevalence of CF was reported as 1 in 3000 in white Americans, 1 in 4000–10,000 in Latin Americans, and 1 in 15,000–20,000 in African Americans (Walters and Mehta al. 2007). Cystic fibrosis was earlier reported to be a rare disorder in Africa and Asia, with a frequency of 1 in 350,000 in Japan (Yamashiro et al. 1997). Frequency F508del mutation was higher in northwest region of Europe than southeast. Similarly, Trp1282X is the most common mutation reported in Israel (O'Sullivan and Freedman 2009).

There could be multiple reasons for lower reporting of CF and CF-related disorders in developing countries in South Asian subcontinent. Majority among them is the lack of awareness about CF and CF-related disorders, limited clinical expertise for diagnosis and management, and limited molecular diagnostic facilities. There has been a good progress in the past few years, and evidence is emerging on *CFTR* gene mutations in CF and CFTR-RD from the South Asian population. A heterogeneous spectrum of *CFTR* gene variants was identified in Asians (Fig. 9.2), with a lower frequency of F508del in Asians as compared to Caucasians (Sharma et al. 2009). Evidence is very limited to prove whether low incidence of CF in Asian populations is due to genetic drift or it is due to misdiagnosis. This needs to be thoroughly investigated especially in South Asian countries. Earlier reports indicated the incidence of CF in immigrant Asians residing in Canada as 1/9200, 1/10,000 in the UK, and 1/40,000 in the USA (Powers et al. 1996; Mei-Zahav et al. 2005). Researchers now hypothesize that India may hold the largest CF and CFTR-RDs population in world with up to 1,00,000 undiagnosed CF patients (CFRI News 2013).

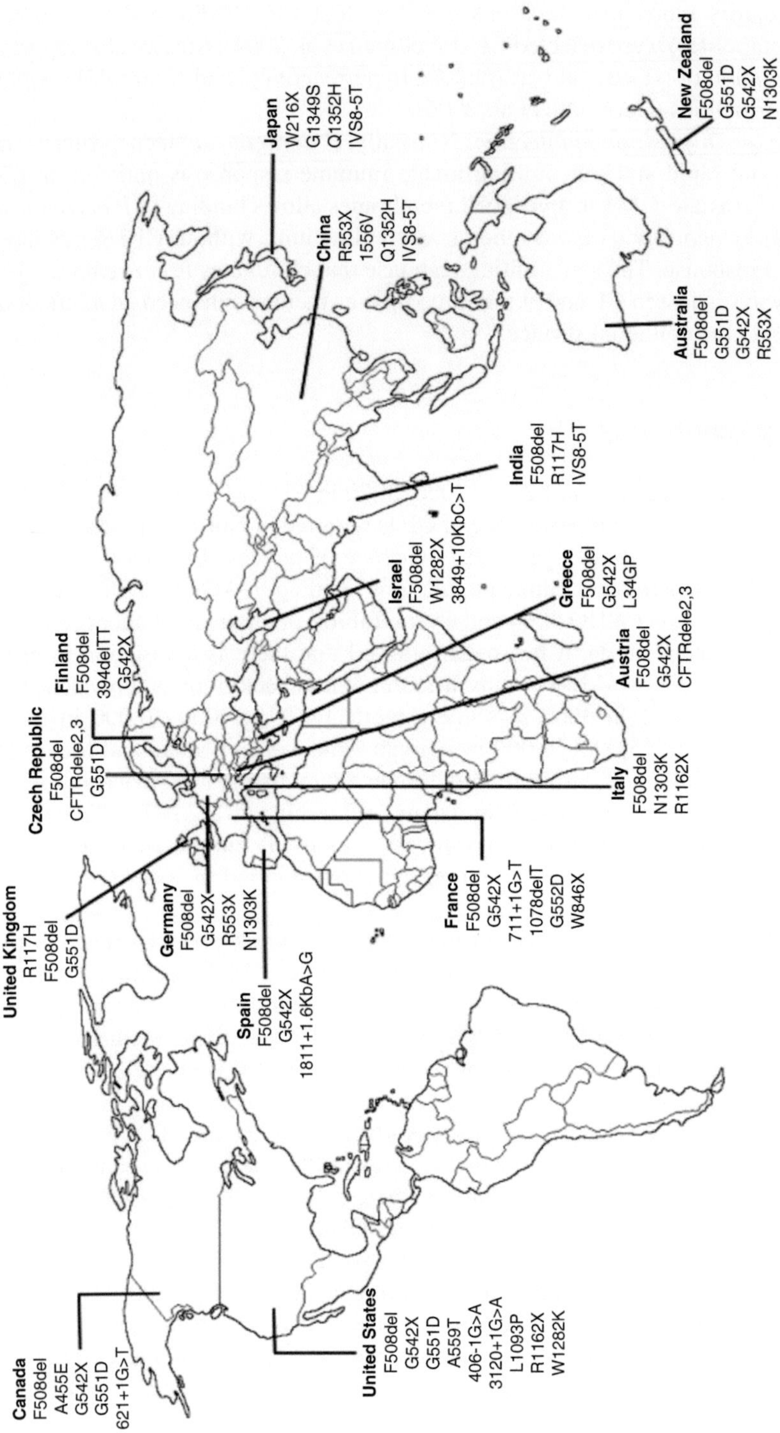

Fig. 9.2 Map showing the most common *CFTR* gene mutations identified across different populations

9.4 Diagnosis

The preliminary diagnosis of CF is based on elevated sweat chloride level (>60 mmol/L). In a more classical condition, CF is diagnosed if the sweat chloride levels are in the intermediate range (for infants >6 months, 30–59 mmol/L and for old individuals, 40–59 mmol/L) and two severe disease-causing mutations are identified in an individual (Rosenstein et al. 1998; De Boeck et al. 2006; Welsh et al. 2001). Patients with intermediate range (30–60 mmol/L) of sweat chloride levels might have CFTR genotype combining two CF-causing mutations. The American College of Medical Genetics recommended a panel of 23 CF-causing mutations. More than 1800 *CFTR* gene mutations have been reported and included in the CF registries. The number of novel mutations is also exponentially increasing (Table 9.1). The diagnosis of CF becomes more problematic when sweat chloride levels are intermediate and patient still has symptoms suggestive of CF. A more severe lung disease is observed among the patients with abnormalities in NPD

Table 9.1 List of high prevalence of *CFTR* gene mutations

Mutation cDNA name	Mutation protein name	Mutation legacy name	Significance
c.54-5940_273+10250del21kb	p.Ser18ArgfsX16	CFTRdele2,3	CF-causing
c.178G>T	p.Glu60X	E60X	CF-causing
c.223C>T	p.Arg75X	R75X	CF-causing
c.224G>A	p.Arg75Gln	R75Q	Non-CF-causing
c.254G>A	p.Gly85Glu	G85E	CF-causing
c.262_263delTT	p.Leu88llefsX22	394delTT	CF-causing
c.328G<C	p.Asp110His	D110H	CF-causing
c.350G>A	p.Arg117His	R117H	Varying clinical significance
c.489+1G>T	No protein name	621+1G>T	CF-causing
c.579+1G>T	No protein name	711+1G>T	CF-causing
c.1040G>C	p.Arg347Pro	R374P	CF-causing
c.1364C>A	p.Ala455Glu	A455E	CF-causing
c.1519_1521delATC	p.lle507del	I507del	CF-causing
c.1521_1523delCTT	p.Phe508del	F508del	CF-causing
c.1585-1G>A	No protein name	1717-1G>A	CF-causing
c.1624G>T	p.Gly542X	G542X	CF-causing
c.1652G>A	p.Gly551Asp	G551D	CF-causing
c.1657C>T	p.Arg553X	R553X	CF-causing
c.2052_2053insA	p.Gln685ThrfsX4	2184insA	CF-causing
c.2052delA	p.Lys684AsnfnsX38	2184delA	CF-causing
c.2657+5G>A	No protein name	2789+5G>A	CF-causing
c.3196C>T	p.Arg1066Cys	R1066C	CF-causing

(continued)

Table 9.1 (continued)

Mutation cDNA name	Mutation protein name	Mutation legacy name	Significance
c.3454G>C	p.Asp1152His	D1152H	Varying clinical consequence
c.3484C>T	p.Arg1162X	R1162X	CF-causing
c.3528delC	p.Lys1177SerfsX15	3659delC	CF-causing
c.3717+12191C>T	No protein name	3849+10kbC>T	CF-causing
c.3846G>A	p.Try1282X	W1282X	CF-causing
c.3909C>G	p.Asn1303Lys	N1303K	CF-causing

measurement or two *CFTR* gene mutations (Goubau et al. 2009). However, their disease symptoms are milder as compared to those with a sweat chloride levels above 60 mmol/L. In children with multiple organ involvement with marginal levels of sweat chloride concentration and/or presence of at least one *CFTR* gene mutation of unknown clinical significance, a terminology of "nonclassical" or "atypical" CF is applicable (Rosenstein et al. 1998). Due to the varied spectrum of clinical phenotypes, now the new terminology of "CFTR-related disorders" (CFTR-RDs) is gaining wider acceptance (Dequeker et al. 2009; Castellani et al. 2008). It is very essential to understand the complete clinical phenotype along with biochemical and molecular tests for reaching out the correct diagnosis of CF or CFTR-RD.

9.5 Fertility in Men Having CF

Spermatogenesis is a well-orchestrated process by which the totipotent primordial spermatogonia undergo meiosis to produce daughter cells called spermatozoa. In order to form a mature sperm, the spermatozoa undergoes a series of morphological and functional differentiation processes under the influence of hormones including follicle-stimulating hormone (FSH), luteinizing hormone (LH), and testosterone. These processes occur within the seminiferous tubules, which are supported by Sertoli cells that are in close contact with the germ cells. Defects at any stage of spermatogenesis may cause male infertility including azoospermia, oligospermia, and teratospermia. However, the importance of CFTR in spermatogenesis is still controversial, even after its experimental evidence of expression in the testis (Trezíse and Buchwald 1991). Histological studies with testicular tissues of men with CF and CBAVD tried to resolve this controversy but resulted in contradictory findings such as normal spermatogenesis (Tuerlings et al. 1998) to severely decreased spermatogenesis with abnormal sperm and a reduced sperm count (Larriba et al. 1998).

Puberty in men having classic CF and chronic lung disease, malnutrition is usually delayed due to lower levels of follicle-stimulating hormone (FSH) and luteinizing hormones (LH). In spite of the delayed onset of puberty, majority of CF patients

(>90%) achieve normal height. Around 2–5% of CF men are fertile. There is a normal production of immature sperm in testes. Bilateral vas deferens is either atrophied or absent in approximately 95% of CF males. Seminal vesicles are hypoplastic or absent and normal maturation of sperm is impaired. As a result of this, there is reduced seminal volume, no mature sperm, and high acid content, absent or low fructose in semen.

9.6 CFTR-Related Disorders Associated with Male Infertility

A CFTR-related disorder (CFTR-RD) is a separate clinical condition associated with *CFTR* gene abnormalities and does not fulfill the diagnostic criteria of CF. Four main clinical entities illustrate these phenotypes:

- Congenital bilateral absence of the vas deferens (CBAVD) with CFTR dysfunction
- CBAVD having renal anomalies
- Congenital unilateral absence of the vas deferens (CUAVD)
- Ejaculatory duct obstruction (EDO)

9.6.1 Congenital Bilateral Absence of the Vas Deferens (CBAVD)

CBAVD is a condition in which there is a complete or partial failure of development of vasa deferens before birth. CBAVD in otherwise healthy men also known as isolated CBAVD accounts for ~3% of male infertility. The incidence of CBAVD is ~1:1000 men (Holsclaw et al. 1971; Oates and Amos 1993; Mak and Jarvi 1996). Isolated CBAVD (MIM#277180) is an autosomal recessive genetic disorder known to be associated with *CFTR* gene abnormalities. Milder phenotype such as CBAVD is due to the *CFTR* gene variants that retain the CFTR function to its minimum. CBAVD is either due to the one inherited *CFTR* gene mutation (Dumur et al. 1990; Anguiano et al. 1992; Patrizio et al. 1993) or due to the inheritance of mutations in both the copies of the *CFTR* gene (70–90% of cases) (Bombieri et al. 2011). CBAVD and CF are now considered as two different spectrums of CFTR due to the distinct genotype and phenotype (Colin et al. 1996).

The diagnosis of CBAVD is based on scrotal examination—bilateral absence of the vas deferens and normal testicular volume (>15 mL) and absence of body and tail of epididymis. Semen analysis is very important in diagnosis as it reveals azoospermia with low seminal volume (<1.0 mL), low pH (average <6.8), and low or absent fructose levels (Casals et al. 1995, Holsclaw et al. 1971). The abnormal CFTR protein could affect the multiple organs including reproductive tract. Transrectal ultrasonography (TRUS) reveals the morphology and size of the seminal vesicles, prostate, and ejaculatory ducts. In CBAVD, body and tail of the epididymis are either atrophic or absent or the epididymis remnants are distended, whereas the head or caput of the epididymis is usually present (McCallum et al.

2000a, b). The sweat chloride levels are usually normal, and testicular biopsy shows normal spermatogenesis in majority of CBAVD cases.

Due to the compound heterozygosity (either one severe and one mild mutation or two mild mutations) in CBAVD, the spectrum of *CFTR* gene mutations differs from that of the classical CF. In CF patients two severe *CFTR* gene mutations (88%) or one severe and one mild or variable CFTR mutations (12%) are detected. In CBAVD, one severe and one mild or variable (88%) or two mild *CFTR* gene mutations (12%) are detected (Bombieri et al. 2011). p.F508del along with IVS8-5T (28%) and p.F508del in trans with p.R117H (6%) are the most common compound heterozygous genotypes found in CBAVD. A significant difference in the frequency is found in the most CF-causing mutations. The frequency of p.F508del is found to be 21–33% in the USA, Canada, and Northern Europe (Oates and Amos 1993; Jarvi et al. 1998; Dork et al. 1997; Claustres et al. 2000; Jarvi et al. 1998) and 12–18% in Southern Europe and India (Kanavakis et al. 1998; Grangeia et al. 2004; Sharma et al. 2009). However, p.F508del is found in lower frequencies in CBAVD men from non-European populations. The IVS8-5T allele is found in similar frequency in Indian (25%) and Japanese (30%) (Sharma et al. 2009; Anzai et al. 2000) or higher frequencies in Egyptians (44%) and Taiwanese (44%) population (Lissens et al. 1995; Wu et al. 2004). IVS8-5T is seen in 5% of general population and is reported in many countries where CF was once considered as a rare disorder. Due to the limited studies in South Asian populations, many of the common *CFTR* gene mutations are yet to be reported in these populations. IVS8-5T allele is 5–8 times higher in CBAVD men than the general population. Hence, it is the most common "mild" CFTR allele, present in at least 5% of general population worldwide (Bombieri et al. 2011). Studies have found that 34% of CBAVD men from European descent inherit at least one IVS8-5T allele (Casals et al. 1992). However, due to mild pathogenicity, IVS8-5T allele alone or in combination with other *CFTR* gene mutation cannot result in severe CF phenotype. IVS8-5T causes alternative splicing of exon 9 of the *CFTR* gene and leads to decreased levels of functional CFTR protein to develop isolated CBAVD phenotype (Casals et al. 1992). It has been reported that Wolffian tissues are the most prone tissues to splicing of exon 9, resulting in reduced full-length CFTR mRNAs as compared to other tissues. IVS8-5T splicing variant also produces low transcript level of full-length CFTR protein which is necessary for normal Wolffian tissues phenotype (Teng et al. 1997). The vas deferens is most sensitive to reduced functional CFTR protein due to the above mentioned mechanisms.

The IVS8-5T allele is known as a genetic modifier of p.R117H mutation when associated *in cis* position. The IVS8-5T allele is considered a CBAVD mutation with partial or incomplete penetrance. The efficiency of exon 9 splicing is influenced by the $(TG)_m$ repeat which lies immediately upstream of the IVS8-T_n tract (Cuppens et al. 1998). Thus, chances of exon 9 skipping is higher in the presence of longer IVS8-TG_m and shorter IVS8-T_n repeats leading to misfolded and/or nonfunctional CFTR protein (Cuppens et al. 1998). It has been found that CBAVD men have longer IVS8-TG repeats (12 or 13) as compared to healthy men, who have shorter IVS8-TG repeats (10 or 11) (Cuppens et al. 1998). Longer IVS8-TG repeats (IVS8-TG12 or TG13) *in cis* with IVS8-5T were found to correlate with CBAVD or CFTR-RD disease status. Therefore, the polymorphic dinucleotide $(TG)_m$ repeats could be the reliable predictor for the penetrance of IVS8-5T as a disease-causing allele. So far, the pathogenicity of TG12-5T and TG13-5T is much higher than that of TG11-5T allele. The TG_mT_n allele represents a model of CBAVD "polyvariant mutant CFTR."

Point mutations are extensively identified in the *CFTR* gene of CBAVD men. Often, large rearrangements such as deletions or duplications within the CFTR locus are also identified in 6–10% CBAVD cases, which is lower than the rearrangements found in CF patients (15–25%). Overall, large rearrangements (null mutations, classified as "severe") represent <1% of CBAVD alleles, a lower proportion than in CF, which reflects the higher contribution of severe alleles to the pathogenesis of CF (Bombieri et al. 2011).

9.6.2 CBAVD Having Renal Anomalies (CBAVD-URA)

CBAVD is associated with congenital malformations or agenesis of the upper urinary tract in 12–21% of cases. The association of *CFTR* gene mutations with CBAVD-URA is controversial as majority of cases failed to detect *CFTR* gene mutation (Anguiano et al. 1992; Augarten et al. 1994; Casals et al. 1995; Mickle et al. 1995; Schlegel et al. 1996; Dörk et al. 1997; de la Taille et al. 1998; Claustres et al. 2000; McCallum et al. 2001). As a result, CBAVD with renal malformation was considered as a distinct clinical phenotype termed "CBAVD-URA" (McCallum et al. 2001). There was no statistically significant difference in physical, laboratory, and radiographic findings of the reproductive derivatives as well as in fertilization and pregnancy rates between CF/CBAVD and CBAVD-URA (Robert et al. 2002). The hypothesis that CBAVD-URA could be a separate clinical disorder is further supported by the marked difference between the renal portions of the mesonephric duct in the two cohorts. The physical separation between the two mesonephric duct derivatives (seminal and renal) occurs by week 7 of gestation (Oates and Amos 1993). During embryonic development, the mesonephric duct gives rise to the vas deferens, seminal vesicle, ejaculatory duct, and distal two-thirds of the epididymis, while the ureteric part induces renal development. The genital ridge extends to form the caput of the epididymis (which is present in men with CBAVD or CF) and the testis. Any abnormalities at the embryonic developmental phase before week 7 could lead to abnormal development of the entire mesonephric duct resulting in CBAVD-URA phenotype (Hall and Oates 1993; McCallum et al. 2001) or CUAVD-URA phenotype (Donohue and Fauver 1989). By contrast, the genetic defect in CBAVD-URA appears to affect the embryo after the division of the mesonephric parts in the seventh week of gestation, so that only the seminal tract will be altered. A few number of patients with CBAVD and URA have now been reported to be heterozygous for a *CFTR* gene mutations (Mak and Jarvi 1996); the significance of these mutations is undetermined as it could be in conjunction with the IVS8-5T carrier status found in the general population and the lack of investigations in large number of CBAVD-URA patients. Hence, a complete family studies are required in both the CBAVD and CBAVD-URA cohort to determine the genetic causes, the mode of inheritance, and the penetrance of genetic factors in CBAVD and nephrogenesis. More studies are required to prove or disprove the association of *CFTR* gene with CBAVD and renal anomalies.

9.6.3 Congenital Unilateral Absence of the Vas Deferens (CUAVD)

Congenital unilateral absence of the vas deferens (CUAVD) occurs in less than 1/1000 men and hence is a rare condition. Mickle et al. (1995) defined CUAVD as the absence of one of the scrotal vasa deferentia and considered as a clinically and genetically distinct phenotype. The frequency of ipsilateral renal agenesis is higher (40–80%) in CUAVD and no *CFTR* gene mutations were detected (Mickle et al. 1995; Mak and Jarvi 1996; Weiske et al. 2000; McCallum et al. 2001; Kolettis and Sandlow 2002). A large variation is observed in the clinical presentation of CUAVD. Surprisingly, patients could be diagnosed of CUAVD during a clinical evaluation for vasectomy or other urologic conditions. Others may be diagnosed due to infertility and azoospermia because of contralateral testicular or Wolffian duct abnormalities. CUAVD exists as two different forms with and without renal anomalies suggesting different pathophysiological processes.

9.6.4 Ejaculatory Duct Obstruction

It was suggested that azoospermia not related to vas aplasia may be in some cases associated with *CFTR* gene mutations, including idiopathic forms of epididymal obstruction (Jarvi et al. 1998; Mak and Jarvi 1996). Bilateral ejaculatory duct obstruction (BEDO) was associated with a higher frequency of *CFTR* gene mutations (Meschede et al. 1997; Mak and Jarvi 1996). In a study involving 16 men with isolated anomalies of the seminal vesicles (IASV), only one was found to be heterozygous for a missense mutation and one for the 5T allele, with a frequency not different from the general population, so that IASV was not considered a CFTR-related entity (Meschede et al. 1997). The association of chronic bronchopulmonary disease with azoospermia due to a complete bilateral obstruction of the epididymis characterize Young's syndrome, but, in contrast to CBAVD or CF, there is no anatomical malformation of the seminal ducts.

9.7 Infertility Management in CBAVD

9.7.1 Assisted Reproduction

Obstructive azoospermia (OA) is due to the blockage in sperm delivery pathway occurring anywhere in the reproductive tract including the vas deferens, epididymis, and ejaculatory duct. The most common etiology of OA is CBAVD, vasectomy, failed vasoepididymostomy, post-infective epididymitis, and other irreparable obstructions (Chen et al. 1995; Mansour et al. 1997). Intracytoplasmic sperm injection (ICSI) with percutaneous epididymal sperm aspiration (PESA) is the treatment of choice in men with OA due to CBAVD (Celikten et al. 2013).

It is a well-established fact that spermatogenesis is usually normal in majority of CBAVD men (Meng et al. 2001). There are various techniques of sperm retrieval such as microsurgical epididymal sperm aspiration (MESA) and testicular sperm extraction (TESE) allowing biological paternity to CBAVD patients.

There is an increased risk of having a child with CF or CFTR-RD if female partner of CBAVD is CF carrier. The percentage of 5T alleles in intron 8 of *CFTR* gene was reported as 26.25% in CBAVD, 20% in CUAVD, and 5% in controls in Indian population (Sharma et al. 2009). Thus, a male with CBAVD and F508del/7T alleles, if partnered with a female of normal phenotype possessing the 9T/5T, according to Mendelian expectation, provides a ratio of one in four embryos with F508del/5T genotype, which would result in CF phenotype. The other three predictions would be F508del/9T male having same phenotype as their father, i.e., CBAVD, 7T/9T offspring of normal phenotype, and 7T/5T offspring having normal phenotype (in females) and CBAVD (in males) (Persson et al. 1996). The evidence from follow-up study of children born after ICSI in CBAVD couples suggested 16% increased risk of CF or CBAVD suggesting the mandatory screening for *CFTR* gene mutations in both the partners prior to ICSI (Bonduelle et al. 1998). The first pregnancy for a couple in which the male partner was having CBAVD was reported in 1987 (Silber et al. 1988). The initial IVF cycles yielded poor oocyte fertilization rates. Since 1993, ICSI is the treatment of choice for CBAVD patients. Although lower fertilization (Patrizio et al. 1993) or lower embryo implantation (Hirsh et al. 1994) rates have been reported in couples with CBAVD, the presence of CFTR mutations in men with CBAVD does not seem to affect sperm function during IVF with micromanipulation (Schlegel et al. 1996; Silber et al. 1995). The success rate of ICSI in CBAVD was reported to be around 31% per cycle and a "take-home baby rate" was 23% (Silber et al. 1990). The meta-analysis of the ICSI outcome suggested that ICSI outcome is independent of whether retrieved spermatozoon is fresh, frozen, epididymal, or testicular. However, it suggested a lower fertilization rate and high miscarriage in CBAVD–CFTR as compared to acquired causes of obstructive azoospermia (Nicopoullos et al. 2004). Liu et al. (1994) reported the first successful PGD for a couple with CBAVD (both partners F508del heterozygous). Three carrier embryos were transferred and a healthy boy was born. The data suggested that the presence of CF- or CBAVD-causing *CFTR* gene mutations in CBAVD does not compromise significantly in fertilization rates, embryo implantation rates, or the successful delivery of asymptomatic child after PGD (McCallum et al. 2000; Phillipson et al. 2000).

9.7.2 Genetic Counseling

Genetic counseling prior to ICSI provides an estimated risk of transmitting the CF mutation from each of the parents. The probable CF or CFTR-RD phenotype of the offspring is calculated based upon the female partner's genotype, the severity of the mutation identified in the male partner, and the presence of intron 8 splice site variant. Even if the female partner is not detected to be a CF carrier by available CF mutation panels, the risk of being a carrier of a missed mutation is 0.1%. The genetic risk for couples having *CFTR* gene mutations to have a CF child is 1/4000 and 1/2000. The main rationale for CFTR testing in CBAVD, irrespective of the fact that they will be using their sperm for ICSI, is that this information is important from the point of genetic counseling regarding future health impacts of CFTR mutations as well as counseling of the siblings regarding their risk of being CF carriers. Therefore, men with CBAVD should be offered genetic counseling and CFTR testing. The CFTR screening should also be carried out in female partner before undergoing ICSI that utilizes the sperm of CBAVD partner.

9.7.3 Sperm Collection Techniques

9.7.3.1 Percutaneous Epididymal Sperm Aspiration (PESA)

PESA is used in obstructive azoospermia due to CBAVD. A small needle is inserted in the scrotum and sperm are collected from the epididymis. Obstructive azoospermia cases can be greatly benefitted from PESA as it is a useful technique to find sperm in the male partner.

This can be done by two methods: (1) testicular sperm extraction (TESE), surgical biopsy of the testis, or (2) testicular sperm aspiration (TESA), sticking a needle in the testis and aspirating fluid and tissue with negative pressure.

9.7.3.2 Microsurgical Epididymal Sperm Aspiration (MESA)

MESA is a highly advanced sperm retrieval technique. The optimal area of the epididymis is selected using operating microscope. The retrieved sperm are used for intracytoplasmic sperm injection (ICSI). MESA is now considered as a gold standard for sperm retrieval obstructive azoospermia cases. High fertilization and pregnancy rates and low risk of complications are some of the advantages of MESA (Bernie et al. 2013).

9.8 Our Experience

9.8.1 *CFTR* Gene Variants in Isolated CBAVD in Indian Population

Due to the limited information in Indian population, studies were initiated through NIRRH-ICMR, Mumbai. The andrology clinic at NIRRH is providing regular clinical and laboratory services to males with obstructive azoospermia due to vas aplasia. Currently, the clinic has one of the largest cohorts of obstructive azoospermia cases due to congenital absence of the vas deferens in India. Studies in Indian population observed heterogeneous spectrum of *CFTR* gene mutations suggesting the need to develop population-specific *CFTR* gene mutation panel. We detected ten novel and nine reported *CFTR* gene mutations in Indian CBAVD men (Gajbhiye et al., unpublished data). Further studies are ongoing to carry out screening of larger cohorts of CAVD representing different ethnic groups in India. Studies are also being undertaken to functionally characterize the novel *CFTR* gene mutations reported in Indian CBAVD.

9.8.2 CBAVD-URA

At NIRRH-ICMR, Mumbai, out of 85 CBAVD men, ten patients (11.76%) were found to have unilateral renal anomalies (URA). We detected *CFTR* gene variants in CBAVD having renal malformations. Congenital bilateral absence of seminal vesicles (CASV) and CBAVD are uncommon anomalies, and such patients usually have normal kidneys. Direct DNA sequencing of the *CFTR* gene in five CBAVD-URA men detected c.1210-12[5] (IVS8-5T) mutation in four out of five CBAVD males having renal anomalies with an allelic frequency of 40%. Four novel *CFTR* gene variants (c.2751+85_88delTA, c.2752+106A>T, c.3120+529InsC, c.4375-69C>T); four coding SNPs, V470M, T854T, P1290P, and Q1463Q; and ten previously reported *CFTR* gene variants were also detected in CBAVD males having renal anomalies (Gajbhiye et al. 2016). Normally, in addition to prostatic secretions, seminal vesicular secretions also contribute to the alkalinity of the ejaculate and make up approximately 90% of fluid in ejaculate. Thus, patients having CASV and CBAVD present with history of infertility, and usually patients with URA remain undiagnosed until there is some pathology in the contralateral kidney.

Two CBAVD-URA patients in our study were found to have longer IVS8-TG repeats (TG12 or TG13) in *cis* with 5T and M470V polymorphism. Previous studies reported that M470V along with short poly-T (5T) and long TG-repeat tracks (TG12, TG13) may contribute to CBAVD risk. This genotype was not detected in normal male participants suggesting that longer TG-short T repeats in association with M470V and other variants might be responsible for CBAVD-URA phenotype.

9.9 Future Perspectives

Evidence suggests that CFTR-related male infertility is now well established. The epidemiologic data also suggested variation in CF and CBAVD incidence by ethnic groups indicating that population-specific *CFTR* gene mutation database and mutation panels should be used for CF or CBAVD men undergoing ICSI. The major challenge is to identify disease-causing *CFTR* gene mutations in CFTR-related male infertility. This would help us to understand the genotype–phenotype correlation and provide accurate genetic counseling to the CF or CBAVD men undergoing ICSI. Further research should be focused on screening large number of infertile men due to vas aplasia and also to detect the CF carrier frequency in populations where CF or CF-related disorders (CFTR-RDs) were considered to be low. Studies are also required to functionally characterize the novel ethnic-specific mutations. There is a great need to create awareness about the CF and CFTR-RD worldwide. The genetic screening and counseling should be made available through public health care, especially in low- and middle-income countries. The global network of clinicians, scientists, and policy makers shall be established to provide standard care to patients having CF and CFTR-RD. The international experts and NGOs should come forward and empower the healthcare providers in developing countries to diagnose and provide treatment to CF and CFTR-RD patients.

References

Anguiano A, Oates RD, Amos JA, Dean M, Gerrard B, Stewart C, Maher TA, White MB, Milunsky A (1992) Congenital bilateral absence of the vas deferens: a primarily genital form of cystic fibrosis. JAMA 267(13):1794–1797

Augarten A, Yahav Y, Laufer J, Szeinberg A, Dor J, Mashiach S, Madgar I, Halle D, Gazit E, Kerem BS (1994) Congenital bilateral absence of vas deferens in the absence of cystic fibrosis. Lancet 344(8935):1473–1474

Anzai C, Morokawa N, Okada H, Kamidono S, Eto Y, Yoshimura K (2003) CFTR gene mutations in Japanese individuals with congenital bilateral absence of the vas deferens. J Cyst Fibros 2:14–18.

Bonduelle M, Aytoz A, Van Assche E, Devroey P, Liebaers I, Van Steirteghem A (1998) Incidence of chromosomal aberrations in children born after assisted reproduction through intracytoplasmic sperm injection. Hum Reprod 13(4):781–782

Bombieri C, Claustres M, De Boeck K, Derichs N, Dodge J, Girodon E, et al. (2011) Recommendations for the classification of diseases as CFTR-related disorders. J Cyst Fibros, 10, pp.86–102.

Boucher RC (2007) Airway surface dehydration in cystic fibrosis: pathogenesis and therapy. Annu Rev Med 58:157–170

Carrabino S, Carpani D, Livraghi A, Di Cicco M, Costantini D, Copreni E, Colombo C, Conese M (2006) Dysregulated interleukin-8 secretion and NF-κB activity in human cystic fibrosis nasal epithelial cells. J Cyst Fibros 5(2):113–119

Casals T, Vazquez C, Lazaro C, Girbau E, Gimenez FJ, Estivill X (1992) Cystic fibrosis in the Basque country: high frequency of mutation delta F508 in patients of Basque origin. Am J Hum Genet 50(2):404

Casals T, Bassas L, Ruiz-Romero J, Chillon M, Gimenez J, Ramos MD, Tapia G, Narvaez H, Nunes V, Estivill X (1995) Extensive analysis of 40 infertile patients with congenital absence of the vas deferens: in 50% of cases only one CFTR allele could be detected. Hum Genet 95(2):205–211

Castellani C, Cuppens H, Macek M, Cassiman JJ, Kerem E, Durie P, Tullis E, Assael BM, Bombieri C, Brown A, Casals T (2008) Consensus on the use and interpretation of cystic fibrosis mutation analysis in clinical practice. J Cyst Fibros 7(3):179–196

Celikten A, Batioglu S, Gungor ANC, Ozdemir E (2013) Intracytoplasmic sperm injection outcomes of obstructive and nonobstructive azoospermic men. Arch Gynecol Obstet 288(3):683–686

CFRI news 2013. http://www.cfri.org/pdf/2013SpringCFRInews.pdf

Chen CS, Chu SH, Soong YK, Lai YM (1995) Andrology: epididymal sperm aspiration with assisted reproductive techniques: difference between congenital and acquired obstructive azoospermia? Hum Reprod 10(5):1104–1108

Chu CS, Trapnell BC, Curristin S, Cutting GR, Crystal RG (1993a) Genetic basis of variable exon 9 skipping in cystic fibrosis transmembrane conductance regulator mRNA. Nat Genet 3(2):151–156

Chu CS, Trapnell BC, Curristin S, Cutting GR, Crystal RG (1993b) Genetic basis of variable exon 9 skipping in cystic fibrosis transmembrane conductance regulator mRNA. Nat Genet 3(2):151–156

Claustres M, Guittard C, Bozon D, Chevalier F, Verlingue C, Ferec C, Girodon E, Cazeneuve C, Bienvenu T, Lalau G, Dumur V (2000) Spectrum of CFTR mutations in cystic fibrosis and in congenital absence of the vas deferens in France. Hum Mutat 16(2):143

Colin AA, Sawyer SM, Mickle JE, Oates RD, Milunsky A, Amos JA (1996) Pulmonary function and clinical observations in men with congenital bilateral absence of the vas deferens. Chest 110:440–445

Cuppens H, Lin W, Jaspers M, et al (1998) Polyvariant mutant cystic fibrosis transmembrane conductance regulator genes. The polymorphic (Tg)m locus explains the partial penetrance of the T5 polymorphism as a disease mutation. J Clin Invest 101:487–496.

Dumur V, Gervais R, Rigot JM, Lafitte JJ, Manouvrier S, Biserte J, Mazeman E, Roussel P (1990) Abnormal distribution of CF delta F508 allele in azoospermic men with congenital aplasia of epididymis and vas deferens. Lancet, pp. 336:512

De Boeck K, Wilschanski M, Castellani C, Taylor C, Cuppens H, Dodge J, Sinaasappel M (2006) Cystic fibrosis: terminology and diagnostic algorithms. Thorax 61(7):627–635

de la Taille A (1998) Correlation between genito-urinary anomalies, semen analysis and CFTR genotype in patients with congenital bilateral absence of the vas deferens. Br J Urol 81(4):614–619

Dequeker E, Stuhrmann M, Morris MA, Casals T, Castellani C, Claustres M, Cuppens H, Des Georges M, Ferec C, Macek M, Pignatti PF (2009) Best practice guidelines for molecular genetic diagnosis of cystic fibrosis and CFTR-related disorders–updated European recommendations. Eur J Hum Genet 17(1):51–65

Donohue RE, Fauver HE (1989) Unilateral absence of the vas deferens: a useful clinical sign. JAMA 261(8):1180–1182

Dörk T, Dworniczak B, Aulehla-Scholz C, Wieczorek D, Böhm I, Mayerova A, Seydewitz HH, Nieschlag E, Meschede D, Horst J, Pander HJ (1997) Distinct spectrum of CFTR gene mutations in congenital absence of vas deferens. Hum Genet 100(3–4):365–377

Estivill, X., Bancells, C. and Ramos, C., 1997. Geographic distribution and regional origin of 272 cystic fibrosis mutations in European populations. *Human mutation*, *10*(2), p. 135.

Freedman SD, Blanco PG, Zaman MM, Shea JC, Ollero M, Hopper IK, Weed DA, Gelrud A, Regan MM, Laposata M, Alvarez JG (2004) Association of cystic fibrosis with abnormalities in fatty acid metabolism. N Engl J Med 350(6):560–569

Gajbhiye R, Kadam K, Khole A, Gaikwad A, Kadam S, Shah R, Kumaraswamy R, Khole V (2016) Cystic fibrosis transmembrane conductance regulator (CFTR) gene abnormalities in Indian males with congenital bilateral absence of vas deferens & renal anomalies. Indian J Med Res 143(5):616

Goldman MJ, Anderson GM, Stolzenberg ED, Kari UP, Zasloff M, Wilson JM (1997) Human β-defensin-1 is a salt-sensitive antibiotic in lung that is inactivated in cystic fibrosis. Cell 88(4):553–560

Goubau C, Wilschanski M, Skalická V, Lebecque P, Southern KW, Sermet I, Munck A, Derichs N, Middleton PG, Hjelte L, Padoan R (2009) Phenotypic characterisation of patients with intermediate sweat chloride values: towards validation of the European diagnostic algorithm for cystic fibrosis. Thorax 64(8):683–691

Grangeia A, Niel F, Carvalho F, et al (2004) Characterization of cystic fibrosis conductance transmembrane regulator gene mutations and IVS8 poly(T) variants in Portuguese patients with congenital absence of the vas deferens. Hum Reprod 19:2502–2508

Hall S, Oates RD (1993) Unilateral absence of the scrotal vas deferens associated with contralateral mesonephric duct anomalies resulting in infertility: laboratory, physical and radiographic findings, and therapeutic alternatives. J Urol 150(4):1161–1164

Hirsh AV, Mills C, Bekir J, Dean N, Yovich JL, Tan SL (1994) Factors influencing the outcome of in-vitro fertilization with epididymal spermatozoa in irreversible obstructive azoospermia. Hum Reprod 9(9):1710–1716

Holsclaw DS, Perlmutter AD, Jockin H, Shwachman H (1971) Genital abnormalities in male patients with cystic fibrosis. J Urol 106(4):568

Jarvi K, McCallum S, Zielenski J, Durie P, Tullis E, Wilchanski M, Margolis M, Asch M, Ginzburg B, Martin S, Buckspan MB (1998) Heterogeneity of reproductive tract abnormalities in men with absence of the vas deferens: role of cystic fibrosis transmembrane conductance regulator gene mutations. Fertil Steril 70(4):724–728

Kanavakis E, Tzetis M, Antoniadi T, Pistofidis G, Milligos S, Kat- tamis C (1998) Cystic fibrosis mutation screening in CBAVD patients and men with obstructive azoospermia or severe oligozoospermia. Mol Hum Reprod 4:333–337.

Kolettis PN, Sandlow JI (2002) Clinical and genetic features of patients with congenital unilateral absence of the vas deferens. Urology 60(6):1073–1076

Larriba S, Bassas L, Gimenez J, Ramos MD, Segura A, Nunes V, Estivill X, Casals T (1998) Testicular CFTR splice variants in patients with congenital absence of the vas deferens. Hum Mol Genet 7(11):1739–1744

Lissens W, Mahmoud KZ, El-Gindi E, et al. (1999) Molecular analysis of the cystic fibrosis gene reveals a high frequency of the intron 8 splice variant 5T in Egyptian males with congenital bilateral absence of the vas deferens. Mol Hum Reprod 5: 10–13

Liu J, Lissens W, Silber SJ et al. (1994) Birth after preimplantation diagnosis of the cystic fibrosis ΔF508 mutation by polymerase chain reaction in human embryos resulting from intracytoplasmic sperm injection with epididymal sperm. J. Am. Med. Assoc 272:1858–1860

Mak V, Jarvi KA (1996) The genetics of male infertility. J Urol 156(4):1245–1257

Mansour RT, Kamal A, Fahmy I, Tawab N, Serour GI, Aboulghar MA (1997) Intracytoplasmic sperm injection in obstructive and non-obstructive azoospermia. Hum Reprod 12(9):1974–1979

Matsui H, Grubb BR, Tarran R, Randell SH, Gatzy JT, Davis CW, Boucher RC (1998) Evidence for periciliary liquid layer depletion, not abnormal ion composition, in the pathogenesis of cystic fibrosis airways disease. Cell 95(7):1005–1015

McCallum TJ, Milunsky JM, Cunningham DL, Harris DH, Maher TA, Oates RD (2000) Fertility in men with cystic fibrosis: an update on current surgical practices and outcomes. Chest 118: 1059–1062

McCallum TJ, Milunsky JM, Cunningham DL, Harris DH, Maher TA, Oates RD (2000a) Fertility in men with cystic fibrosis: an update on current surgical practices and outcomes. Chest J 118(4):1059–1062

McCallum TJ, Milunsky JM, Cunningham DL, Harris DH, Maher TA, Oates RD (2000b) Fertility in men with cystic fibrosis: an update on current surgical practices and outcomes. Chest J 118(4):1059–1062

Mei-Zahav M, Durie P, Zielenski J, Solomon M, Tullis E, Tsui LC, Corey M (2005) The prevalence and clinical characteristics of cystic fibrosis in south Asian Canadian immigrants. Arch Dis Child 90(7):675–679

Meng MV, Black LD, Cha I, Ljung BM, Pera RAR, Turek PJ (2001) Impaired spermatogenesis in men with congenital absence of the vas deferens. Hum Reprod 16(3):529–533

Meschede D, Dworniczak B, Behre HM, Kliesch S, Claustres M, Nieschlag E, Horst J (1997) CFTR gene mutations in men with bilateral ejaculatory-duct obstruction and anomalies of the seminal vesicles. Am J Hum Genet 61(5):1200

Mickle J, Milunsky A, Amos JA, Oates RD (1995) Immunology: congenital unilateral absence of the vas deferens: a heterogeneous disorder with two distinct subpopulations based upon aetiology and mutational status of the cystic fibrosis gene. Hum Reprod 10(7):1728–1735

Nicopoullos JDM, Gilling-Smith C and Ramsay JWA. 2004. Does the cause of obstructive azoospermia affect the outcome intracytoplasmic sperm injection: a meta-analysis. BJU Int 93:1282–1286

Nicopoullos J, Gilling-Smith C, Almeida P, Ramsay J (2005) The predictive value of sperm chromatin structure assay. Hum Reprod 20(3):839–839

Nielsen OH, Thomsen BL, Green A, Andersen PK, Hauge M, Schiøtz PO (1988) Cystic fibrosis in Denmark 1945 to 1985. Acta Paediatr 77(6):836–841

Oates RD, Amos JA (1993) Congenital bilateral absence of the vas deferens and cystic fibrosis. World J Urol 11(2):82–88

O'Sullivan BP, Freedman SD (2009) Cystic fibrosis. Lancet 373(9678):1891–1904

Patrizio P, Asch RH, Handelin B, Silber SJ (1993) Aetiology of congenital absence of vas deferens: genetic study of three generations. Hum Reprod 8(2):215–220

Persson J, Peters G, Saunders D (1996) Genetic consequences of ICSI. Hum Reprod 11:921–932

Phillipson GTM, Petrucco OM, Matthews CD (2000) Congenital bilateral absence of the vas deferens, cystic fibrosis mutation analysis and intracytoplasmic sperm injection. Hum Reprod 15(2):431–435

Powers CA, Potter EM, Wessel HU, Lloyd-Still JD (1996) Cystic fibrosis in Asian Indians. Arch Pediatr Adolesc Med 150(5):554–555

Quinton PM (1999) Physiological basis of cystic fibrosis: a historical perspective. Physiol Rev 79(1):S3–S22

Quinton PM (2007) Cystic fibrosis: lessons from the sweat gland. Physiology (Bethesda) 22(3):212–225

Reddy MM, Light MJ, Quinton PM (1999) Activation of the epithelial Na+ channel (ENaC) requires CFTR Cl-channel function. Nature 402(6759):301–304

Robert F, Bey-Omar F, Rollet J, Lapray JF, Morel Y (2002) Relation between the anatomical genital phenotype and cystic fibrosis transmembrane conductance regulator gene mutations in the absence of the vas deferens. Fertil Steril 77:889–896

Rosenstein BJ, Cutting GR (1998) The diagnosis of cystic fibrosis: a consensus statement. J Pediatr 132(4):589–595

Schlegel PN, Shin D, Goldstein M (1996) Urogenital anomalies in men with congenital absence of the vas deferens. J Urol 155(5):1644–1648

Sharma N, Acharya N, Singh SK, Singh M, Sharma U, Prasad R (2009) Heterogenous spectrum of CFTR gene mutations in Indian patients with congenital absence of vas deferens. Hum Reprod 24(5):1229–1236

Silber SJ, Asch R, Balmaceda J et al. (1988) Pregnancy with sperm aspiration from the proximal head of the epididymis: a new treatment for congenital absence of the vas deferens. Fertil Steril 50:525–528

Silber SJ, Ord T, Balmaceda J et al. (1990) Congenital absence of the vas deferens. The fertilizing capacity of human epididymal sperm. N. Engl. J. Meet 323;788–1792

Silber SJ, Nagy Z, Liu J, Tournaye H, Lissens W, Ferec C, Liebaers I, Devroey P, Van Steirteghem AC (1995) Genetics: the use of epididymal and testicular spermatozoa for intracytoplasmic sperm injection: the genetic implications for male infertility. Hum Reprod 10(8):2031–2043

Teng H, Jorissen M, Van Poppel H, Legius E, Cassiman JJ, Cuppens H (1997) Increased proportion of exon 9 alternatively spliced CFTR tran- scripts in vas deferens compared with nasal epithelial cells. Hum Mol Genet 6:85–90

Tournaye H, Liu J, Nagy PZ, Camus M, Goossens A, Silber S, Van Steirteghem AC, Devroey P (1996) Correlation between testicular histology and outcome after intracytoplasmic sperm injection using testicular spermatozoa. Hum Reprod 11(1):127–132

Trezíse, A.E. and Buchwald, M., 1991. In vivo cell-specific expression of the cystic fibrosis transmembrane conductance regulator.

Tuerlings JH, Mol B, Kremer JA, Looman M, Meuleman EJ, te Meerman GJ, Buys CH, Merkus HM, Scheffer H (1998) Mutation frequency of cystic fibrosis transmembrane regulator is not increased in oligozoospermic male candidates for intracytoplasmic sperm injection. Fertil Steril 69(5):899–903

Walters S, Mehta A (2007) Epidemiology of cystic fibrosis. In: Hodson M, Geddes DM, Bush A (eds) Cystic fibrosis. Edward Arnold, London, pp 21–45

Weiske WH, Sälzler N, Schroeder-Printzen I, Weidner W (2000) Clinical findings in congenital absence of the vasa deferentia. Andrologia 32(1):13–18

Welsh MJ, Fick RB (1987) Cystic fibrosis. J Clin Investig 80(6):1523–1526

Welsh MJ, Smith AE (1993) Molecular mechanisms of CFTR chloride channel dysfunction in cystic fibrosis. Cell 73(7):1251–1254

Welsh MJ, Ramsey BW, Accurso F, Cutting JI (2001) Cystic fibrosis. In: Scriver CR, Beaudet AL, Sly WS, Valle D (eds) The metabolic and molecular basis of inherited diseases, 8th edn. McGraw-Hill, New York, pp 5121–5188

WHO 2004 and Cystic Fibrosis Foundation Patient Registry 2012 Annual Data Report.

Wu CC, Hsieh-Li HM, Lin YM, Chiang HS (2004). Cystic fibrosis trans- membrane conductance regulator gene screening and clinical correla- tion in Taiwanese males with congenital bilateral absence of the vas deferens. Hum Reprod 19:250–253

Yamashiro Y, Shimizu T, Oguchi S, Shioya T, Nagata S, Ohtsuka Y (1997) The estimated incidence of cystic fibrosis in Japan. J Pediatr Gastroenterol Nutr 24(5):544–547

Oxidative Stress and Male Infertility

10

Rima Dada and Shilpa Bisht

Abstract

Male infertility accounts for up to half of the infertility cases and affects 13–15% couples worldwide. An optimal level of reactive oxygen species is crucial for maintaining spermatogenesis and sperm functions. However, excessive production of reactive oxygen species may cause oxidative stress. Oxidative stress has been identified as one of the major risk factors which affects the fertilizing potential of spermatozoa. Oxidative stress occurs due to excessive production of ROS and causes germ cell DNA damage, sperm fragility and defects in motility, culminating in infertility. Poor sperm quality and DNA damage may also result in pregnancy loss. This article highlights the significance of ROS in human male fertility and that of oxidative stress in infertility.

Keywords

Reactive oxygen species • Oxidative stress • DNA damage • Male infertility Antioxidants

Key Points

- A large number of biological reactions are catalysed or mediated by the reactive oxygen species.
- Reactive oxygen species serve essential functions in capacitation, acrosome reaction, sperm hypermotility and hence fertility.
- Excessive production of ROS and not the decline in antioxidant capacity is generally the cause of oxidative stress.

R. Dada (✉) • S. Bisht
Laboratory for Molecular Reproduction and Genetics,
Department of Anatomy, AIIMS, New Delhi, India
e-mail: rima_dada@rediffmail.com

R. Singh, K. Singh (eds.), *Male Infertility: Understanding, Causes and Treatment*, DOI 10.1007/978-981-10-4017-7_10

- Aimless and robust antioxidant therapy can be detrimental rather than beneficial for spermatogenesis and fertility.
- Careful antioxidant therapy or other natural ways of alleviating oxidative stress such as yoga and meditation are suitable measures to keep the ROS levels in check.

10.1 Introduction

Free radicals were first described more than a century ago (Gomberg 1900), and several years after that, it was proposed that all oxidation-reduction reactions involving organic molecules are mediated by free radicals (Michaelis 1939). Free radicals are short-lived chemical intermediates that have one or more unpaired electrons. Free radicals are regulated by a number of mechanisms including hormones, and they regulate various metabolic pathways (Spagnoli et al. 1995; Tafuri et al. 2015; Alfadda and Sallam 2012). They cannot be considered to be only harmful as a number of biological reactions at the chemical level are catalysed by ROS. From fertility point of view, an optimal concentration of reactive oxygen species (ROS) is required for sperm maturation, capacitation, hyperactivation, acrosome reaction, zona pellucida binding and sperm-oocyte interaction (Mishra et al. 2016; de Lamirande and Lamothe 2009; Agarwal et al. 2014a, b).

Oxidative stress (OS) is defined as a condition when the antioxidant scavenging system of the cell is overwhelmed by the overproduction of ROS, which causes cellular damage and affects essential metabolic processes (Valko et al. 2007). OS has long been considered a potential risk factor for impaired spermatogenesis and male infertility (Aitken 2014; Hampl et al. 2012; McLachlan and de Kretser 2001). MacLeod in 1943 was the first to describe the association of elevated ROS levels with male infertility. Elevated ROS level affects the male reproductive functions via two mechanisms: first, it damages sperm membrane, affecting motility (De Lamirande and Gagnon 1992) and fertilizing potential, and second, it causes germ cell DNA damage, resulting in increased apoptosis and compromised paternal genomic contribution to the embryo (Tremellen 2008; Aitken and Curry 2011). It is, thus, considered to be the prime contributor to the aetiology of male factor infertility, especially in unexplained (idiopathic) infertile male patients (Saalu 2010; Ray et al. 2012). The present chapter illustrates the significance of ROS in terms of male reproductive functions and fertility outcomes.

10.2 Significance of ROS in Sperm Function

An optimal level of ROS is required for maintaining dynamic functions such as sperm hyperactivation, capacitation, acrosome reaction, zona pellucida binding and sperm-oocyte interaction. Regulated release of ROS during capacitation initiates various molecular modifications within the cell, which start with an increase in cyclic adenosine 3′,5′-monophosphate (cAMP) level. Activation of cAMP pathway is involved in the phosphorylation of tyrosine moieties in fibrous sheath of sperm membrane, causing an increase in sperm motility and hyperactivation (Aitken et al. 1998; de Lamirande and O'Flaherty 2008; Kothari et al. 2010). Significance of ROS in sperm hyperactivation was further supported by a study which demonstrated that in vitro incubation of spermatozoa with low concentration of OH^- triggered sperm hyperactivation (Makker et al. 2009).

The hyperactivated sperm undergo acrosome reaction through a series of events involving protein tyrosine phosphorylation, Ca^{2+} influx and increase in cAMP and Protein kinase A (PKA) levels. Functions of ROS in initiating acrosome reaction involve phosphorylation of three plasma membrane proteins (Agarwal et al. 2014a, b). The significance of ROS is further supported by the induction of acrosome reaction by in vitro supplementation of O_2^-, H_2O_2 and NO in the seminal plasma (Bansal and Bilaspuri 2010). Further, high concentration of polyunsaturated fatty acids (PUFAs) such as docosahexaenoic acid (DHA) is required for maintaining membrane fluidity in sperm, which is essential for zona pellucida binding and fertilization. ROS has been shown to facilitate the process of sperm-oocyte fusion by increasing the membrane fluidity during capacitation and acrosome reaction (Khosrowbeygi and Zarghami 2007).

10.3 Sources of ROS in Semen

ROS represents a collection of a broad range of radicals (e.g. hydroxyl ion [OH^-], superoxide ion [O_2^-], nitric oxide [NO], peroxyl [RO_2], lipid peroxyl [LOO] and Thiyl [RS^-]) and nonradical molecules (singlet oxygen [$-1O_2$], hydrogen peroxide [H_2O_2], hypochloric acid [HOCL], lipid peroxide [LOOH] and ozone [O_3]). The major sources of ROS production in semen include activated leukocytes mainly neutrophils and macrophages in the seminal plasma. Semen leukocytes produce 1000 times more ROS than spermatozoa (Plante et al. 1994). Immature and morphologically abnormal spermatozoa are other important source of ROS in semen (Griveau and Le Lannou 1997). However, the chief cause of ROS production in human spermatozoa is oxidative phosphorylation reaction in sperm mitochondria (Koppers et al. 2008).

ROS targets the PUFAs, particularly DHA present on the sperm plasma membrane. PUFAs are essential for maintaining the sperm plasma membrane fluidity and physiological homeostasis within the sperm. ROS initiates a cascade of reactions by attacking PUFAs in the sperm plasma membrane. Malondialdehyde (MDA), a by-product of lipid peroxidation, is used for indirect estimation of peroxidative damage in sperm (Colagar et al. 2009). Due to heavy energy requirements, sperm mitochondrion is the major source of ROS production in infertile men and often the major target as well.

10.4 Oxidative Stress: Potential Origins

The term "oxidative stress" began to be used frequently in the 1970s, but its conceptual origin can be traced back to the 1950s, when researchers pondered over the toxic effects of ionizing radiations, free radicals, peroxides and similar toxic effects

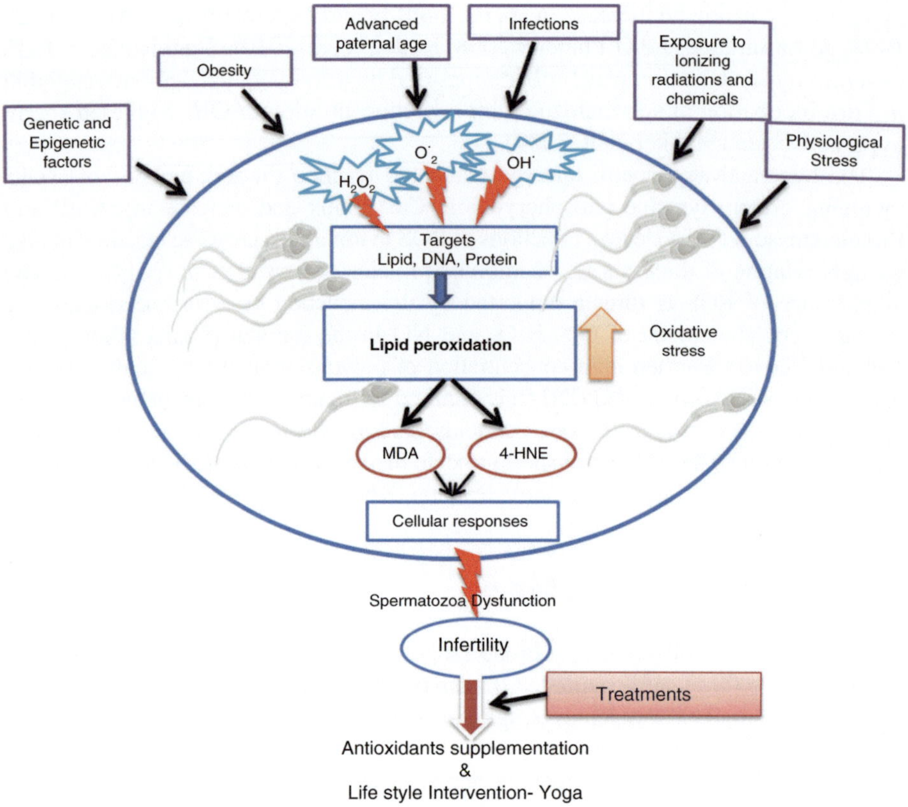

Fig. 10.1 ROS production and its augmentation leading to oxidative stress and male infertility

of molecular oxygen. Oxygen is one of the most ubiquitous elements present on the earth, and aerobic organisms produce energy through various metabolic processes. OS in sperm can arise from intrinsic sources such as testicular or sperm-borne sources to exogenous or extrinsic sources such as lifestyle and environmental sources (Fig. 10.1). Intrinsic origins of ROS can be attributed to damaged or deficient sperm (Aitken and Clarkson 1987) and several other aetiologics such as infection/inflammation, varicocele, cryptorchidism, testicular torsion and ageing.

Varicocele, a pathological condition defined by an abnormal enlargement of the pampiniform plexus of the spermatic veins, is associated with elevated ROS level in infertile men. It occurs in around 30–81% of infertile men (Saypol 1981). Varicocele-induced OS results from increased retrograde flow leading to elevated testicular temperature (Goldstein and Eid 1989; Hendin et al. 1999; Santoro and Romeo 2001). Varicocele is also associated with reduced seminal plasma antioxidant activity (Hendin et al. 1999). Similarly, cryptorchidism is associated with increased ROS levels and OS, particularly due to inactivation of superoxide dismutase activity and decreased catalase activity at elevated temperatures (Ahotupa and Huhtaniemi 1992).

Ischaemia-induced testicular torsion and its repair increase the level of ROS, leading to germ cell-specific apoptosis (Da Ros et al. 1998; Lysiak et al. 2001; Turner et al. 2004). Elevated production of ROS from leukocytes during inflammation and infection in the genital tract creates OS in spermatozoa, which causes various functional defects in sperm. Recently, a study demonstrated a significant positive correlation between OS and leukocytospermia in 88 men (Aggarwal et al. 2015). Overproduction of ROS during testicular inflammation occurs due to elevated levels of inducible nitric oxide synthase (iNOS), cyclooxygenase-2 and interleukin-1b with a corresponding decrease in antioxidant level (Reddy et al. 2006; Henkel 2011). Ageing has been demonstrated to create testicular OS by reducing the antioxidant efficiency, specifically in the Leydig cells (Cao et al. 2004; Luo et al. 2006).

Further, several extrinsic or environmental factors, such as ionizing radiations, toxins and chemotherapy can induce testicular ROS causing abnormal spermatogenesis (Agarwal et al. 2003). Some of these extrinsic factors that affect male fertility by inducing ROS include methoxyethanol from brake fluid and paints (Syed and Hecht 1998), toluene by-products, sulphur dioxide from petroleum, cadmium or lead exposure and cigarette smoking (Koizumi and Li 1992; Hsu et al. 1997). Chemotherapy often involves gonadotoxic elements such as cisplatin (Santos et al. 2008), doxorubicin (Asmis et al. 2006) and cyclophosphamide (Sudharsan et al. 2005) that affect spermatogenesis via OS.

10.5 Oxidative Stress: A Major Contributor to the Disease Pathology

As introduced above, ROS serve beneficial as well as detrimental effects on the body depending upon their levels. In low or moderate levels, ROS serve essential functions such as host defence system, regulation of various intracellular signalling cascades (Nathan 2003), induction of mitogenic response and transcription of various genes etc. High levels of ROS generate OS, which in turn has detrimental effects on cell physiology and disrupts cell membrane permeability and fluidity, disrupts junctional complexes especially connexons, initiates lipid peroxidation cascade, damages mitochondrial and nuclear DNA integrity (Mishra et al. 2014), affects protein functions and is involved in mutagenesis and carcinogenesis, most of which are mediated by hydroxyl radicals (·OH) (Aitken et al. 2012; Datta et al. 2000).

OS is in general bad for overall health. High level of ROS can cause the overall metabolic insult, resulting in a state of poor overall health and defence against other factors that affect health adversely. Accordingly, excessive ROS production has been found to be associated with several disorders like neurodegenerative disorders, autoimmune diseases, cardiovascular dysfunctions, accelerated ageing, cancer, (Carmignani and Bozzini 2006), disease of the reproductive system (male and female infertility), etc. (Droge 2002; Valko et al. 2006; Pacher et al. 2007; Dada et al. 2012). Hence, OS is central to the pathophysiology of "oxidative stress-associated diseases" that affect the whole system to little or large extent (Singh et al. 2004).

10.6 Oxidative Stress and Declining Semen Quality

Uncontrolled production of ROS (25–40%) has detrimental effect on the function and quality of sperm (Sikka 1996), and around 80% of infertile men present elevated level of ROS in semen (Agarwal and Sekhon 2011). Thus, an elevated level of ROS, rather than a compromised antioxidant activity, is the major cause of OS-induced male infertility (Agarwal and Said 2003). Increase in ROS can cause male infertility predominantly by two ways: by damaging sperm plasma membrane and sperm DNA (Agarwal et al. 2014a). A large number of studies have reported the association of OS with abnormal sperm morphology (Aziz et al. 2004), sperm DNA damage (Duru et al. 2000; Desai et al. 2009), elevated apoptosis levels (Buttke and Sandstrom 1994; Agarwal and Said 2005), declined motility (Athayde et al. 2007) and low sperm concentration (Zini et al. 1993; Athayde et al. 2007).

A recent study demonstrated significantly high level of seminal ROS in infertile men as compared to fertile donors (Agarwal et al. 2014b). Further, high level of ROS correlated positively with abnormal seminal parameters such as low sperm concentration, abnormal motility and morphology (Agarwal et al. 2014b). High ROS level has been shown to significantly impair the sperm-oocyte interaction (Agarwal et al. 2008; Peña 2015).

10.7 Oxidative Stress Correlates with Erectile Dysfunction

The factors such as hypercholesterolemia, atherosclerosis, hypertension and diabetes mellitus that affect vascular functions may also affect erectile function by disturbing intricate neurovascular mechanisms (Wang et al. 2004). A number of other stressful conditions such as hypoxaemia, sleep apnea and oxygen supply are now recognized as causes of erectile dysfunction (ED) (Brow et al. 2000). Most of these conditions affect normal body functions by inducing OS as a result of increased ROS generation (Slater 1984). Further, OS in testis can leave the cellular membranes damaged, affecting the functioning of testicular cells and production of testosterone. Decreased testosterone level not only affects spermatogenesis but also affects libido and sexual function. Therefore, OS can have more than one mechanism of action causing loss of spermatogenesis and fertility. A study on a rabbit model found that arteriogenic ED accumulated products in the erectile tissue, and the authors concluded that OS may be of great importance in the pathophysiology of arteriogenic ED (Azadzoi et al. 2005). The study also evaluated various antioxidant regimens and found that alleviating OS helped decrease ED. Similarly, another study suggested that arteriogenic ED is related to OS as it was found that arteriogenic ED cases had higher ROS and lower total antioxidant capacity (Barassi et al. 2009). It is now known that the interaction between nitric oxide and ROS is one of the important mechanisms contributing to the pathophysiology of ED (Nunes and Webb 2012).

10.8 Oxidative Stress: Clinical Perspectives and Laboratory Assessment

In a large number of idiopathic infertility cases, the semen parameters are normal as per the WHO criteria. Infertility in these individuals must have other reasons that are beyond the regular assessment of semen parameters. A number of functional sperm parameters such as ability to undergo capacitation, acrosome reaction, penetrate the zona pellucida and ultimately initiate post-fertilization development remain untested. These tests if adopted could explain the cause of infertility in a significant number of idiopathic infertility cases. However, none of these parameters is in regular screening programs for infertility evaluation. As mentioned above, an optimal level of ROS is required for capacitation and acrosome reaction; therefore, ROS evaluation could serve as a potential marker for functional competence of sperm.

Further OS damages both mitochondrial and nuclear DNA, and therefore DNA damage tests could also be a part of functional sperm tests. Sperm DNA integrity serves as an essential parameter for the assessment of sperm quality and has a predictive value in the outcomes of idiopathic infertility treatment, embryo quality, implantation, spontaneous abortion, congenital malformations and childhood diseases (Gautam et al. 2015) following natural or assisted conception. A number of tests have been devised for the detection of sperm DNA damage in human spermatozoa. These are classified as direct tests or indirect tests depending upon whether DNA fragmentation/oxidation is measured directly by incorporating probes at the site of DNA damage or DNA fragmentation is measured indirectly by measuring chromatin compaction (Table 10.1).

Routine semen analysis involving assessment of semen parameters (motility, morphology, sperm count) is still used by various laboratories to find out the possible presence of sperm OS. A reduction in any of the semen parameters is more commonly seen in men with OS, and asthenozoospermia is considered as probably the best surrogate marker for OS in a routine semen analysis (Keskes-Ammar et al. 2003; Kao et al. 2008). Hyperviscosity of seminal plasma is associated with increased seminal plasma MDA levels (Aydemir et al. 2008) and reduced seminal plasma antioxidant status, making impaired seminal viscosity a reasonable marker for OS evaluation (Siciliano et al. 2001). Other laboratory parameters which could predict the possible involvement of OS-induced oxidative DNA damage and hence male infertility include poor sperm motility, high number of round cells (leucocytospermia), poor sperm membrane integrity on hypo-osmolar swelling test, poor blastocyst development in the absence of a clear female factor, poor sperm motility after overnight incubation with the oocyte, poor fertilization on routine IVF, etc. (Alvarez et al. 2002; Tremellen 2008).

Damaged or defective spermatozoa can also affect the pregnancy outcome and health of the offspring in a successful pregnancy. For example, OS can be the cause of paternally mediated increase in miscarriages, disrupted preimplantation growth, implantation failure, congenital malformations, complex neuropsychiatric disorders

Table 10.1 Tests for detection of DNA damage in human spermatozoa

Test	Detection	Features
TUNEL	Direct quantification of sperm DNA breaks (single and double stranded)	Direct assay, based on flow cytometry
Comet	Double-stranded DNA breaks	Direct assay, based on single-cell gel electrophoresis, implies use of staining dyes such as propidium iodide, SYBR-Green and YOYO-1 iodide
ISNT	Single-stranded DNA break	Direct assay, utilizes template-dependent DNA polymerase I, less sensitive
DNA oxidation	DNA base adduct 8-hydroxy-2′-deoxyguanosine	Direct assay, ELISA based, labour intensive
Sperm chromatin structure assay	Based on susceptibility of sperm DNA to acid-induced denaturation in situ	Indirect assay, flow cytometer based, utilizes intercalating dye acridine orange
Nuclear protein composition	Protamine to histone ratio	Indirect assay, assessed by protein extraction, gel separation, immunoblotting with specific antibodies
Sperm nuclear maturity test	Chromatin integrity, protamine composition of sperm DNA	Indirect assay, simple, inexpensive slide based
Sperm chromatin dispersion	Sperm DNA fragmentation, single-stranded DNA fragments	Indirect assay, simple, based on characteristic halo produced by sperm depending upon sperm DNA integrity

and a wide range of diseases in the offspring, including dominant genetic disorders (Aitken et al. 2003, 2014). Therefore, it is vital and very essential to explore the factors that cause decline in sperm function and disrupt genomic integrity, of which OS is the predominant one (Aitken and Baker 2006; Aitken 2006).

10.9 Management of Oxidative Stress-Induced Male Infertility

As mentioned above, there is an indispensable need of optimal ROS concentration for proper enzymatic activities, and the undesirable effect of ROS in inducing cellular damage is prevented by the scavenging system of antioxidants (Ganestra 2007). This system involves enzymatic and non-enzymatic pathways, which maintain a proper balance between oxidants and antioxidants (Agarwal et al. 2003, 2014a, b). Seminal plasma contains a rich level of scavengers and antioxidants to prevent the damaging effects of ROS in sperm.

Enzymatic antioxidants include glutathione peroxidase, glutathione transferase, ceruloplasmin, catalase and superoxide dismutase. The non-enzymatic antioxidants contain vitamin C (ascorbic acid), vitamin E (alpha-tocopherol), carnitine, pyruvate, ubiquinol, hypotaurine, zinc, β-carotenes and glutathione (Sies et al. 1992; Saleh and Agarwal 2002; Agarwal and Sekhon 2010, 2011). These antioxidants work by terminating the oxidative chain reactions and alleviate the OS-induced damage (Young and Woodside 2001; Bansal and Bilaspuri 2010).

10.9.1 Antioxidants

On the basis of their mode of action, antioxidants can be (1) preventive, for example, metal chelators such as lactoferrin and transferrin which limit the formation of ROS or (2) scavenging antioxidants, such as vitamins C and E, which eliminate the existing ROS (Lampiao et al. 2012). A group of researchers recently suggested that a supplementation of antioxidants such as green tea, white tea, fish oil and melatonin might be beneficial in treating OS-induced male infertility (Cemile Merve and Elmas 2016). However, the latest research shows that an improper antioxidant therapy may result in physiological distress disturbing the intricate balance between various pro-oxidants and antioxidants in the cell/body. Aimless and robust antioxidant therapy may do more harm than good. A detailed description of the vitamins and antioxidants in male infertility treatment can be found in Chap. 20.

Sperm DNA damage or total antioxidant capacity gives a rough estimation of the OS level in spermatozoa. Hence, antioxidant supplementation could be considered as a plausible therapy to reduce OS level and to improve the reproductive outcome. Many studies have already been published discussing the role of various antioxidant therapies in the context of male infertility. However, the results of these studies are not in concordance with each other because of small sample size, too many variables used in these studies (Sikka et al. 1995) and the fact that the impact of antioxidants on DNA at therapeutic doses is still not known. Determination of the most preferred active doses of antioxidants needs further research. The effectiveness of antioxidant therapy remains controversial as it has not been confirmed by other studies.

10.9.2 Other Therapies

Yoga, essentially described as a psychosomatic-spiritual discipline, aims at achieving union and harmony between our mind, body and soul and brings balance to all aspects of one's being from physical, mental, emotional to spiritual spectrum. This ancient Indian discipline includes all aspects of an individual from health to self-realization. It caters to self-management of life and includes regulation of diet, mental attitude and the practice of specific techniques such as asanas (postures), breathing practices (pranayamas) and meditation, to attain the highest level of consciousness (Balaji et al. 2012). Various randomized controlled trials have been previously conducted citing the significant positive impact of yoga in the management of several diseases like bronchial asthma, cardiovascular disorders, diabetes mellitus, attention-deficit hyperactivity disorders, depression (Tolahunase et al. 2016), primary open-angle glaucoma (Mittal et al. 2015; Dada et al. 2016) and infertility (Dada et al. 2015), etc.

It is important to note that infertility itself causes stress, which leads to further increase in ROS level. Stress combined with adverse quality of life leads to elevation of cortisol which, in turn, leads to elevation in ROS. ROS causes oxidative damage in spermiogenesis leading to the production of damaged spermatozoa which again leads to elevation in ROS. Taken together, these factors accelerate progression in the severity of ROS and increase in abnormal or non-viable spermatozoa. Yoga and meditation (encompassing physical postures, breathing practices, relaxation techniques and meditation) are known to modulate neural, endocrine and immune functions at the cellular level through influencing cell cycle control, ageing, OS, apoptosis and several pathways of stress signalling (Dada et al. 2015; Kumar et al. 2015).

In our recent studies, we have also observed upregulation in genes involved in cellular repair and nerve growth maintenance while observing a downregulation of pro-apoptotic and pro-inflammatory genes (Dada et al. 2016; Mittal et al. 2015; Mohanty et al. 2016) in the central as well as peripheral tissues. Therefore, yoga has multisystem effects and is an ideal practice in treatment of infertility and reversing testicular ageing.

Conclusion

In physiological conditions, normal male reproductive system has an optimum oxidative status, which is maintained by equilibrium between ROS production and the antioxidant capacity. As ROS catalyse a number of biological reactions, they are indispensable for spermatogenesis and sperm fertility. Sperm functions such as capacitation, acrosome reaction and hypermotility are dependent on ROS production. Nevertheless, ROS overproduction in many pathological or stressful conditions may lead to oxidative stress, rendering the spermatozoa dysfunctional. Heavy antioxidant therapy is not recommended as it may lead to further damage by scavenging even the physiological level of ROS. Since oxidative stress has a strong lifestyle and environmental component, adoption of healthy lifestyle and interventions like yoga and meditation may help reduce psychological and oxidative stress and alleviate the ill effects of oxidative stress.

References

Agarwal A, Gupta S, Sekhon L, Shah R (2008) Redox considerations in female reproductive function and assisted reproduction: from molecular mechanisms to health implications. Antioxid Redox Signal 10(8):1375–1404

Agarwal A, Prabakaran S, Allamaneni S (2006) What an andrologist/urologist should know about free radicals and why. Urology 67:2–8

Agarwal A, Said TM (2003) Role of sperm chromatin abnormalities and DNA damage in male infertility. Hum Reprod Update 9(4):331–345

Agarwal A, Said TM (2005) Oxidative stress, DNA damage and apoptosis in male infertility: a clinical approach. BJU Int 95(4):503–507
Agarwal A, Saleh RA, Bedaiwy MA (2003) Role of reactive oxygen species in the pathophysiology of human reproduction. Fertil Steril 79(4):829–843
Agarwal A, Sekhon LH (2010) The role of antioxidant therapy in the treatment of male infertility. Hum Fertil 13(4):217–225
Agarwal A, Sekhon LH (2011) Oxidative stress and antioxidants for idiopathic oligoasthenoteratospermia: Is it justified? Indian J Urol 27(1):74
Agarwal A, Sharma RK, Sharma R, Assidi M, Abuzenadah AM, Alshahrani S, Durairajanayagam D, Sabanegh E (2014b) Characterizing semen parameters and their association with reactive oxygen species in infertile men. Reprod Biol Endocrinol 12(1):33
Agarwal A, Virk G, Ong C, du Plessis SS (2014a) Effect of oxidative stress on male reproduction. World J Mens Health 32(1):1–17
Aggarwal R, Puri M, Dada R, Saurabh G (2015) Correlation between leukocytospermia and oxidative stress in male partners of infertile couples with leukocytospermia. Int J Reprod Contracept Obstet Gynecol 4:168–172
Aitken RJ (2006) Sperm function tests and fertility. Int J Androl 29:69–75. doi:10.1111/J.1365-2605.2005.00630.x
Aitken RJ (2014) Age, the environment and our reproductive future: bonking baby boomers and the future of sex. Reproduction 147:S1–S11. doi:10.1530/REP-13-0399
Aitken RJ, Baker MA (2006) Oxidative stress, sperm survival and fertility control. Mol Cell Endocrinol 250:66–69
Aitken RJ, Baker MA, Sawyer D (2003) Oxidative stress in the male germ line and its role in the aetiology of male infertility and genetic disease. Reprod Biomed Online 7:65–70
Aitken RJ, Clarkson JS (1987) Cellular basis of defective sperm function and its association with the genesis of reactive oxygen species by human spermatozoa. J Reprod Fertil 81(2):459–469
Aitken RJ, Curry BJ (2011) Redox regulation of human sperm function: from the physiological control of sperm capacitation to the etiology of infertility and DNA damage in the germ line. Antioxid Redox Signal 14:367–381
Aitken RJ, Gibb Z, Baker MA, Drevet J, Gharagozloo P (2016) Causes and consequences of oxidative stress in spermatozoa. Reprod Fertil Dev 28(1–2):1–10. doi:10.1071/RD15325. PubMed PMID: 27062870
Aitken RJ, Gibb Z, Mitchell LA, Lambourne SR, Connaughton HS et al (2012) Sperm motility is lost in vitro as a consequence of mitochondrial free radical production and the generation of electrophilic aldehydes but can be significantly rescued by the presence of nucleophilic thiols. Biol Reprod 87:110
Aitken R, Harkiss D, Knox W, Paterson M, Irvine D (1998) A novel signal transduction cascade in capacitating human spermatozoa characterised by a redox-regulated, cAMP-mediated induction of tyrosine phosphorylation. J Cell Sci 111:645–656
Aitken RJ, Smith TB, Jobling MS, Baker MA, De Iuliis GN (2014) Oxidative stress and male reproductive health. Asian J Androl 16:31–38. doi:10.4103/1008-682X.122203
Alfadda AA, Sallam RM (2012) Reactive oxygen species in health and disease. Biomed Res Int 2012:936486
Alvarez JG, Sharma RK, Ollero M et al (2002) Increased DNA damage in sperm from leukocytospermic semen samples as determined by the sperm chromatin structure assay. Fertil Steril 78:319–329
Asmis R, Qiao M, Rossi RR, Cholewa J, Xu L, Asmis LM (2006) Adriamycin promotes macrophage dysfunction in mice. Free Radic Biol Med 41(1):165–174
Athayde KS, Cocuzza M, Agarwal A, Krajcir N, Lucon AM, Srougi M, Hallak J (2007) Development of normal reference values for seminal reactive oxygen species and their correlation with leukocytes and semen parameters in a fertile population. J Androl 28(4):613–620
Aydemir B, Onaran I, Kiziler AR, Alici B, Akyolcu MC (2008) The influence of oxidative damage on viscosity of seminal fluid in infertile men. J Androl 29:41–46

Azadzoi KM, Schulman RN, Aviram M, Siroky MB (2005) Oxidative stress in arteriogenic erectile dysfunction: prophylactic role of antioxidants. J Urol 174(1):386–393. PubMed PMID: 15947695

Aziz N, Saleh RA, Sharma RK, Lewis-Jones I, Esfandiari N, Thomas AJ, Agarwal A (2004) Novel association between sperm reactive oxygen species production, sperm morphological defects, and the sperm deformity index. Fertil Steril 81(2):349–354

Balaji PA, Varne SR, Ali SS (2012) Physiological effects of yogic practices and transcendental meditation in health and disease. N Am J Med Sci 4(10):442

Bansal, A. K., Bilaspuri, G. S. (2010). Impacts of oxidative stress and antioxidants on semen functions. Vet Med Int 2010. pii: 686137. doi:10.4061/2011/686137

Barassi A, Colpi GM, Piediferro G, Dogliotti G, D'Eril GV, Corsi MM (2009) Oxidative stress and antioxidant status in patients with erectile dysfunction. J Sex Med 6(10):2820–2825. doi:10.1111/j.1743-6109.2009.01279.x. Epub 2009 Apr 23

Brow SL, Seftel AD, Strohl KP, Herbener T (2000) Vasculogenic impotence and cavernosal oxygen tension. Int J Impot Res 12:19

Buttke TM, Sandstrom PA (1994) Oxidative stress as a mediator of apoptosis. Immunol Today 15(1):7–10

Cao L, Leers-Sucheta S, Azhar S (2004) Aging alters the functional expression of enzymatic and non-enzymatic anti-oxidant defense systems in testicular rat Leydig cells. J Steroid Biochem Mol Biol 88(1):61–67

Carmignani L, Bozzini G (2006) Re: increased incidence of testicular cancer in men presenting with infertility and abnormal semen analysis. Raman JD, Nobert CF, Goldstein M. J Urol 175(4):1574. Author reply 1574. PubMed PMID: 16516050

Cemile Merve S, Elmas Ç (2016) The effects of oxidative stress and some of the popular antioxidants on reproductive system: a mini review. J Nutr Food Sci 6:464

Colagar AH, Pouramir M, Marzony ET, Jorsaraei SGA (2009) Relationship between seminal malondialdehyde levels and sperm quality in fertile and infertile men. Braz Arch Biol Technol 52(6):1387–1392

Da Ros CT, Teloken C, Tannhauser M, Hartmann A (1998) Does intratesticular testosterone administration modify the evolution of transitory testicular ischemia in prepubertal rats? J Urol 159(5):1752–1754

Dada T, Faiq MA, Mohanty K, Mittal D, Bhat M, Yadav RK, Sihota R, Dada R (2016) Effect of yoga and meditation based intervention on intraocular pressure, quality of life, oxidative stress and gene expression pattern in primary open angle glaucoma: a randomized controlled. Association of Research on Vision and Ophthalmology (ARVO). ARVO 2016, p 2

Dada R, Kumar M, Jesudasan R, Fernández JL, Gosálvez J, Agarwal A (2012) Epigenetics and its role in male infertility. J Assist Reprod Genet 29(3):213–223

Dada R, Kumar SB, Tolahunase M, Mishra S, Mohanty K, Mukesh T (2015) Yoga and meditation as a therapeutic intervention in oxidative stress and oxidative DNA damage to paternal genome. J Yoga Phys Ther 5:4

Datta K, Sinha S, Chattopadhyay P (2000) Reactive oxygen species in health and disease. Natl Med J India 13(6):304–310

De Lamirande E, Gagnon C (1992) Reactive oxygen species and human spermatozoa: I. Effects on the motility of intact spermatozoa and on sperm axonemes. J Androl 13:368–368

Desai N, Sharma R, Makker K, Sabanegh E, Agarwal A (2009) Physiologic and pathologic levels of reactive oxygen species in neat semen of infertile men. Fertil Steril 92(5): 1626–1631

Droge W (2002) Free radicals in the physiological control of cell function. Review. Physiol Rev 82:47–95

Duru NK, Morshedi M, Oehninger S (2000) Effects of hydrogen peroxide on DNA and plasma membrane integrity of human spermatozoa. Fertil Steril 74(6):1200–1207

Gautam S, Chawla B, Kumar SB, Bisht S, Dada R (2015) Sperm DNA damage in non-familial sporadic heritable retinoblastoma (NFSHRb). Clin Epidemiol Glob Health 3:S20–S25

Genestra M (2007) Oxyl radicals, redox-sensitive signalling cascades and antioxidants. Review. Cell Signal 19:1807–1819
Goldstein M, Eid JF (1989) Elevation of intratesticular and scrotal skin surface temperature in men with varicocele. J Urol 142(3):743–745
Gomberg M (1900) An instance of trivalent carbon: triphenylmethyl. J Am Chem Soc 22:757–771
Griveau JF, Le Lannou D (1997) Reactive oxygen species and human spermatozoa: physiology and pathology. Int J Androl 20:61–69
Hampl R, Drábková P, Kanďár R, Stěpán J (2012) Impact of oxidative stress on male infertility. Ceska Gynekol 77:241–245
Hendin BN, Kolettis PN, Sharma RK, Thomas AJ, Agarwal A (1999) Varicocele is associated with elevated spermatozoal reactive oxygen species production and diminished seminal plasma antioxidant capacity. J Urol 161(6):1831–1834
Henkel RR (2011) Leukocytes and oxidative stress: dilemma for sperm function and male fertility. Asian J Androl 13(1):43
Hsu PC, Liu MY, Hsu CC, Chen LY, Guo YL (1997) Lead exposure causes generation of reactive oxygen species and functional impairment in rat sperm. Toxicology 122(1):133–143
Kao SH, Chao HT, Chen HW, Hwang TI, Liao TL, Wei YH (2008) Increase of oxidative stress in human sperm with lower motility. Fertil Steril 89(5):1183–1190
Keskes-Ammar L, Feki-Chakroun N, Rebai T, Sahnoun Z, Ghozzi H, Hammami S, Zghal K, Fki H, Damak J, Bahloul A (2003) Sperm oxidative stress and the effect of an oral vitamin E and selenium supplement on semen quality in infertile men. Arch Androl 49:83–94
Koppers AJ, De Iuliis GN, Finnie JM, McLaughlin EA, Aitken RJ (2008) Significance of mitochondrial reactive oxygen species in the generation of oxidative stress in spermatozoa. J Clin Endocrinol Metab 93:3199–3207
Kumar SB, Yadav R, Yadav RK, Tolahunase M, Dada R (2015) Telomerase activity and cellular aging might be positively modified by a yoga-based lifestyle intervention. J Altern Complement Med 21(6):370–372
de Lamirande E, Lamothe G (2009) Reactive oxygen-induced reactive oxygen formation during human sperm capacitation. Free Radic Biol Med 46(4):502–510
de Lamirande E, O'Flaherty C (2008) Sperm activation: role of reactive oxygen species and kinases. Biochim Biophys Acta 1784(1):106–115
Lampiao F, Opperman CJ, Agarwal A, du Plessis SS (2012) Oxidative stress. In: Parekattil SJ, Agarwal A (eds) Male infertility: contemporary clinical approaches, andrology, ART & antioxidants. Springer, New York, pp 225–235
Luo L, Chen H, Trush MA, Show MD, Anway MD, Zirkin BR (2006) Aging and the brown Norway rat leydig cell antioxidant defense system. J Androl 27(2):240–247
Lysiak JJ, Turner SD, Nguyen QAT, Singbartl K, Ley K, Turner TT (2001) Essential role of neutrophils in germ cell-specific apoptosis following ischemia/reperfusion injury of the mouse testis. Biol Reprod 65(3):718–725
Makker K, Agarwal A, Sharma R (2009). Oxidative stress & male infertility. Indian J Med Res 129(4):357–367
McLachlan RI, de Kretser DM (2001) Male infertility: the case for continued research. Med J Aust 174:116–117
Michaelis L (1939) Free radicals as intermediate steps of oxidation–reduction. Cold Spring Harb Symp Quant Biol 7:33–49
Mishra S, Kranthi V, Kumar R, Malhotra N, Mohanty K, Pathak V, Dada R (2014) Oxidative damage to sperm DNA: clinical implications. Androl Open Access 3(116):2167–0250
Mishra SS, Kumar R, Malhotra N et al (2016) Mild oxidative stress is beneficial for sperm telomere length maintenance. World J Methodol 6(2):163–170
Mittal D, Sihota R, Dada R, Sharma R, Faiq MA, Mohanty K, Dada T (2015) Impact of meditation and yoga based intervention on quality of life and stress markers in glaucoma patients: a prospective randomized controlled study. Int Glaucoma Rev 16(3):237

Nathan C (2003) Specificity of a third kind: reactive oxygen and nitrogen intermediates in cell signalling. J Clin Investig 111:769–778
Nunes KP, Webb RC (2012) Mechanisms in erectile function and dysfunction: an overview. Georgia Health Sciences University, Augusta, Georgia. In: Nunes K (ed) Erectile dysfunction—disease-associated mechanisms and novel insights into therapy. InTech, Rijeka
Pacher P, Beckman JS, Liaudet L (2007) Nitric oxide and peroxynitrite in health and disease. Physiol Rev 87:315–424
Peltola V, Huhtaniemi I, Ahotupa M (1992) Antioxidant enzyme activity in the maturing rat testis. J Androl 13:450–450
Peña F (2015) New aspects of sperm physiology and sperm oocyte interactions. Anim Reprod 12:359–365
Plante M, de Lamirande E, Gagnon C (1994) Reactive oxygen species released by activated neutrophils, but not by deficient spermatozoa, are sufficient to affect normal sperm motility. Fertil Steril 62(2):387–393
Ray PD, Huang BW, Tsuji Y (2012) Reactive oxygen species (ROS) homeostasis and redox regulation in cellular signaling. Cell Signal 24(5):981–990
Reddy MM, Mahipal SV, Subhashini J, Reddy MC, Roy KR, Reddy GV, Reddy PR, Reddanna P (2006) Bacterial lipopolysaccharide-induced oxidative stress in the impairment of steroidogenesis and spermatogenesis in rats. Reprod Toxicol 22(3):493–500
Saalu LC (2010) The incriminating role of reactive oxygen species in idiopathic male infertility: an evidence based evaluation. Pak J Biol Sci 13:413–422
Saleh RA, Agarwal A (2002) Oxidative stress and male infertility: from research bench to clinical practice. J Androl 23:737–752
Santoro G, Romeo C (2001) Normal and varicocele testis in adolescents. Asian J Androl 3(4):259–262
Santos NAG, Bezerra CC, Martins NM, Curti C, Bianchi MLP, Santos AC (2008) Hydroxyl radical scavenger ameliorates cisplatin-induced nephrotoxicity by preventing oxidative stress, redox state unbalance, impairment of energetic metabolism and apoptosis in rat kidney mitochondria. Cancer Chemother Pharmacol 61(1):145
Saypol DC (1981) Varicocele. J Androl 2(2):61–71
Siciliano L, Tarantino P, Longobardi F, Rago V, De Stefano C, Carpino A (2001) Impaired seminal antioxidant capacity in human semen with hyperviscosity or oligoasthenozoospermia. J Androl 22:798–803
Sies H, Stahl W, Sundquist AR (1992) Antioxidant functions of vitamins. Ann N Y Acad Sci 669(1):7–20
Sikka SC (1996) Oxidative stress and role of antioxidants in normal and abnormal sperm function. Front Biosci 1:78–86
Sikka SC, Rajasekaran M, Hellstrom WJ (1995) Role of oxidative stress and antioxidants in male infertility. J Androl 16:464–481
Singh RP, Sharad S, Kapur S (2004) Free radicals and oxidative stress in neurodegenerative diseases: relevance of dietary antioxidants. JIACM 5:218–225
Slater TF (1984) Free-radical mechanisms in tissue injury. Biochem J 222:1
Spagnoli A, Spadoni GL, Sesti G, Del Principe D, Germani D, Boscherini B (1995) Effect of insulin on hydrogen peroxide production by human polymorphonuclear leukocytes. Studies with monoclonal anti-insulin receptor antibodies, and an agonist and an inhibitor of protein kinase C. Horm Res 43:286–293
Sudharsan PT, Mythili Y, Selvakumar E, Varalakshmi P (2005) Cardioprotective effect of pentacyclic triterpene, lupeol and its ester on cyclophosphamide-induced oxidative stress. Hum Exp Toxicol 24(6):313–318
Syed V, Hecht NB (1998) Rat pachytene spermatocytes down-regulate a polo-like kinase and up-regulate a thiol-specific antioxidant protein, whereas sertoli cells down-regulate a phosphodiesterase and up-regulate an oxidative stress protein after exposure to methoxyethanol and methoxyacetic acid 1. Endocrinology 139(8):3503–3511

Tafuri S, Ciani F, Iorio EL, Esposito L, Cocchia N (2015) Reactive oxygen species (ROS) and male fertility. In: Wu B (ed) New discoveries in embryology. Rijcka, InTcch, pp 19–33

Tolahunase M, Bisht S, Sagar R, Yadav RK, Khan S, & Dada R (2016). Effect of yoga on quality of life and seminal quality in infertile male patients with major depressive disorder (MDD): trial. In: ASA 41st Annual Conference 2016, Suppl. 112

Tremellen K (2008) Oxidative stress and male infertility—a clinical perspective. Hum Reprod Update 14:243–258

Turner TT, Bang HJ, Lysiak JL (2004) The molecular pathology of experimental testicular torsion suggests adjunct therapy to surgical repair. J Urol 172(6):2574–2578

Valko M, Leibfritz D, Moncol J, Cronin MT, Mazur M, Telser J (2007) Free radicals and antioxidants in normal physiological functions and human disease. Int J Biochem Cell Biol 39:44–84

Valko M, Rhodes CJ, Moncol J, Izakovic M et al (2006) Free radicals, metals and antioxidants in oxidative stress-induced cancer. Mini-review. Chem Biol Interact 160:1–40

Wang T, Soker S, Atala A, Siroky MB, Azadzoi KM (2004) Alterations in angiogenic growth factors and neuronal nitric oxide synthase expression in chronic cavernosal ischemia. Int J Impot Res 16:403

Young I, Woodside J (2001) Antioxidants in health and disease. J Clin Pathol 54:176–186

Zini A, Lamirande E, Gagnon C (1993) Reactive oxygen species in semen of infertile patients: levels of superoxide dismutase—and catalase-like activities in seminal plasma and spermatozoa. Int J Androl 16(3):183–188

Obesity, Spermatogenesis, and Male Infertility

11

Joseph R.D. Fernandes and Arnab Banerjee

Abstract

Obesity is one of the major global health problems concerning people from almost all age groups. Obesity has profound adverse effects on the reproductive health and may lead to metabolic disturbances and syndromes. Obese couples have less fertility potential and often opt for assisted reproductive technologies for conception. It has been observed that obese men have an altered adipokine profile, increased serum estrogen levels, and poor sperm quality. These alterations correlate with impaired spermatogenesis and may subsequently lead to subfertility or infertility. The increasing incidence of obesity calls for molecular, genetic, and epigenetic research to elucidate the underlying risk factors for loss of fertility. This chapter focuses on the factors responsible for obesity and provides an account of its effects on spermatogenesis and fertility in males. Finally, potential reversibility measures and management options for obesity-associated infertility have been discussed.

Keywords

Obesity • Metabolic syndrome • Spermatogenesis • Oligozoospermia • Male infertility

Key points

- Sedentary lifestyle and urbanization can lead to metabolic syndrome including obesity.
- Obesity affects almost all the physiological processes in the body, including reproduction.

J.R.D. Fernandes • A. Banerjee (✉)
Department of Biological Sciences, BITS Pilani KK Birla Goa Campus, Sancoale, Goa, India
e-mail: arnabb@goa.bits-pilani.ac.in

R. Singh, K. Singh (eds.), *Male Infertility: Understanding, Causes and Treatment*, DOI 10.1007/978-981-10-4017-7_11

- With increasing incidence of male obesity, there is a rise in male infertility worldwide.
- Obese males have reduced spermatogenesis, poor sperm quality, abnormal sperm morphology, reduced serum testosterone level, and altered adipokine profile.
- Paternal obesity also has transgenerational effects, thus affecting the health and reproduction of the coming generations.
- The most effective way to manage obesity would be following a good lifestyle, weight loss, healthy diet, and adequate physical activity.

11.1 Introduction

Excessive use of escalators and elevators, less sidewalks, sedentary lifestyle, poor eating habits, and the lack of physical activity can lead to metabolic disorders or metabolic syndromes (MetS), such as non-insulin-dependent diabetes mellitus (NIDDM), cardiovascular diseases (CVDs), and other groups of abnormalities including overweight (Kasturi et al. 2008). The following criteria have been defined for an individual to be diagnosed with metabolic syndrome in accordance with the International Diabetes Federation in 2006; central obesity measured by waist circumference plus two additional factors such as reduced high-density lipoprotein (HDL) cholesterol level (<40–50 mg/dL), raised triglycerides level (>150 mg/dL), increased blood pressure (>130 mmHg systolic or >85 mmHg diastolic), or raised level of fasting plasma glucose (>100 mg/dL) (Grundy et al. 2004). Obesity is one of the most important aspects of MetS as it opens the gateway to many associated abnormalities to develop (Deedwania and Gupta 2006). Exposure to endocrine disruptors and environmental toxins with estrogenic effects causes reproductive disorders as these toxins are fat soluble and accumulate in adipose tissue. The etiology of obesity is indeed highly complex which includes genetic, environmental, physiological, and psychological factors, which coalesce to promote the development of obesity (Aronne et al. 2009).

Obesity introduces a deficient overall health, which convolutes with time and further affects almost every vital organ in body. Various studies suggest that obesity negatively affects the reproductive potential in males, usually associated with erectile dysfunction and reduced semen quality (Hammoud et al. 2012). Evidenced by the increase in couples seeking assisted reproductive technologies (ART), especially intracytoplasmic sperm injection (ICSI) confirms the increase in male infertility due to a number of reasons, including obesity. There is also heightening awareness about male obesity and its effects on spermatogenesis, thereby reducing sperm quality, count, and viability, in particular by affecting the germ cells in testes. There is a high prevalence of obese men with poor semen quality than normal men (Magnusdottir et al. 2005). Increased adipose tissue around scrotal area raises gonadal temperature which affects the heat sensitive spermatogenesis process. This chapter brings together the mechanisms by which obesity contributes to the loss of fertility and the methods to overcome some of these.

11.2 Obesity and Reproduction

Ancient human lifestyle demanded them to eat and store food to thrive through fast and famines; however, that lifestyle is not applicable in the modern society. People now dwell in sedentary behavior and suffer from severe obesity that may also be linked to subfertility and infertility. As evidenced by an increase in the number of overweight and obese couples seeking assisted reproductive technology (ART), obesity is found to be in direct correlation with male infertility. Not all individuals in a population are able to reproduce, and one of the factors that inhibit the ability to reproduce is obesity. Reproduction is an energy demanding process that has a fundamental effect on fat metabolism, which is the major form of energy stored in animals (Bronson 1989). Lipid metabolism and reproduction are intricately related. Several studies in animals suggest that the fat reserves are mobilized during reproduction; hence, reduced or abolished reproductive activity can elevate the lipid storage and may lead to weight gain in various species (McElroy and Wade 1987; Corona et al. 2009; Judd et al. 2011).

Obesity and overweight are consorted with severe reproductive consequences in women as well as men. Excess of body fat has been associated with increased risk of polycystic ovarian syndrome, miscarriages, infertility and infertility treatment failure, menstrual cycle disturbances, multiple complications in pregnancy, gestational diabetes, preeclampsia, cesarean delivery, and macrosomic fetus. It has been observed that girls with delayed puberty are relatively thin during their adolescence; a critical body weight of 47 kg or ~22% body fat content has been suggested for the onset of cyclical ovarian activity (Pasquali et al. 2007). A study showed that around 1–5% women suffer from weight-related amenorrhea (Laughlin et al. 1998). Another study suggested that it is the regional fat loss that might trigger amenorrhea rather than the total fat loss. Reduction of fat from thighs, hips, and buttocks, which provide much of energy during pregnancy and lactation, may severely disrupt reproductive function (Brownell and Jeffery 1987).

Obesity affects the reproductive parameters in men to the same degree as in women. Obesity is directly linked to reduced spermatogenesis, poor sperm quality, and altered sperm morphology. The incidence of obesity is increasing; likewise the number of men with compromised sperm quality is rising. The effects of obesity creep in steadily, thus making it difficult to comprehend due to a wide variation in semen parameters in humans. Nevertheless, interesting studies over the last few decades have identified a significant correlation between obesity and fertility. A number of mechanisms have been put forward to explain the impact of obesity on male fertility, some of which are supported by animal experimental studies and human observational studies.

11.3 Obesity Compromises Testosterone Production

Altered hormonal profile in obese males changes the microenvironment of germ cells, affecting spermatogenesis at the molecular level. Males suffering from severe obesity generally have decreased plasma concentration of testosterone and the level decreases as obesity increases (Kley et al. 1980; Wang et al. 2011). Free testosterone level might be at normal in men with an appropriate weight or even in men with moderate obesity, but in men suffering from massive obesity, there is a significant decrease in the level of free testosterone (Zumoff et al. 1990). Increased insulin level in obese men has been shown to be responsible for reduced sex hormone-binding globulin (SHBG) production in liver, further decreasing the availability of free unbound testosterone in blood, which reflects in poor sperm count in obese patients (Haffner et al. 1992; Laing et al. 1998; Wang et al. 2011). Levels of free testosterone have been reported to be lower among diabetic obese men in comparison to nondiabetic obese men (Corona et al. 2011). Increasing evidence suggest that decreased serum testosterone can further induce metabolic syndrome. Hence, one can hypothesize that obesity in males may be promoted further by decreased testosterone level (Kupelian et al. 2006; Akishita et al. 2010).

The impact of fat on reproductive function can also be attributed to the endocrine disturbances and mechanisms as suggested by a study showing that higher BMI values are related to lower inhibin B levels (Pauli et al. 2008). Lower inhibin level is associated with decreased number of Sertoli cells, thereby lowering the sperm count in obesity (Ramaswamy et al. 2000; Cabler et al. 2010). Several studies suggest that metabolic parameters such as high levels of cholesterol and triglycerides that are associated with obesity have a direct and drastic effect on the testicular function, which may lead to poor semen quality and infertility (Padron et al. 1997). This observation is further supported and expanded by a study that reported a 65% incidence of dyslipidemia (defined by isolated hypercholesterolemia and triglyceridemia) in infertile men (Ramirez-Torres et al. 2000). LH pulse remains undisturbed in obese men; however, the LH amplitude is severely attenuated as compared to nonobese men (Vermeulen et al. 1993). The decrease in LH level depends strongly on the degree of obesity and is observed more often in massive obese men with a BMI greater than 40 kg/m^2. Increased serum glucose in obese men has been shown to be responsible for decreased LH levels due to altered HPG activity (Clarke et al. 1990).

11.4 Obesity Disturbs Testosterone: Estrogen Ratio

Testosterone and estradiol exert a negative feedback on GnRH release through the activation of kisspeptin 1 (KISS1) neurons present in large numbers in the arcuate nucleus (ARC) of the hypothalamus, which has receptors for androgens and estrogens (Tng 2015). KISS1 also stimulates the release of GnRH. Besides stimulating GnRH levels, estradiol and testosterone also have a negative effect on the release of FSH and LH by the pituitary. In comparison to testosterone, obese men exhibit a considerable increase in estradiol, estrone, and defective estrogen receptors that further leads to decreased testosterone-estradiol binding globulin (TeBG) (Schneider et al. 1979). Unlike testosterone levels, estrogen levels in obese individuals are elevated, predominantly due to the aromatization of free testosterone levels by aromatase in the adipose tissue. The overall rate of aromatization of testosterone to estradiol increases with age and fat mass (Vermeulen et al. 2002). The conversion of testosterone to estradiol enhances fat deposition and contributes to a greater degree of testosterone deficiency, which may also cause secondary hypogonadotropic hypogonadism and infertility in obese men. Obese men when treated with aromatase inhibitors, such as anastrozole or letrozole that inhibit the conversion of testosterone to estradiol, show normalization of testosterone levels, spermatogenesis, and fertility (De Boer et al. 2005; Roth et al. 2008).

11.5 Obesity Disturbs Scrotal Thermal Regulation

The key role of the scrotal sac is to keep the temperature of testes 2–4 °C below the core body temperature (Ivell 2007). Spermatogenesis is highly sensitive to increase in temperature, and a small rise in testicular temperature may induce testicular stress, germ cell apoptosis, and hypogonadism. In case of obese men, increased mass in thighs results in close proximity between thighs and testicles, causing testicular heating. High scrotal temperature in obese individuals may also be due to increased fat deposition in the scrotum (Wise et al. 2011). Obese men have been reported to have high scrotal temperatures with altered seminal parameters, hypogonadism, and increased sperm aneuploidies, suggesting that increased testicular heating might be the cause of impaired testicular activity in these subjects (Garolla et al. 2015). The same study pointed out those obese individuals show circadian rhythm in the changes in scrotal temperature, with high fluctuations between day and night from the observed mean scrotal temperature of 34.73 °C in healthy subjects (Garolla et al. 2015).

11.6 Obesity Increases DNA Damage

Appropriate sperm concentration and motility as well as the molecular entities are vital to generate a healthy pregnancy. To have a successful fertilization and embryonic development, it is important to have adequate sperm DNA integrity as sperm with poor DNA integrity negatively correlate with pregnancy outcome (Benchaib et al. 2003). It has been shown that sperm with high DNA damage are more frequent in obese men as compared to nonobese men (Chavarro et al. 2010). Another study showed a positive association of increased BMI with DNA fragmentation using sperm chromatin structure assay, wherein obese and overweight men showed a higher percentage of DNA fragmentation index (27 and 25.8%) in comparison to normal men (19.9%) (Kort et al. 2006). Earlier studies have shown that with increase in BMI, the oxidative stress increases mainly due to the rise in seminal macrophage activation. This may lead to decreased sperm motility, increased sperm DNA damage, decreased acrosome reaction, and lower embryo implantation rates (Palmer et al. 2012; Katib 2015). Excessive generation of reactive oxygen species may attack the phospholipid membranes, causing disruption and inhibition of oxidative phosphorylation, ultimately leading to decreased production of ATP (Bourgeron 2000; Turner 2003).

11.7 Obesity Leads to Transgenerational Epigenetic Effects

An interesting study on rats showed that paternal obesity compromised the metabolic and reproductive health of the first and second generation offsprings, indicating that sperm is the potent carrier for any such physiological change (Palmer et al. 2012). Such molecular mechanisms include altered epigenetic modifications and sperm noncoding RNA content (Youngson and Whitelaw 2011). Reduced pregnancy success and increased sperm DNA damage in males undergoing infertility treatment are associated with hypomethylation of imprinted genes and repeat elements in sperm (Palmer et al. 2012). Interestingly, a study showed that male mice fed with high-fat diet displayed altered acetylation status in the late round spermatids, which correlated with DNA damage in the germ cells. Disruptions to the sperm histone acetylation lead to increased DNA damage in mature sperm and potentially correspond to poor sperm parameters that are observed in obese males (Jenkins and Carrell 2012). Daughters of male rats fed a high-fat diet were shown to have abnormal DNA methylation in the pancreas (Ng et al. 2010), whereas offsprings of male mice fed a low-protein diet showed altered liver expression of cholesterol genes (Carone et al. 2010).

11.8 microRNAs (miRNAs), Obesity, and Male Infertility

miRNAs are small noncoding RNAs of about 18–25 nucleotides long playing a crucial role in gene regulation and in silencing or repressing thousands of genes at the posttranscriptional levels (He and Hannon 2004). miRNAs are known to express in human adipose tissue and show significant modulation in obese individuals (Hilton et al. 2013; Oger et al. 2014). It is evident from human and animal studies that obesity alters microRNA (miRNA) expression in metabolically important organs and that miRNAs are involved in changes of normal physiology, acting as mediators of disease. miRNAs regulate multiple pathways including insulin signaling, immune-mediated inflammation, adipokine expression, adipogenesis, lipid metabolism, and food intake regulation (Oger et al. 2014).

Though spermatozoa are transcriptionally inactive, now it has been shown that mature sperm contain mRNA, noncoding RNA, and piwi-interacting RNA (piRNAs) (Lalancette et al. 2008; Dadoune 2009). These RNAs have roles during fertilization and embryo development (Palmer et al. 2012). There is a report suggesting an altered sperm miRNA profile in obese rodents (Lane et al. 2012). Obese men have also been reported to have altered circulatory miRNA content, which can be restored by diet and exercise (Ortega et al. 2013). It has been recently shown that diet alteration or exercise intervention in obese fathers might prevent female offspring from being predisposed to metabolic syndrome by regulating miRNA profile (McPherson et al. 2015).

11.9 Obesity Correlates with Erectile Dysfunction

The inability to have and maintain a penile erection adequate for sexual intercourse is known as erectile dysfunction (ED), which has organic and psychogenic components (Muneer et al. 2014). ED is reported to be one of the consequences of obesity in males. Obesity-associated ED cases are more common than age-related ED (Skrypnik et al. 2014). Various studies have suggested that obesity negatively affects the reproductive potential in males, which is usually associated with erectile dysfunction and reduced sperm quality (Hammoud et al. 2012). Central obesity is associated with both arteriogenic ED and reduced testosterone (T) levels (Corona et al. 2014). Hypogonadism is prevalent among obese men, which justifies the higher prevalence of ED among them (Corona et al. 2009). A study involving the Massachusetts male aging cohort found that men who were overweight at baseline were at a higher risk of developing ED regardless if they lost weight during the follow-up (Derby et al. 2000). However, a report suggests that obese men with EDs showed better erectile function within 2 years of incorporating changes in their dietary habits and increased physical activity (Esposito et al. 2004). The link between obesity and ED might be a useful motivation for men to improve their health-related lifestyle choices.

11.10 Obesity, Adipokines, and Male Infertility

Adipose tissue is no longer considered to be an inert tissue for fat storage. This tissue is considered an endocrine organ that actively affects the whole body metabolism (Coelho et al. 2013). Adipose tissue secretes adipokines or adipocytokines, which directly influence insulin sensitivity along with several other physiological events including fertility (Hutley and Prins 2005). Adipose tissue releases chemokines like adiponectin, visfatin, resistin, tumor necrosis factor (TNF)-α, interleukin 6, etc., and their balance is dysregulated in obesity (Rosen and Spiegelman 2006). In obesity, there is increased level of all the pro-inflammatory adipokines, such as leptin, TNF-α, etc., which in turn causes insulin resistance in these individuals (Hotamisligil et al. 1993; Rotter et al. 2003). Interestingly, the only anti-inflammatory adipokine, adiponectin, shows an opposite trend. Adiponectin expression decreases with increase in obesity, and earlier studies showed that adiponectin protects against several metabolic dysfunctions and ameliorates insulin resistance and glucose tolerance (Maeda et al. 1996).

Increased levels of the pro-inflammatory cytokines, such as leptin, significantly decrease testicular testosterone production, thereby promoting infertility in obese individuals (Caprio et al. 1999). Studies in rodents showed that leptin concentrations at par with obese men directly inhibited the conversion of 17OH-progesterone to testosterone (Caprio et al. 1999; Isidori et al. 1999). The presence of leptin has been demonstrated in human male spermatocytes in the testes, suggesting that increased levels of leptin might disrupt spermatogenesis (Ishikawa et al. 2007). Another study showed that intratesticular delivery of TNF-α reduced the human chorionic gonadotropin (hCG) and stimulated steroidogenic acute regulatory protein expression and testosterone biosynthesis in rats (Morales et al. 2003). Leptin has also been shown to augment the secretion of gonadotropin hormones, thereby impairing testicular activity (Hausman et al. 2012). It has also been reported that persistent insulin resistance condition induced by pro-inflammatory cytokines (TNF-α and IL-6) affects the HPG axis, causing secondary hypogonadism in males (Bhasin et al. 2010).

Interestingly, a recent study has shown that obese male seminal vesicle fluid has increased levels of both leptin and insulin (Leisegang et al. 2014). Since both leptin and insulin have their receptors on spermatozoa (Aquila et al. 2005), it was hypothesized that after ejaculation during the movement through the female reproductive tract, altered levels of these hormones in obese males may affect sperm functions (Binder et al. 2015).

11.11 Impact of Childhood Obesity on Puberty

In the recent past, the incidence of childhood obesity has increased tremendously (Kiess et al. 2015). One of the important concerns for these obese children is the effect of obesity on pubertal development, i.e., whether pubarche will be at a faster pace. However, this is a point of debate, mainly because of diverse views generated by several studies and due to scarcity of data. A study from Belgium pointed out that in boys, adult median height and weight have increased by 1.2 cm and 0.9 kg per decade; however, the timing of puberty has not advanced as evident from pubertal onset and enlargement of scrotum and testes in boys (Roelants et al. 2009). Conversely, other studies from Denmark have shown that while body mass index is inversely proportional to the age at puberty and that there is a general trend of attaining puberty at early age, but the trend cannot be attributed exclusively to BMI (Juul et al. 2006; Aksglaede et al. 2009). Another report showed that short stature children had delayed puberty, portraying an inverse relationship between weight status and pubertal development and suggesting that over nutrition may accelerate development in boys (Juul et al. 2007). Yet another study showed that boys with a higher childhood BMI attained puberty earlier, and the childhood BMI correlated positively with adult adiposity (Nathan et al. 2006).

In a very recent study, it was shown that rapid infancy weight gain had a strong correlation with increased risk of childhood obesity. These individuals had increased insulin-like growth factor I and adrenal androgen levels, upregulated aromatase activity, and decreased sex hormone-binding globulin levels, which in turn increased free serum steroid levels promoting the activity of the GnRH pulse generator. In addition, obese children have been reported to have higher leptin level, which triggers LH pulsatility and early onset of puberty (Pintana et al. 2015). A recent study demonstrated that early pubarche predicts a central fat mass distribution, while a predominantly subcutaneous obese phenotype is strongly predicted by a high prepubertal body mass index (Kindblom et al. 2006).

11.12 Effect of Maternal Obesity on Fetal Health

Maternal obesity not only increases the risk of complications during pregnancy, including preeclampsia and gestational diabetes mellitus (Bautista-Castaño et al. 2013; reviewed in Guelinckx et al. 2008) but also can affect child's health later in life. High birth weight of fetuses has been hypothesized to have several complications later in life for the young one (Barker 2001). According to the Barker hypothesis, a link has been established between maternal obesity during the first trimester and obesity in the offspring. Maternal obesity has been associated with increased risk of neural tube defects in offspring. Further, maternal obesity has also been shown to be closely associated with semen abnormalities in male offspring by affecting the testicular development during the fetal life in utero (Teerds et al. 2011). Animal studies have also shown that paternal obesity might cause the offspring to be more susceptible to obesity through epigenetic modifications (Ozanne 2015). Thus, evidence suggest that obesity at adult or prepuberty and maternal obesity before birth all affect fertility, often negatively.

11.13 Obesity and Quality of Sexual Life

Sexual health had been described by the WHO as "a state of physical, emotional, mental, and social well-being in relation to sexuality." Clinical syndromes such as sexual aversion, dysfunctional sexual arousal, erectile dysfunction, and premature ejaculation in males are the major signs of sexual dysfunction in males as marked by the WHO. Earlier studies have highlighted that both obese men and women have more problems in their sexual life in comparison to their lean counterparts (Marchesini et al. 2002; Esposito and Giugliano 2005). Altered adipokine profile, obstructive sleep apnea syndrome, physical disability, and social and psychosocial problems in obese males may explain the association between obesity and sexual dysfunction (Poggiogalle et al. 2014). However, obesity associated with lack of enjoyment of sexual activity and sexual desire and difficulties with sexual performance is common in obese individuals (Kolotkin et al. 2006).

11.14 Animal Studies on Obesity and Fertility

To understand and delineate the molecular mechanisms and pathways relating obesity with fertility, the best approach had been to create animal models that mimic the obese condition of humans. Animal models of obesity include loss of function mutation-based and diet-induced models. Recent studies have demonstrated that obese animal models have frequently displayed reproductive problems as observed in humans. Loss of function mutations in obesity (ob) gene in mice display similar effects as observed in humans, such as low sex steroid and gonadotropin levels. In ob/ob mice testes, multinucleated spermatids, few spermatozoa, and abnormal Leydig cells were observed (Bhat et al. 2006). An earlier study has suggested that treatment with leptin in ob/ob mice significantly reduced the neuropeptide Y (NPY) mRNA expression and normalized reproductive functions (Lutz and Woods 2012). Treatment with leptin in ob/ob mice significantly decreased the body weight, increased the weight of testis and seminal vesicle, and normalized plasma LH, Leydig cell morphology, and spermatogenesis (Stephens et al. 1995). In another interesting study on the molecular effects of obesity, Palmer et al. (2012) undertook a study on SIRT6 expression in mice fed with normal and high-fat diet for 16 weeks. The authors observed that SIRT6 protein was localized to the nucleus of transitional spermatics and the acrosome of mature spermatozoan with the levels significantly reduced in high-fat diet-fed male mice. Further, this study showed that decrease in SIRT6 level in sperm was mainly due to altered acetylation status of the H3K9 in the nucleus (Palmer et al. 2012).

Just like the loss of function mutations in the leptin receptor, the Zucker rats also show many similarities after leptin receptor mutations with that of db/db mouse (Wang et al. 2014). These mutations cause severe hyperphagia, early onset of obesity, insulin resistance, and infertility as observed in humans. Male Zucker rats have been shown to have increased sperm DNA damage (Vendramini et al. 2013). Along with rodent models to study obesity, there are some wild animals showing seasonal adiposity during winter, which also coincides with suppressed reproductive activity during this period. A study in male *Scotophilus heathi* showed that increase in circulating leptin level during winter decreases testicular activity by inhibiting testicular steroidogenesis (Roy and Krishna 2010). In this animal model, there is a period of decreased spermatogenesis, which coincides with peak body mass due to increased accumulation of white adipose tissue (WAT) (Roy and Krishna 2010). Similar seasonal adiposity has been investigated in the Siberian or Djungarian hamster (*Phodopus sungorus*), Syrian or golden hamster (*Mesocricetus auratus*), collared lemming (*Dicrostonyx groenlandicus*) (Bartness and Goldman 2002) and in various species of vole (Dark and Zucker 1984; Peacock et al. 2004).

11.15 Management of Obesity-Related Infertility

Obesity affects male fertility negatively through adipokines. Obese men also have high levels of estrogen, and inhibitors of aromatase are in use to lower the estrogen levels, which increase peripheral testosterone levels, spermatogenesis, and fertility. Another alternative might be to manage obesity by nontherapeutic methods, which include lifestyle modifications as the prominent change. There are clusters of options to manage obesity and metabolic disorders; however, the most effective would be having a drastic change in lifestyle, weight loss, eating habits incorporating a regular healthy diet, and adequate physical activity. Daily physical activity of moderate intensity for 30–45 min duration at least four times a week is considered to be a healthy practice. Daily walking for a moderate period is also a good way to avoid overweight and is considered as an adaptation for cardiorespiratory fitness (Poirier and Després 2001). It was reported that obese children and adolescents randomly assigned to a 6-month combined exercise (aerobic and resistance training) and Mediterranean diet program showed improved body parameters (Lisón et al. 2012). In another such trial, a 56-week treatment with liraglutide as an adjunct to diet and exercise to obese individuals showed reduced body weight and improved metabolic control (Pi-Sunyer et al. 2015). Other effective measures used along with low-calorie intake are fat absorption blockers, which inhibit the gastric and pancreatic lipases (Curran and Scott 2004).

Noninvasive methods have also been adopted; for example, metformin has been widely used against hyperglycemia to lower hepatic glucose through partial AMPK activation (Zhou et al. 2001). Although some drugs might have side effects, they are prescribed, such as thiazolidinediones that are PPAR-α agonists, which have an effect on type II diabetes as they enhance insulin sensitivity by modulating adipose tissue (Edgerton et al. 2009). Recently, GLP-1 analogs which also stimulate insulin secretion are used, and DPP4 inhibitors that help to prolong GLP-1 action are prescribed without many side effects. Sibutramine (Meridia), which works by inhibiting noradrenergic and serotonergic reuptake in the hypothalamus, is prescribed for the treatment of obesity in spite of the fact that it has serious side effects.

Invasive surgeries may be the option for severely obese individuals who cannot resort to physical workout due to very high weight, but are willing to adopt a good lifestyle once mobile. Bariatric surgeries including gastroplasty, gastric bypass, biliopancreatic diversion, etc. are performed to reduce weight, which is also an effective way to control glucose and insulin resistance. However, there are reports that weight loss through surgeries are associated with a death rate of 0.3% and may start a series of serious complications in 4.1% patients (Lim et al. 2010).

Conclusion and Future Directions

Obesity and fat accumulation are important risk factors for the development of type 2 diabetes, hypertension, cardiovascular disease, and infertility. Obesity brings a state of poor overall health, which disturbs glucose metabolism, physical fitness, and the quality of sexual life. In addition to these indirect effects on male fertility, obesity also affects spermatogenesis and fertility by a number of means discussed above. Obesity is now established as a risk factor for loss of spermatogenesis and male infertility. Weight once gained is difficult to loose; therefore, avoiding weight gain is the best method of prevention against the ill effects of obesity. The molecular mechanisms by which obesity contributes to male infertility are being unraveled, which may contribute to the development of better therapeutics in severely obese individuals. Targeting miRNA and the small noncoding RNAs might open up new arenas for obesity therapeutics. For example, initially brown adipose tissue (BAT) was thought to be present only in infants; now there is increasing evidence that these are also present in adult human (Saito et al. 2009), and these could be potential target for obesity management. BAT activation releases excess of energy in the form of heat. A better understanding of adipogenesis along with the associated adipokine profile is a prerequisite for managing obesity-associated infertility issues as these fat released hormones form a connection between metabolism and reproduction.

References

Akishita M, Fukai S, Hashimoto M, Kameyama Y, Nomura K, Nakamura T, Ogawa S, Iijima K, Eto M, Ouchi Y (2010) Association of low testosterone with metabolic syndrome and its components in middle-aged Japanese men. Hypertens Res 33(6):587–591

Aksglaede L, Juul A, Olsen LW, Sørensen TI (2009) Age at puberty and the emerging obesity epidemic. PLoS One 4(12):e8450

Aquila S, Gentile M, Middea E, Catalano S, Andò S (2005) Autocrine regulation of insulin secretion in human ejaculated spermatozoa. Endocrinology 146(2):552–557

Aronne LJ, Nelinson DS, Lillo JL (2009) Obesity as a disease state: a new paradigm for diagnosis and treatment. Clin Cornerstone 9(4):9–29

Bartness TJ, Demas GE, Song CK (2002). Seasonal changes in adiposity: the roles of the photoperiod, melatonin and other hormones, and sympathetic nervous system. Experimental Biology and Medicine, 227(6):363–376.

Bautista-Castaño I, Henriquez-Sanchez P, Alemán-Perez N, Garcia-Salvador JJ, Gonzalez-Quesada A, García-Hernández JA, Serra-Majem L (2013) Maternal obesity in early pregnancy and risk of adverse outcomes. PLoS One 8:e80410

Barker DJ, Forsén T, Uutela A, Osmond C, Eriksson JG (2001). Size at birth and resilience to effects of poor living conditions in adult life: longitudinal study. Bmj, 323(7324):1273

Benchaib M, Braun V, Lornage J, Hadj S, Salle B, Lejeune H, Guérin JF (2003) Sperm DNA fragmentation decreases the pregnancy rate in an assisted reproductive technique. Hum Reprod 18(5):1023–1028

Bhasin S, Cunningham GR, Hayes FJ, Matsumoto AM, Snyder PJ, Swerdloff RS, Montori VM (2010) Testosterone therapy in men with androgen deficiency syndromes: an Endocrine Society clinical practice guideline. J Clin Endocrinol Metab 95(6):2536–2559

Bhat GK, Sea TL, Olatinwo MO, Simorangkir D, Ford GD, Ford BD, Mann DR (2006) Influence of a leptin deficiency on testicular morphology, germ cell apoptosis, and expression levels of apoptosis-related genes in the mouse. J Androl 27(2):302–310

Binder NK, Sheedy JR, Hannan NJ, Gardner DK (2015) Male obesity is associated with changed spermatozoa Cox4i1 mRNA level and altered seminal vesicle fluid composition in a mouse model. Mol Hum Reprod 21(5):424–434

Bourgeron T (2000) Mitochondrial function and male infertility. In: McElreavey K (ed) The genetic basis of male infertility. Springer, Berlin, pp 187–210

Bronson FH (1989) Mammalian reproductive biology. University of Chicago Press, Chicago

Brownell KD, Jeffery RW (1987) Improving long-term weight loss: pushing the limits of treatment. Behav Ther 18(4):353–374

Cabler S, Agarwal A, Flint M, Du Plessis SS (2010) Obesity: modern man's fertility nemesis. Asian J Androl 12(4):480–489

Caprio M, Isidori AM, Carta AR, Moretti C, Dufau ML, Fabbri A (1999) Expression of functional leptin receptors in rodent Leydig cells 1. Endocrinology 140(11):4939–4947

Carone BR, Fauquier L, Habib N, Shea JM, Hart CE, Li R, Bock C, Li C, Gu H, Zamore PD Meissner A (2010). Paternally induced transgenerational environmental reprogramming of metabolic gene expression in mammals. Cell, 143(7):1084–1096.

Chavarro JE, Toth TL, Wright DL, Meeker JD, Hauser R (2010) Body mass index in relation to semen quality, sperm DNA integrity, and serum reproductive hormone levels among men attending an infertility clinic. Fertil Steril *93*(7):2222–2231

Clarke IJ, Horton RJE, Doughton BW (1990) Investigation of the mechanism by which insulin-induced hypoglycemia decreases luteinizing hormone secretion in ovariectomized ewes. Endocrinology 127(3):1470–1476

Coelho M, Oliveira T, Fernandes R (2013) Biochemistry of adipose tissue: an endocrine organ. Arch Med Sci 9(2):191–200

Corona G, Mannucci E, Forti G, Maggi M (2009) Hypogonadism, ED, metabolic syndrome and obesity: a pathological link supporting cardiovascular diseases. Int J Androl 32(6): 587–598

Corona G, Monami M, Rastrelli G, Aversa A, Sforza A, Lenzi A, Forti G, Mannucci E, Maggi M (2011) Type 2 diabetes mellitus and testosterone: a meta-analysis study. Int J Androl 34(6 pt 1):528–540

Corona G, Rastrelli G, Filippi S, Vignozzi L, Mannucci E, Maggi M (2014) Erectile dysfunction and central obesity: an Italian perspective. Asian J Androl 16(4):581

Curran MP and Scott LJ (2004) Orlistat. Drugs, 64(24):2845–2864.

Dadoune JP (2009) Spermatozoal RNAs: what about their functions? Microsc Res Tech 72(8):536–551

Dark J, Zucker I (1984) Gonadal and photoperiodic control of seasonal body weight changes in male voles. Am J Phys Regul Integr Comp Phys 247(1):R84–R88

De Boer H, Verschoor L, Ruinemans-Koerts J, Jansen M (2005) Letrozole normalizes serum testosterone in severely obese men with hypogonadotropic hypogonadism. Diabetes Obes Metab 7(3):211–215

Deedwania P, Gupta R (2006) Management issues in the metabolic syndrome. JAPI 54:797–810

Derby CA, Mohr BA, Goldstein I, Feldman HA, Johannes CB, McKinlay JB (2000) Modifiable risk factors and erectile dysfunction: can lifestyle changes modify risk? Urology 56(2):302–306

Edgerton DS, Ramnanan CJ, Grueter CA, Johnson KM, Lautz M, Neal DW, Williams PE, Cherrington AD (2009) Effects of insulin on the metabolic control of hepatic gluconeogenesis in vivo. Diabetes 58(12):2766–2775

Esposito K, Giugliano D (2005) Obesity, the metabolic syndrome, and sexual dysfunction. Int J Impot Res 17(5):391–398

Esposito K, Giugliano F, Di Palo C, Giugliano G, Marfella R, D'Andrea F, D'Armiento M, Giugliano D (2004) Effect of lifestyle changes on erectile dysfunction in obese men: a randomized controlled trial. JAMA 291(24):2978–2984

Garolla A, Torino M, Miola P, Caretta N, Pizzol D, Menegazzo M, Bertoldo A, Foresta C (2015) Twenty-four-hour monitoring of scrotal temperature in obese men and men with a varicocele as a mirror of spermatogenic function. Hum Reprod 30(5):1006–1013

Grundy SM, Brewer HB, Cleeman JI, Smith SC, Lenfant C (2004) Definition of metabolic syndrome report of the National Heart, Lung, and Blood Institute/American Heart Association conference on scientific issues related to definition. Circulation 109(3):433–438

Guelinckx I, Devlieger R, Beckers K, Vansant G (2008) Maternal obesity: pregnancy complications, gestational weight gain and nutrition. Obes Rev 9(2):140–150

Haffner SM, Valdez RA, Hazuda HP, Mitchell BD, Morales PA, Stern MP (1992) Prospective analysis of the insulin-resistance syndrome (syndrome X). Diabetes 41(6):715–722

Hammoud AO, Meikle AW, Reis LO, Gibson M, Peterson CM, Carrell DT (2012) Obesity and male infertility: a practical approach. Semin Reprod Med 30(6):486–495

Hausman GJ, Barb CR, Lents CA (2012) Leptin and reproductive function. Biochimie 94(10):2075–2081

He L, Hannon GJ (2004) MicroRNAs: small RNAs with a big role in gene regulation. Nat Rev Genet 5(7):522–531

Hilton C, Neville MJ, Karpe F (2013) MicroRNAs in adipose tissue: their role in adipogenesis and obesity. Int J Obes (Lond) 37(3):325–332

Hotamisligil GS, Shargill NS, Spiegelman BM (1993) Adipose expression of tumor necrosis factor- alpha: direct role in obesity-linked insulin resistance. Science 259:87–91

Hutley L, Prins JB (2005) Fat as an endocrine organ: relationship to the metabolic syndrome. Am J Med Sci *330*(6):280–289

Ishikawa T, Fujioka H, Ishimura T, Takenaka A, Fujisawa M (2007) Expression of leptin and leptin receptor in the testis of fertile and infertile patients. Andrologia 39(1):22–27

Isidori AM, Caprio M, Strollo F, Moretti C, Frajese G, Isidori A, Fabbri A (1999) Leptin and androgens in male obesity: evidence for leptin contribution to reduced androgen levels 1. J Clin Endocrinol Metab 84(10):3673–3680

Ivell R (2007) Lifestyle impact and the biology of the human scrotum. Reprod Biol Endocrinol 5(1):1

Jenkins TG, Carrell DT (2012) The sperm epigenome and potential implications for the developing embryo. Reproduction 143(6):727–734
Judd ET, Wessels FJ, Drewry MD, Grove M, Wright K, Hahn DA, Hatle JD (2011) Ovariectomy in grasshoppers increases somatic storage, but proportional allocation of ingested nutrients to somatic tissues is unchanged. Aging Cell 10(6):972–979
Juul A, Teilmann G, Scheike T, Hertel NT, Holm K, Laursen EM, Main KM, Skakkebaek NE (2006) Pubertal development in Danish children: comparison of recent European and US data. Int J Androl 29(1):247–255
Juul A, Magnusdottir S, Scheike T, Prytz S, Skakkebæk NE (2007) Age at voice break in Danish boys: effects of pre-pubertal body mass index and secular trend. Int J Androl 30(6):537–542
Kasturi SS, Tannir J, Brannigan RE (2008). The metabolic syndrome and male infertility. Journal of andrology, 29(3):251–259.
Katib A (2015) Mechanisms linking obesity to male infertility. Cent European J Urol 68(1):79–85
Kiess W, Wagner IV, Kratzsch J, Körner A (2015) Male obesity. Endocrinol Metab Clin North Am 44(4):761–772
Kindblom JM, Lorentzon M, Norjavaara E, Lönn L, Brandberg J, Angelhed JE, Hellqvist Å, Nilsson S, Ohlsson C (2006) Pubertal timing is an independent predictor of central adiposity in young adult males the Gothenburg osteoporosis and obesity determinants study. Diabetes 55(11):3047–3052
Kley HK, Edelmann P, Krüskemper HL (1980) Relationship of plasma sex hormones to different parameters of obesity in male subjects. Metabolism 29(11):1041–1045
Kolotkin RL, Binks M, Crosby RD, Østbye T, Gress RE, Adams TD (2006) Obesity and sexual quality of life. Obesity 14(3):472–479
Kort HI, Massey JB, Elsner CW, Mitchell-Leef D, Shapiro DB, Witt MA, Roudebush WE (2006) Impact of body mass index values on sperm quantity and quality. J Androl 27(3):450–452
Król E, Speakman JR (2007) Regulation of body mass and adiposity in the field vole, *Microtus agrestis*: a model of leptin resistance. J Endocrinol 192(2):271–278
Kupelian V, Page ST, Araujo AB, Travison TG, Bremner WJ, McKinlay JB (2006) Low sex hormone-binding globulin, total testosterone, and symptomatic androgen deficiency are associated with development of the metabolic syndrome in nonobese men. J Clin Endocrinol Metab 91(3):843–850
Laing I, Olukoga AO, Gordon C, Boulton AJM (1998) Serum sex-hormone-binding globulin is related to hepatic and peripheral insulin sensitivity but not to beta-cell function in men and women with type 2 diabetes mellitus. Diabet Med 15(6):473–479
Lalancette C, Miller D, Li Y, Krawetz SA (2008) Paternal contributions: new functional insights for spermatozoal RNA. J Cell Biochem 104(5):1570–1579
Lane M, Owens JA, Teague EMO, Palmer NO, Fullston T (2012) You are what your father ate—paternal programming of reproductive and metabolic health in offspring. Biol Reprod 87(Suppl 1):95–95
Laughlin SB, van Steveninck RRDR, Anderson JC (1998) The metabolic cost of neural information. Nat Neurosci 1(1):36–41
Leisegang K, Bouic PJ, Menkveld R, Henkel RR 2014 Obesity is associated with increased seminal insulin and leptin alongside reduced fertility parameters in a controlled male cohort. Reprod Biol Endocrinol 12(1):p.1
Lim S, Kim JH, Yoon JW, Kang SM, Choi SH, Park YJ, Kim KW, Lim JY, Park KS, Jang HC (2010) Sarcopenic obesity: prevalence and association with metabolic syndrome in the Korean longitudinal study on health and aging (KLoSHA). Diabetes Care 33(7):1652–1654
Lisón JF, Real-Montes JM, Torró I, Arguisuelas MD, Álvarez-Pitti J, Martínez-Gramage J, Aguilar F, Lurbe E (2012) Exercise intervention in childhood obesity: a randomized controlled trial comparing hospital-versus home-based groups. Acad Pediatr 12(4):319–325
Lutz TA, Woods SC (2012) Overview of animal models of obesity. Curr Protoc Pharmacol 58:5–61
Maeda K, Okubo K, Shimomura I, Funahashi T, Matsuzawa Y, Matsubara K (1996) cDNA cloning and expression of a novel adipose specific collagen-like factor, apM1 (AdiPoseMost abundant Gene transcript 1). Biochem Biophys Res Commun 221(2):286–289

Magnusdottir EV, Thorsteinsson T, Thorsteinsdottir S, Heimisdottir M, Olafsdottir K (2005) Persistent organochlorines, sedentary occupation, obesity and human male subfertility. Hum Reprod 20(1):208–215

Marchesini G, Natale S, Tiraferri F, Tartaglia A, Moscatiello S, Marchesini RL, Villanova N, Forlani G, Melchionda N (2002) The burden of obesity on everyday life: a role for osteoarticular and respiratory diseases. Diabetes Nutr Metab 16(5–6):284–290

McElroy JF, Wade GN (1987) Short-and long-term effects of ovariectomy on food intake, body weight, carcass composition, and brown adipose tissue in rats. Physiol Behav 39(3):361–365

McPherson NO, Owens JA, Fullston T, Lane M (2015) Preconception diet or exercise intervention in obese fathers normalizes sperm microRNA profile and metabolic syndrome in female offspring. Am J Physiol Endocrinol Metab 308(9):E805–E821

Morales V, Santana P, Díaz R, Tabraue C, Gallardo G, Blanco FL, Hernández I, Fanjul LF, Ruiz de Galarreta CM (2003) Intratesticular delivery of tumor necrosis factor-α and ceramide directly abrogates steroidogenic acute regulatory protein expression and Leydig cell steroidogenesis in adult rats. Endocrinology 144(11):4763–4772

Muneer A, Kalsi J, Nazareth I, Arya M (2014) Erectile dysfunction. BMJ 348:g129

Nathan BM, Sedlmeyer IL, Palmert MR (2006) Impact of body mass index on growth in boys with delayed puberty. J Pediatr Endocrinol Metab 19(8):971–978

Ng SF, Lin RC, Laybutt DR, Barres R, Owens JA, Morris MJ (2010) Chronic high-fat diet in fathers programs [bgr]-cell dysfunction in female rat offspring. Nature 467(7318):963–966

Oger F, Gheeraert C, Mogilenko D, Benomar Y, Molendi-Coste O, Bouchaert E, Caron S, Dombrowicz D, Pattou F, Duez H, Eeckhoute J (2014) Cell-specific dysregulation of microRNA expression in obese white adipose tissue. J Clin Endocrinol Metab 99(8):2821–2833

Ortega FJ, Mercader JM, Catalán V, Moreno-Navarrete JM, Pueyo N, Sabater M, Gómez-Ambrosi J, Anglada R, Fernández-Formoso JA, Ricart W, Frühbeck G (2013) Targeting the circulating microRNA signature of obesity. Clin Chem 59(5):781–792

Ozanne SE (2015) Epigenetic signatures of obesity. N Engl J Med 372(10):973–974

Padron OF, Brackett NL, Sharma RK, Lynne CM, Thomas AJ, Agarwal A (1997) Seminal reactive oxygen species and sperm motility and morphology in men with spinal cord injury. Fertil Steril 67(6):1115–1120

Palmer NO, Bakos HW, Fullston T, Lane M (2012) Impact of obesity on male fertility, sperm function and molecular composition. Spermatogenesis 2(4):253–263

Pasquali R, Patton L, Gambineri A (2007) Obesity and infertility. Curr Opin Endocrinol Diabetes Obes 14(6):482–487

Pauli EM, Legro RS, Demers LM, Kunselman AR, Dodson WC, Lee PA (2008). Diminished paternity and gonadal function with increasing obesity in men. Fertility and sterility, 90(2):346–351.

Peacock WL, Król E, Moar KM, McLaren JS, Mercer JG, Speakman JR (2004) Photoperiodic effects on body mass, energy balance and hypothalamic gene expression in the bank vole. J Exp Biol 207(1):165–177

Pintana H, Chattipakorn N, Chattipakorn S (2015) Testosterone deficiency, insulin-resistant obesity and cognitive function. Metab Brain Dis 30(4):853–876

Pi-Sunyer X, Astrup A, Fujioka K, Greenway F, Halpern A, Krempf M, Lau DC, le Roux CW, Violante Ortiz R, Jensen CB, Wilding JP (2015) A randomized, controlled trial of 3.0 mg of liraglutide in weight management. N Engl J Med 373(1):11–22

Poggiogalle E, Di Lazzaro L, Pinto A, Migliaccio S, Lenzi A, Donini LM (2014) Health-related quality of life and quality of sexual life in obese subjects. Int J Endocrinol 2014:847871

Poirier P, Després JP (2001) Exercise in weight management of obesity. Cardiol Clin 19(3):459–470

Ramaswamy S, Marshall GR, McNeilly AS, Plant TM (2000) Dynamics of the follicle-stimulating hormone (FSH)-inhibin B feedback loop and its role in regulating spermatogenesis in the adult male rhesus monkey (*Macaca mulatta*) as revealed by unilateral orchidectomy 1. Endocrinology 141(1):18–27

Ramirez-Torres MA, Carrera A, Zambrana M (2000) High incidence of hyperestrogenemia and dyslipidemia in a group of infertile men. Ginecol Obstet Mex 68:224–229

Roelants M, Hauspie R, Hoppenbrouwers K (2009) References for growth and pubertal development from birth to 21 years in Flanders, Belgium. Ann Hum Biol 36(6):680–694

Rosen ED, Spiegelman BM (2006) Adipocytes as regulators of energy balance and glucose homeostasis. Nature 444(7121):847–853

Roth MY, Amory JK, Page ST (2008) Treatment of male infertility secondary to morbid obesity. Nat Clin Pract Endocrinol Metab 4(7):415–419

Rotter V, Nagaev I, Smith U (2003) Interleukin-6 (IL-6) induces insulin resistance in 3 T3-L1 adipocytes and is, like IL-8 and tumor necrosis factor-α, overexpressed in human fat cells from insulin-resistant subjects. J Biol Chem 278(46):45777–45784

Roy VK, Krishna A (2010) Role of leptin in seasonal adiposity associated changes in testicular activity of vespertilionid bat, *Scotophilus heathi*. Gen Comp Endocrinol 168(1):160–168

Saito M, Okamatsu-Ogura Y, Matsushita M, Watanabe K, Yoneshiro T, Nio-Kobayashi J, Iwanaga T, Miyagawa M, Kameya T, Nakada K and Kawai Y (2009). High incidence of metabolically active brown adipose tissue in healthy adult humans. Diabetes, 58(7):1526–1531.

Schneider G, Kirschner MA, Berkowitz R, Ertel NH (1979) Increased estrogen production in obese men. J Clin Endocrinol Metab 48(4):633–638

Skogen JC, Øverland S (2012) The fetal origins of adult disease: a narrative review of the epidemiological literature. JRSM Short Rep 3(8):59

Skrypnik D, Bogdański P, Musialik K (2014) Obesity—significant risk factor for erectile dysfunction in men. Pol Merkur Lekarski 36(212):137–141

Stephens TW, Basinski M, Bristow PK, Bue-Valleskey JM, Burgett SG, Craft L, Hale J, Hoffmann J, Hsiung HM, Kriauciunas A, MacKellar W (1995) The role of neuropeptide Y in the antiobesity action of the obese gene product. Nature 377(6549):530–532

Teerds KJ, De Rooij DG, Keijer J (2011) Functional relationship between obesity and male reproduction: from humans to animal models. Hum Reprod Update 17(5):667–683

Tng EL (2015) Kisspeptin signalling and its roles in humans. Singapore Med J *56*(12):649

Turner RM (2003) Tales from the tail: what do we really know about sperm motility? J Androl 24(6):790–803

Vendramini V, Cedenho AP, Miraglia SM, Spaine DM (2013) Reproductive function of the male obese Zucker rats alteration in sperm production and sperm DNA damage. Reprod Sci 21(2):221–229

Vermeulen A, Kaufman JM, Deslypere JP, Thomas G (1993) Attenuated luteinizing hormone (LH) pulse amplitude but normal LH pulse frequency, and its relation to plasma androgens in hypogonadism of obese men. J Clin Endocrinol Metab 76(5):1140–1146

Vermeulen A, Kaufman JM, Goemaere S, Van Pottelberg I (2002) Estradiol in elderly men. Aging Male 5(2):98–102

Wang C, Jackson G, Jones TH, Matsumoto AM, Nehra A, Perelman MA, Swerdloff RS, Traish A, Zitzmann M, Cunningham G (2011) Low testosterone associated with obesity and the metabolic syndrome contributes to sexual dysfunction and cardiovascular disease risk in men with type 2 diabetes. Diabetes Care 34(7):1669–1675

Wang B, Charukcshi Chandrasekera P, Pippin J (2014) Leptin-and leptin receptor-deficient rodent models: relevance for human type 2 diabetes. Curr Diabetes Rev 10(2):131–145

Watkins ML, Rasmussen SA, Honein MA, Botto LD, Moore CA (2003) Maternal obesity and risk for birth defects. Pediatrics 111(Supplement 1):1152–1158

Wise LA, Cramer DW, Hornstein MD, Ashby RK, Missmer SA (2011) Physical activity and semen quality among men attending an infertility clinic. Fertil Steril 95(3):1025–1030

Youngson NA, Whitelaw E (2011) The effects of acquired paternal obesity on the next generation. Asian J Androl 13(2):195–196

Zhou G, Myers R, Li Y, Chen Y, Shen X, Fenyk-Melody J, Wu M, Ventre J, Doebber T, Fujii N, Musi N (2001) Role of AMP-activated protein kinase in mechanism of metformin action. J Clin Invest 108(8):1167–1174

Zumoff B, Strain GW, Miller LK, Rosner W, Senie R, Seres DS, Rosenfeld RS (1990) Plasma free and non-sex-hormone-binding-globulin bound testosterone are decreased in obese men in proportion to their degree of obesity. J Clin Endocrinol Metab 71(4):929–931

Sexually Transmitted Infections and Male Infertility: Old Enigma, New Insights

12

Bhavana Kushwaha and Gopal Gupta

Abstract

Sexually transmitted diseases (STDs) are caused by bacteria, viruses, or parasites that are transmitted through venereal contact. The major STDs causing bacterial infections include *Chlamydia trachomatis* (chlamydiasis), *Neisseria gonorrhoeae* (gonorrhea), *Treponema pallidum* (syphilis), *Mycoplasma, and Ureaplasma* species. On the other hand, the major viral STD infections include herpes simplex virus (HSV), human papillomavirus (HPV), human immunodeficiency virus (HIV), human cytomegalovirus (HCV), and hepatitis B and C viruses. Similarly, the major parasite infecting the genital tract is the protozoan *Trichomonas vaginalis*, which causes trichomoniasis. In males, these STDs may either be asymptomatic or cause urethritis, epididymitis, orchitis, vasiculitis, and prostatitis. Most of these infections have been shown to affect male fertility by affecting semen parameters like sperm count, motility, and morphology; however, their exact mechanism of action is still not known. The presence of infection(s) on sperm and/or in the seminal plasma causes their horizontal transmission to sexual partners and vertical transmission to offsprings. This chapter briefly reviews some of the published literature on major STDs in relation to male infertility and relevant treatment strategies. (CDRI communication number 9456)

Keywords

Sexually transmitted diseases • Male fertility • Gonorrhea • Chlamydiasis • Syphilis • Human immunodeficiency virus • Human papilloma virus • Herpes simplex virus

B. Kushwaha • G. Gupta (✉)
Male Reproductive Biology Laboratory, CSIR-Central Drug Research Institute, Lucknow, Uttar Pradesh, India
e-mail: g_gupta@cdri.res.in

R. Singh, K. Singh (eds.), *Male Infertility: Understanding, Causes and Treatment*, DOI 10.1007/978-981-10-4017-7_12

Key Points

- Sexually transmitted diseases (STDs) are caused by pathogenic bacteria, viruses, or protozoa that are transmitted by venereal contact.
- The infections trigger inflammatory processes, which may lead to obstruction of the seminal tract and deterioration of spermatogenesis.
- STDs in males result in urethritis, epididymitis, orchitis, vasiculitis, and prostatitis and have been associated with reduced sperm quality, concentration, and motility.
- STDs are often asymptomatic in males and spread horizontally to sexual partners and vertically to offsprings.
- In males, sexually transmitted infections require repetitive screening for the prognosis of disease.

12.1 Introduction

Sexually transmitted diseases (STDs) are caused by bacteria, viruses, or parasites and transmitted through the venereal contact. These microorganisms, which colonize the male and female genital tracts, pose a serious threat to the normal well-being of mankind. Morbidity, mortality, and stigma related to STDs make it a major global health problem (Maan et al. 2014). More than 1 million sexually transmitted infections (STIs) are acquired every day worldwide (Okamura et al. 1986). These infections are often asymptomatic, which holds a greater risk of passing the disease on to others, if left untreated. These infections pose serious threat to one's immediate and long-term reproductive health and well-being, as well as that of one's partner (Marconi et al. 2009).

Male fertility involves a systematic process of germ cell division and maturation aimed at producing competent gametes with normal fertilization potential. Sperm have to pass through an elaborate labyrinth of tubules where they are bathed in fluids of special composition added by the accessory glandular organs for their full functionalization. Several factors secreted by male accessory glands (viz., epididymis, seminal vesicles, prostate, and the bulbourethral glands) such as fructose, ascorbic acid, prostaglandins, polyamines, ergothioneine, L-carnitine, glycerylphosphorylcholine, alpha-glucosidase, bicarbonate, zinc, and citric acid are crucial for normal sperm physiology. The seminal vesicles produce factors that act as reducing agents and prevent sperm agglutination (Schneede et al. 2003; Apari et al. 2014).

Sperm production and delivery involves a delicate interplay between various organs with unobstructed ducts. However, the process is susceptible to various inflammatory and other pathologies caused by infectious instigators. Bacteria, viruses, protozoa, and epizoa include causative organisms of STDs. These pathogens cause acute and chronic diseases. The most common sexually transmitted infections/diseases are chlamydial, mycoplasmal, ureaplasmal infections and syphilis, gonorrhea, hepatitis, genital herpes, human immunodeficiency virus, trichomoniasis, chancroid, lymphogranuloma venereum, and donovanosis (La Vignera et al. 2011).

In males, these infectious instigators colonize particularly the genital region and cause genital injury, prostatitis, urethritis, epididymitis, and orchitis, resulting in fertility impairment due to organ damage, cell and gamete damage, and obstruction (Marconi et al. 2009). These pathogens are very sensitive to physical and chemical factors and do not cause immunity. Therefore, multiple infections with different organisms may occur at the same time (Baird et al. 2007).

12.2 STD Pathogens: Locus of Infection and Resultant Pathology in the Male Urogenital Tract

Male urogenital tract pathologies include urethritis, epididymitis, orchitis, prostatitis, and vesiculitis and are mostly caused by STD/RTI pathogens. Depending on the site of inflammation, these may contribute to fertility impairment of various magnitudes. Chronic infections often result in transient or permanent infertility, impairing hormones, testicular function, and spermatogenesis (Bignell et al. 2011). Testicular damage by orchitis directly hampers sperm production (Weidner et al. 1999). However, impairment of male accessory glands (epididymis, prostate, and seminal vesicles) and urethral infections exert a negative effect on male reproductive function and fertility owing to obstruction or sub-obstruction, altered secretory functions, and release of inflammatory mediators (Bignell et al. 2011).

12.2.1 Urethritis

Urethritis refers to infection-induced inflammation of the urethra, which can be classified into gonococcal urethritis (GU) and non-gonococcal urethritis (NGU). GU is caused by *Neisseria gonorrhoea*, while the two most common organisms implicated in NGU are *Chlamydia trachomatis*, and *Mycoplasma genitalium*. Infections are often asymptomatic as seen in 90–95% of men with gonorrhea (Weidner et al. 1999) and 50% of the patients with chlamydial infections (Cosentino and Cockett 1986). Non-gonococcal urethritis infections have long-term consequences, when left untreated, and include painful infection of the testicles with reduced fertility (Fig. 12.1).

12.2.2 Epididymitis and Orchitis

The epididymis contains high order of vascularization, which not only provides a nourishing environment for sperm maturation but also serves as a fertile ground for bacterial growth. The epididymis collects sperm formed in the testes and undertakes final maturation of the spermatozoa (Trojian et al. 2009). Epididymitis is caused by infectious bacteria such as chlamydia and gonorrhea that reach the epididymis and develop a painful inflammation. This condition may affect some parts or the entire organ depending upon the severity of inflammation (Fig. 12.1).

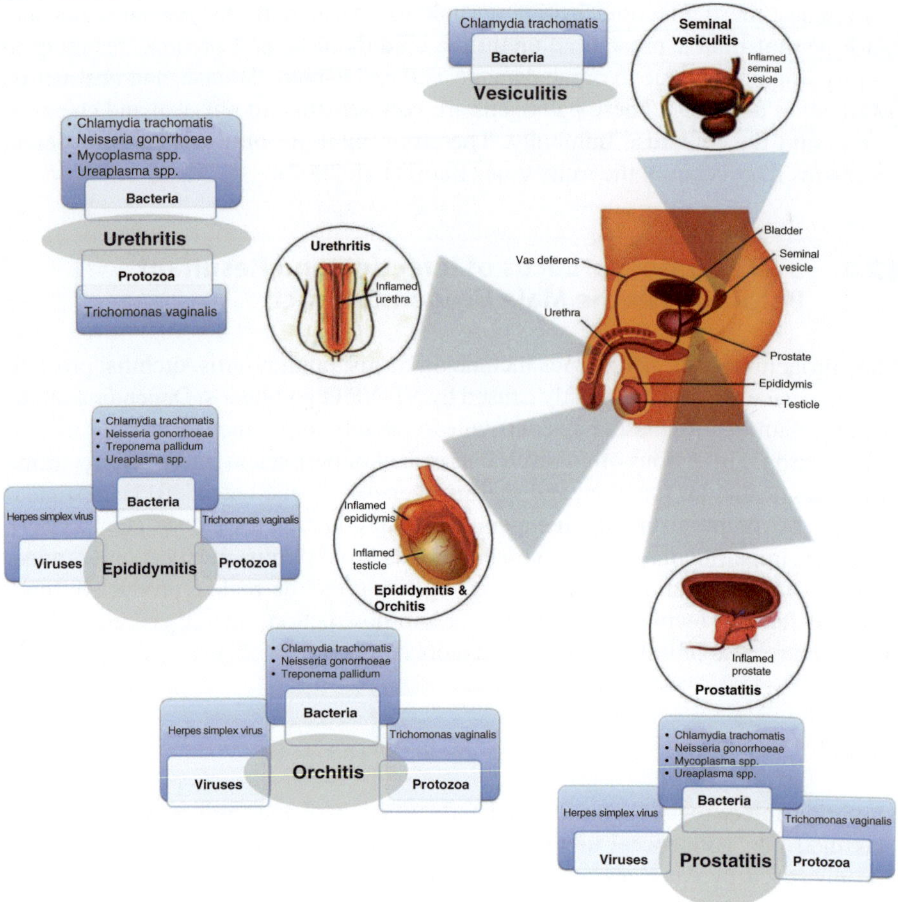

Fig. 12.1 Locus of infection by sexually transmitted infections and the resultant pathology in the male genital tract

As the infection progresses, it may lead to complete obstruction of the duct system. Even after the culmination of symptomatic phase, the scarred areas can still be a source of bacterial infections leading to asymptomatic bacteriospermia. When the inflammation spreads from the epididymis to the adjacent testicle, it develops epididymo-orchitis (Ness et al. 1997), a condition which reduces sperm count and motility, resulting in high rates of infertility (Krieger et al. 1999; Mazzoli et al. 2010).

12.2.3 Prostatitis

Prostatitis refers to an infection or inflammation of the prostate gland. Prostatitis syndromes can be categorized as follows: acute, chronic, non-bacterial prostatitis, chronic pelvic pain syndrome, and asymptomatic inflammatory prostatitis (Lepor et al. 1994). A high prevalence of *C. trachomatis* in chronic prostatitis (39.1%) has been reported (Motrich et al. 2005). It is hypothesized that *C. trachomatis* infection in the prostate gland may cause inflammation and impair the normal functionality of the gland that in turn causes male infertility (Martínez-Prado and CamejoBermúdez 2010) (Fig. 12.1).

However, other studies suggest the role of cytokines in prostatitis and male infertility. Increased levels of IL-1β, IL-6, IL-8, IL-12, IL-18, and TNF-α in semen have been correlated with poor semen parameters, with decreased sperm count and motility, and with increased oxidative stress (Allahbadia 2016; Furuya et al. 2004). Cytokines play an important role in male infertility caused by prostatitis and are released locally from stimulated tissues, serving as important components of the innate host defense against infection (Krishnan and Heal 1991).

12.2.4 Vesiculitis

Seminal vesiculitis refers to the inflammation of the seminal vesicles, commonly a secondary outcome of prostatitis, and sometimes occurs independently. Studies suggest that *C. trachomatis* is found predominantly in the seminal vesicle fluid (Ochsendorf 2008). It has also been shown that after antimicrobial treatment, vesiculitis-associated symptoms disappear with an improvement in the symptoms of epididymitis (Ochsendorf 2008). Some studies propose that chlamydial epididymitis may originate from seminal vesiculitis (Rybar et al. 2012). These findings strongly indicate that seminal vesicles are involved in the urogenital inflammation process (Fig. 12.1).

12.3 Bacterial Infections and Male Infertility

Fertile and infertile men contain several species of bacteria in their genital tract and semen, and their prevalence and relevance to the etiology of male infertility varies according to the geographical location. The major pathogenic bacterial strains that affect male fertility include *Chlamydia trachomatis*, *Neisseria gonorrhoeae*, *Mycoplasma* spp., *Ureaplasma* spp., and *Treponema pallidum* (Gimenes et al. 2014). Furthermore, STI instigators present in semen directly affect and alter semen quality, sperm number, motility, and morphology (Rybar et al. 2012; Isaiah et al. 2011) (Fig. 12.2).

12.3.1 *Neisseria gonorrhoeae*

Gonorrhea is a common sexually transmitted disease caused by a gram-negative diplococcus, *Neisseria gonorrhoeae* (Edwards and Apicella 2004), with a global annual incidence of ~78 million new infections (Newman et al. 2015). The incidence of gonorrhea has escalated over the years, and according to the Center for Disease Control and Prevention (CDC), gonorrhea is the second most commonly

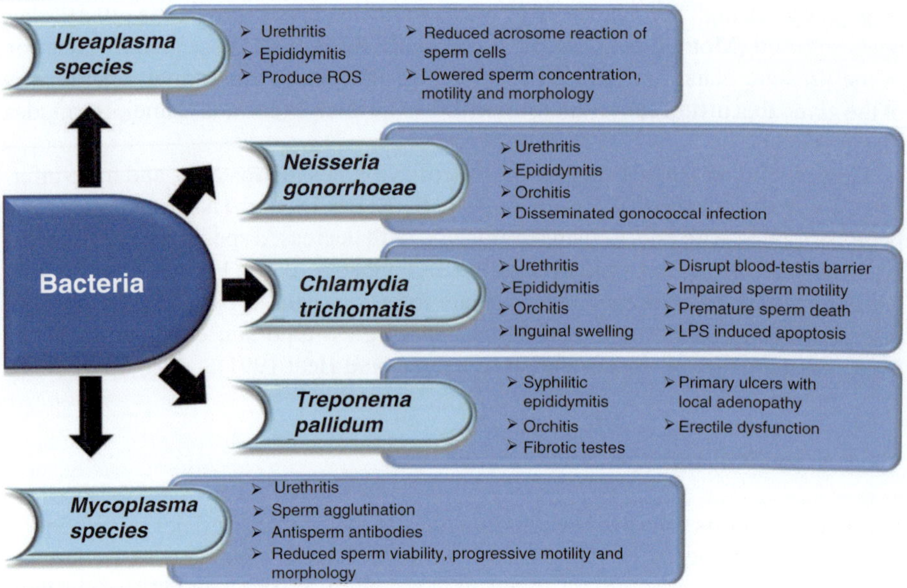

Fig. 12.2 Pathophysiological action of STD bacteria and male infertility

reported STD after chlamydiasis (WHO 2016; CDC 2009). Gonococcal infections in males are manifested by urethritis, orchitis, disseminated gonococcal infections, epididymitis, sexual gland obstruction, and penile discharge (Gimenes et al. 2014). Coinfection of chlamydia is observed in 10% cases with gonorrhoea (Brookings et al. 2013). If left untreated, such infections can cause serious damage, which includes scarring and obliteration of the epididymal canal causing azoospermia in majority of the cases with bilateral involvement (Harkness 1948). The infection is transmitted from asymptomatic carriers rather than the symptomatic patients (Korzeniewski and Juszczak 2015).

Reports have confirmed an adverse effect of *N. gonorrhoeae* infection on fertility (Ness et al. 1997; Gimenes et al. 2014). Despite the global prevalence of the disease, the exact role of *N. gonorrhoeae* in male infertility is still poorly understood. Using polymerase chain reaction, *N. gonorrhoeae* DNA was detected in semen of 6.5% of infertile Jordanian men as compared to 0% in fertile men, highlighting the association of *N. gonorrhoeae* with infertility in men (Abusarah et al. 2013). On the other hand, eradication of gonorrhoea lowered the infertility risk in Swedish men (Akre et al. 1999). Gonococcal interactions with urethral epithelium may initiate cytokine release promoting neutrophil influx and inflammatory response (Edwards et al. 2004), which is a risk factor for male fertility (Henkel et al. 2006). Infections of the male genital tract account for 15% cases of male infertility, and *N. gonorrhoeae* along with *C. trachomatis* form the primary sexually transmitted infections that are involved in male infertility (Pellati et al. 2008).

A range of different virulence factors have been identified in *N. gonorrhoeae*, which enable the bacteria to adapt in the host microenvironment (Edwards and Apicella 2004; Carson et al. 2000). Such adaptations and antigenic shifts allow the bacteria to thrive well in the sole human host and also hinder the development of a vaccine. Several neisserial adhesin proteins (i.e., pilli, Opa, Opc, and P36), additional putative virulence elements, and proteins involved in invasion (i.e., LOS, capsule, PorB) have been identified. Pilus (Dehio et al. 2000), opacity-associated (Opa) outer membrane proteins, and lipooligosaccharide (LOS) undergo phase or antigenic change to avoid the immune cells (Schneider et al. 1988; Danaher et al. 1995). Evidently, the variations in surface as an adaptive mechanism help in immune evasion and survival in the host.

N. gonorrhoeae continues to exhibit elevated resistance to penicillin and tetracycline treatment ever since the 1970s (CDC 2011), hence fluoroquinolones are the recommended therapeutic regimen for uncomplicated gonococcal infections. In 2011, the first resistance of *N. gonorrhoeae* to azithromycin was reported in the United States. Third-generation cephalosporins remain the available class of antibiotics, which can be used effectively against *N. gonorrhoeae* (CDC 2011). However, parasites with abridged sensitivity to oral cephalosporins and cefixime have started to emerge (Wang et al. 2003; Yokoi et al. 2007; Lo et al. 2008). In 2009, first multidrug-resistant gonococcal isolate was identified in Japan (Ohnishi et al. 2011). In order to counter the emerging cephalosporin resistance, the CDC 2015 STD treatment guidelines recommend using a combination of two antibiotics with different mechanisms of action, and accordingly, the recommended treatment for all cases of gonorrhea (uncomplicated urethral, cervical, oropharyngeal, and anal) is ceftriaxone intramuscular 250 mg in a single dose along with a single oral azithromycin dose of 1.0 g.

12.3.2 *Chlamydia trachomatis*

Chlamydia trachomatis is the most common sexually transmitted bacterial infection with global incidence of 131 million new infections annually (Newman et al. 2015), and its prevalence is similar in men and women. It is mostly asymptomatic in humans (Geisler 2010; Taylor and Haggerty 2011). In England, its prevalence in people under 25 years of age was reported to be 10.1% in women and 13.3% in men, and age, sexual behavior, and ethnicity were identified as the primary risk factors (LaMontagne et al. 2004). However, the disease has been reported to have a very high prevalence of 43.3% in asymptomatic infertile men, though its direct correlation with semen parameters could not be established (Gdoura et al. 2008). Other studies similarly reported a high occurrence of chlamydial infection in infertile men (Abusarah et al. 2013; Joki-Korpela et al. 2009; Karinen et al. 2004). The disease may cause acute epididymo-orchitis (Ibrahim et al. 1996), seminal vesiculitis, epididymitis (Furuya et al. 2004), prostatis (Motrich et al. 2006; Ouzounova-Raykova et al. 2010), orchitis, scrotal pain, and even fever (Trojian et al. 2009).

It has been shown that male genital tract inflammation is an important factor for male infertility due to the action of inflammatory mediators. For example, the presence of leukocytes in semen adversely affects sperm motility, viability, and morphology (Henkel et al. 2006). Abnormally high concentrations of leukocytes in semen (pyospermia) caused by *C. trachomatis* infection in infertile men could be treated successfully with antibiotics, which resulted in improved seminal pH, volume, sperm concentration, and motility (Pajovic et al. 2013). The presence of *Chlamydia trachomatis* IgG antibodies in men has been shown to be related with decreased pregnancy rates in asymptomatic infertile couples (Karinen et al. 2004; Joki-Korpela et al. 2009; Idahl et al. 2004), and *C. trachomatis*-infected semen samples have shown poorer sperm morphology, volume, sperm concentration, motility, and velocity (Veznik et al. 2004). Chronic prostatitis and immune-mediated germ cell damage may cause decreased male fertility in males with chlamydial infection (Mazzoli et al. 2010). Though some studies did not find a direct correlation between chlamydial infection and semen parameters (de Barbeyrac et al. 2006), several other studies have demonstrated a direct adverse effect of bacteria on sperm motility, viability, and acrosomal status. Co-incubation of normal human sperm with chlamydia caused decline in percent sperm motility and premature sperm apoptosis (Hosseinzadeh et al. 2001; Satta et al. 2006), caused mainly by the bacterial lipopolysaccharide (Eley et al. 2005; Hosseinzadeh et al. 2003). The elementary bodies of the serovar E infection were most toxic to human sperm (Hosseinzadeh et al. 2001, 2003). Thus *Chlamydia trachomatis*, the most prevalent STI, causes severe inflammation of the male genital tract causing orchitis, epididymitis, vasiculitis, and prostatitis, ensuing innate immune response, which results in increased concentration of leukocytes and antibodies in semen and sperm damage (Fig. 12.4). A direct adverse effect of the presence of *C. trachomatis* infection and sperm parameters has been seen *in vivo* and *in vitro*.

After diagnosis of *Chlamydia trachomatis* infection in men by nucleic acid amplification test (NAAT) of the first catched urine sample or the urethral swab, the treatment recommendations remain essentially the same since 2010: oral azithromycin (1.0 g in a single dose) or 1 week of oral doxycycline (100 mg twice daily). Chlamydia is highly prevalent among adolescents. Azithromycin is also recommended as first line of treatment for *C. trachomatis* infections during pregnancy (Van Vranken 2007). Since most infected persons remain asymptomatic, detection relies on routine screening. The CDC also recommends alternative treatment regimens of erythromycin base 500 mg orally four times a day for 7 days or erythromycin ethylsuccinate 800 mg orally four times a day for 7 days or levofloxacin 500 mg orally once daily for 7 days or ofloxacin 300 mg orally twice a day for 7 days.

12.3.3 *Treponema pallidum*

Syphilis is a chronic sexually transmitted disease, and the etiologic agent is a spirochete, *Treponema pallidum.* Approximately, 6 million new infections are reported annually in the world (Newman et al. 2015). An adverse effect of *Treponema pallidum* on male fertility has not been established or reported, though complications caused by syphilis can affect fertility. In syphilitic epididymitis, obstruction of the epididymis may occur, and in case of chronic obliterative endarteritis, interstitial inflammation can occur in congenital area and lead to small, fibrotic testes (Brookings et al. 2013; Gimenes et al. 2014). Syphilis develops in different stages with varied symptoms at each stage. The primary stage symptoms include sores or chancres on the genitals, rectum, or mouth. The second stage is characterized by rashes on body especially on palms and soles. And the final stage may occur after few years, which includes gummatous syphilis, late neurosyphilis, and cardiovascular syphilis causing severe problems to the brain, nerves, eyes, or heart (CDC 2015). Gummatous lesions are soft and non-cancerous growth, which occurs in tertiary stage of syphilis (Kent and Romanelli 2008). These lesions can cause destruction of the local tissue and, when they are formed in the testicles, may affect the testicular function and fertility. Indirect effects of syphilis can cause erectile dysfunction (Gimenes et al. 2014). The primary mode of syphilis transmission is sexual contact. After *T. pallidum* penetrates through the genital mucosa or abraded skin, it enters the lymphatic and bloodstream and disseminates to various organs including the CNS (Ficarra and Carlos 2009). It has also been seen that syphilis aids in the transmission of HIV.

The parenteral administration of penicillin G (benzathine penicillin G 2.4 million units intramuscular in a single dose) is the preferred treatment for syphilis (all stages); the CDC 2015 STD treatment guidelines recommend doxycycline 100 mg given orally twice a day for 14 days or tetracycline 500 mg given orally four times a day for 14 days to be considered as treatment options in cases of penicillin allergy (Katz et al. 2012). The updated 2015 guidelines warn against the use of single-dose azithromycin due to the emergence of azithromycin-resistant strains of *T. pallidum* (Workowski and Berman 2007). Patients with syphilis should also to be routinely tested for HIV infection in order to exclude the coinfection.

12.3.4 *Mycoplasma* Species

Mycoplasmas are the smallest free-living organisms (bacteria), widespread in nature. Unlike other bacteria, they lack a cell wall and therefore are not affected by common antibiotics that target cell wall synthesis. *Mycoplasma hominis*, *M. primatum*, *M. genitalium*, *M. spermatphilum*, and *M. penetrans* infect and colonize the genital tract of humans (Uuskula and Kohl 2002). *M. genitalium* was first isolated from two men with NGU and was later shown to be an important causative agent for acute and chronic urethritis in men (Deguchi and Maeda 2002; Uuskula and Kohl 2002; Jensen 2004; Manhart et al. 2011). A study conducted in Japan has shown that among 153 patients with NGU, 17% were infected with *M. genitalium* and 2.6% with *M. hominis* (Maeda et al. 2004). The rates of M. genitalium infection were found to be higher than that of other bacterial STI in HIV-positive men who have sex with men (Soni et al. 2010). The prevalence of these infections is high in semen samples of infertile men though a direct relationship of infection with sperm concentration, viability, motility, and morphology, and leukocyte count could not be established (Andrade-Rocha 2003). On the other hand, some studies could not find a significant difference between the rate of *Mycoplasma* infection in fertile and infertile men, but an effect on semen quality was reported in infected patients (Al-Sweih et al. 2012). *Mycoplasma genitalium* has been shown to adhere to the head, midpiece, and tail of human spermatozoa and is carried by motile sperm on its neck and midpiece. Thus, sperms serve as vectors for carrying *Mycoplasma genitalium* infection to women causing female genital disease and infertility (Svenstrup et al. 2003). *M. genitalium* was detected in 5.8% of human immunodeficiency virus (HIV)-positive men in Brazil (da Costa et al. 2010).

Tetracyclines and fluoroquinolones are highly active against Mycoplasmas (Taylor-Robinson and Bebear 1997), whereas β-lactams face resistance due to the absence of cell wall in these microbes. In addition to tetracyclines and erythromycin, some newer macrolides such as clarithromycin, azithromycin, and telithromycin were also found to be highly potent against *M. genitalium* with low minimum inhibitory concentrations (MICs) (~0.01 μg/mL or less) (Renaudin et al. 1992; Bébéar et al. 1999, 2000; Hannan and Woodnutt 2000). Current research has strongly presented azithromycin as the first drug of choice against *M. genitalium* infections.

Despite having such potent antibiotics, treatment against *M. genitalium*-positive urethritis is lagging behind with no well-accepted guidelines or recommendations. Also, only few studies have confirmed antimicrobial chemotherapy in men with *M. genitalium*-positive urethritis, but with limitations like small number of patients, no detection of other potentially important genital mycoplasmas, or other types of organisms (Gambini et al. 2000; Johannisson et al. 2000). Thus, it is difficult to draw any conclusions to devise the best strategy for managing *M. genitalium*-positive non-gonococcal urethritis (Deguchi and Maeda 2002). In-house PCRs are valuable tool to diagnose *M. genitalium* infections, although there is a need of highly accurate internationally validated and approved commercial NAAT.

12.3.5 *Ureaplasma* Species

Ureaplasma are a class of bacteria that can perform urea hydrolysis and belong to the family Mycoplasmataceae. *Ureaplasma urealyticum* is a causative agent for non-gonococcal urethritis, prostatitis, and epididymitis, and its prevalence in semen was found to be more in infertile patients (9%) than in healthy men (1%) (Zeighami et al. 2009). The infected infertile patients presented lower semen parameters like volume, sperm count, and morphology than uninfected patients (Zeighami et al. 2009). Similarly, in another study the frequency of *U. urealyticum* infection was found to be 39% in semen samples of infertile men, who also displayed abnormal semen parameters in terms of viscosity, pH, sperm morphology, motility and concentration, and leukocyte count (Zinzendorf et al. 2008). *U. urealyticum*-infected men also exhibit higher sperm apoptosis rates, which may indicate the effect of the STI on male fertility (Shang et al. 1999). In African men, a study reported *U.*

urealyticum infection in infertile men to be as high as 42% (Bornman et al. 1990), while in Shanghai (China), its reported incidence in infertile men was ~39% against 9% in fertile subjects, with the bacteria found adhering to the membrane of spermatozoa and exfoliated germ cells (Xu et al. 1997). In *in vitro* experiments, incubation of *U. urealyticum* with sperm cells resulted in significant alteration in motility and membrane alterations (Reichart et al. 2001), with marked damage to paternal DNA that could affect embryonic development (Reichart et al. 2000). It has also been shown that although the fertilization parameters in IVF were similar for infected and non-infected semen samples, pregnancy rates were significantly lower in the infected group, indicating the role of *U. urealyticum* at the level of the endometrium (Montagut et al. 1991). Hence it is apparent that *Ureaplasma*, especially *U. urealyticum* infection, may play a significant role in affecting semen parameters and/or pregnancy outcome in case of infertile male patients.

Ureaplasma species infection is diagnosed mainly through culture and polymerase chain reaction (PCR); however commercial assays are also available. For measuring antimicrobial susceptibility of isolates, micro-broth dilution is the routine technique. Biovar strains (that differs physiologically and/or biochemically from other strains) have shown variations in susceptibility with biovar 2 maintaining higher sensitivity rates. Antibiotics such as azithromycin, josamycin, ofloxacin, and doxycycline have been tested against *Ureaplasma* species, and results have shown resistance against macrolides, tetracyclines, and fluoroquinolones. Thus, rapid diagnosis and appropriate antibiotic therapy can prevent long-term complications associated with *Ureaplasma* infections (Sethi et al. 2012).

12.4 Viral Infections and Associated Male Infertility

Viral infections either ascend through the urethra or invade the reproductive tract via the bloodstream to encourage male infertility either by direct toxic effects on the cells of the male reproductive tract or indirectly by causing local inflammatory and/or immunological reactions. These infections are often more disseminated and can occur in epithelial [e.g., human papilloma virus (HPV)] and neuronal cells [e.g., herpes simplex virus (HSV)-2], as well as in WBCs [e.g., human immunodeficiency virus (HIV-1), cytomegalovirus (CMV), Epstein–Barr virus (EBV)] (Fig. 12.3).

12.4.1 Human Papillomavirus

Human papilloma virus (HPV) is one of the most common sexually transmitted viral infections. Over the last two decades, research has found a strong correlation between HPV infection and cancer, including penile cancer in men (Colon-Lopez et al. 2010). HPV is present in semen of asymptomatic men with a higher prevalence in infertile men seeking treatment for fertility (16%) than in other men (10%) (Laprise et al. 2014). In a study conducted in China, 17.4% of infertile men were HPV positive against 6.7% HPV-positive fertile men (Yang et al. 2013), confirming

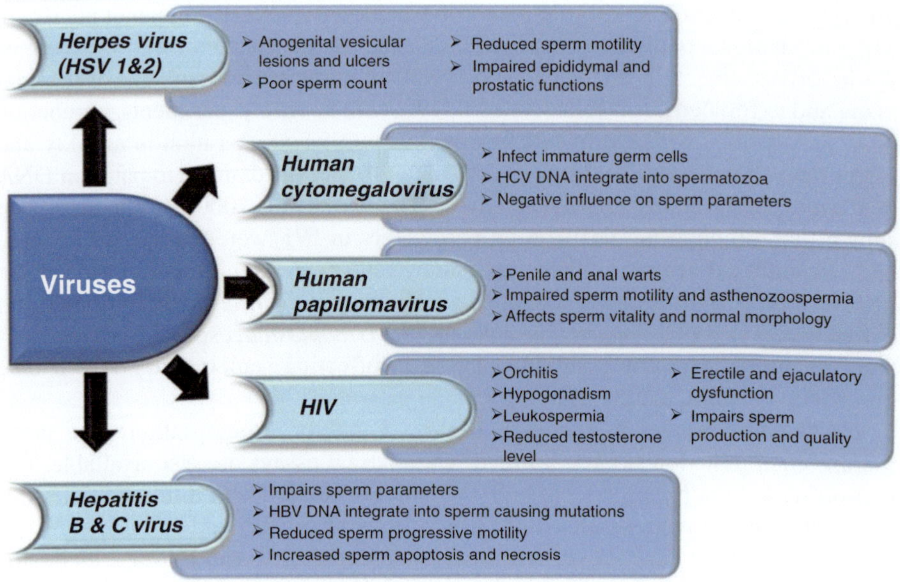

Fig. 12.3 Pathophysiological action of STD viral infections and male infertility

the high infertility rates in infected males. HPV infection was also associated with reduced sperm motility and abnormal sperm morphology. The incidence of asthenozoospermia was significantly higher in HPV-infected sperm samples (75%) versus non-infected samples (8%) (Lai et al. 1997). Another study analyzed the semen samples of 100 young adult sexually experienced men (18 year old) in comparison with 100 sexually inexperienced men of equal age and found HPV infection in 10% of sexually active men who also presented reduced sperm motility, while none among sexually inexperienced men were infected (Foresta et al. 2010). The sexual transfer of HPV is also indicated by HPV detection in the cervix and sperm of 53 married couples, out of which ~50% had at least one partner infected. Nine out of 12 partners of HPV-positive men and 9 out of 23 partners of HPV-positive women were infected, indicating sexual transmission (Kyo et al. 1994). HPV infection is also found to negatively affect the outcome of assisted reproductive technologies (ART) in infertile couples. 66.7% of infertile couples with HPV-infected male partner suffered pregnancy loss as compared to only 15% of infertile couples with HPV-negative males (Perino et al. 2011).

Sperms act as vectors for the transmission of HPV horizontally to sexual partners (as seen above) and vertically to the offsprings. The HPV has been localized in the equatorial region of sperm head from where it is transferred to the oocyte during fertilization (Foresta et al. 2011). HPV capsids bind efficiently to two distinct sites at the equatorial region of sperm head surface, which can be reduced by infection inhibitors like heparin and carrageenan (Perez-Andino et al. 2009). Direct swim-up and modified swim-up techniques can also effectively remove HPV DNA from naturally and artificially infected human sperm for ART (Garolla et al. 2012)

Virus-transmitted sexual diseases are difficult to cure, and the antiviral therapy can only mitigate the symptoms but not eradicate the virus. During the past years, substantial improvement has been made in the discovery and development of antiviral drugs. One of the oldest antiviral drugs, acyclovir (ACV), is approved for initial and recurrent genital herpes infections. Resistance toward ACV and related drugs has been reported among immune-compromised patients. In case of drug resistance, infections can be managed by the second line of drugs that include foscarnet or cidofovir (Mlynarczyk-Bonikowska et al. 2013). In HPV-infected individuals, there is no known specific drug target for the medication, and therefore antimitotics or immunomodulators are used for therapy.

12.4.2 Human Cytomegalovirus

Human cytomegalovirus (HCMV) is a virus of the Herpesviridae family whose infection remains undetected in healthy people but can be life-threatening in the immunocompromised individuals (Ryan and Ray 2004). It has phases of latency and reactivation and has been isolated from the semen and vagina. In males, its presence has been detected in the epididymis (Gimenes et al. 2014), vas deferens (Kimura et al. 1993), prostate (Geder et al. 1976), and seminal vesicles (DeTure et al. 1976).

Among herpes viruses, HCMV harbors maximum genes that are dedicated toward altering (evading) innate and adaptive immunity in the host and represents a lifelong burden of antigenic T cell surveillance and immune dysfunction (Varani and Landini 2011). These infections are life-threatening when the balance maintained by host immune surveillance is disturbed (Khan 2007).

HCMV has been detected in the epididymis, seminal vesicles and vas deferens (Kimura et al. 1993), and prostate (Mastroianni et al. 1996) of men. Significantly lowered sperm quality has been reported in HCMV-infected men, and its incidence was found to be more than that of HSV-2 (Wu et al. 2007). HCMV has been shown to be present in the extracellular fluid of semen (Lang et al. 1974) and in the semen of infertile patients (Levy et al. 1997). HCMV has also been detected both in sperm samples and in testis organotypic culture, and it has been concluded that the virus may infect immature spermatogenic cells, which may later develop into HCMV-carrying spermatozoa (Fig. 12.4). Significant reduction in the number of testicular germ cells indicates that HCMV can produce a direct gametotoxic effect, leading to male infertility (Naumenko et al. 2011).

The appropriate diagnostic test for identifying HCMV infection is the presence of CMV-specific IgG antibodies. The confirmatory tests include serology or detection of HCMV antigen (pp65) or DNA (by PCR) from infected individuals (Ljungman et al. 2002).

The preemptive therapy for CMV disease is to start antivirals for patients in order to halt the proliferation of virus (Griffiths 2002). Drugs available for combating HCMV infection include valganciclovir (VGCV), ganciclovir (GCV), acyclovir (ACV), valacyclovir (VACV), maribavir, foscarnet, and cidofovir.

12.4.3 Human Immunodeficiency Virus

The human immunodeficiency virus (HIV) is a lentivirus (a subgroup of retrovirus) that causes HIV infection and, over a period of time, acquired immunodeficiency syndrome (AIDS) (Weiss 1993; Douek et al. 2009). The major mode of HIV transmission is through sexual contact, though transmission may also occur through contact with contaminated blood and vertical transmission to the offsprings from infected mothers.

HIV infection has been shown to affect male fertility. The testes of AIDS patients are atrophic with decreased spermatogenesis and can be grouped into three major categories, (a) low spermatogenesis, (b) arrested spermatogenesis at primary spermatocyte stage, and (c) Sertoli-cell-only syndrome (Shevchuk et al. 1998). HIV-1 virus has been detected in >90% of testes from HIV-infected adults and in 25–33% of the residual germ cells of these infected testes, though HIV-1 DNA was found to be absent in the testes of all the three preadolescent boys who had acquired HIV-1 *in utero*. (Shevchuk et al. 1998). The presence of HIV-1 proviral DNA has also been detected in the nuclei of germ cells at all stages of spermatogenesis, which suggested that HIV-seropositive men are capable of producing infected spermatozoa and releasing them in the genital tract. A direct infection of the germ cells by cell-free virus in the testis could be supported by the immune privilege of this organ (Muciaccia et al. 1998). Similarly, in the testis of men who deceased of AIDS, hypoplasia and spermatogenic arrest were presented with infected spermatogonia and spermatocytes, demonstrating the testis as one of the major site for early HIV infection. The authors concluded that HIV-positive men release infected spermatozoa (Fig. 12.4) in the genital tract (Muciaccia et al. 1998). In yet another study, HIV-1 DNA was detected in the testis of 11 out of 12 HIV-infected men by PCR *in situ* hybridization technique, which was localized mainly in the testis cells (spermatogonia and spermatocytes, rarely in spermatids), but not in the epithelium of the prostate, epididymis, seminal vesicles, or penis of men with AIDS (Nuovo et al. 1994). It was inferred that spermatogenic cells serve as the primary source for venereal transmission of virus (Nuovo et al. 1994). However, experimental studies on nonhuman primates have demonstrated that HIV may infect the testes (spermatogonia but not more mature spermatogenic cells) as well as the epididymis (Shehu-Xhilaga et al. 2007).

HIV infection has also been shown to adversely affect semen parameters in infected individuals. Semen volume, sperm count, and progressive motility have been reported to be markedly decreased in HIV-infected patients (van Leeuwen et al. 2008; Huang et al. 2002a; Dulioust et al. 2002; Bujan et al. 2007). The above is reported to be accompanied with increased population of immature germ cells and spermiophage cells in semen, suggesting defective epididymal sperm maturation due to reduced testosterone levels (Dondero et al. 1996). Cytokine and chemokine concentrations are elevated in semen upon HIV infection which triggers inflammation and can contribute to male infertility (Gimenes et al. 2014). Undoubtedly, semen is considered as the main vector for HIV-1 transmission, which can transmit HIV-1 as free virions, infected leukocytes, and spermatozoa-associated virus

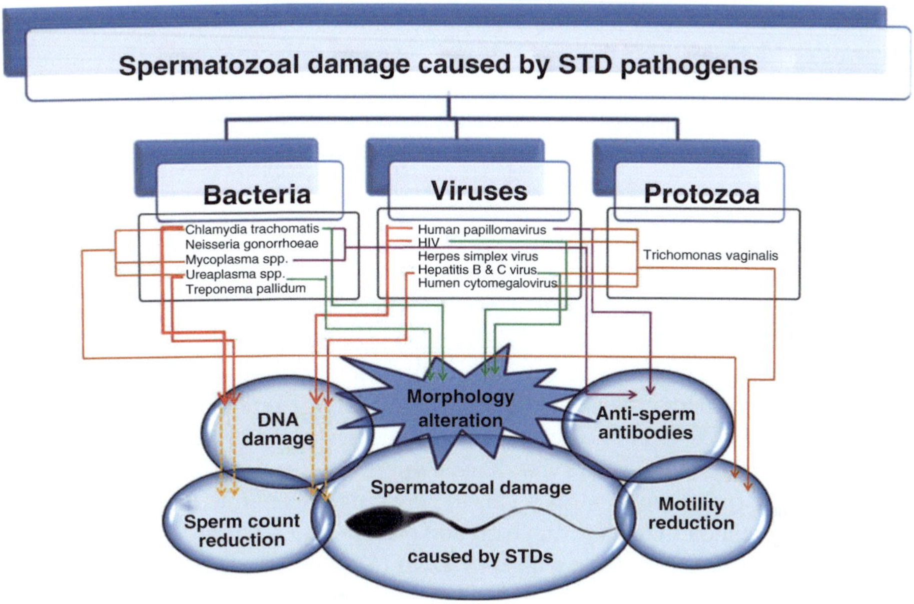

Fig. 12.4 Pathophysiological action of protozoa and male infertility

(Ceballos et al. 2009). HIV-1 proviral DNA has been detected in the seminal cells of 45% HIV-1-infected men and 33% of their sperm samples, with positive staining for HIV-1 in the midpiece and rarely in the head (Bagasra et al. 1994). Electron microscopy has revealed that HIV-1 attaches to the sperm surface and enters the sperm cells through the cell membrane suggesting human sperm as the primary cellular element involved in the transmission of HIV through semen (Bagasra et al. 1988). Sperm express a 165–175 kDA mannose receptor for HIV-1 attachment, resulting in vertical and horizontal transmission of virus (Cardona-Maya et al. 2006). Heparin sulfate expressed in spermatozoa may also play a critical role in capturing HIV-1, as it has been shown that at low vaginal pH, the binding of HIV-1 to spermatozoa and its transmission to the dendritic cells is greatly enhanced (Ceballos et al. 2009). HIV-1 may be carried passively by spermatozoa through attachment of virus to the cell surface, mediated by beta-chemokine receptors (CCR5 and CCR3) present on the acrosomal surface of spermatozoa (Muciaccia et al. 2007).

Development of combinatorial antiretroviral therapy in late 1990s helped in managing fatal HIV illness which was unconquered earlier. More than 25 licensed drugs have been developed since then which block the replication of virus at different steps. A large number of vaginal microbicides have been tested at different stages of preclinical and clinical development, but their development has been delayed perhaps due to lack of support from the pharmaceutical industries (Lederman et al. 2006).

Before antiretroviral therapies, protease inhibitor-based drugs were more in use. However, antiretroviral therapies were developed to decrease the dose and increase the safety and effectiveness of drugs. Standard antiretroviral therapy used currently involves combination of two nucleoside reverse transcriptase inhibitors like emtricitabine or lamivudine along with abacavir, tenofovir, or zidovudine and a non-nucleoside reverse transcriptase inhibitor, a protease inhibitor, or a integrase inhibitor (Maartens et al. 2014).

12.4.4 Herpes Simplex Virus

Herpes simplex virus 1 and 2 (HSV-1 and HSV-2) are known as oral herpes and genital herpes virus, respectively. These two are members of the herpesvirus family Herpesviridae, which infects human beings (Ryan and Ray 2004). Globally, around 3.7 billion people under the age of 50 (67%) have HSV-1 infection, while 417 million people aged 15–49 (11%) have HSV-2 infection (WHO 2016).

Many reports have linked HSV with male infertility (Borai et al. 1997; Kapranos et al. 2003). Herpes simplex virus seems to play a significant role in male infertility and has been related to low sperm count and poor motility. Semen of infertile men has been shown to be populated with HSV DNA (Kapranos et al. 2003; Sheikh et al. 2014). A study on Iranian men detected HSV DNA in 22.9% of men with male factor infertility having lower sperm count (Monavari et al. 2013). HSV-2 has been isolated from the testes and epididymis of male cadavers, suggesting that these organs may serve as reservoirs for the transmission of virus (DeTure et al. 1976). Expression of HSV thymidine kinase (HSV-tk) in transgenic mouse testis led to abnormal spermatogenesis, acrosomal aberrations, spermatogenic arrest, and infertility (Huttner et al. 1993). In transgenic rats, two HSV-tk proteins of 37 and 39 kDa were accumulated in the round spermatids, which increased apoptosis of testicular germ cells. Their sperm had malformed heads and looped tails (Cai et al. 2009). The DNA of herpes viruses is frequently detected in the semen of asymptomatic fertile and infertile male patients (Neofytou et al. 2009). A significant association between the evidence for infertility and HSV-positive test in semen was observed in men ($P = 0.024$) (el Borai et al. 1997). Oligozoospermia is reported to be two times more frequent in HSV-containing ejaculates than in HSV negative with sperm displaying microhead and cytoplasm drops in HSV-infected patients, which indicated that asymptomatic HSV infection may adversely affect male fertility (Abdulmedzhidova et al. 2007; Wu et al. 2007). A study reported that as compared to apparently healthy individuals, men with infertility had a higher incidence of HSV infection of the semen ($P < 0.05$), decreased numbers of actively motile sperm ($P = 0.0001$), and lower percent of morphologically normal sperm ($P = 0.002$), indicating HSV as one of the factors for male infertility (Klimova et al. 2010). However, several other studies could not find an association of HSV infection with male infertility (Kaspersen and Hollsberg 2013).

Screening for HSV is not routinely performed except in those people who are at high risk or have a partner diagnosed with herpes or have developed some symptoms. Diagnosis for HSV is readily available, which includes viral culture, polymerase chain reaction (PCR), and serology testing (Workowski and Berman 2007;

Gupta et al. 2008; Patel and Rompalo 2005). Serology testing is also important to distinguish between HSV-1 from HSV-2. Suppressive therapy is highly encouraged in genital HSV infections to reduce the transmission of infection. One study has reported that valacyclovir (Valtrex) reduces the rate of virus transmission in heterosexual companions (Corey et al. 2004). Some vaginal microbicides like Ushercell and T-PSS have been shown to inhibit transmission of STIs, including HSV (Anderson et al. 2002; Zaneveld et al. 2002), but these could not be developed further as OTC products.

12.4.5 Hepatitis B Virus (HBV)

According to a WHO report, 2 billion people have been infected with HBV globally, and approximately 380 million are chronic carriers (6% of the world population). About 4.5 million new infections are reported worldwide annually with 620,000 deaths per year (Lemoine et al. 2013). HBV prevalence has been found in the reproductive system including semen and vaginal secretions and also in many other body fluids (Alexander and Kowdley 2006). The transmission of the virus occurs mainly from mother to child at time of birth, the use of infected syringes in drug addicts, sexual transmission, and emigration from endemic areas (Toy et al. 2008).

A study of 457 HBV-positive and 459 HBV-negative men seeking fertility assistance at Zhejiang University indicated HBV-infected men exhibiting lower semen volume and lower sperm count, motility, and normal morphology than HBV-negative men ($P < 0.05$). HBV infection increased the incidence of asthenozoospermia, oligozoospermia, or azoospermia ($P < 0.05$) and decreased the rates of implantation/clinical pregnancy in ICSI cycles ($P < 0.05$) (Zhou et al. 2011). Another study reported that HBV patients had significantly worse sperm density, total number, forward motility, morphology, and viability than HCV patients, and additional presence of varicocele further worsened the sperm parameters in HBV patients (Vicari et al. 2006). Perhaps exposure to HBV leads to ROS generation, lipid peroxidation, phosphatidylserine externalization, activation of caspases, and DNA fragmentation, resulting in increased apoptosis of sperm cells and loss of sperm membrane integrity and causing sperm dysfunctions (Kang et al. 2012). The presence of integrated HBV DNA sequences in spermatozoa of 66% and HBV DNA in seminal plasma of 33% of patients has been reported, which could cause vertical and horizontal transmissions, respectively (Hadchouel et al. 1985). Similar integration of HBV DNA in spermatozoa was also observed by Huang et al. (2002). In an interesting study, HBV markers were found in 12 sperm samples, and HBV DNA was detected integrated in 3 sperm samples of infected men whose 66.7% wives and 57.1% of children were also found to be HBV positive. One infant was found infected with HBV-negative mother and HBV-positive father (with HBV DNA-integrated sperm), indicating the possibility of vertical transmission. The study also indicated that men to women transmissions were easier and sperm was the most plausible vector (Xu 1992).

Diagnosis for HBV provides the opportunity to identify those who are positive and recommend vaccination to those who are negative but at risk (Cooke et al. 2010). Hepatitis B vaccines contain inactivated HBsAg and are available from early the 1980s. Initially the vaccines were plasma derived, but currently these have been replaced by vaccines manufactured using recombinant DNA technology (Shouval 2003).

Currently, seven drugs are available for the treatment of chronic hepatitis B, which include five nucleos(t)ide analogues (lamivudine, adefovir, entecavir, tenofovir, and telbivudine) and two interferon-based therapies (conventional interferon and pegylated interferon-alpha) (Aspinall et al. 2011). Nucleoside analogues subdue the viral replication by inhibiting the viral polymerase, whereas interferon therapy works by boosting the host immune response. In addition to vaccination, the risk of HBV transmission can be reduced through other preventive measures such as routine testing of blood, organ and tissue donors, and screening of blood and blood products.

12.4.6 Hepatitis C Virus or HCV

Hepatitis C virus (HCV), a member of the *Hepacivirus* genus of the family Flaviviridae, is a small, enveloped, single-stranded, positive-sense RNA virus (Rosen 2011). HCV infection has been correlated with male infertility. HCV patients show significantly lower sperm concentration, forward motility, normal sperm morphology and mitochondrial membrane potential, and higher DNA fragmentation, apoptosis, and ROS levels and abnormal chromatin in semen (La Vignera et al. 2012; Hofny et al. 2011). Similar results were obtained in another study in which mean sperm motility ($P < 0.001$), viability ($P < 0.001$), and normal morphology ($P < 0.05$) were significantly reduced in HCV patients as compared to controls (Lorusso et al. 2010). Adverse effects of HCV infection have also been shown on spermatogenesis, which could be improved by therapy (Durazzo et al. 2006). Lower serum testosterone and higher serum estradiol and prolactin levels in HCV patients have also been reported (Hofny et al. 2011). HCV RNA and viral particles have been found in seminal plasma but not in spermatozoa (Gimenes et al. 2014).

Hepatitis C infection can be acute or chronic and is usually asymptomatic but can cause significant liver damage before its diagnosis. Treatment recommendations for hepatitis C are based on the condition of the disease (acute vs chronic). The foremost aim of the treatment is to achieve a sustained virologic response, defined as the absence of HCV RNA in serum at least 6 months after the withdrawal of treatment (Modi and Liang 2008). Therapy for hepatitis C infection has improved substantially over the period. The current recommended drugs for chronic HCV infection include the combination of peginterferon and ribavirin and will be in primary use for the next few years (Manns et al. 2001; Fried et al. 2002; Hadziyannis et al. 2004).

12.5 Protozoan Infections and Male Infertility

12.5.1 Trichomonas vaginalis

Trichomonosis, a sexually transmitted disease (STD), is caused by the flagellated parasitic protozoan *Trichomonas vaginalis*. The parasite is a common cause of genitourinary tract infection in both men and women (Mielczarek and Blaszkowska 2015). Worldwide 160–180 million people are affected annually by trichomoniasis, and over half of new *T. vaginalis* infections each year are estimated to occur in men. It is one of the most poorly studied parasites with respect to virulence properties, pathogenesis, and immunopathogenesis (Harp and Chowdhury 2011). The infections are largely neglected, despite being highly prevalent (Bar et al. 2015). About 75% of men harboring *T. vaginalis* are asymptomatic and may not seek treatment (Twu et al. 2014).

Some males infected with *T. vaginalis* show symptoms like inflammation, irritation, urethritis, and urethral discharge. They serve as vectors for the transmission of this STI and other infections including the HIV through disruption of urogenital epithelial layers, which activates local immune cells leading to increased viral replication (Kushwaha et al. 2016; Guenthner et al. 2005). Antimicrobial (Alidina et al. 2016) and cytotoxic agents (Twu et al. 2014) present in male genital secretions and prostatic fluid, viz., pathogenic inhibitory factors (Fair et al. 1976) and zinc, respectively, force the parasite toward asymptomatic infection through various mechanisms.

Recent reports have shown the effect of *T. vaginalis* on sperm morphology, viability, motility, and function through various mechanisms. Some studies have reported reduced sperm function in *T. vaginalis*-infected infertile men than in non-infected fertile controls. *T. vaginalis* produces a proteinaceous substance that kills sperm rapidly (Soper 2004), and *in vitro* mixing of sperm with *Trichomonas* reduced sperm activity (Jarecki-Black et al. 1988). Gopalkrishnan et al. 1990 assessed the effect of *Trichomonas* on sperm and found reduced sperm motility and viability with a reduction in the percentage of sperm with normal morphology (Tuttle et al. 1977; Gopalkrishnan et al. 1990). A significant improvement in disrupted sperm parameters was found in 50% males after treatment with metronidazole, suggesting reversible effect of *Trichomonas* on sperm (Fig. 12.4).

The most sensitive and accessible method for diagnosis of trichomoniasis in women is culture of vaginal secretions, whereas in men, a culture from urethral swab, urine, and semen has the highest sensitivity (Workowski et al. 2006). The 5-nitroimidazole drugs (metronidazole and tinidazole) are the only recommended drugs, of which (FDA)-approved metronidazole is the most prescribed and effective drug to treat trichomoniasis (Dunne et al. 2003). Studies have shown that tinidazole 2 g is equivalent or better than metronidazole 2 g (Fung and Doan 2005). Tinidazole has a longer half-life than metronidazole (Workowski et al. 2006). Oral metronidazole is the preferred way of delivery; metronidazole in vaginal gels cannot reach therapeutic levels in the urethra and vaginal glands and has limited efficacy when used vaginally.

On the other hand, 5-nitroimidazole class compounds are prone to drug resistance. The first case of drug resistance was reported with metronidazole within 2 years of its introduction, and more than 100 cases were reported by 2003 (Crowell et al. 2004). With an increasing incidence of resistance against metronidazole and cross-resistance among the family of 5-nitroimidazole drugs, disease prevention with a safe and effective alternative to nitroimidazoles would clearly be desirable.

A number of studies have been published demonstrating the *in vitro* trichomonicidal activities of non-5-nitroimidazole drugs. These compounds include both drug derivatives synthesized specifically for the treatment of *T. vaginalis* infection and agents currently in use for the treatment of other infectious diseases. The synthesized compounds include imidazole derivatives (Kumar et al. 2013; Anthwal et al. 2014), ammonium salts of carbamodithioic acid (Jain et al. 2011, 2014; Kushwaha et al. 2016), and dithiocarbamate derivatives (Bala et al. 2014, 2015; Mandalapu et al. 2015).

Some synthetic derivatives of benzisothiazolinone showed significantly higher *in vitro* trichomonicidal activities than metronidazole (Ziomko and Kuczyńska 1993). Some other drugs investigated for antitrichomonal activities include sulfimidazole, a 5-nitroimidazole with a functional sulfonamide group (Malagoli et al. 2002); nifuratel, a nitrofuran derivative (Lossick 1990); berberine sulfate, a plant alkaloid (Kaneda et al. 1991); thiadiazine derivatives (Atienza et al. 1992); some 4-nitrobenzimidazole derivatives (Alcalde et al. 1992); specific benzimidazole derivatives (Katiyar et al. 1994); acetylated derivatives of sugar hydrazones (Macickova et al. 1990); disulfiram, a drug often used to treat alcoholism (Bouma et al. 1998); etc. These chemical agents have shown some promising trichomonacidal activity.

Conclusion

This chapter summarizes the current knowledge relating to the major sexually transmitted infections and their influence on sperm and male fertility. STDs are caused by bacterial, viral, and protozoal pathogens and can induce male infertility through multiple pathophysiological mechanisms, including impairment of sperm parameters and functions (Fig. 12.5). Pathogens can be transmitted horizontally to sexual partners and vertically to fetuses and neonates. However, the effect of these pathogens and their role in the manifestation of male infertility are still imprecise and require further validation. While most of the bacterial pathogens and *Trichomonas vaginalis* are curable by the use of appropriate drugs/antibiotics, the asymptomatic nature of some of these diseases results in high rates of transmission. On the other hand, viral infections are difficult to manage and require timely intervention. Inflammation caused by sexually transmitted microbial infections appears to play a major role in male infertility. Lower prevalence of infection reported in earlier studies may be mainly due to the lack of sensitive diagnostic parameters. However, it must also be taken into consideration that most of the studies were performed on men with acute/active disease. Screening and treatment of individuals at risk of acquiring STIs as well as those with infertility due to acute manifestation of the disease will help reduce the incidences of infertility. Readers are encouraged to consult Gimenes et al., Nat Rev. Urol. 2014;11(12), pp. 672–87 and Brookings et al., Alternative formats. Korean J Urol, 2013; 54(3), pp. 149–156 for further reading on this subject.

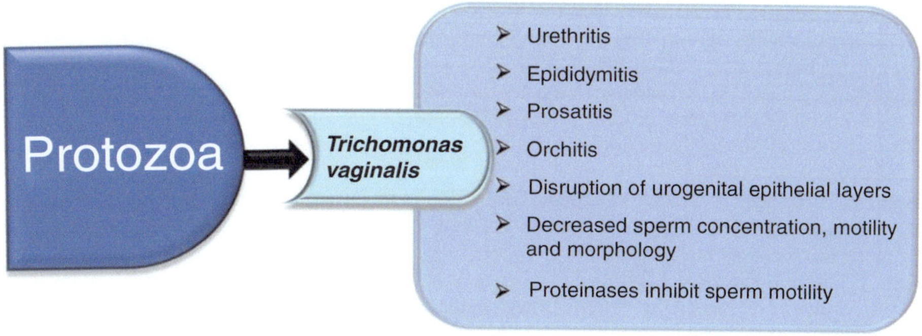

Fig. 12.5 Spermatozoal damage caused by STD pathogens

References

Abdulmedzhidova AG et al. (2007) Asymptomatic genital herpes infection and infertility in males [Russian]. Urologiia 56–59

Abusarah EA, Awwad ZM, Charvalos E, Shehabi AA (2013) Molecular detection of potential sexually transmitted pathogens in semen and urine specimens of infertile and fertile males. Diagn Microbiol Infect Dis 77:283–286

Akre O, Cnattingius S, Bergström R, Kvist U, Trichopoulos D, Ekbom A (1999) Human fertility does not decline: evidence from Sweden. Fertil Steril 71(6):1066–1069

Alcalde E, Perez-Garcia L, Dinares I, Coombs GH, Frigola J (1992) Synthesis and antitrichomonal activity of azinium (azolium) 4-nitrobenzimidazolate betaines and their derivatives. Eur J Med Chem 27(2):171–176

Alexander J, Kowdley KV (2006) Epidemiology of hepatitis B—clinical implications. Medscape general. Medicine 8(2):13

Alidina Z, Wormsbecker AE, Urquia M, MacGillivray J, Tork E, Yudin MH, Campbell DM (2016) HIV prophylaxis in high risk newborns: an examination of sociodemographic factors in an inner city context. Can J Infect Dis Med Microbiol 2016

Allahbadia GN (2016) Yes, it is possible. J Obstet Gynecol India 66(3):139–143

Al-Sweih NA, Al-Fadli AH, Omu AE, Rotimi VO (2012) Prevalence of *Chlamydia trachomatis, Mycoplasma hominis, Mycoplasma genitalium,* and *Ureaplasma urealyticum* infections and seminal quality in infertile and fertile men in Kuwait. J Androl 33:1323–1329

Anderson RA, Feathergill KA, Diao XH, Cooper MD, Kirkpatrick R, Herold BC, Doncel GF, Chany CJ, Waller DP, Rencher WF, Zaneveld LJ (2002) Preclinical evaluation of sodium cellulose sulfate (Ushercell) as a contraceptive antimicrobial agent. J Androl 23(3):426–438

Andrade-Rocha FT (2003) *Ureaplasma urealyticum* and *Mycoplasma hominis* in men attending for routine semen analysis. Prevalence, incidence by age and clinical settings, influence on sperm characteristics, relationship with the leukocyte count and clinical value. Urol Int 71:377–381

Anthwal A, Rajesh UC, Rawat MSM, Kushwaha B, Maikhuri JP, Sharma VL, Gupta G, Rawat DS (2014) Novel metronidazole–chalcone conjugates with potential to counter drug resistance in trichomonas vaginalis. Eur J Med Chem 79:89–94

Apari P, de Sousa JD, Müller V (2014) Why sexually transmitted infections tend to cause infertility: an evolutionary hypothesis. PLoS Pathog 10(8):e1004111

Aspinall EJ, Hawkins G, Fraser A, Hutchinson SJ, Goldberg D (2011) Hepatitis B prevention, diagnosis, treatment and care: a review. Occup Med 61(8):531–540

Atienza J, Diaz RAM, Barrio A, Escario JA, Herrero A, Ochoa C, Rodríguez J (1992) Activity assays of thiadiazine derivatives on trichomonas vaginalis and amastigote forms of Trypanosoma cruzi. Chemotherapy 38(6):441–446

Bagasra O, Freund M, Weidmann J, Harley G (1988) Interaction of human immunodeficiency virus with human sperm in vitro. J Acquir Immune Defic Syndr 1:431–435

Bagasra O et al (1994) Detection of HIV-1 proviral DNA in sperm from HIV-1-infected men. AIDS 8:1669–1674

Baird A, Olarinde O, Talbot M (2007) Evaluation, using two assessment instruments, of the American and British national guidelines for the management of sexually transmissible and genital infections. Sex Health 4(4):255–260

Bala V, Jangir S, Kumar V, Mandalapu D, Gupta S, Kumar L, Kushwaha B, Chhonker YS, Krishna A, Maikhuri JP, Shukla PK (2014) Design and synthesis of substituted morpholin/piperidin-1-yl-carbamodithioates as promising vaginal microbicides with spermicidal potential. Bioorg Med Chem Lett 24(24):5782–5786

Bala V, Jangir S, Mandalapu D, Gupta S, Chhonker YS, Lal N, Kushwaha B, Chandasana H, Krishna S, Rawat K, Maikhuri JP (2015a) Dithiocarbamate–thiourea hybrids useful as vaginal microbicides also show reverse transcriptase inhibition: design, synthesis, docking and pharmacokinetic studies. Bioorg Med Chem Lett 25(4):881–886

Bala V, Mandalapu D, Gupta S, Jangir S, Kushwaha B, Chhonker YS, Chandasana H, Krishna S, Rawat K, Krishna A, Singh M (2015b) N-Alkyl/aryl-4-(3-substituted-3-phenylpropyl) piperazine-1-carbothioamide as dual-action vaginal microbicides with reverse transcriptase inhibition. Eur J Med Chem 101:640–650

Bar AK, Phukan N, Pinheiro J, Simoes-Barbosa A (2015) The interplay of host microbiota and parasitic protozoans at mucosal interfaces: implications for the outcomes of infections and diseases. PLoS Negl Trop Dis 9(12):e0004176

Bébéar CM, Renaudin H, Schaeverbeke T, Leblanc F, Bébéar C (1999) In-vitro activity of grepafloxacin, a new fluoroquinolone, against mycoplasmas. J Antimicrob Chemother 43(5):711–714

Bebear CM, Renaudin H, Bryskier A, Bebear C (2000) Comparative activities of telithromycin (HMR 3647), levofloxacin, and other antimicrobial agents against human mycoplasmas. Antimicrob Agents Chemother 44(7):1980–1982

Bignell C, FitzGerald M, Guideline Development Group (2011) UK national guideline for the management of gonorrhoea in adults, 2011. Int J STD AIDS 22(10):541–547

Borai NE, Inoue M, Lefèvre C, Naumova EN, Sato B, Yamamura M (1997) Detection of herpes simplex DNA in semen and menstrual blood of individuals attending an infertility clinic. J Obstet Gynaecol Res 23(1):17–24

Bornman MS, Mahomed MF, Boomker D, Schulenburg GW, Reif S, Crewe-Brown HH (1990) Die mikrobielle Flora imSperma von kinderlosenafrikanischenMännernausdem Garankuwa-Hospital [Microbial flora in semen of infertile African men at Garankuwa hospital]. Andrologia 22(2):118–121

Bouma MJ, Snowdon D, Fairlamb AH, Ackers JP (1998) Activity of disulfiram (bis (diethylthiocarbamoyl) disulphide) and ditiocarb (diethyldithiocarbamate) against metronidazole-sensitive and-resistant Trichomonas vaginalis and Tritrichomonas foetus. J Antimicrob Chemother 42(6):817–820

Brookings C, Goldmeier D, Sadeghi-Nejad H (2013) Sexually transmitted infections and sexual function in relation to male. Korean J Urol 54(3):149–156

Bujan L et al (2007) Decreased semen volume and spermatozoa motility in HIV-1-infected patients under antiretroviral treatment. J Androl 28

Cai LY et al (2009) HSV type 1 thymidine kinase protein accumulation in round spermatids induces male infertility by spermatogenesis disruption and apoptotic loss of germ cells. Reprod Toxicol 27:14–21

Cardona-Maya W, Lopez-Herrera A, Velilla-Hernandez P, Rugeles MT, Cadavid AP (2006) The role of mannose receptor on HIV-1 entry into human spermatozoa. Am J Reprod Immunol 55:241–245

Carson SDB, Stone B, Beucher M, Fu J, Sparling PF (2000) Phase variation of the gonococcal siderophore receptor FetA. Mol Microbiol 36(3):585–593

Ceballos A et al (2009) Spermatozoa capture HIV-1 through heparan sulfate and efficiently transmit the virus to dendritic cells. J Exp Med 206:2717–2733

Centers for Disease Control (2015). https://www.cdc.gov/std/tg2015/

Centers for Disease Control and Prevention (2009) Sexually transmitted disease surveillance, 2008. US Department of Health and Human Services, Atlanta, GA, 2109755293(01111871), p 10148121180485

Centers for Disease Control and Prevention (2011) Consultation meeting on cephalosporin-resistant gonorrhea outbreak response plan. 2009

Colon-Lopez V, Ortiz AP, Palefsky J (2010) Burden of human papillomavirus infection and related comorbidities in men: implications for research, disease prevention and health promotion among Hispanic men. P R Health Sci J 29:232–240

Cooke GS, Main J, Thursz MR (2010) Treatment for hepatitis B. BMJ 340:87–90

Corey L, Wald A, Patel R, Sacks SL, Tyring SK, Warren T, Douglas JM Jr, Paavonen J, Morrow RA, Beutner KR, Stratchounsky LS (2004) Once-daily valacyclovir to reduce the risk of transmission of genital herpes. N Engl J Med 350(1):11–20

Cosentino MJ, Cockett ATK (1986) Review article: structure and function of the epididymis. Urol Res 14(5):229–240

Crowell AL, Stephens CE, Kumar A, Boykin DW, Secor WE (2004) Activities of dicationic compounds against Trichomonas vaginalis. Antimicrob Agents Chemother 48(9):3602–3605

da Costa FAM, da Silva RC, Arruda LB, Montanheiro P, da Silva Duarte AJ, Casseb J (2010) Prevalence of Mycoplasma genitalium among HIV-infected men in Sao Paulo city detected by realtime polymerase chain reaction. Int J STD AIDS 21(1):23–25

Danaher RJ, Levin JC, Arking D, Burch CL, Sandlin R, Stein DC (1995) Genetic basis of Neisseria gonorrhoeae lipooligosaccharide antigenic variation. J Bacteriol 177(24):7275–7279

de Barbeyrac B et al (2006) *Chlamydia trachomatis* in subfertile couples undergoing an *in vitro* fertilization program: a prospective study. Eur J Obstet Gynecol Reprod Biol 129:46–53

Deguchi T, Maeda SI (2002) Mycoplasma genitalium: another important pathogen of nongonococcal urethritis. J Urol 167(3):1210–1217

Dehio C, Gray-Owen SD, Meyer TF (2000) Host cell invasion by pathogenic Neisseriae. Subcell Biochem 33:61–96

DeTure FA, Drylie DM, Kaufman HE, Centifanto YN (1976) Herpesvirus type 2: isolation from seminal vesicle and testes. Urology 7(5):541–544

Dondero, F, Rossi T, Offizi GD, Mazzilli F, Rosso R, Sarandrea N, Pinter E and Aiuti F (1996) Semen analysis in HIV seropositive men and in subjects at high risk for HIV infection. Hum Reprod. 11(4):765–8.

Douek DC, Roederer M, Koup RA (2009) Emerging concepts in the immunopathogenesis of AIDS. Annu Rev Med 60:471

Dulioust E et al (2002) Semen alterations in HIV-1 infected men. Hum Reprod 17:2112–2118

Dunne RL, Linda AD, Upcroft P, O'donoghue PJ, Upcroft JA (2003) Drug resistance in the sexually transmitted protozoan Trichomonas vaginalis. Cell Res 13(4):239–249

Durazzo M et al (2006) Alterations of seminal and hormonal parameters: an extrahepatic manifestation of HCV infection? World J Gastroenterol 12:3073–3076

Edwards JL, Apicella MA (2004) The molecular mechanisms used by *Neisseria gonorrhoeae* to initiate infection differ between men and women. Clin Microbiol Rev 17:965–981 el Borai N et al (1997) Detection of herpes simplex DNA in semen and menstrual blood of individuals attending an infertility clinic. J Obstet Gynaecol Res 23:17–24

Eley A, Pacey AA, Galdiero M, Galdiero M, Galdiero F (2005) Can *Chlamydia trachomatis* directly damage your sperm? Lancet Infect Dis 5:53–57

Fair WR, Couch J, Wehner N (1976) Prostatic antibacterial factor identity and significance. Urology 7(2):169–177

Ficarra G, Carlos R (2009) Syphilis: the renaissance of an old disease with oral implications. Head Neck Pathol 3(3):195–206

Foresta C et al (2010) Clinical and prognostic significance of human papillomavirus DNA in the sperm or exfoliated cells of infertile patients and subjects with risk factors. Fertil Steril 94:1723–1727

Foresta C, Patassini C, Bertoldo A, Menegazzo M, Francavilla F, Barzon L, Ferlin A (2011) Mechanism of human papillomavirus binding to human spermatozoa and fertilizing ability of infected spermatozoa. PLoS One 6(3):e15036

Fried MW, Shiffman ML, Reddy KR, Smith C, Marinos G, Gonçales FL Jr, Häussinger D, Diago M, Carosi G, Dhumeaux D, Craxi A (2002) Peginterferon alfa-2a plus ribavirin for chronic hepatitis C virus infection. N Engl J Med 347(13):975–982

Fung HB, Doan TL (2005) Tinidazole: a nitroimidazole antiprotozoal agent. Clin Ther 27(12):1859–1884

Furuya R, Takahashi S, Furuya S, Kunishima Y, Takeyama KOH, Tsukamoto T (2004) Is seminal vesiculitis a discrete disease entity? Clinical and microbiological study of seminal vesiculitis in patients with acute epididymitis. J Urol 171(4):1550–1553

Gambini D, Decleva I, Lupica L, Ghislanzoni M, Cusini M, Alessi E (2000) Mycoplasma genitalium in males with nongonococcal urethritis: prevalence and clinical efficacy of eradication. Sex Transm Dis 27(4):226–229

Garolla A et al (2012) Human papillomavirus sperm infection and assisted reproduction: a dangerous hazard with a possible safe solution. Hum Reprod 27:967–973

Gdoura R et al (2008) Assessment of *Chlamydia trachomatis*, *Ureaplasma urealyticum*, *Ureaplasma parvum*, *Mycoplasma hominis*, and *Mycoplasma genitalium* in semen and first void urine specimens of asymptomatic male partners of infertile couples. J Androl 29:198–206

Geder L, Sanford EJ, Rohner TJ, Rapp F (1976) Cytomegalovirus and cancer of the prostate: in vitro transformation of human cells. Cancer Treat Rep 61(2):139–146

Geisler WM (2010) Duration of untreated, uncomplicated *Chlamydia trachomatis* genital infection and factors associated with chlamydia resolution: a review of human studies. J Infect Dis 201(Suppl. 2):S104–S113

Gimenes F, Souza RP, Bento JC, Teixeira JJ, Maria-Engler SS, Bonini MG, Consolaro ME (2014) Male infertility: a public health issue caused by sexually transmitted pathogens. Nat Rev Urol 11(12):672–687

Gopalkrishnan K, Hinduja IN, Kumar TA (1990) Semen characteristics of asymptomatic males affected by Trichomonas vaginalis. J In Vitro Fert Embryo Transf 7(3):165–167

Griffiths PD (2002) The treatment of cytomegalovirus infection. J Antimicrob Chemother 49(2):243–253

Guenthner PC, Secor WE, Dezzutti CS (2005) Trichomonas vaginalis-induced epithelial monolayer disruption and human immunodeficiency virus type 1 (HIV-1) replication: implications for the sexual transmission of HIV-1. Infect Immun 73(7):4155–4160

Gupta R, Warren T, Wald A (2008) Genital herpes. Lancet 370(9605):2127–2137

Hadchouel M et al (1985) Presence of HBV DNA in spermatozoa: a possible vertical transmission of HBV via the germ line. J Med Virol 16:61–66

Hadziyannis SJ, Sette H, Morgan TR, Balan V, Diago M, Marcellin P, Ramadori G, Bodenheimer H, Bernstein D, Rizzetto M, Zeuzem S (2004) Peginterferon-α2a and ribavirin combination therapy in chronic hepatitis C: a randomized study of treatment duration and ribavirin dose. Ann Intern Med 140(5):346–355

Hannan PC, Woodnutt G (2000) In vitro activity of gemifloxacin (SB 265805; LB20304a) against human mycoplasmas. J Antimicrob Chemother 45(3):367–369

Harkness AH (1948) The pathology of gonorrhoea. Br J Vener Dis 24(4):137

Harp DF, Chowdhury I (2011) Trichomoniasis: evaluation to execution. Eur J Obstet Gynecol Reprod Biol 157(1):3–9

Henkel R et al (2006) Chronic pelvic pain syndrome/chronic prostatitis affect the acrosome reaction in human spermatozoa. World J Urol 24:39–44

Hofny ERM, Ali MEM, Taha EA, Nafeh HM, Sayed DS, Abdel-Azeem HG, Abdou EF, Kamal GM, Mostafa T (2011) Semen and hormonal parameters in men with chronic hepatitis C infection. Fertil Steril 95(8):2557–2559

Hosseinzadeh S, Brewis IA, Eley A, Pacey AA (2001) Co-incubation of human spermatozoa with *Chlamydia trachomatis* serovar E causes premature sperm death. Hum Reprod 16:293–299

Hosseinzadeh S, Pacey AA, Eley A (2003) *Chlamydia trachomatis*-induced death of human spermatozoa is caused primarily by lipopolysaccharide. J Med Microbiol 52:193–200

Huang JM, Huang TH, Qiu HY, Fang XW, Zhuang TG, Qiu JW (2002) Studies on the integration of hepatitis B virus DNA sequence in human sperm chromosomes. Asian J Androl 4(3):209–212

Huttner KM, Pudney J, Milstone DS, Ladd D, Seidman JG (1993) Flagellar and acrosomal abnormalities associated with testicular HSV-tk expression in the mouse. Biol Reprod 49:251–261

Ibrahim AA, Refeidi A, El Mekki AA (1996) Etiology and clinical features of acute epididymo-orchitis. Ann Saudi Med 16:171–174

Idahl A, Boman J, Kumlin U, Olofsson JI (2004) Demonstration of *Chlamydia trachomatis* IgG antibodies in the male partner of the infertile couple is correlated with a reduced likelihood of achieving pregnancy. Hum Reprod 19:1121–1126

Isaiah IN, Nche BT, Nwagu IG, Nnanna II (2011) Current studies on bacterospermia the leading cause of male infertility: a protégé and potential threat towards mans extinction. N Am J Med Sci 3(12):562

Jain A, Lal N, Kumar L, Verma V, Kumar R, Kumar L, Singh V, Mishra RK, Sarswat A, Jain SK, Maikhuri JP (2011) Novel trichomonacidal spermicides. Antimicrob Agents Chemother 55(9):4343–4351

Jain A, Kumar L, Kushwaha B, Sharma M, Pandey A, Verma V, Sharma V, Singh V, Rawat T, Sharma VL, Maikhuri JP (2014) Combining a synthetic spermicide with a natural trichomonacide for safe, prophylactic contraception. Hum Reprod 29(2):242–252

Jarecki-Black JC, Lushbaugh WB, Golosov L, Glassman AB (1988) Trichomonas vaginalis: preliminary characterization of a sperm motility inhibiting factor. Ann Clin Lab Sci 18(6):484–489

Jensen JS (2004) *Mycoplasma genitalium*: the aetiological agent of urethritis and other sexually transmitted diseases. J Eur Acad Dermatol Venereol 18:1–11

Johannisson G, Enström Y, Löwhagen GB, Nagy V, Ryberg K, Seeberg S, Welinder-Olsson C (2000) Occurrence and treatment of Mycoplasma genitalium in patients visiting STD clinics in Sweden. Int J STD AIDS 11(5):324–326

Joki-Korpela P et al (2009) The role of Chlamydia trachomatis infection in male infertility. Fertil Steril 91:1448–1450

Kaneda Y, Torii M, Tanaka T, Aikawa M (1991) In vitro effects of berberine sulphate on the growth and structure of *Entamoeba histolytica*, *Giardia lamblia* and Trichomonas vaginalis. Ann Trop Med Parasitol 85(4):417–425

Kang X, Xie Q, Zhou X, Li F, Huang J, Liu D, Huang T (2012) Effects of hepatitis B virus S protein exposure on sperm membrane integrity and functions. PLoS One 7(3):e33471

Kapranos N, Petrakou E, Anastasiadou C, Kotronias D (2003) Detection of herpes simplex virus, cytomegalovirus, and Epstein-Barr virus in the semen of men attending an infertility clinic. Fertil Steril 79:1566–1570

Karinen L et al (2004) Association between *Chlamydia trachomatis* antibodies and subfertility in the Northern Finland Birth Cohort 1966 (NFBC 1966), at the age of 31 years. Epidemiol Infect 132:977–984

Kaspersen MD, Hollsberg P (2013) Seminal shedding of human herpesviruses. Virol J 10:226

Katiyar SK, Gordon VR, McLaughlin GL, Edlind TD (1994) Antiprotozoal activities of benzimidazoles and correlations with beta-tubulin sequence. Antimicrob Agents Chemother 38(9): 2086–2090

Katz AR, Lee MV, Wasserman GM (2012) Sexually transmitted disease (STD) update: a review of the CDC 2010 STD treatment guidelines and epidemiologic trends of common STDs in Hawai'i. Hawaii J Med Public Health 71(3):68–73

Kent ME, Romanelli F (2008) Reexamining syphilis: an update on epidemiology, clinical manifestations, and management. Ann Pharmacother 42(2):226–236

Khan N (2007) The immunological burden of human cytomegalovirus infection. Arch Immunol Ther Exp 55(5):299–308

Kimura M, Maekura S, Satou T, Hashimoto S (1993a) Cytomegaloviral inclusions detected in the seminal vesicle, ductus deferens and lungs in an autopsy case of lung cancer [Japanese]. Rinsho Byori 41:1059–1062

Kimura M, Maekura S, Satou T, Hashimoto S (1993b) [Cytomegaloviral inclusions detected in the seminal vesicle, ductus deferens and lungs in an autopsy case of lung cancer]. Rinsho Byori 41(9):1059–1062

Klimova RR et al (2010) Herpes simplex virus and cytomegalovirus in male ejaculate: herpes simplex virus is more frequently encountered in idiopathic infertility and correlates with the reduction in sperm parameters [Russian]. Vopr Virusol 55:27–31

Korzeniewski K, Juszczak D (2015) Travel-related sexually transmitted infections. Int Marit Health 66(4):238–246

Krieger JN, Nyberg L Jr, Nickel JC (1999) NIH consensus definition and classification of prostatitis. JAMA 282(3):236–237

Krishnan R, Heal MR (1991) Study of the seminal vesicles in acute epididymitis. Br J Urol 67(6):632–637

Kumar L, Lal N, Kumar V, Sarswat A, Jangir S, Bala V, Kumar L, Kushwaha B, Pandey AK, Siddiqi MI, Shukla PK (2013) Azole–carbodithioate hybrids as vaginal anti-Candida contraceptive agents: design, synthesis and docking studies. Eur J Med Chem 70:68–77

Kushwaha B, Mandalapu D, Bala V, Kumar L, Pandey A, Pandey D, Yadav SK, Singh P, Shukla PK, Maikhuri JP, Sankhwar SN (2016) Ammonium salts of carbamodithioic acid as potent vaginal trichomonacides and fungicides. Int J Antimicrob Agents 47(1):36–47

Kyo S et al (1994) Detection of high-risk human papillomavirus in the cervix and semen of sex partners. J Infect Dis 170:682–685

La Vignera S, Vicari E, Condorelli RA, d'Agata R, Calogero AE (2011) Male accessory gland infection and sperm parameters (review). Int J Androl 34(5 Pt 2):e330–e347

La Vignera S, Condorelli RA, Vicari E, D'Agata R, Calogero AE (2012) Sperm DNA damage in patients with chronic viral C hepatitis. Eur J Intern Med 23:e19–e24

Lai YM et al (1997) The effect of human papillomavirus infection on sperm cell motility. Fertil Steril 67:1152–1155

LaMontagne DS, Fenton KA, Randall S, Anderson S, Carter P (2004) Establishing the National Chlamydia Screening Programme in England: results from the first full year of screening. Sex Transm Infect 80:335–341

Lang DJ, Kummer JF, Hartley DP (1974) Cytomegalovirus in semen. Persistence and demonstration in extracellular fluids. N Engl J Med 291:121–123

Laprise C, Trottier H, Monnier P, Coutlee F, Mayrand MH (2014) Prevalence of human papillomaviruses in semen: a systematic review and meta-analysis. Hum Reprod 29:640–651

Lederman MM, Offord RE, Hartley O (2006) Microbicides and other topical strategies to prevent vaginal transmission of HIV. Nat Rev Immunol 6:371–382

Lemoine M, Nayagam S, Thursz M (2013) Viral hepatitis in resource-limited countries and access to antiviral therapies: current and future challenges. Futur Virol 8(4):371–380

Lepor H, Wang B, Shapiro E (1994) Relationship between prostatic epithelial volume and serum prostate-specific antigen levels. Urology 44(2):199–205

Levy R et al (1997) Detection of cytomegalovirus in semen from a population of men seeking infertility evaluation. Fertil Steril 68:820–825

Ljungman P, Griffiths P, Paya C (2002) Definitions of cytomegalovirus infection and disease in transplant recipients. Clin Infect Dis 34(8):1094–1097

Lo JY, Ho KM, Leung AO, Tiu FS, Tsang GK, Lo AC, Tapsall JW (2008) Ceftibuten resistance and treatment failure of Neisseria gonorrhoeae infection. Antimicrob Agents Chemother 52(10):3564–3567

Lorusso F et al (2010) Impact of chronic viral diseases on semen parameters. Andrologia 42:121–126

Lossick JG (1990) Epidemiology of urogenital trichomoniasis. In: Trichomonads parasitic in humans. Springer, New York, pp 311–323

Maan MA, Fatma H, Muhammad J (2014) Epidemiology of hepatitis C viral infection in Faisalabad, Pakistan: a retrospective study (2010–2012). Afr Health Sci 14(4):810–814

Maartens G, Celum C, Lewin SR (2014) HIV infection: epidemiology, pathogenesis, treatment, and prevention. Lancet 384(9939):258–271

Macickova T, Kettner M, Linek K (1990) In vivo effect of 2, 4-dinitro-phenyl-substituted sugar hydrazones and phenylosazone against Trichomonas vaginalis and Tritrichomonas foetus. Pharmazie 45(3):204–206

Maeda S et al (2004) Detection of *Mycoplasma genitalium*, *Mycoplasma hominis*, *Ureaplasma parvum* (biovar 1) and *Ureaplasma urealyticum* (biovar 2) in patients with non-gonococcal urethritis using polymerase chain reaction-microtiter plate hybridization. Int J Urol 11:750–754

Malagoli M, Rossi T, Baggio A, Zandomeneghi G, Zanca A, Casolari C, Castelli M (2002) In vitro study of chemotherapeutic activity of sulphimidazole on some sensitive and metronidazole resistant-Trichomonas vaginalis strains. Pharmacol Res 46(5):469–472

Mandalapu D, Lal N, Kumar L, Kushwaha B, Gupta S, Kumar L, Bala V, Yadav SK, Singh P, Singh N, Maikhuri JP (2015) Innovative disulfide esters of dithiocarbamic acid as women-controlled contraceptive microbicides: a bioisosterism approach. ChemMedChem 10(10):1739–1753

Manhart LE, Broad JM, Golden MR (2011) *Mycoplasma genitalium*: should we treat and how? Clin Infect Dis 53(Suppl. 3):S129–S142

Manns MP, McHutchison JG, Gordon SC, Rustgi VK, Shiffman M, Reindollar R, Goodman ZD, Koury K, Ling MH, Albrecht JK, Group, I.H.I.T (2001) Peginterferon alfa-2b plus ribavirin compared with interferon alfa-2b plus ribavirin for initial treatment of chronic hepatitis C: a randomised trial. Lancet 358(9286):958–965

Marconi M, Pilatz A, Wagenlehner F, Diemer T, Weidner W (2009) Impact of infection on the secretory capacity of the male accessory glands. Int Braz J Urol 35(3):299–309

Martínez-Prado E, CamejoBermúdez MI (2010) Expression of IL-6, IL-8, TNF-α, IL-10, HSP-60, anti-HSP-60 antibodies, and anti-sperm antibodies, in semen of men with leukocytes and/or bacteria. Am J Reprod Immunol 63(3):233–243

Mastroianni A, Coronado O, Manfredi R, Chiodo F, Scarani P (1996) Acute cytomegalovirus prostatitis in AIDS. Genitourin Med 72

Mazzoli S, Cai T, Addonisio P, Bechi A, Mondaini N, Bartoletti R (2010) Chlamydia trachomatis infection is related to poor semen quality in young prostatitis patients. Eur Urol 57(4):708–714

Mielczarek E, Blaszkowska J (2015) Trichomonas vaginalis: pathogenicity and potential role in human reproductive failure. Infection:1–12

Mlynarczyk-Bonikowska B, Majewska A, Malejczyk M, Mlynarczyk G, Majewski S (2013) Antiviral medication in sexually transmitted diseases. Part I: HSV, HPV. Mini Rev Med Chem 13(13):1837–1845

Modi AA, Liang TJ (2008) Hepatitis C: a clinical review. Oral Dis 14(1):10–14

Monavari SH, Vaziri MS, Khalili M, Shamsi-Shahrabadi M, Keyvani H, Mollaei H, Fazlalipour M (2013) Asymptomatic seminal infection of herpes simplex virus: impact on male infertility. J Biomed Res 27(1):56–61

Montagut JM, Lepretre S, Degoy J, Rousseau M (1991) Ureaplasma in semen and IVF. Hum Reprod 6:727–729

Motrich RD, Maccioni M, Molina R, Tissera A, Olmedo J, Riera CM, Rivero VE (2005) Reduced semen quality in chronic prostatitis patients that have cellular autoimmune response to prostate antigens. Hum Reprod 20(9):2567–2572

Motrich RD, Cuffini C, Oberti JP, Maccioni M, Rivero VE (2006) *Chlamydia trachomatis* occurrence and its impact on sperm quality in chronic prostatitis patients. J Infect 53:175–183
Muciaccia B et al (1998) Testicular germ cells of HIV-seropositive asymptomatic men are infected by the virus. J Reprod Immunol 41:81–93
Muciaccia B, Corallini S, Vicini E, Padula F, Gandini L, Liuzzi G, Lenzi A, Stefanini M (2007) HIV-1 viral DNA is present in ejaculated abnormal spermatozoa of seropositive subjects. Hum Reprod 22(11):2868–2878
Naumenko VA et al (2011) Detection of human cytomegalovirus in motile spermatozoa and spermatogenic cells in testis organotypic culture. Herpesviridae 2:7
Neofytou E, Sourvinos G, Asmarianaki M, Spandidos DA, Makrigiannakis A (2009) Prevalence of human herpes virus types 1–7 in the semen of men attending an infertility clinic and correlation with semen parameters. Fertil Steril 91
Ness RB, Markovic N, Carlson CL, Coughlin MT (1997) Do men become infertile after having sexually transmitted urethritis? An epidemiologic examination. Fertil Steril 68(2):205–213
Newman L, Rowley J, Vander Hoorn S, Wijesooriya NS, Unemo M, Low N, Stevens G, Gottlieb S, Kiarie J, Temmerman M (2015) Global estimates of the prevalence and incidence of four curable sexually transmitted infections in 2012 based on systematic review and global reporting. PLoS One 10(12):e0143304
Nuovo GJ et al (1994) HIV-1 nucleic acids localize to the spermatogonia and their progeny. A study by polymerase chain reaction in situ hybridization. Am J Pathol 144:1142–1148
Ochsendorf FR (2008) Sexually transmitted infections: impact on male fertility. Andrologia 40(2):72–75
Ohnishi M, Saika T, Hoshina S, Iwasaku K, Nakayama SI, Watanabe H, Kitawaki J (2011) Ceftriaxone-resistant Neisseria gonorrhoeae, Japan. Emerg Infect Dis 17(1):148–150
Okamura N, Tajima Y, Ishikawa H, Yoshii S, Koiso K, Sugita Y (1986) Lowered levels of bicarbonate in seminal plasma cause the poor sperm motility in human infertile patients. Fertil Steril 45(2):265–272
Ouzounova-Raykova V, Ouzounova I, Mitov IG (2010) May *Chlamydia trachomatis* be an aetiological agent of chronic prostatic infection? Andrologia 42:176–181
Pajovic B, Radojevic N, Vukovic M, Stjepcevic A (2013) Semen analysis before and after antibiotic treatment of asymptomatic Chlamydia-and Ureaplasma-related pyospermia. Andrologia 45:266–271
Patel R, Rompalo A (2005) Managing patients with genital herpes and their sexual partners. Infect Dis Clin N Am 19(2):427–438
Pellati D et al (2008) Genital tract infections and infertility. Eur J Obstet Gynecol Reprod Biol 140:3–11
Perez-Andino J, Buck CB, Ribbeck K (2009) Adsorption of human papillomavirus 16 to live human sperm. PLoS One 4(6):e5847
Perino A et al (2011) Human papillomavirus infection in couples undergoing in vitro fertilization procedures: impact on reproductive outcomes. Fertil Steril 95:1845–1848
Reichart M, Kahane I, Bartoov B (2000) In vivo and in vitro impairment of human and ram sperm nuclear chromatin integrity by sexually transmitted Ureaplasma urealyticum infection. Biol Reprod 63:1041–1048
Reichart M, Levi H, Kahane I, Bartoov B (2001) Dual energy metabolism-dependent effect of Ureaplasma urealyticum infection on sperm activity. J Androl 22:404–412
Renaudin H, Tully JG, Bebear C (1992) In vitro susceptibilities of Mycoplasma genitalium to antibiotics. Antimicrob Agents Chemother 36(4):870–872
Rosen HR (2011) Chronic hepatitis C infection. N Engl J Med 364(25):2429–2438
Ryan KJ, Ray CG (2004) Medical microbiology: an introduction to infectious diseases. Mcgraw-Hill
Rybar R, Prinosilova P, Kopecka V, Hlavicova J, Veznik Z, Zajicova A, Rubes J (2012) The effect of bacterial contamination of semen on sperm chromatin integrity and standard semen parameters in men from infertile couples. Andrologia 44(s1):410–418

Satta A et al (2006) Experimental *Chlamydia trachomatis* infection causes apoptosis in human sperm. Hum Reprod 21:134–137

Schneede P, Tenke P, Hofstetter AG (2003) Sexually transmitted diseases (STDs)—a synoptic overview for urologists. Eur Urol 44(1):1–7

Schneider H, Hammack CA, Apicella MA, Griffiss JM (1988) Instability of expression of lipooligosaccharides and their epitopes in Neisseria gonorrhoeae. Infect Immun 56(4):942–946

Sethi S, Singh G, Samanta P, Sharma M (2012) Mycoplasma genitalium: an emerging sexually transmitted pathogen. Indian J Med Res 136(6):942

Shang XJ et al (1999) Ureaplasma urealyticum infection and apoptosis of spermatogenic cells. Asian J Androl 1:127–129

Shehu-Xhilaga M et al (2007) The testis and epididymis are productively infected by SIV and SHIV in juvenile macaques during the post-acute stage of infection. Retrovirology 4:7

Sheikh SH, Waqqar S, William GP, Aziz S, Bano S, David A, Ahsan M (2014) Herpes simplex & cytomegaloviruses inflicted semen substantiates infertility among men. J Dent Med Sci:74–77

Shevchuk MM, Nuovo GJ, Khalife G (1998) HIV in testis: quantitative histology and HIV localization in germ cells. J Reprod Immunol 41:69–79

Shouval D (2003) Hepatitis B vaccines. J Hepatol 39:70–76

Soni S et al (2010) The prevalence of urethral and rectal *Mycoplasma genitalium* and its associations in men who have sex with men attending a genitourinary medicine clinic. Sex Transm Infect 86:21–24

Soper D (2004) Trichomoniasis: under control or undercontrolled? Am J Obstet Gynecol 190(1):281–290

Svenstrup HF, Fedder J, Abraham-Peskir J, Birkelund S, Christiansen G (2003) *Mycoplasma genitalium* attaches to human spermatozoa. Hum Reprod 18:2103–2109

Taylor BD, Haggerty CL (2011) Management of *Chlamydia trachomatis* genital tract infection: screening and treatment challenges. Infect Drug Resist 4:19–29

Taylor-Robinson D, Bebear C (1997) Antibiotic susceptibilities of mycoplasmas and treatment of mycoplasmal infections. J Antimicrob Chemother 40(5):622–630

Toy M, Veldhuijzen IK, Mostert MC, De Man RA, Richardus JH (2008) Transmission routes of hepatitis B virus infection in chronic hepatitis B patients in the Netherlands. J Med Virol 80(3):399–404

Trojian TH, Lishnak TS, Heiman D (2009) Epididymitis and orchitis: an overview. Am Fam Physician 79:583–587

Tuttle JP Jr, Holbrook TW, Derrick FC (1977) Interference of human spermatozoal motility by Trichomonas vaginalis. J Urol 118(6):1024–1025

Twu O, Dessí D, Vu A, Mercer F, Stevens GC, De Miguel N, Rappelli P, Cocco AR, Clubb RT, Fiori PL, Johnson PJ (2014) Trichomonas vaginalis homolog of macrophage migration inhibitory factor induces prostate cell growth, invasiveness, and inflammatory responses. Proc Natl Acad Sci 111(22):8179–8184

Uuskula A, Kohl PK (2002) Genital mycoplasmas, including *Mycoplasma genitalium*, as sexually transmitted agents. Int J STD AIDS 13:79–85

van Leeuwen E et al (2008) Effects of antiretroviral therapy on semen quality. AIDS 22:637–642

Van Vranken M (2007) Prevention and treatment of sexually transmitted diseases: an update. Am Fam Physician 76(12)

Varani S, Landini MP (2011) Cytomegalovirus-induced immunopathology and its clinical consequences. Herpesviridae 2(1):1

Veznik Z, Pospisil L, Svecova D, Zajicova A, Unzeitig V (2004) Chlamydiae in the ejaculate: their influence on the quality and morphology of sperm. Acta Obstet Gynecol Scand 83:656–660

Vicari E et al (2006) Sperm output in patients with primary infertility and hepatitis B or C virus; negative influence of HBV infection during concomitant varicocele. Minerva Med 97:65–77

Wang SA, Lee MVC, O'Connor N, Iverson CJ, Ohye RG, Whiticar PM, Hale JA, Trees DL, Knapp JS, Effler PV, Weinstock HS (2003) Multidrug-resistant Neisseria gonorrhoeae with decreased susceptibility to cefixime—Hawaii, 2001. Clin Infect Dis 37(6):849–852

Weidner W, Krause W, Ludwig M (1999) Relevance of male accessory gland infection for subsequent fertility with special focus on prostatitis. Hum Reprod Update 5(5):421–432

Weiss RA (1993) How does HIV cause AIDS? Science 260(5112):1273–1279

Workowski KA, Berman SM (2007) Centers for disease control and prevention sexually transmitted diseases treatment guidelines. Clin Infect Dis 44(Suppl 3):S73–S76

Workowski KA, Berman SM, Centers for Disease Control and Prevention (2006) Sexually transmitted diseases treatment guidelines, 2006. MMWR Recomm Rep 55(RR-11):1–94

World Health Organization (2016) Fact sheets, sexually transmitted infections (STIs). http://www.who.int/mediacentre/factsheets/fs110/en/

Wu KH et al (2007) Infection of cytomegalovirus and herpes simplex virus and morphology of the infected spermatogenic cells in infertile men [Chinese]. Zhonghua Nan Ke Xue 13

Xu X (1992) The possible role of sperm in family HBV infection [Chinese]. Zhonghua Liu Xing Bing Xue Za Zhi 13:337–339

Xu C, Sun GF, Zhu YF, Wang YF (1997) The correlation of Ureaplasma urealyticum infection with infertility. Andrologia 29(4):219–226

Yang Y, Jia CW, Ma YM, Zhou LY, Wang SY (2013) Correlation between HPV sperm infection and male infertility. Asian J Androl 15(4):529–532

Yokoi S, Deguchi T, Ozawa T, Yasuda M, Ito SI, Kubota Y, Tamaki M, Maeda SI (2007) Threat to cefixime treatment for gonorrhea. Emerg Infect Dis 13(8):1275–1277

Zaneveld LJ, Waller DP, Anderson RA, Chany C 2nd, Rencher WF, Feathergill K, Diao XH, Doncel GF, Herold B, Cooper M (2002) Efficacy and safety of a new vaginal contraceptive antimicrobial formulation containing high molecular weight poly(sodium 4-styrenesulfonate). Biol Reprod 66(4):886–894

Zeighami H, Peerayeh SN, Yazdi RS, Sorouri R (2009) Prevalence of *Ureaplasma urealyticum* and *Ureaplasma parvum* in semen of infertile and healthy men. Int J STD AIDS 20:387–390

Zhou XP et al (2011) Comparison of semen quality and outcome of assisted reproductive techniques in Chinese men with and without hepatitis B. Asian J Androl 13:465–469

Zinzendorf NY, Kouassi-Agbessi BT, Lathro JS, Don C, Kouadio L (2008) Ureaplasma urealyticum or Mycoplasma hominis infections and semen quality of infertile men in Abidjan. Journal of Reproduction and Contraception. 19(2):65–72.

Ziomko I, Kuczyńska E (1993) [Trichomonacidal activity of newly synthesized derivatives of benzoizothiazolinon (BIT) in vitro]. Wiad Parazytol 40(1):59–64

Cytogenetic Factors in Male Infertility 13

Vertika Singh and Kiran Singh

Abstract

Nearly 15% of the couples worldwide face the problem of infertility. A number of cytogenetic aberrations in the form of somatic chromosome aneuploidies, sperm aneuploidies, chromosomal translocations and inversions, etc. are known to contribute to male infertility. Couples with normal hormonal profile should be evaluated for possible cytogenetic abnormalities before proceeding to treatment. The identification of cytogenetic abnormality cannot only explain infertility but also guide treatment in the affected cases. This chapter summarizes the cytogenetic factors that increase the risk of male infertility. Towards the end, we have provided a glimpse of the contemporary techniques that have revolutionized the classical field of cytogenetics.

Keywords

Cytogenetics • 47,XXY • 46,XX male • 47,XYY • Klinefelter's syndrome • Germ cell aneuploidy • Interchromosomal effects • Cytogenetic techniques

Key Points

- Chromosomal abnormalities and mutations of genes involved in germ cell production and function account for 30% of the infertile cases.
- The prevalence of cytogenetic anomalies in infertile males varies between 2% and 8% and is as high as 20% in azoospermia.
- High frequency of chromosomal abnormalities (42.5%) has been reported in embryos derived from ICSI cycles of males with meiotic aberrations.

V. Singh • K. Singh (✉)
Department of Molecular and Human Genetics, Institute of Science, Banaras Hindu University, Varanasi, Uttar Pradesh, India
e-mail: singhk4@rediffmail.com

R. Singh, K. Singh (eds.), *Male Infertility: Understanding, Causes and Treatment*, DOI 10.1007/978-981-10-4017-7_13

- 66% of abnormal karyotypes obtained from miscarriages have a male factor origin.
- Autosomal structural rearrangements may lead to an increase in the frequency of sex chromosome aneuploidies, called as the interchromosomal effect.
- Next-generation sequencing technique offers the highest accuracy and detection limit till now.

13.1 Introduction

Spermatogenesis is essential for the perpetuation of the male germline. It is marked by highly regulated mechanism of interaction between somatic and germ cells with coordinated hormonal regulation. After the identification of the hormonal dysregulation as a contributor to male infertility, the next major discovery was the identification of the molecular aberrations at the level of chromosomes. With the inception of the field of cytogenetics, chromosomal defects in the form of aneuploidies, breaks or translocations were identified. It is estimated that chromosomal abnormalities and mutations of genes involved in germ cell production and function account for 30% of the infertile cases. Due to deletions or insertions or a large number of genes in the form of big chunks of DNA or the whole chromosome, a number of cytogenetic abnormalities infest in the form of syndromes. Cytogenetic aberrations interfere with the process of spermatogenesis, and the percentage of chromosomal abnormality increases proportionally with the decline in sperm concentrations and increasing severity of the infertility (McLachlan and O'Bryan 2010). The estimated frequency of the overall occurrence of a chromosomal factor in infertile males varies between 2% and 8% (Foresta et al. 2002). The prevalence further increases to 20% in azoospermic males, with sex chromosome most frequently involved (Dohle et al. 2002).

Aneuploidy reflects a change in the chromosome number from a normal diploid complement in somatic cells or haploid complement in the gametes. These defects can be either numerical in nature involving a gain or loss of an entire chromosome or structural involving a gain or loss of a chromosomal segment. Numerical chromosomal anomalies are the most frequently associated chromosomal abnormality in infertile males (Emery and Carrell 2006). High incidences of sex chromosome aneuploidy are reported in men with nonobstructive azoospermia (Palermo et al. 2002; Mateizel et al. 2002). Aneuploidy results from non-disjunction of chromosomes during meiosis. Klinefelter syndrome (KS), a condition characterized by at least a single supernumerary X chromosome is the most common abnormality reported in infertile males (Tüttelmann and Gromoll 2010; Fu et al. 2012) (cumulative 4.9%). The abnormality is characterized by progressive testicular failure resulting in small firm testes, androgen deficiency and azoospermia in males (Klinefelter et al. 1942). It affects around 1:600 males (Lissitsina et al. 2006). Around 80–90% of KS cases show an 'original' compliment of 47,XXY, whereas the remaining display (in decreasing frequency) a varying mosaicism (e.g. 47,XXY/46,XY), additional sex chromosomes (48,XXXY; 48,XXYY; 49,XXXXY) or structurally abnormal X chromosomes (Bojesen et al. 2003; Lanfranco et al. 2004; Tüttelmann and Gromoll 2010). Variants of Klinefelter patients with increasing number of X chromosomes show a female sexual phenotype owing to the X-chromosome dosage effect on the male gonad development (Vogt 2004). Other rare chromosomal abnormalities include XYY syndrome, 46,XX males and Noonan syndrome. The frequency of XYY syndrome is seen in around 0.84/1000 infertile males. This disorder takes place as a result of paternal meiotic non-disjunction events and is characterized by tall, azoo-/oligozoospermia phenotype (Robinson and Jacobs 1999). Noonan syndrome also called as pterygium colli syndrome or male Turner syndrome occurs with a frequency of 1 in 1000 to 2500 live births. The chromosomal constitution of Noonan syndrome shows 46,XO/XY mosaicism. Most of the males with Noonan syndrome typically display cryptorchidism with elevated gonadotrophin levels. 46, XX males are sterile which occur with a frequency of 1 in 20,000. Though these males are devoid of Y chromosome, they show a distinctive presence of *SRY* gene, responsible for male sexual characteristic. Chromosomal translocations account for an additional source of aneuploidy in humans (Gianaroli et al. 2002). Disruption of genes due to translocations leads to a loss of genetic material resulting in transmission of an incorrect genetic message (Carrell 2008). Robertsonian translocations are the most frequent structural chromosomal abnormalities in humans (Therman and Susman 2012; Ferlin et al. 2007). Furthermore, the meiotic abnormalities during spermatogenesis and sperm aneuploidies are the leading cause of infertility, recurrent miscarriage, abnormal embryos and offspring. In this chapter we aim to delineate various chromosomal abnormalities associated with impaired fertility and associated reproductive outcomes.

Most of the evidence of chromosomal abnormalities in male infertility has come from case studies or case series as the individuals bearing these abnormalities acquire them de novo and are generally infertile.

13.2 *SRY* Gene Translocation on X Chromosome or Autosomes

The presence of *SRY* gene on the distal end of the Y chromosome encodes the 'testis-determining factor' responsible for the regression of Mullerian structures and production of testosterone in the Leydig cells. Most of the males (80%) devoid of Y chromosome are 'SRY positive'. Due to unequal crossing over events between homologous regions during paternal meiotic division, the *SRY* from the Y chromosome fragment is translocated to the short arm of an X chromosome or an autosome. Due to *SRY*, the affected individual shows male sexual characteristic, despite the XX sex chromosome pattern. However, a few reports demonstrated the presence of testicular tissue in *SRY*-negative patients. Several hypotheses have been put forward to explain the mechanism. Rajender et al. reported a hidden gonadal mosaicism for *SRY* gene as the reason for the development of testicular tissue and male phenotype in an *SRY*-negative male (Rajender et al. 2006). They suggested that the development of the male phenotype in the absence of *SRY* probably occurred from the loss of function mutation in some unknown gene termed 'Z'. Furthermore, a gain of function mutation in a few genes that might function downstream to the *SRY* gene in the sex-determination pathway may also lead to 46,XX female-to-male sex reversal (Kent et al. 1996; Meeks et al. 2003; Rajender et al. 2006; Maciel-Guerra et al. 2008). In 1999, Kusz et al. described that a preferential inactivation of the Y-bearing X chromosome in XX males with Y-to-X translocations could be the major mechanism triggering a sexually ambiguous phenotype (Kusz et al. 1999). The origin of a male phenotype in XX males thus could be the result of translocation of Y chromosome sequences, including the *SRY* gene, to an X chromosome or to an autosome, a mutation in a yet unknown X-linked or autosomal gene from the testis-determination pathway and cryptic Y chromosome mosaicism (Pandith et al. 2015).

13.3 Somatic Chromosome Aneuploidies

Somatic chromosomal abnormalities are reasonably common in humans. These can be numerical, with an extra chromosome, or structural, such as translocations. Chromosomal abnormalities have been widely accepted as a causative factor associated with infertility, an increased probability of pregnancy loss and various birth defects. These abnormalities are known to occur with a frequency of 3% and 19%, in the cases of subfertility and nonobstructive azoospermia (NOA), respectively (Yoshida et al. 1997). Thus, it is imperative to routinely practice karyotyping in all the unexplained cases of male infertility to provide the estimated probable risk and contribution of the cytogenetic abnormalities.

13.3.1 47,XXY

Klinefelter syndrome (KS) characterized by an extra X-chromosome (karyotype 47,XXY) is seen in about 80–90% of patients, while the remaining show chromosomal mosaicisms (e.g. 47, XXY/46, XY), additional sex chromosomes (e.g. 48, XXXY; 48, XXYY; 49, XXXXY) or X chromosome structural abnormalities (e.g. 47,X,iXq,Y) (Maiburg et al. 2012). The patients with Klinefelter syndrome (47,XXY) or mosaic variants display an impaired spermatogenic phenotype, which includes primary testicular failure with reduced testicular volume, hypergonadotropic hypogonadism and azoospermia or severe oligozoospermia in 90% and 10% of non-mosaic patients, respectively (Ferlin et al. 2007; De Sanctis and Ciccone 2010; Foresta et al. 2012). KS is the most frequent genetic cause of male infertility, which occurs with a frequency of 15% of all azoospermic cases and 3% of all infertile men (Anawalt 2013). These aneuploidies result due to parental and maternal non-disjunction events during meiosis, resulting in an extra X chromosome (Fig. 13.1). Progressive germ cell degeneration and impaired Sertoli cells (SCs) function in association with an extensive fibrosis and hyalinization of the seminiferous tubules, and Leydig cell hyperplasia results in an azoospermic phenotype in KS patients (Aksglæde et al. 2006). The mechanism associated with the global degeneration in these patients is still unclear. However, it has been hypothesized that an

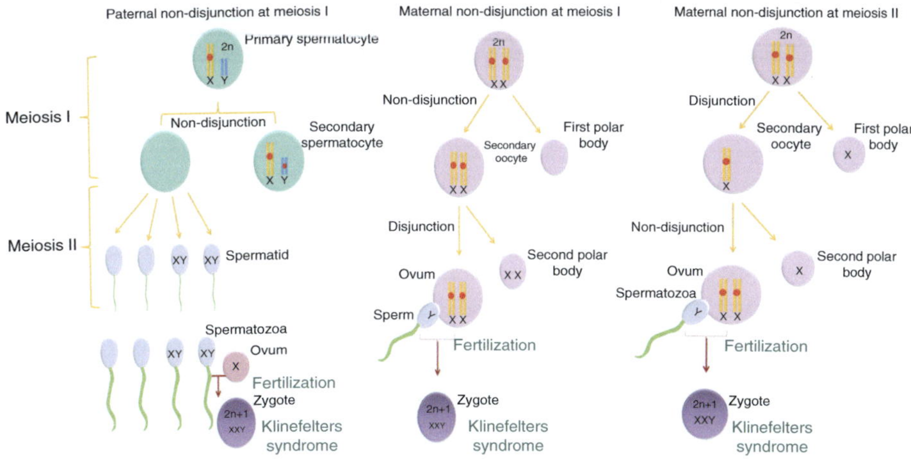

Fig. 13.1 Diagrammatic illustration of the phenomenon of non-disjunction (paternal and maternal) at the time of meiosis, leading to the development of a 47,XXY genotype

altered dosage of the genes which escape the inactivation on the supernumerary X chromosome might be responsible for the induction of germ cell loss during spermatogenesis (Aksglæde et al. 2006). A study recently reported that *SHOX* gene, which encodes for a transcription factor expressed in the developing skeleton and is associated with various skeletal anomalies seen in 47, XXY syndrome, might be responsible for the tall stature regularly seen in KS (Tüttelmann and Gromoll 2010).

It has been well established that in 47,XXY men, the supernumerary X chromosome is inherited with equal probability from the mother and the father (Thomas and Hassold 2003). This may further result in an altered differential expression of paternal versus maternal alleles due to imprinting (Iitsuka et al. 2001). This hypothesis was supported by another study which demonstrated that in a study on 61 KS men, a higher incidence of developmental problems in speech/language and motor impairment was seen when the supernumerary X chromosome was paternally inherited (Stemkens et al. 2006). The X chromosome harbours 9% of the genes (99 out of 1098) specifically expressed in testis which makes the fertility status of XXY patients highly variable (Ross et al. 2005; Aksglæde et al. 2006). The beginning of various testicular sperm retrieval technologies and microdissection technologies provided a testicular sperm retrieval rate ranging between 30% and 70% in the KS patients (Schiff et al. 2005; Koga et al. 2007; Yarali et al. 2009). This augmented the possibility for KS men to become father with the assistance of various assisted reproductive technologies.

13.3.2 47,XYY

47, XYY is the most common sex chromosome aneuploidy after Klinefelter syndrome (47, XXY) (Hook and Hamerton 1977; Gekas et al. 2001; Rives et al. 2005) with a clinical presentation of around 1 in 1000 live male births. Men with 47,XYY karyotype are generally phenotypically normal, but are at greater risk of behavioural problems, mild learning disability, delayed speech and language development (Evans et al. 1991). These men are usually fertile, but are more frequently seen in the infertile populations. Recent studies have reported an association of 47,XYY phenotype with chromosomally abnormal spermatozoa in the semen (Speed et al. 1991; Blanco et al. 1997; Chevret et al. 1997; Lim et al. 1999a, b; Gonzalez-Merino et al. 2007; Wong et al. 2008). However, men with 47,XYY karyotype display a wide range of seminal phenotypic variability from normal sperm count to azoospermia (Lim et al. 1999a, b; Egozcue et al. 2000; Rives et al. 2005; Moretti et al. 2007; Abdel-Razic et al. 2012). A group of researchers analysed 75 sperm karyotypes from a 47,XYY male and reported that none of the karyotypes were abnormal for sex chromosomes. They suggested that elimination of the extra chromosome during spermatogenesis can result in production of sperm lacking sex chromosome disomy in 47,XYY men (Benet and Martin 1988). They further analysed 10,000 sperm from the same men by fluorescence in situ hybridization and reported a small but significant increase of XY disomy (Martin et al. 1999). Since men with 47,XYY syndrome present a heterogeneous phenotype with a diverse spectrum of clinical presentation, it becomes really difficult to diagnose especially in the cases where there is normal fertility.

13.4 Meiotic Abnormalities and Sperm Aneuploidies

In the recent years, the use of intra-cytoplasmic sperm injection (ICSI) in assistance with the new methods used for recovering testicular spermatozoa has significantly improved the fertility dimensions for infertile patients (Van Steirteghem et al. 1993). However, the preimplantation genetic diagnosis and careful observations following ICSI have shown a significant increase in de novo sex chromosome abnormalities and structural abnormalities in sperm (Bonduelle et al. 2002; Van Steirteghem et al. 2002). These abnormalities may arise due to meiotic abnormalities during the process of spermatogenesis. High frequency of sperm aneuploidies has been shown to be associated with altered meiosis in infertile men. Meiotic errors can even result in the production of abnormal sperm, which retain fertilization capabilities, but increase the risk of recurrent miscarriage and abnormal embryos or offspring. High frequency of chromosomal abnormalities (42.5%) has been reported in embryos derived from ICSI cycles of males with meiotic aberrations (Aran et al. 2004).

In 1990s, fluorescence in situ hybridization (FISH) was developed as a faster, easier and economic method to detect aneuploidies in human sperm (Martin 2008a, b). This technique utilizes fluorescent-tagged probes complementary to the DNA to visualize the region of interest. Since then, it is the most widely used technique for the detection of sperm aneuploidies. FISH analyses have revealed a higher manifestation of numerical chromosomal abnormalities in infertile men, particularly in sex chromosomes (Moosani et al. 1995; Bernardini et al. 1998; Aran et al. 1999; Colombero et al. 1999; Pang et al. 1999; Pfeffer et al. 1999; Nishikawa et al. 2000; Ushijima et al. 2000; Vegetti et al. 2000; Calogero et al. 2001a, b; Rubio et al. 2001;

Martin et al. 2003). However, a higher incidence of sex chromosome aneuploidy was observed in sperm derived from oligoasthenoteratozoospermic men as compared to the normozoospermic, asthenozoospermic, teratozoospermic and asthenoteratozoospermic males (Rubio et al. 2001). The percentage of chromosome 21 disomic sperm was also higher in OAT patients in comparison with other infertile phenotypes (Rubio et al. 2001). A gradual increase in aneuploidy rates was observed with declining sperm concentration in infertile men. Some studies, however, reported the highest percentage of aneuploidy in severe oligozoospermia patients having a sperm concentration of $<1 \times 10^6$ sperm/mL (Rubio et al. 2001). Individuals with karyotype abnormalities present an obvious predisposition towards chromosomally abnormal conceptions as a result of which they fail to achieve a successful pregnancy often due to repeated spontaneous abortions (Harton and Tempest 2012). The FISH analysis demonstrated that the sex chromosome aneuploidies range between 2% (Rives et al. 2000) and 45% (Estop et al. 1999) in sperm derived from men with Klinefelter syndrome and from 1.5% (Lim et al. 1999a, b) to 7% (Kruse et al. 1998) in sperm from Klinefelter mosaics. Interestingly, some studies have reported that sperm chromosomes derived from men with Klinefelter syndrome have a tendency to eliminate the extra sex chromosome during the process of spermatogenesis.

The majority of children born to 47,XXY men have been normal although chromosomally abnormal foetuses have been reported (Ron-el et al. 2000; Friedler et al. 2001). Staessen et al. in 2003 studied 113 embryos by preimplantation genetic diagnosis (PGD) and found a significantly increased frequency of autosomal and sex chromosomal abnormalities (Staessen et al. 2003). Thus, there appears to be a small increased risk for these men. This group found that the frequency of sperm aneuploidy was concordant with the frequency of aneuploidy in preimplantation embryos (32%). Since many 47,XYY men have normal semen parameters, severe oligozoospermia observed in these men may indicate more perturbations during meiotic pairing, subsequent loss of germ cells and the production of aneuploid sperm.

13.5 Chromosomal Translocations and Inversions

A balanced chromosomal translocation occurs when two chromosomes break followed by abnormal repair of their chromosome fragments, resulting in the transposition of genetic material from one chromosome to the other without loss of any genetic material. In most of the cases, the carriers of balanced translocation display normal phenotypic characters, but may experience various birth defects and reduced fertility outcomes (Tempest and Simpson 2010). Robertsonian translocation involves the fusion of long arms of two acrocentric chromosomes. In this instance, the short arm is generally lost, which results in chromosomal constitution of 45 chromosomes. Reciprocal translocations, however, occur when there is an exchange of material between two or more chromosomes with the involvement of at least one non-acrocentric chromosome. Reduced fertility in translocation carriers may result due to the formation of quadrivalent (in case of reciprocal translocation or trivalent in case of Robertsonian translocation) structures during the pairing of homologous chromosomes. The probable explanation behind the reduced fertility due to the formation of quadrivalent and divalent are the time restrictions and mechanics involved in the formation of such complexes and disjunction of these structures which may result in the production of unbalanced gametes (Tempest and Simpson 2010). The frequency of unbalanced gametes has been reported to vary between 3% and 36% in a study performed on 20 carriers of balanced Robertsonian translocations (Ferlin et al. 2005; Sarrate et al. 2005). However, an analysis of 30 balanced reciprocal translocation carriers reported a high frequency of genetically unbalanced sperm (29–81%) as compared to the Robertsonian translocation carriers. Similarly, the chromosomal inversions may also result in fertility- and birth-related complications due to the formation of inversion loops to enable the chromosomes pair at the time of meiosis (Brown et al. 1998). Various molecular studies have reported that the recombination event between these loops is restrained due to mechanics and time constraints associated with the formation of loop. A restricted recombination further results in meiotic breakdown and germ cell apoptosis, thus reducing sperm count. Moreover, the recombination between the inversions loops augments the possibility of a production of unbalanced gametes. Nevertheless, the frequency of the production of an unbalanced gamete depends on the chromosomes involved, the length of the region involved and the probability of recombination to occur within the inverted sequences. Very few studies have investigated the frequency of unbalanced gametes in the carriers of balanced inversions; however, the reported frequency ranges from 1% to 54% of unbalanced sperm.

13.6 The Interchromosomal Effects

The effect of inversions and translocations on the synapsis and disjunction of heterologous chromosomes is called as the interchromosomal effect or the 'Schultz–Redfield effect' (Schultz and Redfield 1951; Averhoff and Richardson 1974). The concept of interchromosomal effect in humans was first postulated by Lejeune in 1963 (Lejeune 1963). The logical explanation behind this phenomenon is based on the hypothetical formation of heterologous pairing among the rearranged chromosomes, which frequently adopt configurations with asynaptic regions and other chromosomes (Guichaoua et al. 1990). The chromosome of infertile males shows a high degree of variability in chromosomal segregation patterns during meiosis, which can be attributed to complex chromosomal rearrangements and aneuploidies that could affect the meiotic synapsis (Miharu et al. 1994; Moosani et al. 1995). Many reports have demonstrated an increase in the frequency of sex chromosome aneuploidies in patients with autosomal structural rearrangements (Morel et al. 2001; Anton et al. 2004; Roux et al. 2005). This can be attributed to alterations in the meiotic process produced by the rearrangement. A research group recently analysed the meiotic segregation pattern in sperm of a patient who was a double Robertsonian carrier with karyotype 45,XY,der(13;13)/45,XY,der(13;14). The patient with this rare translocation reflected a high rate of unbalanced gametes with disomic spermatozoa (de Vozzi et al. 2009). Studies performed on sperm obtained from carriers of structural chromosomal rearrangements have reported an ICE of 58% in Robertsonian translocation carriers and 64% in reciprocal translocation carriers (Martin 2008a, b). In addition, the PGD studies have also reported an increase in chromosomal aneuploidy rates in embryos for chromosomes which are not involved in rearrangements (Martin 2008a, b).

13.7 Sperm Aneuploidies and Adverse Reproductive Outcomes

Sperm aneuploidies can influence the reproductive outcomes at various stages of development, right from the process of fertilization to the embryo development, pregnancy and birth (Vera et al. 2012). A large number of studies have identified the clinical concerns of sperm aneuploidies during in vitro fertilization (IVF) cycles. Sperm aneuploidy rates were associated with repeated ICSI failures in a prospective study, and the frequency was even higher for aneuploidies of chromosome 18 and the sex chromosomes (Nicopoullos et al. 2008). In an analysis, Martin et al. identified an infertile men who had a ninefold higher frequency of 24,XY sperm than controls (Martin 1986). Later, this man was identified to induce a pregnancy through ICSI that resulted in a 47,XXY foetus (Moosani et al. 1999). Increase in sperm aneuploidy levels is also shown to correlate with the abnormal development of embryo. FISH analyses have identified that abnormal sperm significantly increase the number of mosaic embryos (Rodrigo et al. 2003). A decline in sperm concentration to <5 million sperm/mL has also been shown to result in the development of abnormal embryos with sex chromosome aneuploidies (Pehlivan et al. 2003). High embryonic mosaicism rate (53%) is reported in nonobstructive azoospermia patients (Silber et al. 2003). Moreover, an increase in diploid sperm resulted in triploid embryos, which resulted in spontaneous abortions (Rodrigo et al. 2010). Chromosomally abnormal sperm is related to poor implantation rates and miscarriage. An increase in sex chromosome disomy was reported in 31.6% of males with three or more implantation failures (Rubio et al. 2001). A study recently identified that around 66% of abnormal karyotypes obtained from miscarriages have a male factor origin (Kim et al. 2010).

13.8 Advances in Human Molecular Cytogenetics: From Chromosomes to SNPs

It was in 1956 when Tjio and Levan established the human diploid chromosome number as 46. From then began the area of human cytogenetics, which became even more popular with the advent of G-banding technique, which utilized trypsin digestion followed by Giemsa staining to identify each chromosome. This technique could identify chromosomal rearrangements of 5–10 Mb in size. With the discovery of fluorescence in situ hybridization (FISH), it became possible to localize target

DNA sequence in the regions of interest. This technique was breakthrough in bringing cytogenetics towards clinical centres to investigate the cytogenetic cause of various disorders.

With complete genome sequencing of human, the field gained enormous expansions in technological advancements and research applications. Now we know that in addition to aneuploidies and large structural rearrangements, human genome is susceptible to much wider range of variations in form of single nucleotide polymorphisms (SNPs), copy number variations (CNVs) and large-scale deletions. Some of these variations may be tolerated, while others can be pathogenic in nature. The availability of human genome sequence brought forward a number of cytogenetic high-resolution techniques that allow the detection of submicroscopic alterations and single nucleotide variations. Some of these techniques include modified FISH techniques, called multiplex-FISH and combining binary and ratio-labelling (COBRA-FISH) and high-throughput techniques such as array CGH, SNP array and massive parallel sequencing, popularly known as next-generation sequencing.

13.8.1 Multiplex-Fluorescence In Situ Hybridization (M-FISH) and Spectral Karyotyping (SKY)

M-FISH is used to stain human chromosomes with distinctive colours for karyotyping. It is based on combinatorial labelling of fluorescent probes to produce 24 colours in order to identify complex chromosomal rearrangements and the presence of marker chromosomes. Thus, it allows the detection of 22 autosomes, X and Y chromosomes. A similar technique called as spectral karyotyping (SKY) permits the classification of chromosomes on the basis of different emission spectra (Schrock et al. 1996). SKY uses the combination of chromosome painting and multicolour fluorescence to paint each of the 24 chromosomes of human with different colours. In this technique new colours are developed by mixing a pair of different fluorescent dyes among spectrum orange, Cy5, Texas red, spectrum green and Cy5.5.

13.8.2 Combining Binary and Ratio Labelling (COBRA-FISH)

COBRA-FISH is a modification of classical MFISH technique that combines the use of combinatorial labelling and ratio labelling (Tanke et al. 1999). The ratio labelling procedure allows different ratios of fluorescent labels to differentiate between the probes. Thus, a fewer number of fluorochromes could generate more number of pseudocolours, allowing 48 colour combinations for differential recognition of human chromosome arm (Wiegant et al. 2000). Depending upon applications, some other modifications of FISH include Comet-FISH (for detection of DNA damage), Halo-FISH (for detection of DNA/chromatin organization) and Flow-FISH (For identification of telomeric repeats).

13.8.3 Array-Based CGH

Array-based comparative genomic hybridization (aCGH) is a powerful method used to detect aneuploidies, uniparental disomy and genome-wide imbalances/submicroscopic alterations in the form of copy number changes (gains/loss) in target DNA sample. In this technique millions of oligonucleotide probes from the human genome are immobilized on a glass slide by photolithography in the form of an array. Differentially labelled/fluorescent-tagged genomic DNA from sample of interest and reference DNA are co-hybridized on the array to detect deferential hybridization in the form of relative fluorescence signals. Cyanine 3 (Cy3) and cyanine 5 (Cy5) are the two most commonly used fluorescent labels in aCGH. The inability to detect balanced translocations and other rearrangements that do not allow a detectable change in copy number are some of the limitations of aCGH.

13.8.4 Single-Nucleotide Polymorphism Array (SNP Array)

Around ten million SNPs are present in the human genome, which can be pathogenic or nonpathogenic in nature (Kruglyak et al. 2001). SNP array allows the identification of various SNPs across the genome. It is based on the complementary binding of target DNA base to unique reference oligonucleotide probes spotted on a chip. Each probe is designed for a specific DNA region. The detection is made on the basis of differential signal intensity produced depending on affinity between the target and the probe. It offers genotyping accuracy over 99.5% and is capable of interrogating millions of SNPs per run.

13.8.5 Next-Generation Sequencing (NGS)

Massive parallel sequencing or next-generation sequencing is one of the highly used techniques nowadays to analyse novel SNPs, copy number variations and transcriptomic and epigenetic alterations. NGS technologies use a number of different chemistries which allows parallel sequencing of a number of DNA fragments. It is based on the concept of incorporation fluorescently labelled dNTPs by DNA polymerase during consecutive cycles of DNA synthesis and the identification of nucleotide incorporation by signal detection and strength. This technique offers the highest accuracy and detection limit till now.

With these techniques, we can hope for better detection, diagnosis and management of various complex cytogenetic disorders with unknown aetiology.

Conclusion

Chromosomal aberrations are seen in around 0.6% of the general population (Berger 1975). However, the frequency of karyotype abnormalities increases to 2–14% in males presenting with infertility (Shi and Martin 2000). Chromosomal aberrations may have profound effects on fertility outcomes. The advent of ICSI procedures has provided new dimensions and hopes to the infertile men to father a biological child. However, since their inception, there has been a growing concern about the risk of utilizing sperm from infertile men with chromosomal defect. This becomes particularly important in order to reduce the transmission of various fertility defects (chromosomal rearrangements and aneuploidies) to the offspring, the probability of which increases with increasing severity of the disease. Thus, it is of substantial interest to analyse various chromosomal defects in the gametes before going ahead with assisted reproductive procedures. Screening of the gametes will further allow the patients to be appropriately counselled regarding various repercussions of the ART. However, we are far from being able to answer the mechanisms that cause various meiotic errors and disjunctions driving chromosomal rearrangements.

References

Abdel-Razic MM, Abdel-Hamid IA, ElSobky ES (2012) Nonmosaic 47, XYY syndrome presenting with male infertility: case series. Andrologia 44(3):200–204

Aksglæde L, Wikström AM, Rajpert-De Meyts E, Dunkel L, Skakkebæk NE, Juul A (2006) Natural history of seminiferous tubule degeneration in Klinefelter syndrome. Hum Reprod Update 12(1):39–48

Anawalt BD (2013) Approach to male infertility and induction of spermatogenesis. J Clin Endocrinol Metabol 98(9):3532–3542

Anton E, Blanco J, Egozcue J, Vidal F (2004) Sperm FISH studies in seven male carriers of Robertsonian translocation t (13; 14)(q10; q10). Hum Reprod 19(6):1345–1351

Aran B, Blanco J, Vidal F, Vendrell JM, Egozcue S, Barri PN, Egozcue J, Veiga A (1999) Screening for abnormalities of chromosomes X, Y, and 18 and for diploidy in spermatozoa from infertile men participating in an in vitro fertilization-intracytoplasmic sperm injection program. Fertil Steril 72(4):696–701

Aran B, Veiga A, Vidal F, Parriego M, Vendrell JM, Santal J, Egozcue J, Barri PN (2004) Preimplantation genetic diagnosis in patients with male meiotic abnormalities. Reprod Biomed Online 8(4):470–476

Averhoff WW, Richardson RH (1974) Pheromonal control of mating patterns in Drosophila melanogaster. Behav Genet 4(3):207–225

Benet J, Martin RH (1988) Sperm chromosome complements in a 47, XYY man. Hum Genet 78(4):313–315

Berger R (1975) The incidence of constitutional chromosome aberrations. J Hum Genet 23:42–49

Bernardini L, Borini A, Preti S, Conte N, Flamigni C, Capitanio GL, Venturini PL (1998) Study of aneuploidy in normal and abnormal germ cells from semen of fertile and infertile men. Hum Reprod 13(12):3406–3413

Blanco J, Rubio C, Simon C, Egozcue J, Vidal F (1997) Increased incidence of disomic sperm nuclei in a 47, XYY male assessed by fluorescent in situ hybridization (FISH). Hum Genet 99(3):413–416

Bojesen A, Juul S, Gravholt CH (2003) Prenatal and postnatal prevalence of Klinefelter syndrome: a national registry study. J Clin Endocrinol Metabol 88(2):622–626

Bonduelle M, Van Assche E, Joris H, Keymolen K, Devroey P, Van Steirteghem A, Liebaers I (2002) Prenatal testing in ICSI pregnancies: incidence of chromosomal anomalies in 1586 karyotypes and relation to sperm parameters. Hum Reprod 17(10):2600–2614

Brown GM, Leversha M, Hulten M, Ferguson-Smith MA, Affara NA, Furlong RA (1998) Genetic analysis of meiotic recombination in humans by use of sperm typing: reduced recombination within a heterozygous paracentric inversion of chromosome 9q32-q34. 3. Am J Hum Genet 62(6):1484–1492

Calogero AE, De Palma A, Grazioso C, Barone N, Burrello N, Palermo I, Gulisano A, Pafumi C, D'Agata R (2001a) High sperm aneuploidy rate in unselected infertile patients and its relationship with intracytoplasmic sperm injection outcome. Hum Reprod 16(7):1433–1439

Calogero AE, De Palma A, Grazioso C, Barone N, Romeo R, Rappazzo G, D'Agata R (2001b) Aneuploidy rate in spermatozoa of selected men with abnormal semen parameters. Hum Reprod 16(6):1172–1179

Carrell DT (2008) Contributions of spermatozoa to embryogenesis: assays to evaluate their genetic and epigenetic fitness. Reprod Biomed Online 16(4):474–484

Chevret E, Rousseaux S, Monteil M, Usson Y, Cozzi J, Pelletier R, Sele B (1997) Meiotic behaviour of sex chromosomes investigated by three-colour FISH on 35 142 sperm nuclei from two 47, XYY males. Hum Genet 99(3):407–412

Colombero LT, Hariprashad JJ, Tsai MC, Rosenwaks Z, Palermo GD (1999) Incidence of sperm aneuploidy in relation to semen characteristics and assisted reproductive outcome. Fertil Steril 72(1):90–96

De Sanctis V, Ciccone S (2010) Fertility preservation in adolescents with Klinefelter's syndrome. Pediatr Endocrinol Rev 8:178–181

de Vozzi MSJ, Santos SA, Pereira CS, Cuzzi JF, Laureano LA, Franco JG Jr, Martelli L (2009) Meiotic segregation and interchromosomal effect in the sperm of a double translocation carrier: a case report. Mol Cytogenet 2(1):1

Dohle GR, Halley DJJ, Van Hemel JO, Van Den Ouwel AMW, Pieters MHEC, Weber RFA, Govaerts LCP (2002) Genetic risk factors in infertile men with severe oligozoospermia and azoospermia. Hum Reprod 17(1):13–16

Egozcue S, Blanco J, Vendrell JM, Garcia F, Veiga A, Aran B, Barri PN, Vidal F, Egozcue J (2000) Human male infertility: chromosome anomalies, meiotic disorders, abnormal spermatozoa and recurrent abortion. Hum Reprod Update 6(1):93–105

Emery BR, Carrell DT (2006) The effect of epigenetic sperm abnormalities on early embryogenesis. Asian J Androl 8(2):131–142

Estop AM, Cieply KM, Wakim A, Feingold E (1999) Meiotic products of two reciprocal translocations studied by multicolor fluorescence in situ hybridization. Cytogenet Genome Res 83(3–4):193–198

Evans J, Hamerton J, Robinson A (1991) Children and young adults with sex chromosome aneuploidy: follow-up, clinical, and molecular studies, vol 26, no 4. Wiley-Liss

Ferlin A, Garolla A, Foresta C (2005) Chromosome abnormalities in sperm of individuals with constitutional sex chromosomal abnormalities. Cytogenet Genome Res 111(3–4):310–316

Ferlin A, Raicu F, Gatta V, Zuccarello D, Palka G, Foresta C (2007) Male infertility: role of genetic background. Reprod Biomed Online 14(6):734–745

Foresta C, Ferlin A, Gianaroli L, Dallapiccola B (2002) Guidelines for the appropriate use of genetic tests in infertile couples. Eur J Hum Genet 10(5):303–312

Foresta C, Caretta N, Palego P, Ferlin A, Zuccarello D, Lenzi A, Selice R (2012) Reduced artery diameters in Klinefelter syndrome. Int J Androl 35(5):720–725

Friedler S, Raziel A, Strassburger D, Schachter M, Bern O, Ron-El R (2001) Outcome of ICSI using fresh and cryopreserved–thawed testicular spermatozoa in patients with non-mosaic Klinefelter's syndrome. Hum Reprod 16(12):2616–2620

Fu L, Xiong DK, Ding XP, Li C, Zhang LY, Ding M, Nie SS, Quan Q (2012) Genetic screening for chromosomal abnormalities and Y chromosome microdeletions in Chinese infertile men. J Assist Reprod Genet 29(6):521–527

Gekas J, Thepot F, Turleau C, Siffroi JP, Dadoune JP, Briault S, Rio M, Bourouillou G, Carre-Pigeon F, Wasels R, Benzacken B (2001) Chromosomal factors of infertility in candidate couples for ICSI: an equal risk of constitutional aberrations in women and men. Hum Reprod 16(1):82–90

Gianaroli L, Magli MC, Ferraretti AP, Munne S, Balicchia B, Escudero T, Crippa A (2002) Possible interchromosomal effect in embryos generated by gametes from translocation carriers. Hum Reprod 17(12):3201–3207

Gonzalez-Merino E, Hans C, Abramowicz M, Englert Y, Emiliani S (2007) Aneuploidy study in sperm and preimplantation embryos from nonmosaic 47, XYY men. Fertil Steril 88(3):600–606

Guichaoua MR, Quack B, Speed RM, Noel B, Chandley AC, Luciani JM (1990) Infertility in human males with autosomal translocations: meiotic study of a 14; 22 *Robertsonian translocation*. Hum Genet 86(2):162–166

Harton GL, Tempest HG (2012) Chromosomal disorders and male infertility. Asian J Androl 14(1):32–39

Hook EB, Hamerton JL (1977) The frequency of chromosome abnormalities detected in consecutive newborn studies—differences between studies—results by sex and by severity of phenotypic involvement. Population cytogenetics, pp 63–79

Iitsuka Y, Bock A, Nguyen DD, Samango-Sprouse CA, Simpson JL, Bischoff FZ (2001) Evidence of skewed X-chromosome inactivation in 47, XXY and 48, XXYY Klinefelter patients. Am J Med Genet 98(1):25–31

Kent J, Wheatley SC, Andrews JE, Sinclair AH, Koopman P (1996) A male-specific role for SOX9 in vertebrate sex determination. Development 122(9):2813–2822

Kim JW, Lee WS, Yoon TK, Seok HH, Cho JH, Kim YS, Lyu SW, Shim SH (2010) Chromosomal abnormalities in spontaneous abortion after assisted reproductive treatment. BMC Med Genet 11(1):1

Klinefelter HF Jr, Reifenstein EC Jr, Albright F Jr (1942) Syndrome characterized by gynecomastia, aspermatogenesis without A-Leydigism, and increased excretion of follicle-stimulating hormone 1. J Clin Endocrinol Metabol 2(11):615–627

Koga M, Tsujimura A, Takeyama M, Kiuchi H, Takao T, Miyagawa Y, Takada S, Matsumiya K, Fujioka H, Okamoto Y, Nonomura N (2007) Clinical comparison of successful and failed microdissection testicular sperm extraction in patients with nonmosaic Klinefelter syndrome. Urology 70(2):341–345

Kruglyak L, Nickerson DA (2001) Variation is the spice of life. Nature genetics 27(3):234–235

Kruse R, Guttenbach M, Schartmann B, Schubert R, van der Ven H, Schmid M, Propping P (1998) Genetic counseling in a patient with XXY/XXXY/XY mosaic Klinefelter's syndrome: estimate of sex chromosome aberrations in sperm before intracytoplasmic sperm injection. Fertil Steril 69(3):482–485

Kusz K, Kotecki M, Wojda A, Szarras-Czapnik M, Latos-Bielenska A, Warenik-Szymankiewicz A, Ruszczynska-Wolska A, Jaruzelska J (1999) Incomplete masculinisation of XX subjects carrying the SRY gene on an inactive X chromosome. J Med Genet 36(6):452–456

Lanfranco F, Kamischke A, Zitzmann M, Nieschlag E (2004) Klinefelter's syndrome. Lancet 364(9430):273–283

Lejeune J (1963) Autosomal disorders. Pediatrics 32(3):326–337

Lim AST, Fong Y, Yu SL (1999a) Analysis of the sex chromosome constitution of sperm in men with a 47, XYY mosaic karyotype by fluorescence in situ hybridization. Fertil Steril 72(1):121–123

Lim AST, Fong Y, Yu SL (1999b) Estimates of sperm sex chromosome disomy and diploidy rates in a 47, XXY/46, XY mosaic Klinefelter patient. Hum Genet 104(5):405–409

Lissitsina J, Mikelsaar R, Punab M (2006) Cytogenetic analyses in infertile men. Arch Androl 52(2):91–95

Maciel-Guerra AT, de Mello MP, Cocli FB, Ribeiro ML, Miranda ML, Marques-de-Faria AP, Baptista MTM, Moraes SG, Guerra-Junior G (2008) XX maleness and XX true hermaphroditism in SRY-negative monozygotic twins: additional evidence for a common origin. J Clin Endocrinol Metabol 93(2):339–343

Maiburg M, Repping S, Giltay J (2012) The genetic origin of Klinefelter syndrome and its effect on spermatogenesis. Fertil Steril 98(2):253–260

Martin RH (1986) Sperm chromosome analysis in a man heterozygous for a paracentric inversion of chromosome 7 (q11q22). Hum Genet 73(2):97–100

Martin RH (2008a) Cytogenetic determinants of male fertility. Hum Reprod Update 14(4):379–390

Martin RH (2008b) Meiotic errors in human oogenesis and spermatogenesis. Reprod Biomed Online 16(4):523–531

Martin RH, McInnes B, Rademaker AW (1999) Analysis of aneuploidy for chromosomes 13, 21, X and Y by multicolour fluorescence in situ hybridisation (FISH) in a 47, XYY male. Zygote 7(2):131–134

Martin RH, Greene C, Rademaker AW, Ko E, Chernos J (2003) Analysis of aneuploidy in spermatozoa from testicular biopsies from men with nonobstructive azoospermia. J Androl 24(1):100–103

Mateizel I, Verheyen G, Van Assche E, Tournaye H, Liebaers I, Van Steirteghem A (2002) FISH analysis of chromosome X, Y and 18 abnormalities in testicular sperm from azoospermic patients. Hum Reprod 17(9):2249–2257

McLachlan RI, O'Bryan MK (2010) State of the art for genetic testing of infertile men. J Clin Endocrinol Metabol 95(3):1013–1024

Meeks JJ, Weiss J, Jameson JL (2003) Dax1 is required for testis determination. Nat Genet 34(1)

Miharu N, Best RG, Young SR (1994) Numerical chromosome abnormalities in spermatozoa of fertile and infertile men detected by fluorescence in situ hybridization. Hum Genet 93(5): 502–506

Moosani N, Pattinson HA, Carter MD, Cox DM, Rademaker AW, Martin RH (1995) Chromosomal analysis of sperm from men with idiopathic infertility using sperm karyotyping and fluorescence in situ hybridization. Fertil Steril 64(4):811–817

Moosani N, Chernos J, Lowry RB, Martin RH, Rademaker A (1999) A 47, XXY fetus resulting from ICSI in a man with an elevated frequency of 24, XY spermatozoa. Hum Reprod 14(4):1137–1138

Morel F, Roux C, Bresson JL (2001) FISH analysis of the chromosomal status of spermatozoa from three men with 45, XY, der (13; 14)(q10; q10) karyotype. Mol Hum Reprod 7(5):483–488

Moretti E, Anichini C, Sartini B, Collodel G (2007) Sperm ultrastructure and meiotic segregation in an infertile 47, XYY man. Andrologia 39(6):229–234

Nicopoullos JDM, Gilling-Smith C, Almeida PA, Homa S, Nice L, Tempest H, Ramsay JWA (2008) The role of sperm aneuploidy as a predictor of the success of intracytoplasmic sperm injection? Hum Reprod 23(2):240–250

Nishikawa N, Murakami I, Ikuta K, Suzumori K (2000) Sex chromosomal analysis of spermatozoa from infertile men using fluorescence in situ hybridization. J Assist Reprod Genet 17(2):97–102

Palermo GD, Colombero LT, Hariprashad JJ, Schlegel PN, Rosenwaks Z (2002) Chromosome analysis of epididymal and testicular sperm in azoospermic patients undergoing ICSI. Hum Reprod 17(3):570–575

Pandith AA, Akbar S, Faheem S, Malla TM, Zargar MH, Shah ZA, Lateef A, Qasim I, Dar FA, Azad NA, Baba S (2015) Molecular and cytogenetic evaluation of gender in patients born with ambiguous genitalia from different regions of the valley of Kashmir, North India. J Genet Syndr Gene Ther 6

Pang MG, Hoegerman SF, Cuticchia AJ, Moon SY, Doncel GF, Acosta AA, Kearns WG (1999) Detection of aneuploidy for chromosomes 4, 6, 7, 8, 9, 10, 11, 12, 13, 17, 18, 21, X and Y by fluorescence in-situ hybridization in spermatozoa from nine patients with oligoasthenoteratozoospermia undergoing intracytoplasmic sperm injection. Hum Reprod 14(5):1266–1273

Pehlivan TS, Rodrigo L, Simón C, Remohí J, Pellicer A, Rubio C (2003) Chromosomal abnormalities in day 3 embryos from severe oligozoospermic patients. Proceedings in 59th annual meeting of the American Society for Reproductive Medicine, Texas, USA, October 2003

Pfeffer J, Pang MG, Hoegerman SF, Osgood CJ, Stacey MW, Mayer J, Oehninger S, Kearns WG (1999) Aneuploidy frequencies in semen fractions from ten oligoasthenoteratozoospermic patients donating sperm for intracytoplasmic sperm injection. Fertil Steril 72(3):472–478

Rajender S, Rajani V, Gupta NJ, Chakravarty B, Singh L, Thangaraj K (2006) SRY-negative 46, XX male with normal genitals, complete masculinization and infertility. Mol Hum Reprod 12(5):341–346

Rives N, Joly G, Machy A, Simeon N, Leclerc P, Mace B (2000) Assessment of sex chromosome aneuploidy in sperm nuclei from 47, XXY and 46, XY/47, XXY males: comparison with fertile and infertile males with normal karyotype. Mol Hum Reprod 6(2):107–112

Rives N, Milazzo JP, Miraux L, North MO, Sibert L, Mace B (2005) From spermatocytes to spermatozoa in an infertile XYY male. Int J Androl 28(5):304–310

Robinson DO, Jacobs PA (1999) The origin of the extra Y chromosome in males with a 47, XYY karyotype. Hum Mol Genet 8(12):2205–2209

Rodrigo L, Rubio C, Gil-Salom M, Mateu E, Pérez-Cano I, Herrer R, Simon C, Remohí J, Pellicer A (2003) Sperm chromosomal abnormalities: a new indication for preimplantational genetic diagnosis? Proceedings in 19th annual meeting of the European Society of Human Reproduction and Embryology, Madrid, Spain, June–July 2003

Rodrigo L, Peinado V, Mateu E, Remohí J, Pellicer A, Simón C, Gil-Salom M, Rubio C (2010) Impact of different patterns of sperm chromosomal abnormalities on the chromosomal constitution of preimplantation embryos. Fertil Steril 94(4):1380–1386

Ron-El R, Strassburger D, Gelman-Kohan S, Friedler S, Raziel A, Appelman Z (2000) A 47, XXY fetus conceived after ICSI of spermatozoa from a patient with non-mosaic Klinefelter's syndrome: case report. Hum Reprod 15(8):1804–1806

Ross MT, Grafham DV, Coffey AJ, Scherer S, McLay K, Muzny D, Platzer M, Howell GR, Burrows C, Bird CP, Frankish A (2005) The DNA sequence of the human X chromosome. Nature 434(7031):325–337

Roux C, Tripogney C, Morel F, Joanne C, Fellmann F, Clavequin MC, Bresson JL (2005) Segregation of chromosomes in sperm of Robertsonian translocation carriers. Cytogenet Genome Res 111(3–4):291–296

Rubio C, Gil-Salom M, Simon C, Vidal F, Rodrigo L, Minguez Y, Remohi J, Pellicer A (2001) Incidence of sperm chromosomal abnormalities in a risk population: relationship with sperm quality and ICSI outcome. Hum Reprod 16(10):2084–2092

Sarrate Z, Blanco J, Anton E, Egozcue S, Egozcue J, Vidal F (2005) FISH studies of chromosome abnormalities in germ cells and its relevance in reproductive counseling. Asian J Androl 7(3):227–236

Schiff JD, Palermo GD, Veeck LL, Goldstein M, Rosenwaks Z, Schlegel PN (2005) Success of testicular sperm injection and intracytoplasmic sperm injection in men with Klinefelter syndrome. J Clin Endocrinol Metabol 90(11):6263–6267

Schrock EDMS, Du Manoir S, Veldman T, Schoell B (1996) Multicolor spectral karotyping of human chromosomes. Science 273(5274):494

Schultz J, Redfield H (1951) Interchromosomal effects on crossing over in drosophila. In: Cold Spring Harbor symposia on quantitative biology, vol 16. Cold Spring Harbor Laboratory Press, pp 175–197

Shi Q, Martin RH (2000) Aneuploidy in human sperm: a review of the frequency and distribution of aneuploidy, effects of donor age and lifestyle factors. Cytogenet Genome Res 90(3–4):219–226

Silber S, Escudero T, Lenahan K, Abdelhadi I, Kilani Z, Munne S (2003) Chromosomal abnormalities in embryos derived from testicular sperm extraction. Fertil Steril 79(1):30–38

Speed RM, Faed MJW, Batstone PJ, Baxby K, Barnetson W (1991) Persistence of two Y chromosomes through meiotic prophase and metaphase I in an XYY man. Hum Genet 87(4):416–420

Staessen C, Tournaye H, Van Assche E, Michiels A, Van Landuyt L, Devroey P, Liebaers I, Van Steirteghem A (2003) PGD in 47,XXY Klinefelter's syndrome patients. Hum Reprod Update 9(4):319–330

Stemkens D, Roza T, Verrij L, Swaab H, Van Werkhoven MK, Alizadeh BZ, Sinke RJ, Giltay JC (2006) Is there an influence of X chromosomal imprinting on the phenotype in Klinefelter syndrome? A clinical and molecular genetic study of 61 cases. Clin Genet 70(1):43–48

Tanke HJ, Wiegant J, Van Gijlswijk RPM, Bezrookove V, Pattenier H, Heetebrij RJ, Talman EG, Raap AK, Vrolijk J (1999) New strategy for multi-colour fluorescence in situ hybridisation: COBRA: COmbined Binary RAtio labelling. Eur J Hum Genet 7(1):2–11

Tempest HG, Simpson JL (2010) Role of preimplantation genetic diagnosis (PGD) in current infertility practice. Int J Infertil Fetal Med 1:1–10

Therman E, Susman M (2012) Human chromosomes: structure, behavior, and effects. Springer Science & Business Media

Thomas NS, Hassold TJ (2003) Aberrant recombination and the origin of Klinefelter syndrome. Hum Reprod Update 9(4):309–317

Tüttelmann F, Gromoll J (2010) Novel genetic aspects of Klinefelter's syndrome. Mol Hum Reprod 16(6):386–395

Ushijima C, Kumasako Y, Kihaile PE, Hirotsuru K, Utsunomiya T (2000) Analysis of chromosomal abnormalities in human spermatozoa using multi-colour fluorescence in-situ hybridization. Hum Reprod 15(5):1107–1111

Van Steirteghem AC, Nagy Z, Joris H, Liu J, Staessen C, Smitz J, Wisanto A, Devroey P (1993) High fertilization and implantation rates after intracytoplasmic sperm injection. Hum Reprod 8(7):1061–1066

Van Steirteghem A, Bonduelle M, Devroey P, Liebaers I (2002) Follow-up of children born after ICSI. Hum Reprod Update 8(2):111–116

Vegetti W, Van Assche E, Frias A, Verheyen G, Bianchi MM, Bonduelle M, Liebaers I, Van Steirteghem A (2000) Correlation between semen parameters and sperm aneuploidy rates investigated by fluorescence in-situ hybridization in infertile men. Hum Reprod 15(2):351–365

Vera M, Peinado V, Al-Asmar N, Gruhn J, Rodrigo L, Hassold T, Rubio C (2012) Human male meiosis and sperm aneuploidies. Aneuploidy in health and disease, p 141.

Vogt PH (2004) Genomic heterogeneity and instability of the AZF locus on the human Y chromosome. Mol Cell Endocrinol 224(1):1–9

Wiegant J, Bezrookove V, Rosenberg C, Tanke HJ, Raap AK, Zhang H, Bittner M, Trent JM, Meltzer P (2000) Differentially painting human chromosome arms with combined binary ratio-labeling fluorescence in situ hybridization. Genome Res 10(6):861–865

Wong EC, Ferguson KA, Chow V, Ma S (2008) Sperm aneuploidy and meiotic sex chromosome configurations in an infertile XYY male. Hum Reprod 23(2):374–378

Yarali H, Polat M, Bozdag G, Gunel M, Alpas I, Esinler I, Dogan U, Tiras B (2009) TESE–ICSI in patients with non-mosaic Klinefelter syndrome: a comparative study. Reprod Biomed Online 18(6):756–760

Yoshida A, Miura K, Shirai M (1997) Cytogenetic survey of 1,007 infertile males. Urol Int 58(3):166–176

Autosomal Genes in Male Infertility

14

Vertika Singh, Sandeep Kumar Bansal, Rajender Singh, and Kiran Singh

Abstract

Spermatogenesis is driven by the master genes present on the Y chromosome. These driver genes need support from numerous other genes spread across the genome for a number of actions such as energy metabolism, cell death and apoptosis, protein turnover, synthesis of new proteins and garbage disposal. Preliminary studies on infertility focused on the Y chromosome genes due to their primary and indispensable role in spermatogenesis. A number of other studies on human infertility and mouse knockouts have identified several spermatogenically important genes present on chromosomes other than X and Y. For some of these genes, molecular pathways they participate in have also been worked out. This chapter summarizes the genes present on the autosomes that facilitate the process of spermatogenesis and fertility.

Keywords

Autosomal genes • Spermatogenesis • Male infertility • Gonadal development Germ cell apoptosis • Metabolic pathways

V. Singh • K. Singh (✉)
Department of Molecular and Human Genetics, Institute of Science, Banaras Hindu University, Varanasi, Uttar Pradesh, India
e-mail: singhk4@rediffmail.com

S.K. Bansal • R. Singh
Male Reproductive Biology Laboratory, CSIR-Central Drug Research Institute, Lucknow, Uttar Pradesh, India
e-mail: rajender_singh@cdri.res.in

R. Singh, K. Singh (eds.), *Male Infertility: Understanding, Causes and Treatment*, DOI 10.1007/978-981-10-4017-7_14

Key Points

- Y-chromosomal spermatogenic driver genes need autosomal genes for spermatogenesis.
- Mutations in the transcription factor gene *WT1* result in a series of genitourinary anomalies in humans, including gonadal dysgenesis.
- Deletions/mutations of *Sox9* in humans and mice result in male-to-female phenotypic sex reversal, whereas *Sox9* gain-of-function causes testis formation in XX individuals.
- Men with ataxia-telangiectasia (AT) display gonadal atrophy and azoospermia due to meiotic arrest at the zygotene-pachytene stage.
- *PROP1* mutations cause combined pituitary hormone deficiency, including hypogonadotropic hypogonadism and infertility.
- INSL3 serves as an excellent marker in monitoring the treatment of hypogonadal patients.
- Deletions in the *CATSPER* genes, which encode cation channel of sperm, are associated with human male infertility.

14.1 Introduction

The *SRY* gene on the Y chromosome regulates the development of maleness. Logically, nature chooses to integrate a number of spermatogenic genes on the Y chromosome. From the initial studies that identified the deletions on Y chromosome in infertile individuals, the dissection of the Y chromosome in infertile male individuals has identified a number of major, minor and partial deletions that result in male infertility. This signifies the presence of a number of spermatogenic genes on Y chromosome, which are indispensable for spermatogenesis. Interestingly, most of these genes exist in multiple copies, and accordingly, deletions of various lengths show a continuous spectrum of spermatogenic loss.

Spermatogenic genes of Y chromosome cannot work in isolation and need assistance from numerous other genes spread across the whole human genome. The identification of these genes has progressed slowly due to complex nature of spermatogenic loss and vast size of human genome. Mouse knockout studies have identified a

Table 14.1 List of autosomal gene knockouts and their effect on spermatogenesis leading to infertility

Gene knockout and male infertility			
Gene symbol	Gene name	Reproductive phenotype	Reference
Mlh1	MutL homologue 1	Apoptosis of pachytene spermatocytes and infertility	Edelmann et al. (1996)
A-myb	Proto-oncogene A-myb	Arrest at pachytene spermatocyte stage	Toscani et al. (1997)
		Complete absence of postmeiotic cells such as spermatids or spermatozoa and infertility	
Fkbp6	FK506-binding protein	Apoptosis of pachytene spermatocytes and infertility	Crackower et al. (2003)

Table 14.1 (continued)

Gene knockout and male infertility			
Gene symbol	Gene name	Reproductive phenotype	Reference
Rxrb	Retinoid X receptor beta	Accumulation of lipids in Sertoli cells, testicular degeneration and infertility	Mascrez et al. (2004)
H1t2	Testis-specific and histone H1 variant	Abnormal cell restructuring and DNA condensation during the elongation phase of spermiogenesis and reduced fertility	Martianov et al. (2005)
MFP-2	Multifunctional protein-2	Sertoli cell apoptosis and infertility	Huyghe et al. (2006)
Spo11	Meiosis-specific protein Spo11	Failure of spermatocytes synapsis and progress beyond zygotene stage and infertility	Smirnova et al. (2006)
TSLC1	Tumour suppressor of lung cancer 1	Apoptosis of spermatid and infertility	van der Weyden et al. (2006)
Rara	Retinoic acid receptor A protein	Apoptosis of early meiotic prophase spermatocytes, degeneration or germ cells and infertility	Doyle et al. (2007)
Akap4	A-kinase anchor protein 4	Loss of sperm progressive motility and infertility	Miki et al. (2002)
Csnk2a2	Casein kinase 2 alpha 2	Oligospermia and globozoospermia with male infertility	Xu et al. (1999)
hook1	Hook microtubule-tethering protein 1	Ectopic positioning of microtubular structures within the spermatid and infertility	Mendoza-Lujambio et al. 2002
PCI	Protein C inhibitor	Abnormal spermatogenesis, sperm malformation and infertility	Uhrin et al. (2000)
Dazl	Deleted in azoospermia like	Spermatogenic arrest and infertility	Schrans-Stassen et al. (2001)
CcnA1	Cyclin A1	Arrest of spermatogenesis and infertility	Liu et al. (1998)
Cks2	Cyclin-dependent kinases regulatory subunit 2	Spermatogenesis blockage at the metaphase of meiosis I and infertility	Spruck et al. (2003)
Tnp2	Transition protein 2	Teratozoospermia and infertility	Adham et al. (2001)
DMRT1	Doublesex and mab3-related transcription factor 1	Disorganized seminiferous tubules, absence of germ cells and infertility	Raymond et al. (2000)
Insl3	Insulin-like 3	Cryptorchidism and male infertility	Gorlov et al. (2002)

number of genes that are indispensable for spermatogenesis (Table 14.1). Candidate gene studies on infertile human patients have further confirmed the importance of a number of these genes in spermatogenesis and male infertility. Interestingly, further research has even identified the biological pathways and roles of some of these genes. The present chapter discusses candidate autosomal genes important for spermatogenesis and fertility with a glimpse of the biological functions they facilitate.

14.2 Genes in Gonadal Development and Fertility: Establishing Fertility

Apart from the sex chromosomal genes that are exclusively involved in gonadal functions and regulations, a number of genes from autosomes are reported to play an indispensable role in gonadal development, testis differentiation and spermatogenesis. These include genes such as *SF1*, *WT1*, *GATA4*, *SOX9*, *SOX8*, *FGF9* and *DMRT1*. The steroidogenic factor 1 (SF1) protein is encoded by the *NR5A1* gene located on chromosome 9. Steroidogenic factor-1 (SF-1) transcription factor is widely expressed throughout the reproductive axis, including the hypothalamus, gonadotropic cells of the pituitary, gonads and adrenal gland. It plays a key role in the regulation of adrenal and functional development of gonads (Parker and Schimmer 1997; Lin and Achermann 2008; Schimmer and White 2010). Genetic studies, in mice and humans, have demonstrated its significance in male fertility. Male and female Nr5a1 null mice show adrenal agenesis, internal genitalia and gonadal agenesis.

WT1 gene is located on chromosome 11 and encodes for a transcription factor that plays an essential role in cell survival and development. Mutations in the transcription factor gene *WT1* result in a series of genitourinary anomalies in humans, including gonadal dysgenesis, suggesting its critical role in sex determination. In mice, Wt1 is demonstrated to play an essential role in cell survival and proliferation at genital ridge. The genital ridge fails to thicken in *Wt1−/−* animals and completely disappears by E14 (Kreidberg et al. 1993). One of the studies reported that *Wt1* gene plays a very crucial role in spermatogenesis by regulating the polarity of Sertoli cells via Wnt signalling pathway and is one of the genetic causes of nonobstructive azoospermia in humans (Wang et al. 2013).

The role of transcription factor *GATA4* has been well established in sustaining the development and function of the mammalian testis (Viger et al. 2008). The expression of GATA4 at foetal stage is observed in pre-Sertoli cells, Sertoli cells, Leydig cells, fibroblast-like interstitial cells and peritubular myoid cells (Bielinska et al. 2007; Viger et al. 1998). GATA4 expression is postnatally seen in the Sertoli cells and adult Leydig cells only (Ketola et al. 2002; Oréal et al. 2002; LaVoie et al. 2004). *Gata4* knockout mice are lethal and die by embryonic day 9.5 due to abnormal ventral morphogenesis and developmental heart anomalies (Kuo et al. 1997; Molkentin et al. 1997). Due to this reason, it becomes difficult to ascertain the role of this transcription factor in postnatal gonadal development. A study performed using adult transgenic mice with small interfering RNA directed against Gata4 revealed poor breeding capacity with reduced testicular expression of Gata4 target genes, such as Amh and StAR (Thurisch et al. 2009). Recently, a group of investigators generated mice with conditionally deleted Gata4 in Sertoli cells using Cre-LoxP recombination with Amhr2-Cre. The knockout (cKO) mice displayed age-dependent testicular atrophy and loss of fertility with decrease in sperm concentration and motility. The histological analysis further showed Sertoli cell vacuolation, impaired spermatogenesis and altered permeability of the blood-testis barrier. These findings highlight the importance of this gene in spermatogenesis and fertility.

SOX proteins are transcription factors containing a high-mobility group (HMG) domain which facilitates binding and bending of DNA, allowing the transactivation of target genes (Giese et al. 1994; Pontiggia et al. 1994). *SOX9* (Sry-related HMG box gene 9) is present at cytogenetic locus 17q24.3. It is one of the important genes that help in Sertoli cell differentiation. It performs its functions initiated by direct interaction with SRY. Along with male sexual development, *SOX9* is involved in the regulation and maintenance of other male specific factors. Deletion of *SOX9* in human and mice results in male-to-female phenotypic sex reversal, whereas *Sox9* gain-of-function and *Sry*-independent upregulation of *Sox9* cause testis formation in XX individuals (Wagner et al. 1994; Foster 1996; Vidal et al. 2001; Chaboissier et al. 2004; Barrionuevo et al. 2006; Lavery et al. 2011). Similarly, the loss of *SOX8* has been demonstrated to result in progressive degeneration of the seminiferous epithelium through impaired communication between the Sertoli cells and the developing germ cells. A study performed to analyse the copy number variations in an infertile dog revealed a copy number difference in *TEKT1*, *DNM2* and *SOX8* genes (Cassatella et al. 2013). The growth factor Fgf9 is shown to express initially in gonads of both sexes, but the expression shoots up in the developing testis shortly after the activation of *SRY* and *SOX9* in pre-Sertoli cells (Colvin et al. 2001; Nef et al. 2005). Deletion of *Fgf9* results in male-to-female phenotypic sex reversal (Colvin et al. 2001; Schmahl et al. 2004). Using in vitro and in vivo models, it has been demonstrated that Fgf9 acts directly on germ cells to inhibit meiosis (Bowles et al. 2010).

DMRT1 gene maps to the chromosomal region 9p24.3 having in common a zinc finger-like DNA-binding motif referred to as the "DM domain" and a nuclear localization signal (Raymond et al. 1998; Ying et al. 2007). DMRT1 is chiefly expressed in the testis and plays a central role in testis differentiation (Raymond et al. 1998). DMRT1 expression is highly upregulated in undifferentiated spermatogonia, while it is downregulated in differentiating spermatogonia. In addition, DMRT1 represses male germ cell meiosis and stimulates germ cells to enter mitosis (Matson et al. 2010). *Dmrt1−/−* mice showed testicular hypoplasia, disorganized seminiferous tubules and undifferentiated Sertoli cells. However, *Dmrt1+/−* males were fertile with normal testicular development (Raymond et al. 2000). Deletions on the short arm of chromosome 9 are associated with XY gonadal dysgenesis. A study utilized a high-resolution array CGH (comparative genomic hybridization) approach to observe the genomic imbalances in patients with XY gonadal dysgenesis. The study revealed very small partial deletions of *DMRT1* gene as an unexpected finding of XY ovotesticular disorder of sexual development (DSD) (Vinci et al. 2007; Ledig et al. 2012). Deletions of *DMRT1* were identified in impaired spermatogenesis cases (Lopes et al. 2013).

14.3 Autosomal Pathways in Spermatogenesis

14.3.1 Infertility and Apoptosis: Eliminating the Unfit

Spermatogenesis is a dynamic process intricately regulated by germ cell proliferation and differentiation. The testis is an organ having the highest rate of all multiplications in a normal male individual. Mitosis and meiosis take place in a tightly regulated manner to supply millions of spermatozoa without turning cancerous. Therefore, a process of death is required to eliminate the unwanted cells and to match the ratio of the germ cells with supporting Sertoli cells. Excessive proliferation of germ cells is counterbalanced by selective apoptosis of their progenies (Allan et al. 1992; Bartke 1995; Billig et al. 1995). The process of death can occur via several processes such as necrosis, apoptosis, autophagy and entosis. Apoptosis is the most studied and well-elucidated process of cell death in spermatogenesis. Apoptotic mechanisms in the testis are governed by complex interactions between the diverse kind of cells and their unique ability to respond to various types of stimuli (Hikim et al. 2003). As high as 75% of the germ cells from various stages undergo apoptosis in the testis (Huckins 1978).

Germ cell apoptosis within the testis is regulated by both intrinsic or mitochondrial pathway and extrinsic or death receptor pathway (Kawamura et al. 2004; Said et al. 2004; Theas et al. 2006; Yin et al. 2007; Sofikitis et al. 2008; Aitken et al. 2011), whose execution is governed by the activation of caspase family of proteins. However, in the past few years, some caspase-independent pathways of apoptosis have also been reported (Coureuil et al. 2006). This includes the perforin/granzyme pathway that induces apoptosis via either granzyme B or granzyme A (Martinvalet et al. 2005) and the p53 pathway involved in the regulation of apoptosis induced by genotoxic and non-genotoxic stresses (Vogelstein et al. 2000).

Testicular germ cell apoptosis is a natural phenomenon that occurs normally and continuously throughout life (Bartke 1995; Billig et al. 1995). The testis of a 4-week-old rat is reported to have large number of spermatocytes undergoing apoptosis; however, in adult rat, spermatogonia become the principle cells undergoing process of apoptosis (Billig et al. 1995). The initiation of apoptotic signalling within the cell is mediated via extrinsic or intrinsic pathway. Extrinsic pathway is executed through the stimulation of transmembrane death receptors such as Fas receptors localized on cell membrane. However, the intrinsic pathway is governed by the release of various signalling factors by the mitochondria. Candidate genes from both the extrinsic and intrinsic pathway are involved in maintaining the balance between germ cell proliferation and death (Fig. 14.1).

14.3.1.1 Intrinsic Pathway

During the migration of primordial germ cells to the developing gonad, the cells with abnormal migration undergo apoptosis mediated by Bcl-xl and Bax (Rucker et al. 2000). Heterozygous *Bcl-x* knockout mice [*Bcl-x* (+/−)] exhibit severe defects in male germ cell development (Kasai et al. 2003). Transgenic mice with Bax "knockout" or Bcl-2 or Bcl-x overexpression show an accumulation of spermatogonial and spermatocyte population due to the elimination of the first wave of apoptosis resulting in the development of an infertile phenotype (Knudson et al. 1995). Mice with Bcl-xl overexpression show an increase in germ cell death (Russell et al. 2001). Thus, a fine-tuning is required in order to maintain a proper balance between apoptosis-inducing and apoptosis-protecting proteins in the testis (Russell

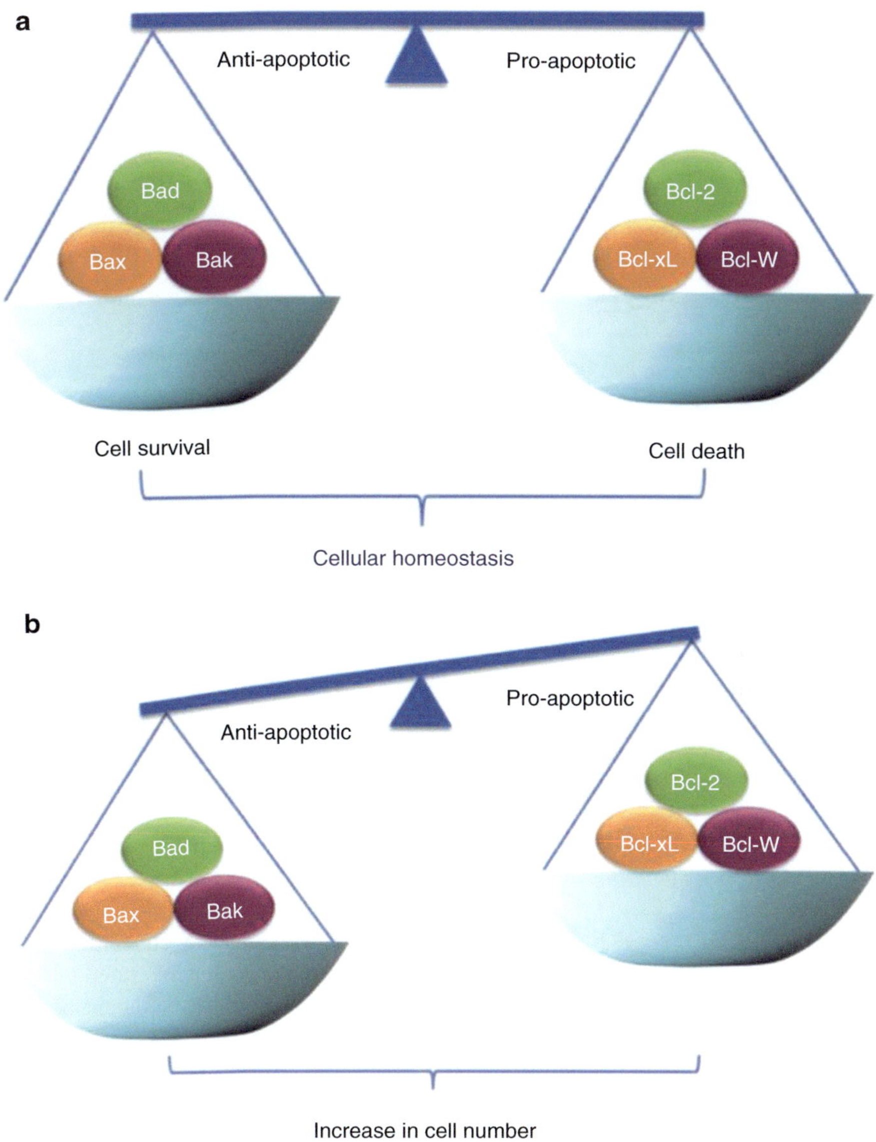

Fig. 14.1 Diagrammatic illustration of the role key apoptotic proteins in sustaining a physiological balance between cell proliferation and cell death. (**a**) A balance between anti-apoptotic and pro-apoptotic proteins are vital for maintaining a cellular homeostasis. (**b**) An excess of anti-apoptotic proteins may result in increase in cell number (**c**) An excess of proapoptotic proteins may result in increase in cell death via apoptosis

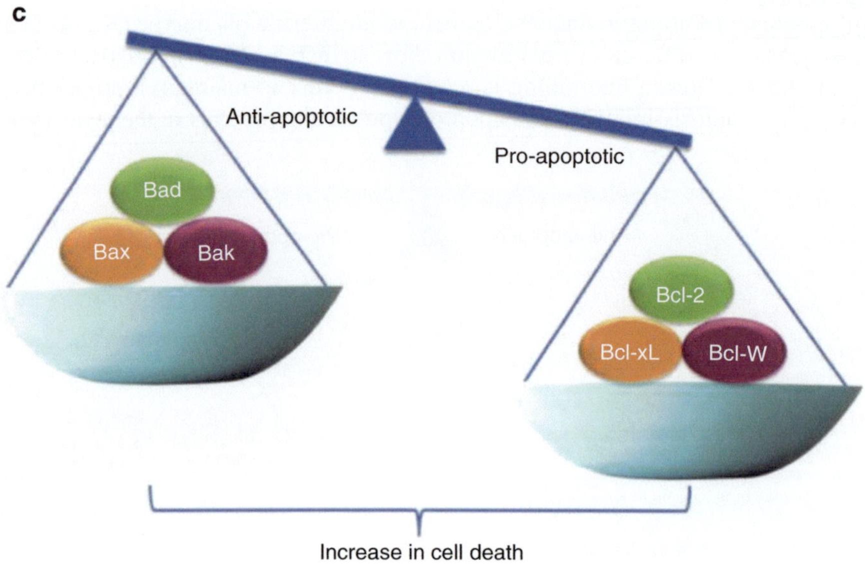

Fig. 14.1 (continued)

et al. 2001). In our lab, we performed a protein profiling of candidate genes from apoptosis pathway in impaired spermatogenesis cases. The analysis indicated a significant expression of pro-apoptotic proteins BAX, BAD and BAK and a low expression of anti-apoptotic BCL2 and BCLW, the anti-apoptotic proteins. A ratio between pro- and anti-apoptotic genes was disturbed, which might be the reason for altered apoptosis in impaired spermatogenesis cases (Jaiswal et al. 2015). The testis of a *bik* (−/−) or *bim* (−/−) male mice develops normally but displays an infertile phenotype (Shaha et al. 2010). A recent report demonstrated significantly increased and decreased level of seminal BAX and BCL2, respectively, in infertile men with varicocele (Mostafa et al. 2014).

14.3.1.2 Extrinsic Pathway

FasL and its corresponding receptor, Fas, interact to form an activated Fas receptor complex that initiates a pro-apoptotic death signal in the receptor-bearing cell (Janssen et al. 2003). Fas ligand and Fas receptor expression are well studied in the testis (Guazzone et al. 2009). Upregulation of Fas receptor is demonstrated in association with spermatocyte apoptosis during the first round of spermatogenesis in rat (Lizama et al. 2007). The altered expression of Fas/FasL system is also associated with germ cell apoptosis in humans. Elevated level of FasL is associated with SCO (Sertoli cell-only) syndrome and MA (maturation arrest) (Kim et al. 2004), indicating a role of altered apoptosis through caspase-3 activation. Studies from our lab have demonstrated an increased expression of FasL in compromised spermatogenesis (Jaiswal et al. 2015).

14.3.2 DNA Damage, Replication and Repair Pathways: Keeping It Correct

Repair of DNA lesions is an indispensable requirement of a cell for maintaining genomic stability and to safeguard healthy propagation of species. The DNA repair and recombination mechanisms are highly conserved across species. Alteration of any of these machineries may result in repair and recombination errors, which might precipitate in the form of reproductive failure. A number of reports suggest the involvement of a series of meiotic checkpoints that may cause a spermatogenic arrest due to defective recombination, leading to male infertility. A study performed by Reijo-Pera, Martin and colleagues identified that nearly half of the infertile patients show measurable defects in recombination (Gonsalves et al. 2004, 2005). SPO11 is a topoisomerase involved in homologous recombination repair. Mice with disrupted spo11 show defective meiosis in both males and females (Baudat et al. 2000; Romanienko and Camerini-Otero 2000).

The synaptonemal complex is a proteinaceous structure involved in linking homologous chromosomes during recombination. This aggregate is composed of SYCP1, SYCP2 and SYCP3 proteins known as SC. Targeted deletion of *Sycp3* resulted in male sterility and synaptic failure (Yuan et al. 2000, 2002). Mutations of *SYCP3*, the human homologue of *SCP3*, are reported in azoospermic infertile men (Miyamoto et al. 2003). FK506-binding protein 6 (Fkbp6) plays an essential role in maintaining fidelity of homologous chromosome pairing during meiosis and is important for sex-specific fertility (Crackower et al. 2003). Targeted inactivation of *Fkbp6* mice showed lack of spermatids and absence of spermatozoa in caudal epididymis and seminiferous tubules. Further analysis of nucleotide sequence revealed that a 93 bp region corresponding to exon 8 of the *Fkbp6* gene was deleted in these animals, which was thus suggested as a causative factor for aspermic phenotype (Crackower et al. 2003).

RAD51, a recombinase protein, plays an important role in meiotic prophase by co-localizing with DMC1. RAD51 and DMC1 proteins are crucial during homology and heteroduplex formation along with other associated proteins (Zenvirth et al. 2003). *Dmc1* null mice display an infertile phenotype showing gross defects in chromosome pairing (Yoshida et al. 1998; Pittman et al. 1998). Similarly, *Rad51* knockout mice exhibit embryonic lethality demonstrating an indispensable role of these proteins in meiosis and development (Thacker 1999). Analysis on a man with spermatocyte arrest showed an abnormal presence of BRCA1 with RAD51 absence in early and late spermatocytes (Sciurano et al. 2006). In case of failure of DNA repair, the cell cycle checkpoint genes get activated resulting in apoptosis. Numerous genes are elaborated in the DNA damage-induced regulation of cell cycle control. Mutation in one such gene, *ATM*, causes ataxia-telangiectasia, a genetic disorder characterized by radiosensitivity, defective cell cycle checkpoint activation, genomic instability and infertility (Meyn 1999). Men with ataxia-telangiectasia (AT) display gonadal atrophy and azoospermia due to meiotic arrest at zygotene-pachytene stage (Xu and Baltimore 1996).

Fanconi anaemia genes such as *BRCA1* and *BRCA2* play an important role in male and female fertility. Targeted deletion of the *FANCA* genes results in germ cell deficiency due to defective proliferation of germ cells (Chen and Tomkinsz 1996; Whitney et al. 1996; Nadler and Braun 2000; Yang et al. 2001; Meetei et al. 2003). *Brac1-p53* double-knockout male mice were infertile due to meiotic failure (Cressman et al. 1999). Tp53 and Ercc1 play a fundamental role in DNA damage response and repair during spermatogenesis. DNA mismatch repair protein family is involved in DNA repair mismatches that arise predominantly during DNA replication. Their function is to ensure chromosomal integrity during meiotic recombination in most of the sexually reproducing organisms (Svetlanov and Cohen 2004). *Msh2* and *Tp53* genes display compromised germ cell function and sperm production (Paul et al. 2007). *Msh4* or *Msh5* knockout mice exhibit defects in synapsis resulting in a failure of primary spermatocytes at the zygotene/pachytene checkpoint (de Vries et al. 1999; Svetlanov and Cohen 2004). Using case-control association approach, another study reported that rs4647269 SNP in *MLH1*, rs1059060 SNP in *PMS2* and rs2075789 in *MSH5* may act as risk factors for azoospermia or oligozoospermia.

14.3.3 Hormonal/Endocrine Pathways

The hormonal control of spermatogenesis is governed by the hypothalamic-pituitary-testicular axis. This axis functions in a highly regulated and coordinated manner to produce optimal concentrations of circulating steroids that are essential for normal male sexual development, spermatogenesis and fertility. The hypothalamus secretes gonadotropin-releasing hormone (GnRH), which further stimulates the pituitary gonadotrophs to secrete follicle-stimulating hormone (FSH) and luteinizing hormone (LH). These hormones play pivotal roles in the process of spermatogenesis. Low level of GnRH results in decreased levels of FSH and LH, which result in hypogonadotropic hypogonadism (HH) (Seminara et al. 2000). Idiopathic congenital hypogonadotropic hypogonadism (CHH) is a reproductive disorder characterized by impaired pubertal development caused by gonadotropin-releasing hormone (GnRH) deficiency. This disorder is often characterized by low plasma luteinizing hormone (LH) and follicle-stimulating hormone (FSH) levels along with undetectable concentrations of circulating sex steroids. About 50% of CHH patients possess a reduced (hyposmia) or deficient (anosmia) sense of smell, termed as Kallmann syndrome (KS). Anosmia is associated with hypoplasia of olfactory bulbs and tracts. This defect is largely associated with abnormal GnRH neuron ontogenesis (Juan A. 1856, Kallmann 1944). Recent report suggests five KS genes that are associated with the Kallmann syndrome, namely, *FGFR1* (Dodé et al. 2003), *FGF8* (Falardeau et al. 2008), *PROKR2*, *PROK2* (Dodé et al. 2006) and *KAL1* (Franco et al. 1991; Legouis et al. 1991; Hardelin et al. 1992). Insulin-like peptide 3 (INSL3) represents an additional regulator of the HPG axis. INSL3 belongs to insulin-like hormone superfamily (which also includes relaxin). In mammalian testis, INSL3 is a chief secreted product of the interstitial Leydig cells. This Leydig cell hormone interacts with specific receptors, called RXFP2, to modulate the process of steroidogenesis and support spermatogenesis. INSL3 receptor is predominantly found on spermatocytes and to a great extent on germ cells (Anand-Ivell et al. 2006). RXFP2 is a G-protein-coupled receptor normally linked to Gs, activating adenylyl cyclase (Bathgate et al. 2006). In mice, the complete loss of *INSL3* (*Insl3−/−*) results in abnormal gubernacular development with intra-abdominal gonads (Zimmermann et al. 1999; Nef and Parada 1999). Global ablation of its receptor *RXFP2* results in cryptorchidism and infertility in male mice (Gorlov et al. 2002). Beta-catenin and NOTCH1 signalling pathways are the major targets of INSL3 signalling during gubernacular development (Huang et al. 2013). INSL3 serves as an excellent marker in monitoring the treatment of hypogonadal patients.

Loss-of-function mutations in the pituitary-expressed *FSHβ* genes did not cause infertility; however, they resulted in reduced testes size and reduced sperm count (Kumar et al. 1997). One of the essential regulatory genes for pituitary gland ontogeny is *PROP1*, which encodes a paired-like homeodomain transcription factor Prop1 (prophet of Pit1) (Sornson et al. 1996). Its expression appears early in embryonic development and is essential for somatotroph, thyrotroph, gonadotroph and lactotroph function and differentiation. *Prop1* (mouse) and *PROP1* (human) gene mutations reveal its significance during pituitary gland organogenesis. A homozygous missense mutation (S83P) in the *Prop1* gene exhibits growth insufficiency, hypothyroidism and infertility in Ames dwarf mice (Wu et al. 1998a, b). PROP1-related combined pituitary hormone deficiency (CPHD) is associated with more than 11 different loss-of-function and null mutations identified in humans. *PROP1* mutations cause combined pituitary hormone deficiency, including HHG and infertility (Cogan et al. 1998; Dattani and Robinson 2000).

Members of the steroid receptor superfamily along with their transcriptional coactivators such as AR, ER, PR, RXRβ, SF1, DAX1 and SRC1 play crucial roles in regulating and maintaining the testicular development and spermatogenesis. Disruption of any of these genes may subsequently affect male development, spermatogenesis and fertility. Mutations in the *AR* gene (X linked) cause male infertility with a frequency of 1:60,000 of live deliveries (Hiort et al. 2000). The functions of oestrogens (OS) in regulating testis development and spermatogenesis are well known (Rochira et al. 2005). In addition to roles in spermatogenesis, the presence of oestrogen receptors (ORα and ORβ) on germ cells shows their importance in spermatogenesis (Carreau et al. 2011). Oestrogen maintains the function of sperm by facilitating capacitation and fertilization (Carreau et al. 2007). Physical functions of oestrogens are mediated through the oestrogen receptors (OR) (Ellmann et al. 2009).

The functional significance of oestrogen has been investigated using genetically modified mice that lack the OR (Korach 1994). *ORα*, *ORβ* and *ORαβ* knockout male mice displayed reproductive incompetence. *ORαKO* mice were infertile with defects in epididymal fluid reabsorption (Delbès et al. 2006). *ORβ* knockout mice presented a 50% increase in the number of gonocytes caused by an increased proliferation and decreased apoptosis. However, *ORα* gene increased testosterone production without affecting the number of gonocytes during foetal life (Delbès et al. 2005).

ORβ is involved in regulating neonatal gametogenesis; however, ORα controls the foetal and neonatal steroidogenesis. Genetic screening for the *ORα* and *ORβ* gene has shown several polymorphic sites associated with the pathogenesis of male infertility (Gennari et al. 2005). A recent study has performed a meta-analysis on the single-nucleotide polymorphisms (SNPs) in oestrogen receptor genes in association with the risk of male infertility. The study revealed a significant association of rs2234693C allele with a decreased risk for male infertility; however, the rs9340799AA and the rs1256049GA genotypes showed an increased risk for male infertility (Li et al. 2014). The steroidogenic factor 1 (SF1) protein, encoded by the *NR5A1* gene, is a member of the nuclear receptor superfamily. It is one of the key regulatory genes of the hypothalamic-pituitary-steroidogenic axis (Morohashi et al. 1992; Luo et al. 1994). The SF1 protein plays a vital role in gonadal development and steroidogenesis. Mutations in *NR5A1* have been shown to be associated with primary adrenal insufficiency, 46, XY gonadal dysgenesis and boys with hypospadias, micropenis and bilateral anorchia (Ferraz-de-Souza et al. 2011). A group of researchers has analysed the frequency of *NR5A1* mutations in infertile men. The study demonstrated that 4% of the infertile men ($N = 315$) with reduced sperm counts and sperm concentrations below one million/mL were having the mutation (Bashamboo et al. 2010). Recently, a lab performed a mutation screening of *NR5A1* gene in infertile patients by sequencing all exons. The investigation identified seven novel and one previously described missense mutation in patients with severe spermatogenic impairment (Ferlin et al. 2015).

14.4 Standalone Drivers from the Autosomal Store

A number of genes which we have discussed so far are involved in the regulation of a multitude of pathways influencing the process of spermatogenesis. However, there are a few autosomal genes which either function in isolation or have not yet been linked to biological pathways. These include *DAZL*, *PRM1*, *PRM2* and *CATSPER* genes that are known to be important for spermatogenesis or sperm functions.

Deletions in the long arm of Y chromosome at Yq11.2 region are found in approximately 5–15% of males with spermatogenic failure. Among these cases, deletions involving the *DAZ* (deleted in azoospermia) gene family are the most frequent (Reijo et al. 1996; Vogt 1998).

The *DAZ* gene has an autosomal homologue, *DAZL* (*DAZ* like), which is located on chromosome 3p24. *DAZL* gene is highly homologous to the *DAZ* gene with 83% similarity cDNA coding region. Both of these genes encode RNA-binding proteins required for germ cell development in diverse organisms (Saxena et al. 1996; Shan et al. 1996; Yen et al. 1996; Chai et al. 1997; Xu et al. 2001). It is believed that the *DAZ* gene evolved around 40 million years ago through a series of events involving transposition, recurring amplification and modification of an ancestral autosomal gene, *DAZL* (Saxena et al. 1996). *Dazl* knockout mice showed a loss of germ cells with complete absence of gamete production (Ruggiu et al. 1997). A number of studies however have reported that Dazl is required for germ cell development in wide range of species (Ruggiu et al. 1997; Yen et al. 1996; Cooke et al. 1996; Reijo et al. 1996; Houston and King 2000; Lin and Page 2005).

The loss of Dazl functionality has been demonstrated to increase the level of germ cell apoptosis along with chromatin configuration changes in the immature germ cells at prenatal stage (Lin and Page 2005). Interestingly, a number of studies have reported the association of *DAZL* gene mutations with male infertility. A recent meta-analysis on the studies of *DAZL* gene polymorphism showed that A260G polymorphism did not correlate with oligo-/azoospermia, while A386G correlated with male infertility (Chen et al. 2016). However, this correlation was found only in China in an ethnicity-specific manner but not in India, Japan and other Caucasian countries (Chen et al. 2016).

Protamines are the most abundant sperm nuclear proteins which help in paternal genome packaging and replace histones during spermatogenesis (Bloch 1969; Calvin 1976; Mezquita and Teng 1977; Subirana 1983; Oliva and Dixon 1991; Lewis et al. 2003; Ando et al. 2012). These proteins contain a high content of positively charged amino acids, predominantly arginine that facilities their binding into the minor groove of DNA. Protamine gene family includes the nucleoprotein genes *PRM1*, *PRM2* and *TNP2* closely linked in a stretch of DNA, 13–15 kb long, one on human chromosome 16p13.3. Mutations in the protamine genes are found to be widely associated with impaired spermatogenesis, defects in imprinting, sperm chromatin abnormalities and DNA breaks (De Yebra et al. 1993; Cho et al. 2001; Miyagawa et al. 2005; Iguchi et al. 2006), which are also shown to affect sperm penetration functions and embryonic development (Ahmadi and Ng 1999; Kempisty et al. 2007). Protamine P1 is synthesized as a mature protein; however, P2 family proteins are formed by proteolysis from a precursor protein. The P1/P2 ratio (content of protamine P1 vs protamine P2) in the human sperm nucleus is approximately one. Alteration of P1 or P2 is shown to significantly affect the DNA integrity and the outcome of various assisted reproduction procedures (Aoki et al. 2005). Another important gene, *TNP2*, encodes for transition nuclear protein 2 that is required for sperm chromatin condensation. These proteins are transition proteins in the sense that they are replaced by protamines in the course of sperm chromatin condensation in the mature sperm nucleus (Steger et al. 2000; Sassone-Corsi 2002; Aoki et al. 2005; Oliva 2006; Tüttelmann et al. 2007). Defects in TNP2 proteins are associated with acrosome deficiencies, defects in sperm movement through the female genital tract and inability of the spermatozoa to penetrate the zona pellucida (Adham et al. 2001). These functional deformities of sperm may explain infertility in a number of normozoospermic cases (Carreras et al. 1990).

The cation channel of sperm (CatSper) is a sperm-specific ion channel, which plays an exclusive role in orchestrating various fertilization events and appears to be entirely evolved for male reproductive functions and fertility (Jaiswal et al. 2014; Singh and Rajender 2015). The CatSper channel is localized to the principal piece of sperm flagellum (Ren et al. 2001) and humans (Cheon et al. 2004). The disruption of CatSper alpha subunits (*CatSper1–4*) by knockout in mouse models results in channel dysfunction and infertility (Qi et al. 2007). *CatSper1* and *CatSper2* mutations have been found to correlate with asthenoteratozoospermia (Avidan et al.

2003; Avenarius et al. 2009). In a recent microarray study, our laboratory reported genomic imbalance/copy number variations in two infertile brothers with reference to control. The analysis demonstrated a common deletion in both the patients at 15q15.3 locus, which harboured several genes including *CATSPER2*. This is the first familial case report from India on the association of *CATSPER* gene deletion in human male infertility (Jaiswal et al. 2014).

Conclusion and Future Prospects: Anticipations from the Next-Generation Era

In the past few decades, candidate gene approach has been used to study the effect of gene mutations/deletions in understanding the mechanism of spermatogenesis and infertility. The progress in the identification of genes important for spermatogenesis and fertility had been slow due to technical limitations. However, the concept of forward genetics has recently taken a big leap in the form of genome-wide scan. SNP microarray technology promises simultaneous detection of a wide range of SNPs across the whole genome. Similarly, massive parallel sequencing allows scanning of the whole genome for genetic and epigenetic variations that can affect spermatogenesis and fertility. These high-throughput techniques allow the documentation of protein-coding mutations, including missense, nonsense, splice site and small deletions or insertions. These powerful techniques have been widely appreciated as efficient strategies for identifying the causes behind the pathophysiology of various complex diseases including male infertility. Well-planned scientific studies aided by powerful genome analysis tools would accelerate the discovery of new autosomal drivers of spermatogenesis and fertility.

References

Adham IM, Nayernia K, Burkhardt-Göttges E, Topaloglu Ö, Dixkens C, Holstein AF, Engel W (2001) Teratozoospermia in mice lacking the transition protein 2 (Tnp2). Mol Hum Reprod 7(6):513–520

Ahmadi A, Ng SC (1999) Destruction of protamine in human sperm inhibits sperm binding and penetration in the zona-free hamster penetration test but increases sperm head decondensation and male pronuclear formation in the hamster-ICSI assay. J Assist Reprod Genet 16(3): 128–132

Aitken RJ, Findlay JK, Hutt KJ, Kerr JB (2011) Apoptosis in the germline. Reproduction 141(2): 139–150

Allan DJ, Harmon BV, Roberts SA (1992) Spermatogonial apoptosis has three morphologically recognizable phases and shows no circadian rhythm during normal spermatogenesis in the rat. Cell Prolif 25(3):241–250

Anand-Ivell RJ, Relan V, Balvers M, Coiffec-Dorval I, Fritsch M, Bathgate RA, Ivell R (2006) Expression of the insulin-like peptide 3 (INSL3) hormone-receptor (LGR8) system in the testis. Biol Reprod 74(5):945–953

Ando T, Yamasaki M, Suzuki K (2012) Protamines: isolation·characterization·structure and function, vol 12. Springer Science & Business Media

Aoki VW, Liu L, Carrell DT (2005) Identification and evaluation of a novel sperm protamine abnormality in a population of infertile males. Hum Reprod 20(5):1298–1306
Avenarius MR, Hildebrand MS, Zhang Y, Meyer NC, Smith LL, Kahrizi K, Najmabadi H, Smith RJ (2009) Human male infertility caused by mutations in the CATSPER1 channel protein. Am J Hum Genet 84(4):505–510
Avidan N, Tamary H, Dgany O, Cattan D, Pariente A, Thulliez M, Borot N, Moati L, Barthelme A, Shalmon L, Krasnov T (2003) CATSPER2, a human autosomal nonsyndromic male infertility gene. Eur J Hum Genet 11(7):497–502
Barrionuevo F, Bagheri-Fam S, Klattig J, Kist R, Taketo MM, Englert C, Scherer G (2006) Homozygous inactivation of Sox9 causes complete XY sex reversal in mice. Biol Reprod 74(1):195–201
Bartke A (1995) Apoptosis of male germ cells, a generalized or a cell type-specific phenomenon? Endocrinology 136(1):3
Bashamboo A, Ferraz-de-Souza B, Lourenço D, Lin L, Sebire NJ, Montjean D, Bignon-Topalovic J, Mandelbaum J, Siffroi JP, Christin-Maitre S, Radhakrishna U (2010) Human male infertility associated with mutations in NR5A1 encoding steroidogenic factor 1. Am J Hum Genet 87(4):505–512
Bathgate RA, Ivell R, Sanborn BM, Sherwood OD, Summers RJ (2006) International Union of Pharmacology LVII: recommendations for the nomenclature of receptors for relaxin family peptides. Pharmacol Rev 58(1):7–31
Baudat F, Manova K, Yuen JP, Jasin M, Keeney S (2000) Chromosome synapsis defects and sexually dimorphic meiotic progression in mice lacking Spo11. Mol Cell 6(5):989–998
Bielinska M, Seehra A, Toppari J, Heikinheimo M, Wilson DB (2007) GATA-4 is required for sex steroidogenic cell development in the fetal mouse. Dev Dyn 236(1):203–213
Billig H, Furuta I, Rivier C, Tapanainen J, Parvinen M, Hsueh AJ (1995) Apoptosis in testis germ cells: developmental changes in gonadotropin dependence and localization to selective tubule stages. Endocrinology 136(1):5–12
Bloch DP (1969) A catalog of sperm histones. Genetics 61(1 Suppl):93–111
Bowles J, Feng CW, Spiller C, Davidson TL, Jackson A, Koopman P (2010) FGF9 suppresses meiosis and promotes male germ cell fate in mice. Dev Cell 19(3):440–449
Calvin HI (1976) Comparative analysis of the nuclear basic proteins in rat, human, Guinea pig, mouse and rabbit spermatozoa. Biochim Biophys Acta 434(2):377–389
Carreau S, Silandre D, Bois C, Bouraima H, Galeraud-Denis I, Delalande C (2007) Estrogens: a new player in spermatogenesis. Folia Histochem Cytobiol 45(Suppl 1):S5–10
Carreau S, Bouraima-Lelong H, Delalande C (2011) Estrogens in male germ cells. Spermatogenesis 1(2):90–94
Carreras A, Palma A, Mendoza C (1990) Hypoosmotic swelling test in normo-, oligo-, astheno- and oligoasthenozoospermic men before and after swim-up separation of spermatozoa. Andrologia 22(4):313–317
Cassatella D, Martino NA, Valentini L, Guaricci AC, Cardone MF, Pizzi F, Dell'Aquila ME, Ventura M (2013) Male infertility and copy number variants (CNVs) in the dog: a two-pronged approach using Computer Assisted Sperm Analysis (CASA) and Fluorescent In Situ Hybridization (FISH). BMC Genomics 14(1):1
Chaboissier MC, Kobayashi A, Vidal VI, Lützkendorf S, van de Kant HJ, Wegner M, de Rooij DG, Behringer RR, Schedl A (2004) Functional analysis of Sox8 and Sox9 during sex determination in the mouse. Development 131(9):1891–1901
Chai NN, Phillips A, Fernandez A, Yen PH (1997) A putative human male infertility gene DAZLA: genomic structure and methylation status. Mol Hum Reprod 3(8):705–708
Chen M, Tomkinsz DI (1996) Inactivation of Fac in mice produces inducible chromosomal instability and. Nat Genet 12
Chen P, Wang X, Xu C, Xiao H, Zhang WH, Wang XH, Zhang XH (2016) Association of polymorphisms of A260G and A386G in DAZL gene with male infertility: a meta-analysis and systemic review. Asian J Androl 18(1):96

Cheon KW, Choi YJ, Byun HK, Hong JY, Choi HK, Seo JT (2004) Expression of sperm-specific cation channel CatSper in human spermatozoa. Korean J Urol 45(4):365–372
Cho C, Willis WD, Goulding EH, Jung-Ha H, Choi YC, Hecht NB, Eddy EM (2001) Haploinsufficiency of protamine-1 or-2 causes infertility in mice. Nat Genet 28(1):82–86
Cogan JD, Wu W, Phillips JA III, Arnhold IJ, Agapito A, Fofanova OV, Osorio MGF, Bircan I, Moreno A, Mendonca BB (1998) The PROP1 2-base pair deletion is a common cause of combined pituitary hormone deficiency 1. J Clin Endocrinol Metabol 83(9):3346–3349
Colvin JS, Green RP, Schmahl J, Capel B, Ornitz DM (2001) Male-to-female sex reversal in mice lacking fibroblast growth factor 9. Cell 104(6):875–889
Cooke HJ, Lee M, Kerr S, Ruggiu M (1996) A murine homologue of the human DAZ gene is autosomal and expressed only in male and female gonads. Hum Mol Genet 5(4):513–516
Coureuil M, Fouchet P, Prat M, Letallec B, Barroca V, Dos Santos C, Racine C, Allemand I (2006) Caspase-independent death of meiotic and postmeiotic cells overexpressing p53: calpain involvement. Cell Death Differ 13(11):1927–1937
Crackower MA, Kolas NK, Noguchi J, Sarao R, Kikuchi K, Kaneko H, Kobayashi E, Kawai Y, Kozieradzki I, Landers R, Mo R (2003) Essential role of Fkbp6 in male fertility and homologous chromosome pairing in meiosis. Science 300(5623):1291–1295
Cressman VL, Backlund DC, Avrutskaya AV, Leadon SA, Godfrey V, Koller BH (1999) Growth retardation, DNA repair defects, and lack of spermatogenesis in BRCA1-deficient mice. Mol Cell Biol 19(10):7061–7075
Dattani MT, Robinson IC (2000) The molecular basis for developmental disorders of the pituitary gland in man. Clin Genet 57(5):337–346
de Vries SS, Baart EB, Dekker M, Siezen A, de Rooij DG, de Boer P, te Riele H (1999) Mouse MutS-like protein Msh5 is required for proper chromosome synapsis in male and female meiosis. Genes Dev 13(5):523–531
De Yebra L, Ballesca JL, Vanrell JA, Bassas L, Oliva R (1993) Complete selective absence of protamine P2 in humans. J Biol Chem 268(14):10553–10557
Delbès G, Levacher C, Duquenne C, Racine C, Pakarinen P, Habert R (2005) Endogenous estrogens inhibit mouse fetal Leydig cell development via estrogen receptor α. Endocrinology 146(5):2454–2461
Delbès G, Levacher C, Habert R (2006) Estrogen effects on fetal and neonatal testicular development. Reproduction 132(4):527–538
Dodé C, Levilliers J, Dupont JM, De Paepe A, Le Dû N, Soussi-Yanicostas N, Coimbra RS, Delmaghani S, Compain-Nouaille S, Baverel F, Pêcheux C (2003) Loss-of-function mutations in FGFR1 cause autosomal dominant Kallmann syndrome. Nat Genet 33(4):463–465
Dodé C, Teixeira L, Levilliers J, Fouveaut C, Bouchard P, Kottler ML, Lespinasse J, Lienhardt-Roussie A, Mathieu M, Moerman A, Morgan G (2006) Kallmann syndrome: mutations in the genes encoding prokineticin-2 and prokineticin receptor-2. PLoS Genet 2(10):e175
Doyle TJ, Braun KW, McLean DJ, Wright RW, Griswold MD, Kim KH (2007) Potential functions of retinoic acid receptor A in Sertoli cells and germ cells during spermatogenesis. Ann N Y Acad Sci 1120(1):114–130
Edelmann W, Cohen PE, Kane M, Lau K, Morrow B, Bennett S, Umar A, Kunkel T, Cattoretti G, Chaganti R, Pollard JW (1996) Meiotic pachytene arrest in MLH1-deficient mice. Cell 85(7):1125–1134
Ellmann S, Sticht H, Thiel F, Beckmann MW, Strick R, Strissel PL (2009) Estrogen and progesterone receptors: from molecular structures to clinical targets. Cell Mol Life Sci 66(15):2405–2426
Falardeau J, Chung WC, Beenken A, Raivio T, Plummer L, Sidis Y, Jacobson-Dickman EE, Eliseenkova AV, Ma J, Dwyer A, Quinton R (2008) Decreased FGF8 signaling causes deficiency of gonadotropin-releasing hormone in humans and mice. J Clin Invest 118(8):2822–2831
Ferlin A, Santa Rocca M, Vinanzi C, Ghezzi M, Di Nisio A, Foresta C (2015) Mutational screening of NR5A1 gene encoding steroidogenic factor 1 in cryptorchidism and male factor infertility and functional analysis of seven undescribed mutations. Fertil Steril 104(1):163–169

Ferraz-de-Souza B, Lin L, Achermann JC (2011) Steroidogenic factor-1 (SF-1, NR5A1) and human disease. Mol Cell Endocrinol 336(1):198–205

Foster JW (1996) Mutations in SOX9 cause both autosomal sex reversal and campomelic dysplasia. Pediatr Int 38(4):405–411

Franco B, Guioli S, Pragliola A, Incerti B, Bardoni B, Tonlorenzi R, Carrozzo R, Maestrini E, Pieretti M, Taillon-Miller P, Brown CJ (1991) A gene deleted in Kallmann's syndrome shares homology with neural cell adhesion and axonal path-finding molecules. Nature 353(6344): 529–536

Gennari L, Merlotti D, De Paola V, Calabro A, Becherini L, Martini G, Nuti R (2005) Estrogen receptor gene polymorphisms and the genetics of osteoporosis: a HuGE review. Am J Epidemiol 161(4):307–320

Giese K, Pagel J, Grosschedl R (1994) Distinct DNA-binding properties of the high mobility group domain of murine and human SRY sex-determining factors. Proc Natl Acad Sci 91(8):3368–3372

Gonsalves J, Sun F, Schlegel PN, Turek PJ, Hopps CV, Greene C, Martin RH, Pera RAR (2004) Defective recombination in infertile men. Hum Mol Genet 13(22):2875–2883

Gonsalves J, Turek PJ, Schlegel PN, Hopps CV, Weier JF, Pera RAR (2005) Recombination in men with Klinefelter syndrome. Reproduction 130(2):223–229

Gorlov IP, Kamat A, Bogatcheva NV, Jones E, Lamb DJ, Truong A, Bishop CE, McElreavey K, Agoulnik AI (2002) Mutations of the GREAT gene cause cryptorchidism. Hum Mol Genet 11(19):2309–2318

Guazzone VA, Jacobo P, Theas MS, Lustig L (2009) Cytokines and chemokines in testicular inflammation: a brief review. Microsc Res Tech 72(8):620–628

Hardelin JP, Levilliers J, del Castillo I, Cohen-Salmon M, Legouis R, Blanchard S, Compain S, Bouloux P, Kirk J, Moraine C (1992) X chromosome-linked Kallmann syndrome: stop mutations validate the candidate gene. Proc Natl Acad Sci 89(17):8190–8194

Hikim APS, Lue Y, Yamamoto CM, Vera Y, Rodriguez S, Yen PH, Soeng K, Wang C, Swerdloff RS (2003) Key apoptotic pathways for heat-induced programmed germ cell death in the testis. Endocrinology 144(7):3167–3175

Hiort O, Holterhus PM, Horter T, Schulze W, Kremke B, Bals-Pratsch M, Sinnecker GH, Kruse K (2000) Significance of mutations in the androgen receptor gene in males with idiopathic infertility 1. J Clin Endocrinol Metabol 85(8):2810–2815

Houston DW, King ML (2000) Germ plasm and molecular determinants of germ cell fate. Curr Top Dev Biol 50:155–IN2

Huang Z, Kaftanovskaya EM, Rivas B, Agoulnik AI (2013) Mechanisms of INSL3 signaling in male reproductive organs. Ital J Anat Embryol 118(1):32–33

Huckins C (1978) The morphology and kinetics of spermatogonial degeneration in normal adult rats: an analysis using a simplified classification of the germinal epithelium. Anat Rec 190(4):905–926

Huyghe S, Mannaerts GP, Baes M, Van Veldhoven PP (2006) Peroxisomal multifunctional protein-2: the enzyme, the patients and the knockout mouse model. Biochim Biophys Acta 1761(9):973–994

Iguchi N, Yang S, Lamb DJ, Hecht NB (2006) An SNP in protamine 1: a possible genetic cause of male infertility? J Med Genet 43(4):382–384

Jaiswal D, Singh V, Dwivedi US, Trivedi S, Singh K (2014) Chromosome microarray analysis: a case report of infertile brothers with CATSPER gene deletion. Gene 542(2):263–265

Jaiswal D, Trivedi S, Agrawal NK, Singh K (2015) Dysregulation of apoptotic pathway candidate genes and proteins in infertile azoospermia patients. Fertil Steril 104(3):736–743

Janssen O, Qian J, Linkermann A, Kabelitz D (2003) CD95 ligand-death factor and costimulatory molecule? Cell Death Differ 10(11):1215–1225

Kallmann FJ (1944) The genetic aspects of primary eunuchoidism. Am J Ment Defic 48:203–236

Kasai S, Chuma S, Motoyama N, Nakatsuji N (2003) Haploinsufficiency of Bcl-x leads to male-specific defects in fetal germ cells: differential regulation of germ cell apoptosis between the sexes. Dev Biol 264(1):202–216

Kawamura K, Kumagai J, Sudo S, Chun SY, Pisarska M, Morita H, Toppari J, Fu P, Wade JD, Bathgate RA, Hsueh AJ (2004) Paracrine regulation of mammalian oocyte maturation and male germ cell survival. Proc Natl Acad Sci U S A 101(19):7323–7328

Kempisty B, Depa-Martynow M, Lianeri M, Jedrzejczak P, Darul-Wasowicz A, Jagodzinski PP (2007) Evaluation of protamines 1 and 2 transcript contents in spermatozoa from asthenozoospermic men. Folia Histochem Cytobiol 45(Suppl 1):S109–S113

Ketola I, Anttonen M, Vaskivuo T, Tapanainen JS, Toppari J, Heikinheimo M (2002) Developmental expression and spermatogenic stage specificity of transcription factors GATA-1 and GATA-4 and their cofactors FOG-1 and FOG-2 in the mouse testis. Eur J Endocrinol 147(3):397–406

Kim SK, Lee HJ, Yang H, Kim HS, Yoon YD (2004) Prepubertal exposure to 4-tert-octylphenol induces apoptosis of testicular germ cells in adult rat. Arch Androl 50(6):427–441

Knudson CM, Tung KS, Tourtellotte WG, Brown GA, Korsmeyer SJ (1995) Bax-deficient mice with lymphoid hyperplasia and male germ cell death. Science 270(5233):96

Korach KS (1994) Insights from the study of animals lacking functional estrogen receptor. Science 266(5190):1524

Kreidberg JA, Sariola H, Loring JM, Maeda M, Pelletier J, Housman D, Jaenisch R (1993) WT-1 is required for early kidney development. Cell 74(4):679–691

Kumar TR, Wang Y, Lu N, Matzuk MM (1997) Follicle stimulating hormone is required for ovarian follicle maturation but not male fertility. Nat Genet 15(2):201–204

Kuo CT, Morrisey EE, Anandappa R, Sigrist K, Lu MM, Parmacek MS, Soudais C, Leiden JM (1997) GATA4 transcription factor is required for ventral morphogenesis and heart tube formation. Genes Dev 11(8):1048–1060

Lavery R, Lardenois A, Ranc-Jianmotamedi F, Pauper E, Gregoire EP, Vigier C, Moreilhon C, Primig M, Chaboissier MC (2011) XY Sox9 embryonic loss-of-function mouse mutants show complete sex reversal and produce partially fertile XY oocytes. Dev Biol 354(1):111–122

LaVoie HA, McCoy GL, Blake CA (2004) Expression of the GATA-4 and GATA-6 transcription factors in the fetal rat gonad and in the ovary during postnatal development and pregnancy. Mol Cell Endocrinol 227(1):31–40

Ledig S, Hiort O, Wünsch L, Wieacker P (2012) Partial deletion of DMRT1 causes 46, XY ovotesticular disorder of sexual development. Eur J Endocrinol 167(1):119–124

Legouis R, Hardelin JP, Levilliers J, Claverie JM, Compain S, Wunderle V, Millasseau P, Le Paslier D, Cohen D, Caterina D, Bougueleret L (1991) The candidate gene for the X-linked Kallmann syndrome encodes a protein related to adhesion molecules. Cell 67(2):423–435

Lewis JD, Song Y, de Jong ME, Bagha SM, Ausió J (2003) A walk though vertebrate and invertebrate protamines. Chromosoma 111(8):473–482

Li TF, Wu QY, Zhang C, Li WW, Li N, Cui YX, Li XJ, Xia XY (2014) Polymorphisms in estrogen receptors predict the risk of male infertility: a meta-analysis. Reprod Biol Endocrinol 12(1):1

Lin L, Achermann JC (2008) Steroidogenic factor-1 (SF-1, Ad4BP, NR5A1) and disorders of testis development. Sex Dev 2(4–5):200–209

Lin Y, Page DC (2005) Dazl deficiency leads to embryonic arrest of germ cell development in XY C57BL/6 mice. Dev Biol 288(2):309–316

Liu D, Matzuk MM, Sung WK, Guo Q, Wang P, Wolgemuth DJ (1998) Cyclin A1 is required for meiosis in the male mouse. Nat Genet 20(4):377–380

Lizama C, Alfaro I, Reyes JG, Moreno RD (2007) Up-regulation of CD95 (Apo-1/Fas) is associated with spermatocyte apoptosis during the first round of spermatogenesis in the rat. Apoptosis 12(3):499–512

Lopes AM, Aston KI, Thompson E, Carvalho F, Gonçalves J, Huang N, Matthiesen R, Noordam MJ, Quintela I, Ramu A, Seabra C (2013) Human spermatogenic failure purges deleterious mutation load from the autosomes and both sex chromosomes, including the gene DMRT1. PLoS Genet 9(3):e1003349

Luo X, Ikeda Y, Parker KL (1994) A cell-specific nuclear receptor is essential for adrenal and gonadal development and sexual differentiation. Cell 77(4):481–490

Martianov I, Brancorsini S, Catena R, Gansmuller A, Kotaja N, Parvinen M, Sassone-Corsi P, Davidson I (2005) Polar nuclear localization of H1T2, a histone H1 variant, required for spermatid elongation and DNA condensation during spermiogenesis. Proc Natl Acad Sci U S A 102(8):2808–2813

Martinvalet D, Zhu P, Lieberman J (2005) Granzyme A induces caspase-independent mitochondrial damage, a required first step for apoptosis. Immunity 22(3):355–370

Mascrez B, Ghyselinck NB, Watanabe M, Annicotte JS, Chambon P, Auwerx J, Mark M (2004) Ligand-dependent contribution of RXRβ to cholesterol homeostasis in Sertoli cells. EMBO Rep 5(3):285–290

Matson CK, Murphy MW, Griswold MD, Yoshida S, Bardwell VJ, Zarkower D (2010) The mammalian doublesex homolog DMRT1 is a transcriptional gatekeeper that controls the mitosis versus meiosis decision in male germ cells. Dev Cell 19(4):612–624

Meetei AR, De Winter JP, Medhurst AL, Wallisch M, Waisfisz Q, Van de Vrugt HJ, Oostra AB, Yan Z, Ling C, Bishop CE, Hoatlin ME (2003) A novel ubiquitin ligase is deficient in Fanconi anemia. Nat Genet 35(2):165–170

Mendoza-Lujambio I, Burfeind P, Dixkens C, Meinhardt A, Hoyer-Fender S, Engel W, Neesen J (2002) The Hook1 gene is non-functional in the abnormal spermatozoon head shape (azh) mutant mouse. Hum Mol Genet 11(14):1647–1658

Meyn MS (1999) Ataxia-telangiectasia, cancer and the pathobiology of the ATM gene. Clin Genet 55(5):289–304

Mezquita CRISTOBAL, Teng CS (1977) Studies on sex-organ development. Changes in nuclear and chromatin composition and genomic activity during spermatogenesis in the maturing rooster testis. Biochem J 164(1):99–111

Miki K, Willis WD, Brown PR, Goulding EH, Fulcher KD, Eddy EM (2002) Targeted disruption of the Akap4 gene causes defects in sperm flagellum and motility. Dev Biol 248(2):331–342

Miyagawa Y, Nishimura H, Tsujimura A, Matsuoka Y, Matsumiya K, Okuyama A, Nishimune Y, Tanaka H (2005) Single-nucleotide polymorphisms and mutation analyses of the TNP1 and TNP2 genes of fertile and infertile human male populations. J Androl 26(6):779–786

Miyamoto T, Hasuike S, Yogev L, Maduro MR, Ishikawa M, Westphal H, Lamb DJ (2003) Azoospermia in patients heterozygous for a mutation in SYCP3. Lancet 362(9397):1714–1719

Molkentin JD, Lin Q, Duncan SA, Olson EN (1997) Requirement of the transcription factor GATA4 for heart tube formation and ventral morphogenesis. Genes Dev 11(8):1061–1072

Morohashi KI, Honda SI, Inomata Y, Handa H, Omura T (1992) A common trans-acting factor, Ad4-binding protein, to the promoters of steroidogenic P-450s. J Biol Chem 267(25):17913–17919

Mostafa T, Rashed L, Nabil N, Amin R (2014) Seminal BAX and BCL2 gene and protein expressions in infertile men with varicocele. Urology 84(3):590–595

Nadler JJ, Braun RE (2000) Fanconi anemia complementation group C is required for proliferation of murine primordial germ cells. Genesis 27(3):117–123

Nef S, Parada LF (1999) Cryptorchidism in mice mutant for Insl3. Nat Genet 22(3):295–299

Nef S, Schaad O, Stallings NR, Cederroth CR, Pitetti JL, Schaer G, Malki S, Dubois-Dauphin M, Boizet-Bonhoure B, Descombes P, Parker KL (2005) Gene expression during sex determination reveals a robust female genetic program at the onset of ovarian development. Dev Biol 287(2):361–377

Oliva R (2006) Protamines and male infertility. Hum Reprod Update 12(4):417–435

Oliva R, Dixon GH (1991) Vertebrate protamine genes and the histone-to-protamine replacement reaction. Prog Nucleic Acid Res Mol Biol 40:25–94

Oréal E, Mazaud S, Picard JY, Magre S, Carré-Eusèbe D (2002) Different patterns of anti-Müllerian hormone expression, as related to DMRT1, SF-1, WT1, GATA-4, Wnt-4, and Lhx9 expression, in the chick differentiating gonads. Dev Dyn 225(3):221–232

Parker KL, Schimmer BP (1997) Steroidogenic factor 1: a key determinant of endocrine development and function. Endocr Rev 18(3):361–377

Paul C, Povey JE, Lawrence NJ, Selfridge J, Melton DW, Saunders PT (2007) Deletion of genes implicated in protecting the integrity of male germ cells has differential effects on the incidence of DNA breaks and germ cell loss. PLoS One 2(10):e989

Pittman DL, Cobb J, Schimenti KJ, Wilson LA, Cooper DM, Brignull E, Handel MA, Schimenti JC (1998) Meiotic prophase arrest with failure of chromosome synapsis in mice deficient for Dmc1, a germline-specific RecA homolog. Mol Cell 1(5):697–705

Pontiggia A, Rimini R, Harley VR, Goodfellow PN, Lovell-Badge R, Bianchi ME (1994) Sex-reversing mutations affect the architecture of SRY-DNA complexes. EMBO J 13(24):6115

Qi H, Moran MM, Navarro B, Chong JA, Krapivinsky G, Krapivinsky L, Kirichok Y, Ramsey IS, Quill TA, Clapham DE (2007) All four CatSper ion channel proteins are required for male fertility and sperm cell hyperactivated motility. Proc Natl Acad Sci 104(4):1219–1223

Raymond CS, Shamu CE, Shen MM, Seifert KJ, Hirsch B, Hodgkin J, Zarkower D (1998) Evidence for evolutionary conservation of sex-determining genes. Nature 391(6668):691–695

Raymond CS, Murphy MW, O'Sullivan MG, Bardwell VJ, Zarkower D (2000) Dmrt1, a gene related to worm and fly sexual regulators, is required for mammalian testis differentiation. Genes Dev 14(20):2587–2595

Reijo R, Alagappan RK, Page DC, Patrizio P (1996) Severe oligozoospermia resulting from deletions of azoospermia factor gene on Y chromosome. Lancet 347(9011):1290–1293

Ren D, Navarro B, Perez G, Jackson AC, Hsu S, Shi Q, Tilly JL, Clapham DE (2001) A sperm ion channel required for sperm motility and male fertility. Nature 413(6856):603–609

Rochira V, Granata AR, Madeo B, Zirilli L, Rossi G, Carani C (2005) Estrogens in males: what have we learned in the last 10 years? Asian J Androl 7(1):3–20

Romanienko PJ, Camerini-Otero RD (2000) The mouse Spo11 gene is required for meiotic chromosome synapsis. Mol Cell 6(5):975–987

Rucker EB III, Dierisseau P, Wagner KU, Garrett L, Wynshaw-Boris A, Flaws JA, Hennighausen L (2000) Bcl-x and Bax regulate mouse primordial germ cell survival and apoptosis during embryogenesis. Mol Endocrinol 14(7):1038–1052

Ruggiu M, Speed R, Taggart M, McKay SJ, Kilanowski F, Saunders P, Dorin J, Cooke HJ (1997) The mouse Dazla gene encodes a cytoplasmic protein essential for gametogenesis. Nature 389(6646):73–77

Russell LD, Warren J, Debeljuk L, Richardson LL, Mahar PL, Waymire KG, Amy SP, Ross AJ, MacGregor GR (2001) Spermatogenesis in Bclw-deficient mice. Biol Reprod 65(1):318–332

Said TM, Paasch U, Glander HJ, Agarwal A (2004) Role of caspases in male infertility. Hum Reprod Update 10(1):39–51

Sassone-Corsi P (2002) Unique chromatin remodeling and transcriptional regulation in spermatogenesis. Science 296(5576):2176–2178

Saxena R, Brown LG, Hawkins T, Alagappan RK, Skaletsky H, Reeve MP, Reijo R, Rozen S, Dinulos MB, Disteche CM, Page DC (1996) The DAZ gene cluster on the human Y chromosome arose from an autosomal gene that was transposed, repeatedly amplified and pruned. Nat Genet 14(3):292–299

Schimmer BP, White PC (2010) Minireview: steroidogenic factor 1: its roles in differentiation, development, and disease. Mol Endocrinol 24(7):1322–1337

Schmahl J, Kim Y, Colvin JS, Ornitz DM, Capel B (2004) Fgf9 induces proliferation and nuclear localization of FGFR2 in Sertoli precursors during male sex determination. Development 131(15):3627–3636

Schrans-Stassen BH, Saunders PT, Cooke HJ, de Rooij DG (2001) Nature of the spermatogenic arrest in Dazl−/− mice. Biol Reprod 65(3):771–776

Sciurano RB, Rahn MI, Pigozzi MI, Olmedo SB, Solari AJ (2006) An azoospermic man with a double-strand DNA break-processing deficiency in the spermatocyte nuclei: case report. Hum Reprod 21(5):1194–1203

Seminara SB, Beranova M, Oliveira LM, Martin KA, Crowley WF Jr, Hall JE (2000) Successful use of pulsatile gonadotropin-releasing hormone (GnRH) for ovulation induction and pregnancy in a patient with GnRH receptor mutations 1. J Clin Endocrinol Metabol 85(2):556–562

Shaha C, Tripathi R, Mishra DP (2010) Male germ cell apoptosis: regulation and biology. Philos Trans R Soc Lond B Biol Sci 365(1546):1501–1515

Shan Z, Hirschmann P, Seebacher T, Edelmann A, Jauch A, Morell J, Urbitsch P, Vogt PH (1996) A SPGY copy homologous to the mouse gene Dazla and the Drosophila gene boule is autosomal and expressed only in the human male gonad. Hum Mol Genet 5(12):2005–2011

Singh AP, Rajender S (2015) CatSper channel, sperm function and male fertility. Reprod Biomed Online 30(1):28–38

Smirnova NA, Romanienko PJ, Khil PP, Camerini-Otero RD (2006) Gene expression profiles of Spo11−/− mouse testes with spermatocytes arrested in meiotic prophase I. Reproduction 132(1):67–77

Sofikitis N, Giotitsas N, Tsounapi P, Baltogiannis D, Giannakis D, Pardalidis N (2008) Hormonal regulation of spermatogenesis and spermiogenesis. J Steroid Biochem Mol Biol 109(3):323–330

Sornson MW, Wu W, Dasen JS, Flynn SE, Norman DJ, O'Connell SM, Gukovsky I, Carriere C, Ryan AK, Miller AP, Zuo L (1996) Pituitary lineage determination by the prophet of Pit-l homeodomain factor defective in Ames dwarfism. Nature 384:327–332

Spruck CH, de Miguel MP, Smith AP, Ryan A, Stein P, Schultz RM, Lincoln AJ, Donovan PJ, Reed SI (2003) Requirement of Cks2 for the first metaphase/anaphase transition of mammalian meiosis. Science 300(5619):647–650

Steger K, Pauls K, Klonisch T, Franke FE, Bergmann M (2000) Expression of protamine-1 and-2 mRNA during human spermiogenesis. Mol Hum Reprod 6(3):219–225

Subirana JA (1983) Nuclear proteins in spermatozoa and their interactions with DNA. In: The sperm cell. Springer, The Hague, pp 197–213

Svetlanov A, Cohen PE (2004) Mismatch repair proteins, meiosis, and mice: understanding the complexities of mammalian meiosis. Exp Cell Res 296(1):71–79

Thacker J (1999) A surfeit of RAD51-like genes? Trends Genet 15(5):166–168

Theas MS, Rival C, Dietrich SJ, Guazzone VA, Lustig L (2006) Death receptor and mitochondrial pathways are involved in germ cell apoptosis in an experimental model of autoimmune orchitis. Hum Reprod 21(7):1734–1742

Thurisch B, Liang SY, Sarioglu N, Schomburg L, Bungert J, Dame C (2009) Transgenic mice expressing small interfering RNA against Gata4 point to a crucial role of Gata4 in the heart and gonads. J Mol Endocrinol 43(4):157–169

Toscani A, Mettus RV, Coupland R, Simpkins H, Litvin J, Orth J, Hatton KS, Reddy EP (1997) Arrest of spermatogenesis and defective breast development in mice lacking A-myb. Nature 386(6626):713–717

Tüttelmann F, Rajpert-De Meyts E, Nieschlag E, Simoni M (2007) Gene polymorphisms and male infertility – a meta-analysis and literature review. Reprod Biomed Online 15(6):643–658

Uhrin P, Dewerchin M, Hilpert M, Chrenek P, Schöfer C, Zechmeister-Machhart M, Krönke G, Vales A, Carmeliet P, Binder BR, Geiger M (2000) Disruption of the protein C inhibitor gene results in impaired spermatogenesis and male infertility. J Clin Invest 106(12):1531–1539

Van Der Weyden L, Wei L, Luo J, Yang X, Birk DE, Adams DJ, Bradley A, Chen Q (2006) Functional knockout of the matrilin-3 gene causes premature chondrocyte maturation to hypertrophy and increases bone mineral density and osteoarthritis. Am J Pathol 169(2):515–527

Vidal VP, Chaboissier MC, de Rooij DG, Schedl A (2001) Sox9 induces testis development in XX transgenic mice. Nat Genet 28(3)

Viger RS, Mertineit C, Trasler JM, Nemer M (1998) Transcription factor GATA-4 is expressed in a sexually dimorphic pattern during mouse gonadal development and is a potent activator of the Mullerian inhibiting substance promoter. Development 125(14):2665–2675

Viger RS, Guittot SM, Anttonen M, Wilson DB, Heikinheimo M (2008) Role of the GATA family of transcription factors in endocrine development, function, and disease. Mol Endocrinol 22(4):781–798

Vinci G, Chantot-Bastaraud S, El Houate B, Lortat-Jacob S, Brauner R, McElreavey K (2007) Association of deletion 9p, 46, XY gonadal dysgenesis and autistic spectrum disorder. Mol Hum Reprod 13(9):685–689

Vogelstein B, Lane D, Levine AJ (2000) Surfing the p53 network. Nature 408(6810):307–310

Vogt PH (1998) Human chromosome deletions in Yq11, AZF candidate genes and male infertility: history and update. Mol Hum Reprod 4(8):739–744

Wagner T, Wirth J, Meyer J, Zabel B, Held M, Zimmer J, Pasantes J, Bricarelli FD, Keutel J, Hustert E, Wolf U (1994) Autosomal sex reversal and campomelic dysplasia are caused by mutations in and around the SRY-related gene SOX9. Cell 79(6):1111–1120

Wang XN, Li ZS, Ren Y, Jiang T, Wang YQ, Chen M, Zhang J, Hao JX, Wang YB, Sha RN, Huang Y (2013) The Wilms tumor gene, Wt1, is critical for mouse spermatogenesis via regulation of Sertoli cell polarity and is associated with non-obstructive azoospermia in humans. PLoS Genet 9(8):e1003645

Whitney MA, Royle G, Low MJ, Kelly MA, Axthelm MK, Reifsteck C, Olson S, Braun RE, Heinrich MC, Rathbun RK, Bagby GC (1996) Germ cell defects and hematopoietic hypersensitivity to gamma-interferon in mice with a targeted disruption of the Fanconi anemia C gene. Blood 88(1):49–58

Wu W, Cogan JD, Pfäffle RW, Dasen JS, Frisch H, O'Connell SM, Flynn SE, Brown MR, Mullis PE, Parks JS, Phillips JA III (1998a) Mutations in PROP1 cause familial combined pituitary hormone deficiency. Nat Genet 18(2):147–149

Wu W, Cogan JD, Pfäffle RW, Dasen JS, Frisch H, O'Connell SM, Flynn SE, Brown MR, Mullis PE, Parks JS, Phillips JA III (1998b) Mutations in PROP1 cause familial combined pituitary hormone deficiency. Nat Genet 18(2):147–149

Xu Y, Baltimore D (1996) Dual roles of ATM in the cellular response to radiation and in cell growth control. Genes Dev 10(19):2401–2410

Xu X, Toselli PA, Russell LD, Seldin DC (1999) Globozoospermia in mice lacking the casein kinase II α′ catalytic subunit. Nat Genet 23(1):118–121

Xu EY, Moore FL, Pera RAR (2001) A gene family required for human germ cell development evolved from an ancient meiotic gene conserved in metazoans. Proc Natl Acad Sci 98(13):7414–7419

Yang Y, Kuang Y, De Oca RM, Hays T, Moreau L, Lu N, Seed B, D'Andrea AD (2001) Targeted disruption of the murine Fanconi anemia gene, Fancg/Xrcc9. Blood 98(12):3435–3440

Yen PH, Chai NN, Salido EC (1996) The human autosomal gene DAZLA: testis specificity and a candidate for male infertility. Hum Mol Genet 5(12):2013–2017

Yin X, Ouyang S, Xu W, Zhang X, Fok KL, Wong HY, Zhang J, Qiu X, Miao S, Chan HC, Wang L (2007) YWK-II protein as a novel Go-coupled receptor for Müllerian inhibiting substance in cell survival. J Cell Sci 120(9):1521–1528

Ying M, Chen B, Tian Y, Hou Y, Li Q, Shang X, Sun J, Cheng H, Zhou R (2007) Nuclear import of human sexual regulator DMRT1 is mediated by importin-β. Biochim Biophys Acta 1773(6):804–813

Yoshida K, Kondoh G, Matsuda Y, Habu T, Nishimune Y, Morita T (1998) The mouse RecA-like gene Dmc1 is required for homologous chromosome synapsis during meiosis. Mol Cell 1(5):707–718

Yuan L, Liu JG, Zhao J, Brundell E, Daneholt B, Höög C (2000) The murine SCP3 gene is required for synaptonemal complex assembly, chromosome synapsis, and male fertility. Mol Cell 5(1):73–83

Yuan L, Liu JG, Hoja MR, Wilbertz J, Nordqvist K, Höög C (2002) Female germ cell aneuploidy and embryo death in mice lacking the meiosis-specific protein SCP3. Science 296(5570): 1115–1118

Zenvirth D, Richler C, Bardhan A, Baudat F, Barzilai A, Wahrman J, Simchen G (2003) Mammalian meiosis involves DNA double-strand breaks with 3′ overhangs. Chromosoma 111(6):369–376

Zimmermann S, Steding G, Emmen JM, Brinkmann AO, Nayernia K, Holstein AF, Engel W, Adham IM (1999) Targeted disruption of the Insl3 gene causes bilateral cryptorchidism. Mol Endocrinol 13(5):681–691

Sex Chromosomal Genes in Male Infertility

15

Sandeep Kumar Bansal, Vertika Singh, Kiran Singh, and Rajender Singh

Abstract

Y chromosome harbors the male-specific region (MSY) that regulates male sex determination and spermatogenesis. Y microdeletions are the most common cause of male infertility. These deletions are found in 15–20% of patients with idiopathic azoospermia and 7–10% of patients with severe oligozoospermia. Apart from microdeletions, partial deletions in the AZFc region result in loss of multiple copies of Y genes and increase the risk of infertility. A few studies have suggested that routine screening of these deletions could help in understanding the etiology, offering counseling and managing infertility by natural or assisted methods. X being a homologue chromosome of Y has drawn attention regarding the presence of spermatogenic genes. A number of theories and speculations have been put forward that are now supported by the identification of a number of testis-specific or testis-predominant genes present on the X chromosome. This chapter provides an overview of the Y deletions and X chromosome genes that affect spermatogenesis or male fertility.

Keywords

Y chromosome • X chromosome • Spermatogenic genes • Y microdeletions • Male infertility

S.K. Bansal • R. Singh (✉)
Male Reproductive Biology Laboratory, CSIR-Central Drug Research Institute, Lucknow, Uttar Pradesh, India
e-mail: rajender_singh@cdri.res.in

V. Singh • K. Singh
Department of Molecular and Human Genetics, Institute of Science, Banaras Hindu University, Varanasi, Uttar Pradesh, India
e-mail: singhk4@rediffmail.com

R. Singh, K. Singh (eds.), *Male Infertility: Understanding, Causes and Treatment*, DOI 10.1007/978-981-10-4017-7_15

Key Points

- Y microdeletions have been established to be a cause of male infertility and are found in high frequency in idiopathic azoospermic patients (18%), severe oligozoospermic (14%), and oligozoospermic (4%) patients.
- Deletions in AZFa and AZFb with or without AZFc have severe consequences, resulting in Sertoli cell-only syndrome or spermatogenic arrest.
- AZFc is made up of ampliconic repeats that make it susceptible to frequent partial deletions by nonallelic homologous recombination (NAHR), resulting in an array of spermatogenic loss phenotypes.
- Meta-analysis and cohort analysis suggest that gr/gr deletions significantly increase the risk of male infertility and that patients with gr/gr deletions have relatively low sperm count in comparison to those without deletions.
- Approximately, 1098 genes are present on the human X chromosome. Out of these, 99 are expressed in the testis and various cancers. Few testis-specific genes are present in multiple copies on the X chromosome.
- Studies have shown that copy number variations (CNVs) on the X chromosome may cause spermatogenic failure because these CNVs are present very close to the genes that show testis-specific expression.

15.1 Introduction

X and Y chromosomes are interesting as males have only one copy of each of these chromosomes. It is believed that the mammalian sex chromosomes evolved from an ordinary pair of autosomes. The divergence of X and Y chromosomes is dated back to about 180 million years, well before the divergence of the marsupial and placental mammalian lineages. The autosomes that became sex chromosomes in mammals still exist in birds as a pair of autosomes. The first step in the process of divergence was the acquisition of the testis-determining gene, now known as SRY (Hughes and Page 2015). Over a period of time, large-scale inversions and deletions on the Y chromosome suppressed recombination with the X chromosome, leading to accumulation of unique differences between the two chromosomes. The male-specific region of the modern Y chromosome (MSY) harbors the genes for male sex determination and spermatogenesis.

Over this course of evolution, the X chromosome remained largely unchanged in size though the gene content on this chromosome also changed significantly. Due to the presence of the MSY region on Y chromosome, most of the initial studies on spermatogenesis and infertility focused on the Y chromosome, leading to the identification of genes that are indispensable for spermatogenesis. Nevertheless, it is believed that X chromosome does not merely serve as a partner for recombination with Y chromosome and harbors genes important for spermatogenesis and fertility. There has been a lot of controversy regarding the presence of male-specific genes on the X chromosome. Some authors believe that a few male-specific genes are present on the X chromosome (Wang 2004), while others claim that X chromosome is enriched for spermatogenesis genes

(Rice 1984). This chapter highlights the role of X and Y chromosome genes that are important for spermatogenesis and fertility.

15.2 Y Deletions Are Common in Infertility

DNA sequencing of the Y chromosome has identified Yq region to contain an array of amplicons that form eight palindromes named as P1 to P8 from distal to proximal (Skaletsky et al. 2003). These ampliconic regions contain most of the multi-copy genes. Intrachromosomal recombination events between these identical repeats give rise to high-frequency de novo Y microdeletions. Eventually, deletions on the Y chromosome are the most common cause of male infertility (McElreavey et al. 2000). These deletions are present in 15–20% of the patients with idiopathic azoospermia and 7–10% of the patients with severe oligozoospermia. Y deletions map to the Yq region of the chromosome and belong to three non-overlapping regions, called as azoospermia factor (AZF).

Tiepolo and Zuffardi (1976) first reported the association between azoospermia and deletions in the Yq region. Subsequently, by using the STS- and YAC-based mapping, Vogt et al. (1996) revealed that deletions in the Yq region correspond to three different regions, which were later termed as AZFa, AZFb, and AZFc. Till date, many studies have been performed on Y microdeletions and male infertility, finding that the frequency distribution of Y microdeletions varies widely with ethnic and geographical affiliations. Foresta et al. (2001) revealed by reviewing the literature that the prevalence of Y microdeletions was highest in idiopathic azoospermic patients (18%) in comparison with severe oligozoospermic (14%) and oligozoospermic (4%). Atia et al. (2015) screened Y microdeletions and revealed that 22% of patients (azoospermic and severe oligozoospermic) had at least one microdeletion in one or the other AZF region. A recent study on Indian population revealed that 3.4% of infertile men had Yq microdeletions (Sen et al. 2013). Combined analysis of all published studies across India revealed that 5.8% of infertile individuals had Yq microdeletions (Sen et al. 2013). They also revealed that the frequency of these deletions in India was 6.4% in azoospermia, 5.8% in oligozoospermia, and 3.2% in oligoasthenozoospermia and teratozoospermia cases.

15.3 Screening of Y deletions

So far, EAA/EMQN guidelines have been recommended and used for detecting Y deletions (Simoni et al. 2004). These deletions are detected by PCR (polymerase chain reaction) amplification of selected regions of the Y chromosome. MSY-specific STS (sequence-tagged site) primers are used for PCR amplification (Skaletsky et al. 2003). Specific sets of these primers amplify both unknown sequences and MSY-specific genes (Skaletsky et al. 2003). However, a microdeletion detected using the STS primers cannot be regarded as a harmful or pathological deletion; it may be a common polymorphism (Repping et al. 2003; Fernandes et al. 2004).

Basically, the analysis of single STS locus in each AZF region is sufficient for detection of deletion in AZFa, AZFb, and AZFc regions; however, analyzing two STS loci in each region increases the diagnostic accuracy (Simoni et al. 2004). Based on the experience of different laboratories and the formats of multiplex PCR, a fixed set of STS primers has been recommended in the guidelines for detection of Y microdeletions. These primers include for AZFa (sY84 and sY86), for AZFb (sY127 and sY134), and for AZFc (sY254 and sY255; both are in the *DAZ* gene). *SRY* (sY14 marker) should be included as a control for detecting the presence of testis-determining factor (Simoni et al. 2004). Primer sequences of these STS markers are given in Table 15.1.

As mentioned elsewhere in this article, Y chromosome also shows partial deletions in the AZFc region that contribute to male infertility. Y partial deletions (Y subdeletion) are detected by using five STS markers (sY1161, sY1191, sY1291, sY1206, sY1201) specific to the AZFc region (Repping et al. 2003; Lin et al. 2006). Absence of sY1291 marker and the presence of all other markers indicate gr/gr deletions, while absence of sY1191 marker and the presence of all other markers indicate b2/b3 deletions. Absence of three STS markers (sY1161, sY1191, and sY1291) and the presence of other markers indicate b1/b3 deletions. Primer sequences of these STS markers are given in Table 15.1.

Table 15.1 Sequences of STS primers of Y microdeletions and partial deletions

STS primer	Forward primer (5′-3′)	Reverse primer (5′-3′)	Amplicon size
sY84	AGA AGG GTC TGA AAG CAG GT	GCC TAC TAC CTG GAG GCT TC	326
sY86	GTG ACA CAC AGA CTA TGC TTC	ACA CAC AGA GGG ACA ACC CT	320
sY127	GGC TCA CAA ACG AAA AGA AA	CTG CAG GCA GTA ATA AGG GA	274
sY134	GTC TGC CTC ACC ATA AAA CG	ACC ACT GCC AAA ACT TTC AA	301
sY254	GGG TGT TAC CAG AAG GCA AA	GAA CCG TAT CTA CCA AAG CAG C	400
sY255	GTT ACA GGA TTC GGC GTG AT	CTC GTC ATG TGC AGC CAC	126
sY14	GAA TAT TCC CGC TCT CCG GA	GCT GGT GCT CCA TTC TTG AG	214
sY1161	CGACACTTTTGGGAAGTTTCA	TTGTGTCCAGTGGTGGCTTA	377
sY1191	CCAGACGTTCTACCCTTTCG	GAGCCGAGATCCAGTTACCA	385
sY1291	TAAAAGGCAGAACTGCCAGG	GGGAGAAAAGTTCTGCAACG	527
sY1201	CCGACTTCCACAATGGCT	GGGAGAAAAGTTCTGCAACG	677
sY1206	ATTGATCTCCTTGGTTCCCC	GACATGTGTGGCCAATTTGA	394

15.4 Classical Deletions/Microdeletions

Prognostic value of Y microdeletions is not clear and has been a topic of debate since long. Generally, doctors or practitioners do not recommend screening of Y microdeletions before proceeding for ART procedures. TESE/ICSI procedures are usually performed for the treatment of azoospermia or oligozoospermia without complete diagnostic workup. TESE/ICSI procedures are highly invasive and may adversely affect the male (Manning et al. 1998). By these procedures, sperm could only be retrieved in 50% of the patients with nonobstructive azoospermia (Silber et al. 1995). Silber et al. (1998) revealed that complete deletions of AZFb+c or AZFa+b+c show total absence of testicular spermatozoa. On the contrary, Mulhall et al. (1997) reported that sperm could be retrieved from the testis in approximately 50% azoospermic patients with AZFc deletions. Hopps et al. (2003) examined the success rate of testicular sperm retrieval in men with deletions of AZFa, AZFb, and AZFc regions. They reported that sperm retrieval is almost nil or poor in men with microdeletions of complete AZFa or AZFb regions on the Y chromosome, whereas most of the men with AZFc deletions have sperm either in semen or in testis. Therefore, the phenotypic consequences of AZFa or AZFb regions are more severe in comparison with deletions involving only AZFc. This also explains the low frequency of AZFa and AZFb deletions.

Brandell et al. (1998) have suggested that screening of Y microdeletions has clinical importance and can be used as a potential prognostic test. By literature search, Krausz et al. (2000) analyzed the correlation between Y microdeletions and infertility phenotypes. They reported that different subtypes of AZFb deletions may exist and complete deletion of AZFb may lead to spermatogenic arrest at spermatocyte or spermatid stage (Vogt et al. 1996; Krausz et al. 2000). In addition to this, complete deletions of AZFb with AZFa and/or AZFc are associated with Sertoli cell-only syndrome. Krausz et al. (2000) have also revealed that if azoospermic patients are found with complete AZFb deletions, the possibility of sperm retrieval by TESE is completely nil; however, in case of partial AZFb deletions, round spermatids could be retrieved (Brandell et al. 1998). Further, Krausz and McElreavey (1999) reported that complete AZFa deletion is associated with Sertoli cell-only syndrome type I. Krausz et al. (2000) have also revealed that AZFc deletions are usually associated with hypospermatogenesis and Sertoli cell-only syndrome type II. Screening of Y microdeletions could help the patients in prediction of sperm retrieval from the testis and the success of ART procedures.

Y microdeletions are now well established as a cause of male infertility. In individuals with the above microdeletions, the chances of fertility are almost nil; however, hidden islands of normal spermatogenesis could be found in multiple biopsies of patients, for example, in Sertoli cell-only syndrome type II (Brandell et al. 1998). Oligozoospermic patients could directly proceed to ART procedures; however, screening of Y microdeletions could help in prediction of success of ART procedures and offer counseling. However, partial AZFb and complete AZFc deletions may be associated with oligozoospermia. Simoni et al. (1997) have reported a progressive decrease in sperm count over several months in patients with AZFc deletions.

We have also come across a few patients with AZFc deletions who showed progressive loss of sperm count. Early detection of Y microdeletions in these oligozoospermic cases could be useful in overcoming the problem of infertility. This way, an individual could escape from TESE-like invasive procedures in the future.

15.5 Partial Deletions

Among several genetic factors, Y partial deletions are a major cause of male infertility (Repping et al. 2003; Tüttelmann et al. 2007; Stouffs et al. 2011). AZFc region is made up of highly repeated sequences, which make this region more susceptible to deletions. A deletion in the b2/b4 region (also referred to as b2/b4 deletion) completely removes the AZFc region (3.5 Mb in size), which contains several important genes in multiple copy numbers (Repping et al. 2003). Besides complete AZFc deletion (B2/b4 deletions), partial deletions such as gr/gr (1.6 Mb), b1/b3 (1.6 Mb), and b2/b3 (1.8 Mb) are known to occur in the AZFc region. The effects of deletions in the AZFc region are quantitatively mild to severe spermatogenic failure (Pryor et al. 1997; Krausz et al. 1999a, b; Foresta et al. 2001). Among partial deletions, gr/gr are the most frequent and major factors for male infertility. One study has revealed that gr/gr deletions do not completely remove important testis-specific genes, but reduce their copy numbers on the Y chromosome (Repping et al. 2003). They reported that dosage of these deleted genes affects sperm production, which may lead to different infertility phenotypes from azoospermia to normozoospermia. Moreover, semen profiles of patients with gr/gr deletions may vary ethnically and geographically. In comparison with gr/gr deletions, b2/b3 deletions (also known as g1/g3 deletions) are less common. B2/b3 deletions do not occur directly but are preceded by an inversion event having occurred in this region on the Y chromosome.

15.5.1 gr/gr Is a Risk Factor for Male Infertility

Screening of gr/gr deletions in male infertility is still questionable. It is now confirmed that gr/gr deletions are present in the general population and could be detected in both cases and controls; however, to what extent these partial deletions contribute to male infertility is still unknown. Nevertheless, most of the studies have shown that gr/gr deletions are significantly more frequent in infertile cases in comparison with controls. Till date, a number of investigators have tried to find out the relevance of partial deletions in male infertility. Several case–control studies have been conducted on the correlation between gr/gr deletions and male infertility. Most of these studies have shown association between the gr/gr deletions (Repping et al. 2003; de Llanos et al. 2005; Lynch et al. 2005) and male infertility, while others have ruled out the possibility of an association (Ravel et al. 2006; Zhang et al. 2006; Stahl et al. 2011). At least three major meta-analyses have been conducted till date to quantify the relationship of gr/gr deletions with infertility. All of these meta-analyses suggested that gr/gr deletions significantly increase the risk of male infertility [Tüttelmann et al. 2007: OR = 1.81, p-value < 0.00001; Stouffs et al. 2011: OR = 1.76, $p < 0.001$; Bansal et al. 2016a, b: OR = 1.821, $p < 0.001$]. Recently, we undertook a meta-analysis on 29 studies with 10,948 cases and 6604 controls and observed that gr/gr deletions are a risk factor for male infertility (Bansal et al. 2016a, b).

In cohort analysis, we found that the gr/gr deletions result in about 25% loss in sperm count (Bansal et al. 2016a, b). This study supported two previous studies suggesting that gr/gr deletions correlate with poor sperm count (Visser et al. 2009; Shahid et al. 2011). Similarly, Sato et al. (2014) also reported a negative correlation between the gr/gr deletions and semen quality. Stouffs et al. (2011) revealed by meta-analysis that gr/gr deletions had significantly higher occurrence in oligozoospermic group/patients in comparison with azoospermic group/patients. This study showed that screening of gr/gr deletions can be useful in oligozoospermic infertile patients. Stouffs et al. (2011) had also reported that gr/gr deletions were present both in normozoospermic presumptive controls and proven fertile controls. This outcome indicates that gr/gr deletions do not invariably lead to fertility problems. Nevertheless, gr/gr deletions increase infertility risk tremendously and can result in infertility or poor sperm count that shows accelerated decline with age. Therefore, routine screening of gr/gr deletions is advised. In a recent meta-analysis and review of literature, we reported that the frequency of gr/gr deletions varies widely across ethnic and geographic affiliations (Bansal et al. 2016a, b).

15.5.2 b2/b3 May Increase Risk in Some Ethnic Groups

A small number of studies have focused on b2/b3 deletions, with most of them suggesting a lack of association between these deletions and male infertility. We recently reviewed the literature on b2/b3 deletions and found that most of the studies reported a higher frequency of b2/b3 deletions in infertile group in comparison with the control group. However, only two studies reported a significant association of b2/b3 deletions with male infertility (Lu et al. 2009; Vijesh et al. 2015). Interestingly, two other studies showed their protective association with fertility (Hucklenbroich et al. 2005; Sen et al. 2015). Eventually, there is a lack of consensus in association between b2/b3 deletions and the risk of male infertility. In a recent meta-analysis, we included 24 studies and found that b2/b3 deletions are significantly associated with spermatogenesis loss/infertility (Bansal et al. 2016a, b). Further in-depth analysis revealed a significant association only in Mongolians and Negro-Caucasians, but not in Caucasians, South Asians, and Dravidian-Indians. The frequency of these deletions in Mongolian populations was almost the same as that of the gr/gr deletions (Bansal et al. 2016a, b). Due to the lack of adequate number of studies in other populations, the contribution of b2/b3 deletions in male infertility risk is still unclear.

15.6 Y Haplotypes

15.6.1 Terminology and Nomenclature of Y Haplotypes

A common nomenclature system was recommended by the Y Chromosome Consortium (YCC) in 2002 (Y Chromosome Consortium 2002). YCC report uses the terminology of de Knijff (2000). According to this terminology, haplogroup (hg) refers to the lineage of NRY (a nonrecombining portion of the Y chromosome), which is defined by binary polymorphisms, while haplotype refers to all sublineages of the haplogroups defined by the variations at STRs (short tandem repeats) on the NRY (Hammer and Zegura 2002). The other terms such as lineage, sublineage, basal lineage, clade, and subclade refer to tree branches at different hierarchical levels. Prefixes "M" and "P" denote to mutation and polymorphism, respectively (Underhill et al. 2000, 2001; Hammer et al. 1998, 2001). The term paragroup refers to lineages that belong to the clade but not to its subclades (Hammer and Zegura 1996).

The YCC (2002) report stated two complementary nomenclature systems (Y Chromosome Consortium 2002; Ferlin et al. 2007). The first system is hierarchical-/lineage-based nomenclature and is dependent on the binary polymorphisms on human NRY. This system uses 19 capital letters (Y and A–R) to denote the major clades. These capital letters are initial letters of all subsequent subclade names. Paragroups are distinguished by the star (*) symbol added after the clade designation. Subclades are named by alternate alphanumeric letters in lowercase.

Alternatively, a second method can be used to name the haplogroups. The format of this nomenclature system is "capital letter–mutation name" where capital letter refers to the major haplogroup and may be a letter from A to R and Y, while the mutation name refers to the name of the terminal mutation. Dash (–) between the capital letter and mutation name distinguishes this system from the lineage based. Due to the simplicity of this system and the widely known "M" and "P" alphanumeric mutational designations, this system is most likely adoptable.

15.6.2 Y Haplotypes and Male Infertility

Y haplotypes represent the genetic diversity linked to the Y chromosome due to single nucleotide polymorphisms (SNPs) and the linkage between them. In the past decades, Y haplotypes have been found to be associated with Y microdeletions or male infertility. Patients with approximately similar Yq microdeletions have shown variability in infertility phenotypes ranging from azoospermia to oligozoospermia suggesting that there should be some modifier genes which might produce a unique Y chromosome constitution that modulate the effect of genes lost due to deletion. Initially, Carlsen et al. (1992) reviewed the papers published in the past 50 years and reported that semen quality including sperm concentration is decreasing and the number of subfertile men is increasing. This study was supported by the work of Auger et al. (1995) and de Mouzon et al. (1996). By then, it was already known that the Y chromosome had crucial genes for spermatogenesis (Tiepolo and Zuffardi 1976; Nakahori et al. 1996). Based on this preliminary work, Kuroki et al. (1999) studied the relationship between Y haplotypes and sperm concentration in fertile Japanese males. They observed that mean sperm concentration correlated with Y chromosome type. They also reported that occurrence of azoospermia was related to a particular Y chromosomal lineage (a branch of D2b haplotype). Thereafter, many studies were conducted on different populations and regions.

Krausz et al. (2001) reported that haplotype hg 26+ [or K*(xP) haplotype according to YCC nomenclature] was associated with male infertility in Danish population. Among various studies, two studies (one on European population and the other on northwestern European population) did not find an association between Y haplotypes and male infertility (Paracchini et al. 2000; Quintana-Murci et al. 2001). Later on, Repping et al. (2003) reported that D2b haplotype contained only gr/gr-deleted chromosomes and may be a risk factor for male infertility in Japanese population where this haplotype is common. They also reported that D2b haplotype is rare in other populations, including European and American populations. On the contrary, Carvalho et al. (2003) reported the lack of association between Y haplotypes and male infertility in Japanese men. These discrepancies in results of different studies may be due to differences in populations/regions.

Further, Fernandes et al. (2004) reported that b2/b3 deletions (more precisely *DAZ3/DAZ4* deletions) are associated with haplotype N, which is an ancient lineage of Y haplotype and is prevalent in northern Europe and Asia. Meanwhile, Arredi et al. (2007) showed an association between AZFc deletions and certain Y haplotypes in northern Italy. Puzuka et al. (2011) reported that Y-hg K* was predominantly present in infertile Latvian men. In a study on Indian population, Shahid et al. (2011) revealed that Y-hg L1 was present in patients with b1/b3 deletions, whereas Y-hg H1a1 and H1b were present in normozoospermic individuals with gr/gr deletions. Choi et al. (2012) reported that gr/gr deletions were significantly associated with impaired spermatogenesis in Korean men with YAP– lineage, but not in YAP+ lineage. Further, Lu et al. (2013) showed that the distribution of Y haplogroups (Y-hgs K* and O3e*) was significantly different between cases and controls in Han Chinese population. They found that Y-hg K* was significantly predisposed to nonobstructive azoospermia, while Y-hg O3e* had a protective effect against the same. After deep analysis, they revealed that overdosage of *DAZ* gene (*DAZ* duplication) was significantly more frequent in Y-hg K* in comparison with Y-hg O3e*.

Sato et al. (2013) revealed that Y-hg d2* lineage is associated with azoospermia in Japanese males. Similarly, Ran et al. (2013) reported that Y-hgs F*, K*, P*, and N1* may be susceptible to spermatogenic impairment in southwest China, while Y-hg O3 may show a protective effect. Recently, Sato et al. (2015) hypothesized that Y haplotypes may be associated with sex hormone levels. By conducting a study on a Japanese population, they observed that haplogroup D2a1 was significantly associated with high LH levels. From the above studies, it is now clear that Y haplogroups are undoubtedly correlated with male infertility, though genetic backgrounds provide various phenotypic responses from none to severe. In the future, Y haplotype analysis, particularly in studies on Y-deletions, is encouraged as genetic background can significantly manipulate the outcome of Y microdeletions/partial deletions.

15.7 Genes on the X Chromosome

There has been a lot of controversy regarding the presence of male-specific genes on the X chromosome. Some authors believe that a few male-specific genes are present on the X chromosome (Wang 2004), while others claim that X chromosome is enriched for spermatogenesis genes (Rice 1984). In this context, some theories have been proposed based on the evolutionary feminization or masculinization of the X chromosome. These theories include hemizygous exposure hypothesis or Rice hypothesis, sexual antagonism-driven X inactivation hypothesis (SAXI), and meiotic sex chromosome inactivation (MSCI) (Stouffs and Lissens 2012). Loss of fertility in Klinefelter's syndrome (KS) patients explains the significance of X chromosome genes in testicular homeostasis and germ cell development. A group of researchers recently performed a transcriptome analysis of testicular biopsies obtained from six non-mosaic KS patients with azoospermia. The analysis revealed a differential upregulation and downregulation of 656 and 247 transcripts, respectively. Majority of the transcripts belonged to Sertoli and Leydig cell functions (D'Aurora et al. 2015).

Rice hypothesis favors the masculinization of the X chromosome. According to this hypothesis, recessive mutations with beneficial or positive effects for men will accumulate on X chromosome. If the mutation is deleterious, it will not affect females and will survive because it would require two copies to show its effect. On the contrary, the male is hemizygous (having only one copy of X chromosome); therefore, beneficial effects of mutation for men will appear immediately. Thus, the genes/alleles beneficial for men or spermatogenesis will enrich on the X chromosome.

Sexual antagonism-driven X inactivation hypothesis (SAXI) favors feminization of the X chromosome. Sexual antagonism refers to a condition, in which a mutation shows positive effects on one sex, but negative on the other. According to this hypothesis, since females have two copies of X chromosome, X-linked mutations with beneficial effects for women are more likely to fix on the X chromosome, even if they are detrimental to men.

Meiotic sex chromosome inactivation (MSCI) is also referred to as meiotic silencing of unsynapsed chromatin (MSUC) and is a process by which X chromosome is inactivated during the male meiosis. This process is different from X chromosome inactivation or lyonization and is independent of the *XIST* gene. MSCI favors the feminization of X chromosome. Due to MSCI, most of the genes on X chromosome are suppressed from meiosis onward. Consequently, it becomes an unfavorable place for meiotic and postmeiotic genes on the X chromosome; therefore, genes relocate themselves from X chromosome to autosomes by retroposition. However, variations in theories on the gene content of the X chromosome may be explained as researchers examine different species and different gene pools.

15.8 X-Linked Testis-Specific or Testis-Enriched Genes

Ross et al. (2005) suggested that 1098 genes are present on the human X chromosome, of which 99 are expressed in the testis and various cancers. These genes are categorized as cancer–testis (CT) group. Several genes have been identified which are linked to X chromosome and specially expressed/enriched in the testis. Out of these, some have homologues/genes in both human and mouse. These genes include *AKAP4/AKAP82*, *ARPT1/ACTRT1*, *CPXCR1*, *DMRT8*, *ESR1*, *FATE1*, *FMR1NB*, *LUZP4*, *NUDT10*, *NUDT11*, *PABPC1L2A/B*, *PAK3*, *RHOXF1*, *RHOXF2*, *TAF7L*, *TEX11*, *TEX13A*, *TEX13B*, *TGIF2LX*, *TKTL1/TKR*, and *USP26*. Some genes are present only in humans: *CXorf61/CT83*, *DDX53/CAGE*, *GLUD2*, *H2BFWT*, *MYCL2*, *PASD1*, and *SAGE1*. Some genes such as *Tex16*, *Tsga8/Halp-X*, and *Tsx* are present only in the mouse.

These enriched genes play various roles in the testis. Some genes work as transcription factors, such as *DMRT8* (Veith et al. 2006), *RHOXF1* (Wayne et al. 2002; Geserick et al. 2002), *RHOXF2* (Wayne et al. 2002), *TAF7L* (Pointud et al. 2003; Cheng et al. 2007), and *TGIF2LX* (Blanco-Arias et al. 2002), while others play different roles during the development, organization, and spermatogenesis, e.g., *AKAP4* is a part of sperm fibrous sheath, *TEX11* plays a role in meiotic recombination, *GLUD2* participates in glutamate metabolism, *USP26* plays a role in deubiquitination, and *FATE1* participates in testicular differentiation/germ cell development.

Some testis-specific genes are found in multiple copies on the X chromosome. Out of these genes, some have homologues in both human and mouse such as *FTHL17* (4 copies), *H2AFB1* (3 copies), *MAGE* family (>24 copies), *NXF* (4 copies), *SPANX* genes (9/11 copies), and *SSX* family (>10 copies). Some are present only in humans, e.g., *CSAG* (4 copies), *CT45* (4 copies), *CT47* (>13 copies), *CTAG* (3 copies), *FAM9* (3 copies), *GAGE* family (13–39 copies), *PAGE* family (7 copies), *VCX* (4 copies), and *XAGE* family (14 copies), while genes such as Cypt (7 copies), Srsx (~14 copies), Sstx (3 copies), and Slx (>25 copies) are present only in mouse.

15.9 Mutation Analysis of Human X-Linked and Testis-Enriched Genes

15.9.1 A-Kinase Anchor Protein 4 (*AKAP4*)

AKAP4 gene belongs to the family of A-kinase anchor proteins. This protein is crucial for tyrosine phosphorylation, which is essential for sperm capacitation. Miki et al. (2002) revealed that disruption of Akap4 protein in mice resulted in infertility. These mice had shortened flagella, which led to immotile spermatozoa. Though no differences have been observed in the expression and location of AKAP4 proteins between patient and control groups (Turner et al. 2001a, b), Bacetti et al. (2005a, b) have reported partial deletions in the *AKAP4* and *AKAP3* genes in one patient with fibrous sheath dysplasia (Baccetti et al. 2005a, b). Further, Visser et al. (2011) reported a mutation (c.887G>A; p.Gly296Asn) in the*AKAP4* gene, which was present only in infertile patients (Visser et al. 2011).

15.9.2 Fetal and Adult Expressed 1 (*FATE1*)

This gene is highly expressed in the testis and to a lesser extent in the brain, heart, kidney, lung, and adrenal gland (Olesen et al. 2003). Cytoplasmic expression of this gene has been seen in spermatogonia, primary spermatocytes, and Sertoli cells, but not in spermatid and spermatozoa. Olesen et al. (2003) reported two mutations (c.185T>C or p.I34T and c.459C>G or p.S125R) in this gene, which were present in two (one in each) patients out of 144 infertile males, but not in controls ($n = 100$).

15.9.3 TATA Box Binding Protein-Associated Factor 7 Like (*TAF7L*)

Homologue of this gene, *Taf7*, is found on the autosome and is expressed ubiquitously. A study has shown that TAF7L protein expression changes from cytoplasm to the nucleus after meiosis (Pointud et al. 2003). Since autosomal TAF7 is expressed in the nucleus, it is expected that function of TAF7 in the nucleus is replaced by TAF7L after meiosis. Cheng et al. (2007) using Taf7l$^{-/Y}$ mice revealed that all females were fertile, while male mice had low sperm count as well as abnormal and immotile sperm cells. Akinloye et al. (2007) found that 1371G>A mutation in this gene was present more often in the patient group (nonobstructive azoospermic; n = 3/41) in comparison with the control group (n = 1/80). They suggested this mutation to be a risk factor for infertility in humans.

15.9.4 Ubiquitin-Specific Peptidase 26 (*USP26*)

This gene encodes a member of ubiquitin-specific processing (UBP) family of proteases and is a deubiquitinating enzyme with His and Cys domains. Lin et al. (2011) revealed that *USP26* is especially expressed in mice testis. They also observed that protein product of this gene is present at the blood–testis barrier (BTB) and on acrosome of round spermatid and spermatozoa where sperm cells attach to the Sertoli cells. Several variations have been observed in this gene; however, only three mutations (c.370_371 insACA, c.494T>C, and c.1423C>T) have been studied extensively. A study performed in 2005 suggested a correlation between this cluster of genes and spermatogenesis defects (Stouffs et al. 2005). However, a meta-analysis involving eight studies ruled out any such association. According to Dirac and Bernards (2010), this gene is involved in the regulation of androgen receptor ubiquitination. These authors investigated the effect of four variants (c494T>C, c.1037T>A, C.1090C>T, and c1423C>T) on transcriptional activity of the androgen receptor gene; however, no change in transcriptional activity of AR with mutant USP26 protein in comparison with the wild type was observed.

15.10 Genes on the X Chromosomes with a Homologue on the Y Chromosome

Some genes on the X chromosome have a homologue on the Y chromosome. These include *AMELX*, *DDX3X*, *EIF1AX*, *KDM5C*, *KDM6A*, *NLGN4X*, *PCDH11X*, *PRKX*, *RBMX*, *RPS4X*, *SOX3*, *TBL1X*, *TGIF2LX*, *TSPYL2*, *USP9X*, *VCX*, and *ZFX genes*. Out of these, two genes (*VCX* and *TGIF2LX*) are expressed in the testis only. Though no mutational study is available on these genes, a study in 2005 revealed that *VCX* gene has a potential role in mental retardation (Van Esch et al. 2005). However, it is less likely that a testis-specific gene may participate in mental retardation; therefore, reinvestigation of the expression pattern of these genes is required.

15.11 X-Linked Copy Number Variations and Male Infertility

Though several genetic factors have been studied extensively for the etiology of male infertility, X-linked copy number variations (CNVs) have been less explored. Copy number variations are structurally variant regions of the genome, which may differ from individual to individual in the number of copies of that region. These CNVs are larger than 1 kilobase (kb) in size and are gains and losses of genomic DNA. These CNVs are either microscopic or submicroscopic and therefore cannot be detected by standard G-banding karyotyping. Latest techniques, such as array-comparative genomic hybridization (a-CGH) and next-generation sequencing (NGS), have provided new insights into the presence of CNVs in the genome and their roles in diseases. At present, CNVs are thought to cover about 10% of the genome and a large part of the X chromosome; however, very little is known about their role and correlation with infertility. Tüttelmann et al. (2011), for the first time, studied CNVs in infertile patients with severe oligozoospermia and Sertoli cell-only syndrome. They revealed that X-linked CNVs were significantly higher in patients with Sertoli cell-only syndrome. They detected a duplication in CXorf48 gene in two oligozoospermic patients and 23 "private" CNVs only in one patient (Tuttelmann et al. 2011). Recently, a case–control study conducted on 276 idiopathic infertile patients and 327 controls revealed that difference in duplication load (CNVs) between patients and controls was highly significant (Chianese et al. 2014). They also concluded that CNVs might cause spermatogenic failure because these CNVs were patient specific and were found very close to the genes that show testis-specific expression (Chianese et al. 2014).

Conclusion

X and Y chromosomes are believed to have originated from a pair of autosomes that later specialized in the function of sex determination and fertility. Males have one copy of each of these chromosomes, making them very penetrating for their variations and interesting for their functions. Since Y harbors the MSY region, deletion analysis of Y chromosome started soon after the identification of the role of Y chromosome in sex determination. Now Y microdeletions have been established as a cause of male infertility. The AZFc partial deletions caught attention a little later, and a number of studies have emphasized on their importance as infertility risk factors. Meta-analyses have established that gr/gr deletions are a significant risk factor, while b2/b3 deletions may confer infertility risk in an ethnic-specific manner. B1/b3 deletions are relatively rare and need further investigation to understand their importance in infertility. X chromosome certainly harbors a number of genes with testis-specific or testis-predominant expression. Knockout studies have revealed the importance of a number of X chromosome genes in spermatogenesis and fertility; however, only a few mutation studies in humans have been conducted. Therefore, X chromosome genes need further investigations in human male infertility that may be guided by mouse knockout studies.

Acknowledgments The authors would like to thank the Council of Scientific and Industrial Research (CSIR), Govt. of India, for financial help under the network project, PROGRAM (BSC0101).

References

Akinloye O, Gromoll J, Callies C, Nieschlag E, Simoni M (2007) Mutation analysis of the X-chromosome linked, testis-specific TAF7L gene in spermatogenic failure. Andrologia 39(5):190–195

Arredi B, Ferlin A, Speltra E, Bedin C, Zuccarello D, Ganz F, Marchina E, Stuppia L, Krausz C, Foresta C (2007) Y-chromosome haplogroups and susceptibility to azoospermia factor c microdeletion in an Italian population. J Med Genet 44(3):205–208

Atia T, Abbas M, Ahmed AF (2015) Azoospermia factor microdeletion in infertile men with idiopathic severe oligozoospermia or non-obstructive azoospermia. Afr J Urol 21(4):246–253

Auger J, Kunstmann JM, Czyglik F, Jouannet P (1995) Decline in semen quality among fertile men in Paris during the past 20 years. N Engl J Med 332(5):281–285

Baccetti B, Collodel G, Gambera L, Moretti E, Serafini F, Piomboni P (2005a) Fluorescence in situ hybridization and molecular studies in infertile men with dysplasia of the fibrous sheath. Fertil Steril 84(1):123–129

Baccetti B, Collodel G, Estenoz M, Manca D, Moretti E, Piomboni P (2005b) Gene deletions in an infertile man with sperm fibrous sheath dysplasia. Hum Reprod 20(10):2790–2794

Bansal SK, Jaiswal D, Gupta N, Singh K, Dada R, Sankhwar SN, Gupta G, Rajender S (2016a) Gr/gr deletions on Y-chromosome correlate with male infertility: an original study, meta-analyses, and trial sequential analyses. Sci Rep 6

Bansal SK, Gupta G, Rajender S (2016b) Y chromosome b2/b3 deletions and male infertility: a comprehensive meta-analysis, trial sequential analysis and systematic review. Mutat Res Rev Mutat Res 768:78–90

Blanco-Arias P, Sargent CA, Affara NA (2002) The human-specific Yp11. 2/Xq21.3 homology block encodes a potentially functional testis-specific TGIF-like retroposon. Mamm Genome 13(8):463–468

Brandell RA, Mielnik A, Liotta D, Ye Z, Veeck LL, Palermo GD, Schlegel PN (1998) AZFb deletions predict the absence of spermatozoa with testicular sperm extraction: preliminary report of a prognostic genetic test. Hum Reprod 13(10):2812–2815

Carlsen E, Giwercman A, Keiding N, Skakkebæk NE (1992) Evidence for decreasing quality of semen during past 50 years. BMJ 305(6854):609–613

Carvalho C, Fujisawa M, Shirakawa T, Gotoh A, Kamidono S, Freitas Paulo T, Santos SE, Rocha J, Pena SD, Santos FR (2003) Lack of association between Y chromosome haplogroups and male infertility in Japanese men. Am J Med Genet A 116(2):152–158

Cheng Y, Buffone MG, Kouadio M, Goodheart M, Page DC, Gerton GL, Davidson I, Wang PJ (2007) Abnormal sperm in mice lacking the Taf7l gene. Mol Cell Biol 27(7):2582–2589

Chianese C, Gunning AC, Giachini C, Daguin F, Balercia G, Ars E, Giacco DL, Ruiz-Castañé E, Forti G, Krausz C (2014) X chromosome-linked CNVs in male infertility: discovery of overall duplication load and recurrent, patient-specific gains with potential clinical relevance. PLoS One 9(6):e97746

Choi J, Song SH, Bak CW, Sung SR, Yoon TK, Lee DR, Shim SH (2012) Impaired spermatogenesis and gr/gr deletions related to Y chromosome haplogroups in Korean men. PLoS One 7(8):e43550

D'Aurora M, Ferlin A, Di Nicola M, Garolla A, De Toni L, Franchi S, Palka G, Foresta C, Stuppia L, Gatta V (2015) Deregulation of sertoli and leydig cells function in patients with Klinefelter syndrome as evidenced by testis transcriptome analysis. BMC Genomics 16(1):1

de Knijff P (2000) Messages through bottlenecks: on the combined use of slow and fast evolving polymorphic markers on the human Y chromosome. Am J Hum Genet 67(5):1055–1061

De Llanos M, Ballescà JL, Gázquez C, Margarit E, Oliva R (2005) High frequency of gr/gr chromosome Y deletions in consecutive oligospermic ICSI candidates. Hum Reprod 20(1):216–220

de Mouzon JACQUES, Thonneau PD, Spira A, Multigner L (1996) Declining sperm count. Semen quality has declined among men born in France since 1950. BMJ 313(7048):43

Dirac AM, Bernards R (2010) The deubiquitinating enzyme USP26 is a regulator of androgen receptor signaling. Mol Cancer Res 8(6):844–854

Ferlin A, Raicu F, Gatta V, Zuccarello D, Palka G, Foresta C (2007) Male infertility: role of genetic background. Reprod Biomed Online 14(6):734–745

Fernandes S, Paracchini S, Meyer LH, Floridia G, Tyler-Smith C, Vogt PH (2004) A large AZFc deletion removes DAZ3/DAZ4 and nearby genes from men in Y haplogroup N. Am J Hum Genet 74(1):180–187

Foresta C, Moro E, Ferlin A (2001) Y microdeletions and alterations of spermatogenesis 1. Endocr Rev 22(2):226–239

Geserick C, Weiss B, Schleuning WD, Haendler B (2002) OTEX, an androgen-regulated human member of the paired-like class of homeobox genes. Biochem J 366(1):367–375

Hammer MF, Zegura SL (2002) The human Y chromosome haplogroup tree: nomenclature and phylogeography of its major divisions. Annu Rev Anthropol 31(1):303–321.

Hammer MF, Karafet T, Rasanayagam A, Wood ET, Altheide TK, Jenkins T, Griffiths RC, Templeton AR, Zegura SL (1998) Out of Africa and back again: nested cladistic analysis of human Y chromosome variation. Mol Biol Evol 15(4):427–441

Hammer MF, Karafet TM, Redd AJ, Jarjanazi H, Santachiara-Benerecetti S, Soodyall H, Zegura SL (2001) Hierarchical patterns of global human Y-chromosome diversity. Mol Biol Evol 18(7):1189–1203

Hopps CV, Mielnik A, Goldstein M, Palermo GD, Rosenwaks Z, Schlegel PN (2003) Detection of sperm in men with Y microdeletions of the AZFa, AZFb and AZFc regions. Hum Reprod 18(8):1660–1665

Hucklenbroich K, Gromoll J, Heinrich M, Hohoff C, Nieschlag E, Simoni M (2005) Partial deletions in the AZFc region of the Y chromosome occur in men with impaired as well as normal spermatogenesis. Hum Reprod 20(1):191–197

Hughes JF, Page DC (2015) The biology and evolution of mammalian Y chromosomes. Annu Rev Genet 49:507–527

Krausz C, McElreavey K (1999) Y chromosome and male infertility. Front Biosci 4:E1–E8

Krausz C, Quintana-Murci L, Barbaux S, Siffroi JP, Rouba H, Delafontaine D, Souleyreau-Therville N, Arvis G, Antoine JM, Erdei E, Taar JP (1999a) A high frequency of Y deletions in males with nonidiopathic infertility 1. J Clin Endocrinol Metabol 84(10):3606–3612

Krausz C, Bussani-Mastellone C, Granchi S, McElreavey K, Scarselli G, Forti G (1999b) Screening for microdeletions of Y chromosome genes in patients undergoing intracytoplasmic sperm injection. Hum Reprod 14(7):1717–1721

Krausz C, Quintana-Murci L, McElreavey K (2000) Prognostic value of Y deletion analysis what is the clinical prognostic value of Y microdeletions analysis? Hum Reprod 15(7):1431–1434

Krausz C, Quintana-Murci L, Rajpert-De Meyts E, Jørgensen N, Jobling MA, Rosser ZH, Skakkebaek NE, McElreavey K (2001) Identification of a Y chromosome haplogroup associated with reduced sperm counts. Hum Mol Genet 10(18):1873–1877

Kuroki Y, Iwamoto T, Lee J, Yoshiike M, Nozawa S, Nishida T, Ewis AA, Nakamura H, Toda T, Tokunaga K, Kotliarova SE (1999) Spermatogenic ability is different among males in different Y chromosome lineage. J Hum Genet 44(5):289–292

Lin YW, Hsu CL, Yen PH (2006) A two-step protocol for the detection of rearrangements at the AZFc region on the human Y chromosome. Mol Hum Reprod 12(5):347–351

Lin YW, Hsu TH, Yen PH (2011) Localization of ubiquitin specific protease 26 at blood–testis barrier and near Sertoli cell–germ cell interface in mouse testes. Int J Androl 34(5 Pt 2):e368–e377

Lu C, Zhang J, Li Y, Xia Y, Zhang F, Wu B, Wu W, Ji G, Gu A, Wang S, Jin L (2009) The b2/b3 subdeletion shows higher risk of spermatogenic failure and higher frequency of complete AZFc deletion than the gr/gr subdeletion in a Chinese population. Hum Mol Genet 18(6):1122–1130

Lu C, Wang Y, Zhang F, Lu F, Xu M, Qin Y, Wu W, Li S, Song L, Yang S, Wu D (2013) DAZ duplications confer the predisposition of Y chromosome haplogroup K* to non-obstructive azoospermia in Han Chinese populations. Hum Reprod 28(9):2440–2449

Lynch M, Cram DS, Reilly A, O'bryan MK, Baker HWG, De Kretser DM, McLachlan RI (2005) The Y chromosome gr/gr subdeletion is associated with male infertility. Mol Hum Reprod 11(7):507–512

Manning M, Jüemann KP, Alken P (1998) Decrease in testosterone blood concentrations after testicular sperm extraction for intracytoplasmic sperm injection in azoospermic men. Lancet 352(9121):37

McElreavey K, Krausz C, Bishop CE (2000) The human Y chromosome and male infertility. In: The genetic basis of male infertility. Springer, Berlin, pp 211–232

Miki K, Willis WD, Brown PR, Goulding EH, Fulcher KD, Eddy EM (2002) Targeted disruption of the Akap4 gene causes defects in sperm flagellum and motility. Dev Biol 248(2):331–342

Mulhall JP, Reijo R, Alagappan R, Brown L, Page D, Carson R, Oates RD (1997) Azoospermic men with deletion of the DAZ gene cluster are capable of completing spermatogenesis: fertilization, normal embryonic development and pregnancy occur when retrieved testicular spermatozoa are used for intracytoplasmic sperm injection. Hum Reprod 12(3):503–508

Nakahori Y, Kuroki Y, Komaki R, Kondoh N, Namiki M, Iwamoto T, Toda T, Kobayashi K (1996) The Y chromosome region essential for spermatogenesis. Horm Res Paediatr 46(Suppl 1):20–23

Olesen C, Silber J, Eiberg H, Ernst E, Petersen K, Lindenberg S, Tommerup N (2003) Mutational analysis of the human FATE gene in 144 infertile men. Hum Genet 113(3):195–201

Paracchini S, Stuppia L, Gatta V, Palka G, Moro E, Foresta C, Mengua L, Oliva R, Ballescà JL, Kremer JAM, Van Golde RJT (2000) Y-chromosomal DNA haplotypes in infertile European males carrying Y-microdeletions. J Endocrinol Investig 23(10):671–676

Pointud JC, Mengus G, Brancorsini S, Monaco L, Parvinen M, Sassone-Corsi P, Davidson I (2003) The intracellular localisation of TAF7L, a paralogue of transcription factor TFIID subunit TAF7, is developmentally regulated during male germ-cell differentiation. J Cell Sci 116(9):1847–1858

Pryor JL, Kent-First M, Muallem A, Van Bergen AH, Nolten WE, Meisner L, Roberts KP (1997) Microdeletions in the Y chromosome of infertile men. N Engl J Med 336(8):534–540

Puzuka A, Pronina N, Grinfelde I, Erenpreiss J, Lejing V, Bars J, Pliss L, Pelnena I, Baumanis V, Krumina A (2011) Y chromosome—a tool in infertility studies of Latvian population. Russ J Genet 47(3):347–353

Quintana-Murci L, Krausz C, Heyer E, Gromoll J, Seifer I, Barton DE, Barrett T, Skakkebaek NE, Rajpert-Dc Meyts E, Mitchell M, Lee AC (2001) The relationship between Y chromosome DNA haplotypes and Y deletions leading to male infertility. Hum Genet 108(1):55–58

Ran J, Han TT, Ding XP, Wei X, Zhang LY, Zhang YP, Li TJ, Nie SS, Chen L (2013) Association study between Y-chromosome haplogroups and susceptibility to spermatogenic impairment in Han People from southwest China. Genet Mol Res 12(1):59–66

Ravel C, Chantot-Bastaraud S, El Houate B, Mandelbaum J, Siffroi JP, McElreavey K (2006) GR/GR deletions within the azoospermia factor c region on the Y chromosome might not be associated with spermatogenic failure. Fertil Steril 85(1):229–231

Repping S, Skaletsky H, Brown L, van Daalen SK, Korver CM, Pyntikova T, Kuroda-Kawaguchi T, de Vries JW, Oates RD, Silber S, van der Veen F (2003) Polymorphism for a 1.6-Mb deletion of the human Y chromosome persists through balance between recurrent mutation and haploid selection. Nat Genet 35(3):247–251

Rice WR (1984) Sex chromosomes and the evolution of sexual dimorphism. Evolution 38:735–742

Ross MT, Grafham DV, Coffey AJ, Scherer S, McLay K, Muzny D, Platzer M, Howell GR, Burrows C, Bird CP, Frankish A (2005) The DNA sequence of the human X chromosome. Nature 434(7031):325–337

Sato Y, Shinka T, Iwamoto T, Yamauchi A, Nakahori Y (2013) Y chromosome haplogroup D2* lineage is associated with azoospermia in Japanese males. Biol Reprod 88(4):107

Sato Y, Iwamoto T, Shinka T, Nozawa S, Yoshiike M, Koh E, Kanaya J, Namiki M, Matsumiya K, Tsujimura A, Komatsu K (2014) Y chromosome gr/gr subdeletion is associated with lower semen quality in young men from the general Japanese population but not in fertile Japanese Men. Biol Reprod 90(6):116

Sato Y, Shinka T, Nozawa S, Yoshiike M, Koh E, Kanaya J, Namiki M, Matsumiya K, Tsujimura A, Komatsu K, Itoh N (2015) Y chromosome haplogroup D2a1 is significantly associated with high levels of luteinizing hormone in Japanese men. Andrology 3(3):520–525

Sen S, Pasi AR, Dada R, Shamsi MB, Modi D (2013) Y microdeletions in infertile men: prevalence, phenotypes and screening markers for the Indian population. J Assist Reprod Genet 30(3):413–422

Sen S, Ambulkar P, Hinduja I, Zaveri K, Gokral J, Pal A, Modi D (2015) Susceptibility of gr/gr rearrangements to azoospermia or oligozoospermia is dependent on DAZ and CDY1 gene copy deletions. J Assist Reprod Genet 32(9):1333–1341

Shahid M, Dhillon VS, Khalil HS, Sexana A, Husain SA (2011) Associations of Y-chromosome subdeletion gr/gr with the prevalence of Y-chromosome haplogroups in infertile patients. Eur J Hum Genet 19(1):23–29

Silber SJ, Van Steirteghem AC, Devroey P (1995) Sertoli cell only revisited. Hum Reprod 10(5):1031–1032

Silber SJ, Alagappan R, Brown LG, Page DC (1998) Y deletions in azoospermic and severely oligozoospermic men undergoing intracytoplasmic sperm injection after testicular sperm extraction. Hum Reprod 13(12):3332–3337

Simoni M, Gromoll J, Dworniczak B, Rolf C, Abshagen K, Kamischke A, Carani C, Meschede D, Behre HM, Horst J, Nieschlag E (1997) Screening for deletions of the Y chromosome involving the DAZ (deleted in AZoospermia) gene in azoospermia and severe oligozoospermia. Fertil Steril 67(3):542–547

Simoni M, Bakker E, Krausz C (2004) EAA/EMQN best practice guidelines for molecular diagnosis of Y-chromosomal microdeletions. State of the art 2004. Int J Androl 27(4):240–249

Skaletsky H, Kuroda-Kawaguchi T, Minx PJ, Cordum HS, Hillier L, Brown LG, Repping S, Pyntikova T, Ali J, Bieri T, Chinwalla A (2003) The male-specific region of the human Y chromosome is a mosaic of discrete sequence classes. Nature 423(6942):825–837

Stahl PJ, Mielnik A, Margreiter M, Marean MB, Schlegel PN, Paduch DA (2011) Diagnosis of the gr/gr Y microdeletions does not help in the treatment of infertile American men. J Urol 185(1):233–237

Stouffs K, Lissens W (2012) X chromosomal mutations and spermatogenic failure. Biochim Biophys Acta 1822(12):1864–1872

Stouffs K, Lissens W, Tournaye H, Van Steirteghem A, Liebaers I (2005) Possible role of USP26 in patients with severely impaired spermatogenesis. Eur J Hum Genet 13(3):336–340

Stouffs K, Lissens W, Tournaye H, Haentjens P (2011) What about gr/gr deletions and male infertility? Systematic review and meta-analysis. Hum Reprod Update 17(2):197–209

Tiepolo L, Zuffardi O (1976) Localization of factors controlling spermatogenesis in the nonfluorescent portion of the human Y chromosome long arm. Hum Genet 34(2):119–124

Turner RM, Musse MP, Mandal ARABINDA, Klotz KEN, Jayes FC, Herr JC, Gerton GL, Moss SB, Chemes HE (2001a) Molecular genetic analysis of two human sperm fibrous sheath proteins, AKAP4 and AKAP3, in men with dysplasia of the fibrous sheath. J Androl 22(2):302–315

Turner RM, Foster JA, Gerton GL, Moss SB, Patrizio P (2001b) Molecular evaluation of two major human sperm fibrous sheath proteins, pro-hAKAP82 and hAKAP82, in stump tail sperm. Fertil Steril 76(2):267–274

Tüttelmann F, Rajpert-De Meyts E, Nieschlag E, Simoni M (2007) Gene polymorphisms and male infertility – a meta-analysis and literature review. Reprod Biomed Online 15(6):643–658

Tüttelmann F, Simoni M, Kliesch S, Ledig S, Dworniczak B, Wieacker P, Röpke A (2011) Copy number variants in patients with severe oligozoospermia and Sertoli-cell-only syndrome. PLoS One 6(4):e19426

Underhill PA, Shen P, Lin AA, Jin L, Passarino G, Yang WH, Kauffman E, Bonné-Tamir B, Bertranpetit J, Francalacci P, Ibrahim M (2000) Y chromosome sequence variation and the history of human populations. Nat Genet 26(3):358–361

Underhill PA, Passarino G, Lin AA, Shen P, Lahr MM, Foley RA, Oefner PJ, Cavalli-Sforza LL (2001) The phylogeography of Y chromosome binary haplotypes and the origins of modern human populations. Ann Hum Genet 65(01):43–62

Van Esch H, Hollanders K, Badisco L, Melotte C, Van Hummelen P, Vermeesch JR, Devriendt K, Fryns JP, Marynen P, Froyen G (2005) Deletion of VCX-A due to NAHR plays a major role in the occurrence of mental retardation in patients with X-linked ichthyosis. Hum Mol Genet 14(13):1795–1803

Veith AM, Klattig J, Dettai A, Schmidt C, Englert C, Volff JN (2006) Male-biased expression of X-chromosomal DM domain-less Dmrt8 genes in the mouse. Genomics 88(2):185–195

Vijesh VV, Nambiar V, Mohammed SI, Sukumaran S, Suganthi R (2015) Screening for AZFc partial deletions in Dravidian men with nonobstructive azoospermia and oligozoospermia. Genet Test Mol Biomarkers 19(3):150–155

Visser L, Westerveld GH, Korver CM, Van Daalen SKM, Hovingh SE, Rozen S, Van Der Veen F, Repping S (2009) Y chromosome gr/gr deletions are a risk factor for low semen quality. Hum Reprod 24(10):2667–2667

Visser L, Westerveld GH, Xie F, van Daalen SK, van der Veen F, Lombardi MP, Repping S (2011) A comprehensive gene mutation screen in men with asthenozoospermia. Fertil Steril 95(3):1020–1024

Vogt PH, Edelmann A, Kirsch S, Henegariu O, Hirschmann P, Kiesewetter F, Köhn FM, Schill WB, Farah S, Ramos C, Hartmann M (1996) Human Y chromosome azoospermia factors (AZF) mapped to different subregions in Yq11. Hum Mol Genet 5(7):933–943

Wang PJ (2004) X chromosomes, retrogenes and their role in male reproduction. Trends Endocrinol Metab 15(2):79–83

Wayne CM, MacLean JA, Cornwall G, Wilkinson MF (2002) Two novel human X-linked homeobox genes, hPEPP1 and hPEPP2, selectively expressed in the testis. Gene 301(1):1–11

Y Chromosome Consortium (2002) A nomenclature system for the tree of human Y-chromosomal binary haplogroups. Genome Res 12(2):339–348

Zhang F, Li Z, Wen B, Jiang J, Shao M, Zhao Y, He Y, Song X, Qian J, Lu D, Jin L (2006) A frequent partial AZFc deletion does not render an increased risk of spermatogenic impairment in East Asians. Ann Hum Genet 70(3):304–313

Male Infertility: An Epigenetic Perspective

16

Sweta Mohan, Sharvari Deshpande, and N.H. Balasinor

Abstract

Besides known genetic and environmental factors, research over the last two decades has shed light on several epigenetic mechanisms and their association with male infertility. The male germ line undergoes extensive epigenetic remodeling throughout fetal to adult life and is thus susceptible to environmental factors that can affect fertility. During fetal life, the primordial germ cells undergo removal of epigenetic marks (demethylation) followed by re-establishment of these marks according to the sex of the fetus, at the time of gonadal differentiation. Extensive programming of the epigenome occurs during the various phases of spermatogenesis, i.e., mitosis, meiosis, and spermiogenesis, leading to haploid-condensed spermatozoa with protamines as the major nucleoproteins. Shortly after fertilization, the sperm chromatin decondenses and the protamines are replaced by histones. The male pronucleus undergoes active demethylation. One such epigenetic phenomenon, genomic imprinting resulting in monoallelic expression of genes depending on the parent of origin, is involved in early embryogenesis. Aberrant methylation pattern of imprinting control region (ICR) of imprinted genes in the spermatozoa is associated with altered sperm morphology, count, and motility. This chapter provides a comprehensive overview of the epigenetic changes affecting spermatogenesis and male fertility.

Keywords

Male infertility • Epigenetics • Genomic imprinting • Methylation • Demethylation
Chromatin compaction

S. Mohan • S. Deshpande • N.H. Balasinor (✉)
Neuroendocrinology, National Institute for Research in Reproductive Health (ICMR),
Parel, Mumbai 400012, India
e-mail: balasinorn@nirrh.res.in; necnirrh@hotmail.com

R. Singh, K. Singh (eds.), *Male Infertility: Understanding, Causes and Treatment*, DOI 10.1007/978-981-10-4017-7_16

Key Points

- The male germ line undergoes extensive epigenetic remodeling throughout fetal to adult life. Thus, it is susceptible to adverse environmental changes affecting germ cell maturation and fertility.
- Extensive epigenetic programming occurs during the various phases of spermatogenesis, i.e., mitosis, meiosis, and spermiogenesis, leading to haploid-condensed spermatozoa with protamines as the major nucleoproteins.
- Aberrant methylation of imprinting control region (ICR) of imprinted genes in spermatozoa is associated with altered sperm count, motility, and morphology.
- Inclusion of male epigenetic diagnostics in routine clinical investigations will be beneficial for infertility management and for selection of cases that will benefit from assisted reproductive technology (ART).
- Extensive research is required to decide on the type of epigenetic tests/parameters that can be included in routine clinical investigations for spermatozoa.

16.1 Introduction

Epigenetics is the study of heritable changes affecting gene expression that are not caused by changes in the underlying DNA sequence. Epigenetics can also be considered as a switch that regulates gene transcription in response to various environmental factors, leading to phenotypic changes essential for survival. Epigenetic modifications are mainly of three types: DNA methylation, histone modifications and noncoding RNAs.

The methylation of cytosine residues in a CpG (5'-C-phosphate-G-3′) dinucleotide in the DNA is the most well-studied epigenetic modification in mammals. Methylation of CpG-rich regions leads to silencing of gene expression, whereas unmethylated CpG regions are transcribed owing to being more accessible to transcription factors. DNA methyltransferases (DNMTs) are a group of enzymes regulating the process of DNA methylation and are involved in establishment and maintenance of methylation marks on the DNA. DNMT3A, DNMT3B and DNMT3L are responsible for establishment of methylation marks, a process which is independent of DNA replication cycles during early embryogenesis, whereas, DNMT1 is a maintenance methyltransferase which restores methylation patterns following DNA replication. DNMT3L lacks catalytic activity and acts as a cofactor to DNMT3A. Recently, 10–11 translocation (TET) family of proteins have been identified, which act as DNA demethylases and are responsible for catalyzing the removal of methyl groups from the DNA. Thus, the expression of both DNMTs and TETs is important to understand the role of DNA methylation in health and disease (Ni et al. 2016).

Another mechanism of epigenetic control of gene transcription is through the modification of histones. The N-terminal tails of histones are modified at specific amino acid residues. The presence of different modified histones associated with specific gene loci is an indicator of the transcriptional activity of genes (Sawan and Herceg 2010). The association of trimethylated histone 3 at lysine 4 (H3K4me3) and histone 3 acetylation (H3ac) at the promoter region marks an actively transcribing gene, while trimethylated histone 3 at lysine 27 (H3K27me3) and at lysine 9 (H3K9me3) is indicative of repression of gene expression (Boyer et al. 2006; Ho et al. 2015).

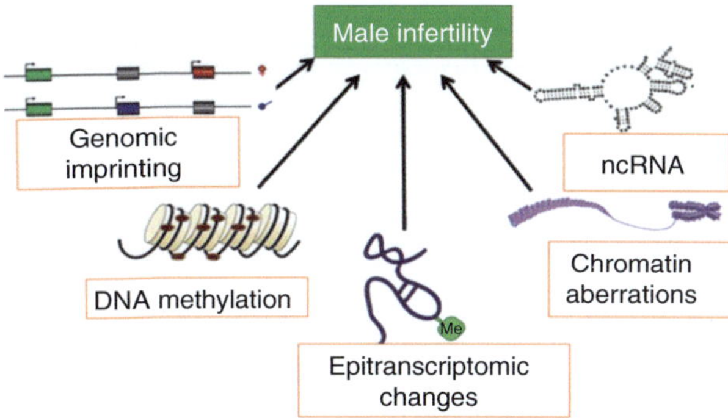

Fig. 16.1 Overview of the epigenetic causes of male infertility

Noncoding RNAs (ncRNAs) forms the third mechanism of epigenetic regulation of gene transcription (Fig. 16.1). Noncoding RNAs are generally classified as small noncoding RNAs: microRNAs and piwi-interacting RNAs (piRNAs) and long noncoding RNAs. These ncRNAs regulate gene expression at the level of transcription, RNA stability and translation (Tsai et al. 2010).

Spermatogenesis is a sequential process, which involves the production of spermatozoa from spermatogonial cells. It takes place in three phases: spermatogonial proliferation by mitosis, meiotic division of primary spermatocytes, and differentiation into haploid sperm cells by the process of spermiogenesis (de Kretser and Kerr 1994). Each of these phases involves extensive epigenetic programming, starting from the primordial germ cells (PGCs) in fetal life. The PGCs undergo genome-wide DNA demethylation and histone modifications such that the PGCs in the males are only ~10% methylated when they enter the genital ridge. Subsequently, *de novo* DNA methylation occurs until meiosis, to establish sperm and oocyte-specific epigenetic marks and silence the retrotransposable elements (Oakes et al. 2007; Tseng and Liao 2015).

Several studies have shown the involvement of both DNA methylation and histone modifications in various aspects of meiosis in males, namely, chromosome condensation, pairing, recombination and XY body formation. Another major epigenetic reorganization occurs post-meiosis, during the process of spermiogenesis, when the histones are replaced by protamines leading to a highly condensed sperm nuclear chromatin (Zamudio et al. 2008). These epigenetic modifications have been found to be involved in the etiology of idiopathic male infertility (Aston et al. 2012). The following sections discuss how the epigenetic modifications affect male fertility.

16.2 DNA Methylation

The role of DNA methylation in the male germ line is evident from a number of studies. The DNMTs are expressed in the male germ cells in a developmentally regulated fashion. Male mice deficient in *Dnmt3l* are infertile due to complete lack of germ cells and a decrease in global DNA methylation (Oakes et al. 2007). In the absence of *Dnmt3l*, chromosomes do not form heterochromatin appropriately, thus failing to pair during meiosis I. This leads to the activation of retrotransposons and repeat elements resulting in a "meiotic catastrophe" (Bourc'his and Bestor 2004). Disruption of *Dnmt3a* in germ cells results in loss of methylation at imprinted genes leading to infertility (Kaneda et al. 2004).

Genome-wide DNA methylation patterns in the spermatozoa are different from other somatic tissues as observed in mice and humans (Oakes et al. 2007). This is due to a reprogramming event that occurs in the PGCs of the developing embryo (Reik et al. 2001; Oakes et al. 2007). In the PGCs of mice, DNA methylation marks on imprinted genes and repetitive elements are erased between days 10.5 and 12.5 of gestation. In males, these marks are re-established around 15.5 dpc and 17.5 dpc on imprinted genes and repeated sequences, respectively. *De novo* DNA methylation and demethylation takes place in the early phases of spermatogenesis and are completed by the end of pachytene stage during meiosis (Oakes et al. 2007).

Studies in mice and humans have demonstrated that male gametogenesis occurs without significant changes in 5mC, but involves a dynamic change in 5-hydroxymethylcytosine (5hmC). This indicates that during spermatogenesis, there is a reduction in DNA demethylation leading to retention of methylation marks in mature sperm (Nettersheim et al. 2013; Ni et al. 2016). Interestingly, hypomethylated promoters in the mature sperm trigger transcription and production of signaling factors required for early embryo development. In mammals, appropriate sperm DNA methylation is essential for both fertilization and early embryo viability. The main sites for methylation in germ cells are the non-CpG island sequences in both distinct loci and repetitive sequences, although CpG islands (CGIs) can also be methylated (Oakes et al. 2007). Recently, Ichiyanagi et al. reported that non-CpG methylation is present within and around B1 retrotransposon sequences in mitotically arrested fetal pro-spermatogonia and reaches its highest level by birth. The level decreases in the neonatal period after the resumption of mitosis and is eventually absent in spermatogonia (Ichiyanagi et al. 2013). However, the biological role of non-CpG methylation remains unknown.

The male germ line undergoes two waves of genome-wide DNA demethylation: the first, in paternal pronucleus shortly after fertilization and, the second, in PGCs (Reik et al. 2001; Hajkova et al. 2002). TET (1–3) dioxygenases are essential for active DNA demethylation in the paternal pronucleus. All three TETs are detectable at the mRNA and protein level in the spermatozoa. Ni et al. reported that normal men exhibited higher levels of TET (1–3) enzymes in spermatozoa as compared to men with oligozoospermia and/or asthenozoospermia. They also reported that levels of TET3 in spermatozoa were significantly associated with high fertilization rates and that of TET2 was significantly associated with healthy pregnancy (Ni et al. 2016).

Evidences pertaining to the involvement of epigenetic mechanisms in male fertility were initially obtained from mice treated with a demethylating agent, 5-aza-20-deoxycytidine. A significant reduction in sperm counts, testis and epididymis weights and litter size was observed (Doerksen and Trasler 1996; Kelly et al. 2003). However, it took several years to clearly demonstrate that besides DNA methylation, other epigenetic mechanisms are also linked to male fertility (Table 16.1) (Boissonnas et al. 2013).

Table 16.1 Epigenetic modifiers and their role in male fertility

Epigenetic modifiers	Function	Substrate	Phenotype	References
DNA methylation				
Dnmt1	Maintenance of methylation marks	DNA	Loss of genomic imprinting; embryonic lethality; enhanced expression of retrotransposons in embryos	Li et al. (1992); Walsh et al. (1998)
Dnmt3a; Dnmt3b	*De novo*; establishment of methylation patterns	DNA	Failure in establishment of imprint patterns; results in male infertility	Kaneda et al. (2004)
Dnmt3l	*De novo*; involved in regulation of Dnmt3a activity	DNA	Failure in establishment of imprint patterns; male mice are infertile; severe hypogonadism; chromatin defects in meiotic cells	Bourc'his and Bestor (2004); Kato et al. (2007); Webster et al. (2005); Oakes et al. (2007)
Histone modifications				
SUV39H1/2	Histone methyl transferase (HMT)	Histone 3 lysine 9 me3 (H3K9me3)	Relaxed heterochromatin organization leading to chromosomal mis-segregation (meiotic defects) in the testis	Peters et al. (2001)
KDM3A/JHDM2A	Histone demethylase (HDM)	Histone 3 lysine 9 me2/1 (H3k9me2/1)	Severe oligospermia, increased apoptosis of pachytene, diplotene spermatocytes, round and elongating spermatids, blocks spermiogenesis	Okada et al. (2007; 2010)
KDM3B	Histone demethylase (HDM)	Histone 3 lysine 9 me2/1 (H3k9me2/1)	Subfertility; reduced litter size; reduced sperm counts and motility; reduced estradiol levels	Liu et al. (2015)
KDM4D	Histone demethylase (HDM)	Histone 3 lysine 9 me3 (H3K9me3)	No effect on fertility; dramatic changes in distribution of H3K9me1, H3K9me2; H3K9me3 in KDM4D-null mice testis	Iwamori et al. (2011)
PRDM9/ MEISETZ	Histone methyl transferase (HMT)	Histone 3 lysine 4 me2 (H3K4me2)	Sterility; reduced testis weights; disruption till pachytene stage of spermatogenesis; impaired pairing of homologous chromosomes; meiotic arrest in spermatocytes	Hayashi et al. (2005)

(continued)

Table 16.1 Continued

Epigenetic modifiers	Function	Substrate	Phenotype	References
MLL2/ KMT2B	Histone methyl transferase (HMT)	Histone 3 lysine 4 me1/2 (H3K4me1/2)	Sterility; disruption in early spermatogonia stage; progressive germ cell loss; dramatic reduction in spermatocytes	Glaser et al. (2009)
LSD1	Histone demethylase (HDM)	Histone 3 lysine 4 me1/2 (H3K4me1/2)	No effect on fertility	Foster et al. (2010)
Noncoding RNAs				
MILI/PIWIL2	Involved in initial processing of piRNAs from retrotransposons	piRNA	Arrest at early pachytene spermatocyte and round spermatid stages	Aravin et al. (2007); Bak et al. (2011)
MIWI/PIWIL1	Key regulator of spermiogenesis	piRNA	Arrest at the beginning of round spermatid stage; spermiogenic defect	Deng and Lin (2002); Bak et al. (2011)
MIWI2/ PIWIL4	Silencer	piRNA	Sterility; arrest at the pachytene spermatocyte stage; meiotic progression defect in the early prophase of meiosis I; germ cell loss	Carmell et al. (2007); Kuramochi-Miyagawa et al. (2008); Bak et al. (2011)
miR-34b/c and miR-449	Silencer	mRNA (FoxJ2, Shkbp1, Uhrf2)	oligoasthenoteratozoospermia in mice; impairs meiosis and sperm maturation	Comazzetto et al. (2014)

Several studies have shown altered sperm DNA methylation patterns in genes associated with spermatogenesis, such as methylene tetrahydrofolate reductase (*MTHFR*), which plays an important role in folate metabolism; cAMP-responsive element modulator (*CREM*), which is associated with spermiogenesis; and deleted in azoospermia-like (*DAZL*) gene, which is involved in germ line establishment and gametogenesis (reviewed by Boissonnas et al. 2013). A study by Jenkins et al. in two participant groups with similar semen characteristics revealed that hypomethylation at two genomic loci, HSPA1L and HSPA1B (members of the heat shock protein family), was associated with decreased fecundity (Jenkins et al. 2016). Urdinguio et al. for the first time studied genome-wide DNA methylation profiles in the spermatozoa of patients with unexplained infertility versus that of fertile individuals and identified about 3000 CpG sites, which displayed aberrant methylation. Among these genes, they found two CpG sites, associated with insulin-like growth factor 2 (*IGF2*) and heat shock 70 kDa protein 6 (*HSPA6*) genes, having altered methylation, which was also observed by Pacheco et al. However, further studies are necessary to elucidate the mechanisms underlying such alterations and their significance for male infertility (Urdinguio et al. 2015; Pacheco et al. 2011).

16.3 Histone Modifications and Chromatin Remodeling

In addition to DNA methylation, histone modifications and chromatin remodeling are important during spermatogenesis. Acetylation of H3 and H4 lysine residues and methylation of H3K4 are required for differentiation of spermatogonial stem cells and later diminish during meiosis. In males, during meiosis, chromosome condensation is correlated to phosphorylation of the histone variant H2AX (γH2AX) and gene silencing (Fernandez-Capetillo et al. 2003). Phosphorylation of H2AX usually occurs in response to DNA double-strand breaks (DSBs). H3K4 mono-, di-, and trimethylation and H3K9me2 are important for chromosome pairing and DNA DSBs formation. Histone methyltransferases Suv39h1, Suv39h2, and G9a which are responsible for tri-, di- and mono-methylation of H3K9 have been implicated to play a role during meiosis. In the Suv39h-deficient mice, spermatocytes undergo apoptosis in the pachytene stage due to incomplete homologous chromosome pairing and synapsis (Peters et al. 2001). In the G9a germ lineage-specific knockout mice, there was failure of synaptonemal complex formation and hence the spermatocytes were unable to progress beyond pachytene stage (Tachibana et al. 2007). Specific lysine demethylases expressed during gametogenesis are important for meiotic progression (Nottke et al. 2009). In males, the sex chromosomes are heterologous and undergo recombination at the pseudoautosomal regions, while the other regions on these chromosomes undergo transcriptional silencing, also known as meiotic sex chromosome inactivation (MSCI), thus forming XY (sex) body. γH2AX phosphorylation brought about by ATX and BRCA1, is involved in this inactivation. To maintain MSCI throughout the pachytene stage, several histone modifications are involved, namely, ubiquitylation of H2A; sumoylation; methylation of H3K27; di-methylation of H3K9, H4K20, H3K79 and H3K27;

trimethylation of H3K9 and H4K20; and deacetylation of H3K9, H4K12 and H4K16. Methylation of H3K9 and H3K27, which are repressing modifications, increases during meiosis and removal of H3K9 methylation at the end of meiosis is essential for the onset of spermiogenesis (reviewed by Zamudio et al. 2008). In addition, hyperacetylation of H4 occurs in elongated spermatids and is an important prerequisite for histone-to-protamine exchange during spermiogenesis.

During spermiogenesis, after the completion of meiosis, the genome of the round spermatid undergoes major changes to ensure efficient packaging of the male genome for its safe travel in the female reproductive tract. The somatic histones get replaced by testis-specific histones that in turn are substituted by transition proteins, which is then followed by tight packaging with protamines. The elongating spermatids also undergo other maturational events that affect motility and fertilizing ability during the period of protamine replacement (reviewed by Stuppia et al. 2015). Recent study has shown that the replacement of histones by protamines is not complete and 5–15% of the sperm chromatin is nucleosomal in humans. This nucleosomal structure is retained at specific gene loci, which are important in early embryogenesis (Hammoud et al. 2009).

The ratio of the two protamines P1 and P2 and their phosphorylation status are important for optimal sperm function. The P1/P2 ratio in fertile men ranges from 0.8 to 1.2 (Carrell and Liu 2001). Higher or lower values are associated with poor semen quality, increased DNA damage, and decreased fertility (Aoki et al. 2005; Stuppia et al. 2015). Protamine deficiency was found to be associated with an increase in methylation and a decrease in hydroxymethylation of the male pronucleus chromatin. Also, the efficiency of fertilization in protamine-deficient sperm cells was less than normal (Rajabi et al. 2016).

16.4 The Epitranscriptome

Abnormal sperm DNA methylation levels are associated with altered semen parameters (Aston et al. 2012). Apart from DNA, several mRNAs, miRNAs, and piwi-interacting RNAs (piRNAs) present in the sperm are also important for male fertility (Hamatani 2012; Liu et al. 2012; Sendler et al. 2013; Johnson et al. 2015). Epigenetic modifications of RNA, including methylation, are being investigated for their contribution in epigenetic regulation. Till date, more than 100 types of RNA modifications, occurring in mRNA, tRNA, rRNA, and small nuclear RNA (snRNA), have been identified (Sun et al. 2016). Among these modifications, N^6-methyladenosine (m^6A) modification is the more prevalent in mammalian mRNA (Yue et al. 2015).

m^6A modification was first reported by Desrosiers et al. (1974). The regulation of m^6A modification is brought about by functional interplay between m^6A methyltransferases and demethylases (Klungland and Dahl 2014). The methyltransferase complex consists of methyltransferase like 3 (METTL3), methyltransferase like 14 (METTL14), and Wilms' tumor 1-associated protein (WTAP). It catalyzes the formation of m^6A with S-adenosyl-L-methionine (SAM) as the methyl donor. METTL14 and METTL3 form a stable heterodimer to mediate mammalian nuclear

m^6A methylation. WTAP itself has no methyltransferase activity, but it can affect cellular m^6A by interacting with the METTL3-METTL14 complex (Liu et al. 2014). The alpha-ketoglutarate and Fe^{2+}- dependent dioxygenase fat mass and obesity-associated protein (FTO) and AlkB family member 5 protein (ALKBH5) are functionally similar to the DNA demethylase enzyme, TET. YTHDF domain family 2 protein (YTHDF2) regulates RNA stability, translation, splicing, transport, and localization through selective recognition of methylated RNA. Thus, methyltransferases act as "writers," demethylases serve as "erasers" and m^6A-selective-binding proteins (YTHDF) represent "readers" of m^6A in mRNA (Ben-Haim et al. 2015).

The expression of the *FTO* gene and m^6A levels are inversely proportional during adipogenesis (Zhao et al. 2014) and in type 2 diabetes mellitus (Shen et al. 2015). In 1997, Bokar et al. first revealed that *METTL3* (*MT-A70*) was expressed in several human tissues, with the highest levels being in the testis (Bokar et al. 1997). A recent study demonstrated that male mice deficient in *Alkbh5* showed an increase in m^6A levels in the mRNA and reduced fertility due to compromised spermatogenesis (Zheng et al. 2013). Recently, Yang et al. showed that m6A levels and expression of *METTL3* and *METTL14* were found to be significantly higher in spermatozoa obtained from asthenozoospemic individuals compared to controls suggesting that it could be one of the risk factor for asthenozoospermic condition (Yang et al. 2016).

16.5 Noncoding RNAs

There are three types of noncoding RNAs: Dicer-dependent miRNAs, long noncoding RNAs and Dicer-independent piRNAs which are expressed in male germ line. The miRNAs are the well-studied noncoding RNAs that are shown to affect spermatogenesis (Moazed 2009). These noncoding RNAs regulate gene expression by degrading their target mRNAs or by either activating or repressing translation (Gangaraju and Lin 2009). Transcription of miRNAs is carried out by polymerase II into large precursors which are later processed by a ribonuclease, DROSHA, and DGCR8, a DiGeorge syndrome critical region 8. These processed forms are then delivered into the cytoplasm, where they are further processed by an endonuclease known as Dicer into functional 20–24 nucleotide mature form and then incorporated into a RNA-induced silencing complex (RISC complex). It has also been observed that testis-specific DROSHA or DICER knockout models show arrested spermatogenesis (Korhonen et al. 2011; Hayashi et al. 2008; Wu et al. 2012).

Several miRNA precursors ranging from 100 to 150 nucleotides, also known as the pri-miRNAs, are found in spermatozoa. Pri-miRNA-181c is the most commonly found immature miRNA in mature human spermatozoa. The targets predicted for this miRNA are known to be involved in early embryonic development (Sendler et al. 2013; Vassena et al. 2011).

Several studies have demonstrated the association between altered miRNA profile and human male infertility. Testicular miRNA profiling identified several miRNAs to be dysregulated in patients with Sertoli cell only (SCO) syndrome, asthenozoospermia, mixed atrophy (MA) and germ cell arrest (GA) compared to healthy fertile males (reviewed in Gou et al. 2014). Interestingly, several SNPs (single nucleotide polymorphisms) are found in binding sites of miRNA of various candidate genes that are significantly associated with male fertility, which in turn may alter the expression of these genes thereby increasing the risk of infertility (reviewed in Gou et al. 2014). miRNAs are known to be a part of several biological fluids, which include seminal plasma. Wang et al. have demonstrated that the miRNA profiles of patients with abnormal morphology/motility or nonobstructive azoospermia were significantly altered compared to miRNA profiles of seminal plasma from healthy donors (Wang et al. 2011). Similarly, Salas-Huetos et al. have demonstrated that spermatozoa from patients with seminal alterations i.e., asthenozoospermia, teratozoospermia and oligozoospermia, exhibited differential miRNA profiles and were able to identify specific microRNAs associated with sperm motility (hasmiR-629-3p) and concentration (has-miR-335-5p, hasmiR-885-5p, and has-miR-152-3p) (Salas-Huetos et al. 2015). However, the function of altered miRNAs in structural integrity, metabolism and motility of spermatozoa is not well understood suggesting that these miRNA signatures need to be functionally characterized in order to be used as diagnostic biomarkers for male infertility.

piRNAs are also present abundantly in the male germ line (Girard et al. 2006). They are found in the spermatozoa of many species (Kawano et al. 2012; Krawetz et al. 2011; Peng et al. 2012). Their genomic organization is in clusters of up to 100 kb. piRNA precursors are processed into mature 23–32 nucleotide form using a mechanism which is dependent on piwi proteins (Ishizu et al. 2012). They are involved in regulation of epigenetic states, RNA stability and protection of germ line genome against transposition (Gangaraju and Lin 2009; Aravin and Hannon 2008). Any alterations in these regulatory RNAs can cause spermatogenic arrest (Carmell et al. 2007; Kuramochi-Miyagawa et al. 2004). The piRNAs may also have protective functions during early embryonic development when the DNA undergoes massive demethylation and remethylation. They also have the ability to maintain and protect DNA integrity by binding to it and prevent the attack of various transposable elements like LINE (long interspersed repeat element), SINE (short interspersed repeat element), LTR (long terminal repeat) and MER (medium reiterated sequence) at several developmental stages of an embryo (Krawetz et al. 2011).

Recent studies have shown that allele-specific differences in DNA methylation in PIWIL2 and PIWIL1 were significantly associated with disturbed spermatogenesis resulting in male infertility suggesting that these allele-specific genetic variations in piRNA and proteins associated with it may also compromise male fertility (reviewed in Gou et al. 2014).

Human spermatozoal small non-coding RNAs (sncRNAs) keeps in check several repetitive/transposable elements of LINE, SINE/ALU and LTR families (Krawetz et al. 2011). Disturbances in LINE1 activity results in arrest at the two- or four-cells stage of embryo (Beraldi et al. 2006). Certain LINE1 RNA fragments, which are poly-purine enriched, form a triple helix in different regions of LINE1 elements and act as a scaffold that will disturb the association of chromatin modifiers with the transcriptional machinery thereby promoting their own transcription (Fadloun et al.

2013). Possibly, large numbers of LINE1 fragments which are delivered by the sperm are capable of activating such a feedback cycle.

Long noncoding RNAs (lncRNAs) are another group of noncoding RNAs, the size of which ranges from 200 to 10,000 nucleotides. lncRNAs regulates gene expression in somatic cells at both transcriptional and post-transcriptional levels (Lee 2012; reviewed in Mercer et al. 2009; Rinn and Chang 2012). During transcription, several lncRNAs activate specific histone modifications. For example, HOX transcript antisense RNA (HOTAIR) regulates transcription by recruitment of a complex called PRC2 (polycomb recruiting complex 2) to HoxD locus, thereby making a repressive mark on histone (H3) (Tsai et al. 2010).

16.6 Genomic Imprinting

Genomic imprinting, an epigenetic phenomenon, is defined as monoallelic expression of genes or chromosomal regions depending on parent of origin of the allele and is reported to occur in mammals and some plant species. There are more than 100 imprinted genes found in mammals, of these many have roles in early embryonic and placental development and also in metabolic and behavioral functions (reviewed by Kitamura et al. 2015) (Table 16.2). The parent-specific expression of imprinted genes is due to differential epigenetic marking, predominantly in the form of DNA methylation in the regions termed as differentially methylated regions (DMRs) on the two parental genomes during gametogenesis when both the parental genomes are physically separated. Besides DNA methylation, histone modification is also known to differentially mark two parental alleles.

One of the most important features of imprinted genes is that they usually occur in clusters. Every cluster of imprinted genes has at least one DMR, where DNA methylation will occur only on one parental allele. One DMR can regulate many imprinted genes within a single cluster. Thus, methylation status of a single DMR can give information about several genes (reviewed by Kitamura et al. 2015). DMRs can be classified into somatic and germ line DMRs. In somatic DMRs, differential DNA methylation is parent-of-origin specific and is acquired only after fertilization, whereas germ line DMRs display differences in DNA methylation states between egg and sperm and are maintained even after fertilization. Once the methylation marks are established, parent-specific imprinted genes escape genome-wide methylome reprogramming after fertilization and tissue differentiation (reviewed by Kitamura et al. 2015).

Genome of the primordial germ cells undergoes extensive methylome reprogramming in order to make sure that it acquires proper sex-specific imprint marks. While inherited maternal and paternal "imprints" in the somatic cells of the embryo are maintained and read, they are erased in the germ line and new imprints are established depending on the sex of the embryo during gametogenesis (Reik et al. 2001) (Fig. 16.2). The erasure and establishment of imprints is initiated in the embryonic gonads and extends till meiosis in the adults. During gametogenesis, establishment of the imprint marks occurs at different time points in both female and male germ

Table 16.2 List of imprinted genes associated with male infertility

Imprinted genes	LOM/ GOM	Condition	Species	References
Human studies				
H19 (P) *MEST* (M)	LOM GOM	Oligozoospermia	Human	Marques et al. (2008)
H19 ICR (P) *SNRPN-ICR* (M)	LOM GOM	Oligozoospermia; asthenozoospermia Asthenozoospermia; teratozoospermia	Human	Dong et al. (2016)
RHOX gene cluster (P), *MEST* (M)	GOM	Idiopathic infertile men	Human	Richardson et al. (2014)
GTL2 (P), *H19* (P) *PEG1* (M), *LIT1* (M), *ZAC* (M), *PEG3* (M), *SNRPN(M)*	LOM GOM	Oligozoospermia	Human	Kobayashi et al. (2007)
H19 (P), *MEG3* (P), *ZDBF2* (M) *PEG1* (M), *PEG3* (M), *PLAGL1* (M), *SNRPN* (M) *KCNQ1OT1* (M)	LOM GOM	Moderate oligozoospermia/severe oligozoospermia	Human	Sato et al. (2011)
SNRPN (M)	LOM	Moderate oligozoospermia/severe oligozoospermia	Human	Manning et al. (2001)
H19 (P)	LOM	Moderate oligozoospermia/severe oligozoospermia	Human	Boissonnas et al. (2010)
H19 (P), *MEG3* (P) *PEG1* (M)	LOM GOM	Obstructive azoospermia/nonobstructive azoospermia/vasectomy reversal	Human	Minor et al. (2011)
Gtl2 (P) *SNRPN* (M)	LOM	Oligozoospermia/asthenozoospermia/ teratozoospermia	Human	El Hajj et al. (2011)
MEST (M), *KCNQ1OT1/LIT1* (M), *PEG1* (M), *ZAC* (M) *SNRPN* (M) *H19* (P)	GOM LOM	Oligozoospermia/abnormal histone–protamine incorporation	Human	Hammoud et al. (2010)
IGF2/H19 ICR1 (P; M) *MEST* (M)	LOM GOM	Idiopathic infertile men	Human	Poplinski et al. (2010)

PLAGL1 (M) *DIRAS3* (M) *MEST* (M)	GOM	Idiopathic infertile men	Human	Houshdaran et al. (2007)
Animal studies				
RTL1 (M), *MEG3 DMR* (P) *DLK1*(M)/*MEG3*(P), *NESP55* (P), *RASGRF1*(M), *PEG10* (M), *WT1* (M), *H19* (P), *IGF2 DMRs*(M)	GOM	Abnormal semen parameters	Boar	Congras et al. (2014)
RASGRF1 DMR (M)	LOM			
MEG3 (P) *MEST* (M), *SNRPN* (M), *PEG3* (M)	LOM GOM	Abnormal sperm parameters on exposure to methoxychlor (MXC)	Mouse	Stouder and Paoloni-Giacobino (2011)
H19 (P), *GTL2* (P) *PEG1* (M), *SNRPN* (M), *PEG3* (M)	LOM GOM	Abnormal sperm parameters on exposure to vinclozolin (VCZ)	Mouse	Stouder and Paoloni-Giacobino (2010)
H19 (P)	LOM	Abnormal sperm parameters on exposure to bisphenol A (BPA)	Rat	Doshi et al. (2013)
H19 (P)	LOM	Abnormal sperm parameters on exposure to alcohol	Mouse	Stouder et al. (2011)

P: paternally imprinted, M: maternally imprinted, LOM: loss of methylation, GOM: gain of methylation

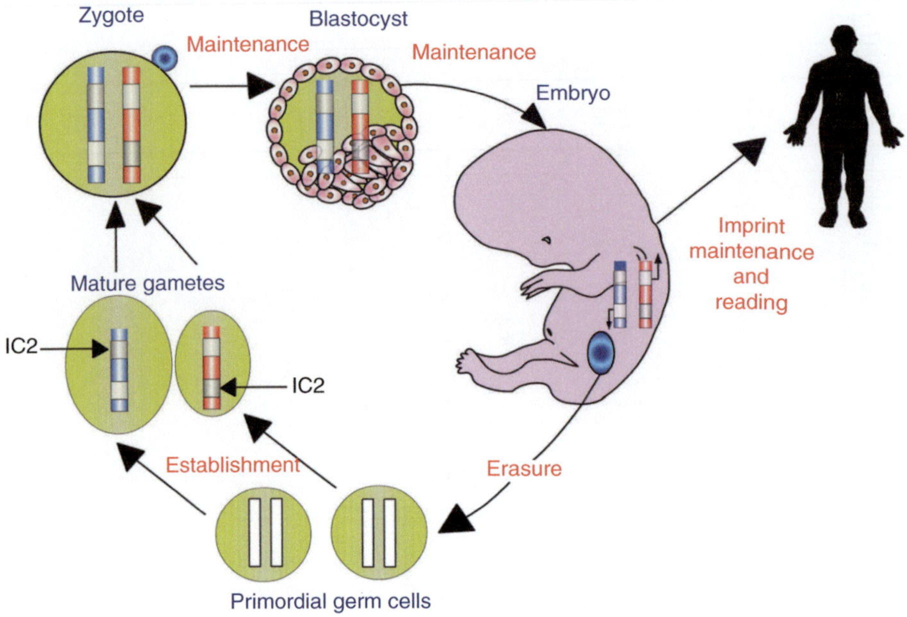

Fig. 16.2 *Life cycle of genomic imprinting*. Erasure, establishment and maintenance of DNA methylation at imprinting clusters during embryonic and germ cell development. Imprinting control regions: IC1 and IC2 are shown as examples. *Gray region* indicates DNA methylation and *white region* indicates the absence of DNA methylation on specific alleles. Parental chromosomes are designated in *blue* which stands for male or *red* which stands for female segregated on the basis of their individual sex (Modified from Reik et al. 2001)

line. The establishment occurs at the embryonic stage in males and is completed before meiosis, whereas in females, imprint acquisition initiates at the time of meiotic division I. In females, maternal DNA methylation of germ line DMRs initiates early in adult oocytes (Sato et al. 2007). In males, methylation occurs at three imprinted genes - *RASGRF1*, *H19* and *GTL2*, which exists prenatally before meiosis and is accomplished by the pachytene stage of spermatogenesis after birth (Davis et al. 2000; Kerjean et al. 2000).

Genomic imprinting involves dynamic remodeling of epigenetic marks that occurs at different phases of growth and development in both males and females. Errors occurring in the process of erasure, establishment or maintenance of imprints can have deleterious effects on the future generations, which are evident in genomic imprinting disorders like Beckwith–Wiedemann, which is an embryo overgrowth syndrome, or Silver–Russell, which is an embryo growth-restriction syndrome.

Abnormal DNA methylation patterns in imprinted genes and genes critical for embryonic development have been observed in the testis and spermatozoa of men suffering from oligozoospermia, azoospermia, and idiopathic infertility and have also been associated with poor semen parameters (Houshdaran et al. 2007;

Hammoud et al. 2010; Marques et al. 2010; Urdinguio et al. 2015). In addition, studies have been done to investigate the increased risk of imprinting disorders in children who were conceived through assisted reproductive technologies (ART). Aberrations in the sperm methylome and defect in other epigenetic factors are suspected to impair the quality of sperm, reduce male fertility in general, cause early developmental problems, and thereby decrease the success rates of ART (Jenkins and Carrell 2012).

It has been observed that the outcome of ART in the form of fertilization or implantation rates is generally poor due to the spermatozoa having altered DNA methylation status (Kobayashi et al. 2007). However, aberrant methylation of two paternally imprinted genes like *GTL2* and *H19* and five maternally imprinted genes like *MEST*, *LIT1*, *PEG3*, *SNRPN* and *NESPAS* demonstrated a significant correlation with poor semen parameters; however, it did not affect the ART outcome (El Hajj et al. 2011). In addition, methylation of ALU elements also showed to have a significant effect on fertilization and live birth rate especially in couples with male factor infertility. El Hajj et al. demonstrated that sperm samples from male partners of women experiencing abortions showed low ALU methylation (El Hajj et al. 2011). Also, studies from our laboratory have shown hypomethylation at the *IGF2-H19* ICR in spermatozoa of male partners of women experiencing recurrent spontaneous abortions (Ankolkar et al. 2012).

In addition to DNA methylation, recent studies have also shown that in somatic cells, DMRs are epigenetically marked by various histone modifications. Locus-specific and genome-wide studies of histone modifications revealed the presence of specific chromatin signatures at both paternal and maternal gametic DMRs. The unmethylated DNA region is usually associated with histone H3 or H4 acetylation and di-methylation of lysine 4 of histone H3 (H3K4me2), which are the hallmarks of activating chromatin, whereas the methylated DNA region is associated with repressing chromatin, i.e. trimethylation on lysine 9 of histone H3 (H3K9me3) and trimethylation on lysine 20 of histone H4 (H4K20me3) (Henckel et al. 2009; reviewed by Arnaud 2010). However, the importance of these activating or repressing chromatin signatures is poorly understood in genomic imprinting.

16.7 Influence of External Factors or Environment and Transgenerational Inheritance

It is well known that spermatogenesis process is under a tight control of gonadotropins and steroid hormones. Disruption of hormonal signaling by endocrine disruptors causes epigenetic disturbances in the germ cells leading to impaired fertility. Many of these endocrine disruptors are present in the environment and are estrogenic or antiandrogenic. For example, pesticides like vinclozolin and phthalates are anti-androgens, whereas plasticizers like bisphenol A are estrogenic and are known to cause several epigenetic disturbances (Zhang and Ho 2011). In the last few decades, poor semen quality, abnormal sperm counts and other reproductive or endocrine disorders in men are shown to have significant association with environmental estrogens and endocrine disrupting compounds exposure (Marques-Pinto and Carvalho 2013).

Similarly, studies in our laboratory on rodents revealed that on paternal subcutaneous administration of tamoxifen drug, which is a selective estrogen receptor modulator (SERM), for 60 days at a dosage of 0.4 mg/kg/day showed a decrease in fecundity with a significant increase in pre- (PIL) and post-implantation loss (POL) without any effect on sperm-fertilizing ability. The PIL was observed at two- and four-cells stage of embryo and POL around mid-gestation (Balasinor et al. 2002). *Igf2* expression, a paternally expressed and maternally imprinted gene, was seen to be significantly downregulated in resorbed embryos, i.e. embryos lost post-implantation (Kedia et al. 2004). The *Igf2-H19* ICR was hypomethylated in resorbing embryos and in spermatozoa obtained from tamoxifen-treated group, suggesting that tamoxifen could be responsible for errors in the acquisition/maintenance of imprint mark during spermatogenesis (Pathak et al. 2009). In addition, microarray experiments using rat whole genome arrays revealed disruption of growth factor signaling pathways and cell cycle arrest in resorbed embryos. Altered expression of imprinted genes important for trophoblast formation and differentiation was observed in resorbed embryos (Kedia-Mokashi et al. 2013; Kedia et al. 2016), suggesting a negative impact on placental development. The study demonstrates that defects in placental development may be caused by paternal drug treatment.

Several environmental factors including diet are known to affect male fertility through influence on epigenetic mechanisms and these factors also affect the health of the offspring. Epidemiological studies have found that paternal diet can influence fertility as well as the health of the offspring. Consumption of low-protein diet by male rats results in abnormal chromatin packaging in spermatozoa and causes aberrant changes in DNA methylation in the offspring (Carone et al. 2010). Paternal insufficiency of folic acid results in increased incidence of progeny suffering from skeletal and muscular defects. To investigate the role of dietary constituents in epigenetic alterations, Lambrot et al. fed female mice diet containing only ~15% of the recommended amount of folate during the preconception period, pregnancy and lactation. The male offsprings were given folate-deficient diet after weaning. It was observed that the pups given deficient diet showed delayed onset of meiosis and DNA damage in spermatocytes. Due to defective sperm function, pregnancy rate of females mated with males put on deficient diet was far lower than that of the control diet group. A number of differentially methylated genes were observed in the spermatozoa from these males (Lambrot et al. 2013). A diet high in fat content also reprograms the sperm epigenome. Barbosa et al. demonstrated that high-fat diet alters the expression of the miRNA let-7c in the sperm of F0 rats and their F1 offspring (Barbosa et al. 2016). These studies indicate how silent effects of the environment, including diet, can significantly alter the epigenome and have grave health consequences for the future generations.

Adverse transgenerational effects on male germ cells, prostate gland, testicular functions, and male fertility, accompanied by aberrant changes in the epigenome either in the form of DNA methylation or gene expression, occurred in Sertoli cells of F3 or F4 generations derived from vinclozolin-exposed F0 dams in a series of studies (Guerrero-Bosagna et al. 2013; Anway and Uzumcu 2006; Skinner 2014). The altered prostate phenotype was accompanied by transgenerational reprogramming of the expression of calcium and WNT signaling pathways. Effects on the sperm and testis were also observed in animal models exposed to a mixture of insect repellent (*N*,*N*-diethyl-meta-toluamide (DEET) and pesticide (permethrin) (Manikkam et al. 2012), insecticide (dichlorodiphenyltrichloroethane (DDT) (Skinner et al. 2013), plasticizer bisphenol A (BPA) (Salian et al. 2009) and di(2-ethylhexyl) phthalate (DEHP) (Doyle et al. 2013), and benzopyrene (Mohamed el et al. 2010).

In addition to the transgenerational inheritance through DNA methylation, male germ cells can transfer functional epigenetic information transgenerationally through RNA, also known as paramutation. An example of RNA-mediated inheritance is mutation of *Kit* gene that codes for a tyrosine kinase receptor involved in melanogenesis, germ cell differentiation and hematopoiesis (Rassoulzadegan et al. 2006). Heterozygous mutation in *Kit* gene in mice has significantly altered *Kit* expression and has distinct white patches on tails and feet. Also, when heterozygotes male or female were crossed with the wild-type counterpart, it was observed that wild-type offspring showed reduced *Kit* expression levels and inherited the mutant phenotype, i.e. white patches (Rassoulzadegan et al. 2006). The altered phenotype of the F1 generation was passed to its subsequent F2 generation suggesting a transgenerational inheritance through RNA. This was further confirmed when *Kit* mRNA and its target miRNA were injected from heterozygotes into zygotes, the white pigmentation was observed in the offspring (Rassoulzadegan et al. 2006).

Conclusion and Future Directions

In the past few years, number of studies has elucidated the involvement of epigenetic mechanisms in spermatogenesis and male fertility. The impact of environment through epigenetic mechanisms on fertility in males as well as transmission to subsequent generations is also well documented. However, these epigenetic factors are not routinely investigated for infertility management. Hence, inclusion of male epigenetic diagnostics in routine clinical investigations will aid in infertility management and selection of cases appropriate for ART. However, more research is required to decide on the type of epigenetic tests/parameters to be included in routine clinical investigations.

Acknowledgment The authors acknowledge the contribution of Ms. Kushaan Dumasia for critically going through the manuscript (IR/420/09-2016).

References

Ankolkar M, Patil A, Warke H, Salvi V, Kedia Mokashi N, Pathak S, Balasinor NH (2012) Methylation analysis of idiopathic recurrent spontaneous miscarriage cases reveals aberrant imprinting at H19 ICR in normozoospermic individuals. Fertil Steril 98:1186–1192

Anway MD, Memon MA, Uzumcu M, Skinner MK (2006) Transgenerational effect of the endocrine disruptor vinclozolin on male spermatogenesis. J Androl 27(6):868–879

Aoki VW, Moskovtsev SI, Willis J, Liu L, Mullen JB, Carrell DT (2005) DNA integrity is compromised in protamine-deficient human sperm. J Androl 26:741–748

Aravin AA, Hannon GJ (2008) Small RNA silencing pathways in germ and stem cells. Cold Spring Harb Symp Quant Biol 73:283–290

Aravin AA, Sachidanandam R, Girard A, Fejes-Toth K, Hannon GJ (2007) Developmentally regulated piRNA clusters implicate MILI in transposon control. Science 316(5825):744–747

Arnaud P (2010) Genomic imprinting in germ cells: imprints are under control. Reproduction 140:411–423

Aston KI, Punj V, Liu L, Carrell DT (2012) Genome-wide sperm deoxyribonucleic acid methylation is altered in some men with abnormal chromatin packaging or poor in vitro fertilization embryogenesis. Fertil Steril 97:285–292

Bak CW, Yoon TK, Choi Y (2011) Functions of PIWI proteins in spermatogenesis. Clin Exp Reprod Med 38(2):61–67

Balasinor N, Gill-Sharma MK, Parte P, D'Souza S, Kedia N, Juneja HS (2002) Effect of paternal administration of an antiestrogen, tamoxifen on embryo development in rats. Mol Cell Endocrinol 190(1–2):159–166

Barbosa TC, Ingerslev LR, Alm PS et al (2016) High-fat diet reprograms the epigenome of rat spermatozoa and transgenerationally affects metabolism of the offspring. Mol Metab 5(3):184–197. doi:10.1016/j.molmet.2015.12.002

Ben-Haim MS, Moshitch-Moshkovitz S, Rechavi G (2015) FTO: linking m6A demethylation to adipogenesis. Cell Res 25(1):3–4

Beraldi R, Pittoggi C, Sciamanna I, Mattei E, Spadafora C (2006) Expression of LINE-1 retroposons is essential for murine preimplantation development. Mol Reprod Dev 73:279–287

Boissonnas CC, Abdalaoui HE, Haelewyn V, Fauque P, Dupont JM, Gut I, Vaiman D, Jouannet P, Tost J, Jammes H (2010) Specific epigenetic alterations of IGF2-H19 locus in spermatozoa from infertile men. Eur J Hum Genet 18(1):73–80

Boissonnas CC, Jouannet P, Jammes H (2013) Epigenetic disorders and male subfertility. Fertil Steril 99:624–631

Bokar JA, Shambaugh ME, Polayes D, Matera AG, Rottman FM (1997) Purification and cDNA cloning of the AdoMet-binding subunit of the human mRNA (N6-adenosine)-methyltransferase. RNA 3:1233–1247

Bourc'his D, Bestor TH (2004) Meiotic catastrophe and retrotransposon reactivation in male germ cells lacking Dnmt3L. Nature 431:96–99

Boyer LA, Plath K, Zeitlinger J, Brambrink T, Medeiros LA, Lee TI, Levine SS, Wernig M, Tajonar A, Ray MK, Bell GW, Otte AP, Vidal M, Gifford DK, Young RA, Jaenisch R (2006) Polycomb complexes repress developmental regulators in murine embryonic stem cells. Nature 441(7091):349–353

Carmell MA, Girard A, van de Kant HJ, Bourc'his D, Bestor TH, de Rooij DG, Hannon GJ (2007) MIWI2 is essential for spermatogenesis and repression of transposons in the mouse male germline. Dev Cell 12:503–514

Carone BR, Fauquier L, Habib N, Shea JM, Hart CE, Li R, Bock C, Li C, Gu H, Zamore PD, Meissner A, Weng Z, Hofmann HA, Friedman N, Rando OJ (2010) Paternally induced transgenerational environmental reprogramming of metabolic gene expression in mammals. Cell 143:1084–1096

Carrell DT, Liu L (2001) Altered protamine 2 expression is uncommon in donors of known fertility, but common among men with poor fertilizing capacity, and may reflect other abnormalities of spermiogenesis. J Androl 22:604–610

Comazzetto S, Di Giacomo M, Rasmussen KD et al (2014) Oligoasthenoteratozoospermia and infertility in mice deficient for miR-34b/c and miR-449 loci. PLoS Genet 10(10):e1004597

Congras A, Yerle-Bouissou M, Pinton A, Vignoles F, Liaubet L, Ferchaud S, Acloque H (2014) Sperm DNA methylation analysis in swine reveals conserved and species-specific methylation patterns and highlights an altered methylation at the GNAS locus in infertile boars. Biol Reprod 91(6):137

Davis TL, Yang GJ, McCarrey JR, Bartolomei MS (2000) The H19 methylation imprint is erased and re-established differentially on the parental alleles during male germ cell development. Hum Mol Genet 9:2885–2894

de Kretser DM, Kerr JB (1994) The cytology of the testis. In: Knobil E, Neill JD (eds) Physiology & reproduction. Raven Press, New York, pp 1177–1290

Deng W, Lin H (2002) Miwi, a murine homolog of piwi, encodes a cytoplasmic protein essential for spermatogenesis. Dev Cell 2(6):819–830

Desrosiers R, Friderici K, Rottman F (1974) Identification of methylated nucleosides in messenger RNA from Novikoff hepatoma cells. Proc Natl Acad Sci U S A 71:3971–3975

Doerksen T, Trasler JM (1996) Developmental exposure of male germ cells to 5-azacytidine results in abnormal preimplantation development in rats. Biol Reprod 55:1155–1162

Dong H, Wang Y, Zou Z, Chen L, Shen C, Xu S, Zhang J, Zhao F, Ge S, Gao Q, Hu H, Song M, Wang W (2016) Abnormal methylation of imprinted genes and cigarette smoking: assessment of their association with the risk of male infertility. Reprod Sci. pii: 1933719116650755

Doshi T, D'souza C, Vanage G (2013) Aberrant DNA methylation at Igf2-H19 imprinting control region in spermatozoa upon neonatal exposure to bisphenol A and its association with post implantation loss. Mol Biol Rep 40(8):4747–4757

Doyle TJ, Bowman JL, Windell VL, McLean DJ, Kim KH (2013) Transgenerational effects of di-(2-ethylhexyl) phthalate on testicular germcell associations and spermatogonial stem cells in mice. Biol Reprod 88(5):112

El Hajj N, Zechner U, Schneider E, Tresch A, Gromoll J, Hahn T, Schorsch M, Haaf T (2011) Methylation status of imprinted genes and repetitive elements in sperm DNA from infertile males. Sex Dev 5:60–69

Fadloun A, Le Gras S, Jost B, Ziegler-Birling C, Takahashi H, Gorab E, Carninci P, Torres-Padilla ME (2013) Chromatin signatures and retrotransposon profiling in mouse embryos reveal regulation of LINE-1 by RNA. Nat Struct Mol Biol 20:7

Fernandez-Capetillo O, Mahadevaiah SK, Celeste A, Romanienko PJ, Camerini-Otero RD, Bonner WM, Manova K, Burgoyne P, Nussenzweig A (2003) H2AX is required for chromatin remodeling and inactivation of sex chromosomes in male mouse meiosis. Dev Cell 4:497–508

Foster CT, Dovey OM, Lezina L, Luo JL, Gant TW, Barlev N, Bradley A, Cowley SM (2010) Lysine specific demethylase 1 (LSD1) regulates the embryonic 2 transcriptome and CoREST stability. Mol Cell Biol 30(20):4851–4863

Gangaraju VK, Lin H (2009) MicroRNAs: key regulators of stem cells. Nat Rev Mol Cell Biol 10:116–125

Girard A, Sachidanandam R, Hannon GJ, Carmell MA (2006) A germline-specific class of small RNAs binds mammalian piwi proteins. Nature 442:199–202

Glaser S, Lubitz S, Loveland KL, Ohbo K, Robb L, Schwenk F, Seibler J, Roellig D, Kranz A, Anastassiadis K, Stewart AF (2009) The histone 3 lysine 4 methyltransferase, Mll2, is only required briefly in development and spermatogenesis. Epigenetics Chromatin 2(5)

Gou L-T, Dai P, Liu MF (2014) Small noncoding RNAs and male infertility. WIREs RNA 5(6):733–745. doi:10.1002/wrna.1252

Guerrero-Bosagna C, Savenkova M, Haque MM, Nilsson E, Skinner MK (2013) Environmentally induced epigenetic transgenerational inheritance of altered Sertoli cell transcriptome and epigenome: molecular etiology of male infertility. PLoS One 8(3):e59922

Hajkova P, Erhardt S, Lane N, Haaf T, El Maarri O, Reik W, Walter J, Surani MA (2002) Epigenetic reprogramming in mouse primordial germ cells. Mech Dev 117:15–23

Hamatani T (2012) Human spermatozoal RNAs. Fertil Steril 97:275–281

Hammoud SS, Nix DA, Zhang H, Purwar J, Carrell DT, Cairns BR (2009) Distinctive chromatin in human sperm packages genes for embryo development. Nature 460:473–478
Hammoud SS, Purwar J, Pflueger C, Cairns BR, Carrell DT (2010) Alterations in sperm DNA methylation patterns at imprinted loci in two classes of infertility. Fertil Steril 94:1728–1733
Hayashi K, Yoshida K, Matsui Y (2005) A histone H3 methyltransferase controls epigenetic events required for meiotic prophase. Nature 438:374–378
Hayashi K, Chuva de Sousa Lopes SM, Kaneda M, Tang F, Hajkova P, Lao K, O'Carroll D, Das PP, Tarakhovsky A, Miska EA et al (2008) MicroRNA biogenesis is required for mouse primordial germ cell development and spermatogenesis. PLoS One 3:e1738
Henckel A, Nakabayashi K, Sanz LA, Feil R, Hata K, Arnaud P (2009) Histone methylation is mechanistically linked to DNA methylation at imprinting control regions in mammals. Hum Mol Genet 18:3375–3383
Ho SM, Cheong A, Lam HM, Hu WY, Shi GB, Zhu X, Chen J, Zhang X, Medvedovic M, Leung YK, Prins GS (2015) Exposure of human prostaspheres to bisphenol a epigenetically regulates SNORD family noncoding RNAs via histone modification. Endocrinology 156(11):3984–3995
Houshdaran S, Cortessis VK, Siegmund K, Yang A, Laird PW, Sokol RZ (2007) Widespread epigenetic abnormalities suggest a broad DNA methylation erasure defect in abnormal human sperm. PLoS One 2:e1289
Ichiyanagi T, Ichiyanagi K, Miyake M, Sasaki H (2013) Accumulation and loss of asymmetric non CpG methylation during male germ cell development. Nucleic Acid Res 41:738–745
Ishizu H, Siomi H, Siomi MC (2012) Biology of PIWI-interacting RNAs: new insights into biogenesis and function inside and outside of germlines. Genes Dev 26:2361–2373
Iwamori N, Zhao M, Meistrich ML, Matzuk MM (2011) The testis-enriched histone demethylase, KDM4D, regulates methylation of histone H3 lysine 9 during spermatogenesis in the mouse but is dispensable for fertility. Biol Reprod 84(6):1225–1234
Jenkins TG, Carrell DT (2012) The sperm epigenome and potential implications for the developing embryo. Reproduction 143:727–734
Jenkins TG, Aston KI, Meyer TD, Hotaling JM, Shamsi MB, Johnstone EB, Cox KJ, Stanford JB, Porucznik CA, Carrell DT (2016) Decreased fecundity and sperm DNA methylation patterns. Fertil Steril 105:51–57
Johnson GD, Mackie P, Jodar M, Moskovtsev S, Krawetz SA (2015) Chromatin and extracellular vesicle associated sperm RNAs. Nucleic Acids Res 43:6847–6859
Kaneda M, Okano M, Hata K, Sado T, Tsujimoto N, Li E, Sasaki H (2004) Essential role for de novo DNA methyltransferase Dnmt3a in paternal and maternal imprinting. Nature 429(6994):900–903
Kato Y, Kaneda M, Hata K, Kumaki K, Hisano M, Kohara Y, Okano M, Li E, Nozaki M, Sasaki H (2007) Role of the Dnmt3 family in de novo methylation of imprinted and repetitive sequences during male germ cell development in the mouse. Hum Mol Genet 16(19):2272–2280
Kawano M, Kawaji H, Grandjean V, Kiani J, Rassoulzadegan M (2012) Novel small noncoding RNAs in mouse spermatozoa, zygotes and early embryos. PLoS One 7:e44542
Kedia N, Gill-Sharma MK, Parte P, Juneja HS, Balasinor N (2004) Effect of paternal tamoxifen on the expression of insulin-like growth factor 2 and insulin-like growth factor type 1 receptor in the post-implantation rat embryos. Mol Reprod Dev 69:22–30
Kedia N, Kadam L, Dumasia K, Balasinor NH (2016) Possible role of paternal aberrant imprinting in placental development: a study in tamoxifen treatment rat model. J Clin Epigenetics 2:1
Kedia-Mokashi N, Kadam L, Ankolkar M, Dumasia K, Balasinor NH (2013) Aberrant methylation of multiple imprinted genes in embryos of tamoxifen-treated male rats. Reproduction 146:155–168
Kelly TL, Li E, Trasler JM (2003) 5-Aza-2′-deoxycytidine induces alterations in murine spermatogenesis and pregnancy outcome. J Androl 24:822–830
Kerjean A, Dupont JM, Vasseur C, Le Tessier D, Cuisset L, Paldi A, Jouannet P, Jeanpierre M (2000) Establishment of the paternal methylation imprint of the human H19 and MEST/PEG1 genes during spermatogenesis. Hum Mol Genet 9:2183–2187

Kitamura A, Miyauchi N, Hamada H, Hiura H, Chiba H, Okae H, Sato A, John RM, Arima T (2015) Epigenetic alterations in sperm associated with male infertility. Congenit Anom 55:133–144

Klungland A, Dahl JA (2014) Dynamic RNA modifications in disease. Curr Opin Genet Dev 26:47–52

Kobayashi H, Sato A, Otsu E et al (2007) Aberrant DNA methylation of imprinted loci in sperm from oligospermic patients. Hum Mol Genet 16:2542–2551

Korhonen HM, Meikar O, Yadav RP, Papaioannou MD, Romero Y, Da Ros M, Herrera PL, Toppari J, Nef S, Kotaja N (2011) Dicer is required for haploid male germ cell differentiation in mice. PLoS One 6:e24821

Krawetz SA, Kruger A, Lalancette C, Tagett R, Anton E, Draghici S, Diamond MP (2011) A survey of small RNAs in human sperm. Hum Reprod 26:3401–3412

Kuramochi-Miyagawa S, Kimura T, Ijiri TW, Isobe T, Asada N, Fujita Y, Ikawa M, Iwai N, Okabe M, Deng W et al (2004) Mili, a mammalian member of piwi family gene, is essential for spermatogenesis. Development 131:839–849

Kuramochi-Miyagawa S, Watanabe T, Gotoh K, Totoki Y, Toyoda A, Ikawa M, Asada N, Kojima K, Yamaguchi Y, Ijiri TW, Hata K, Li E, Matsuda Y, Kimura T, Okabe M, Sakaki Y, Sasaki H, Nakano T (2008) DNA methylation of retrotransposon genes is regulated by piwi family members MILI and MIWI2 in murine fetal testes. Genes Dev 22(7):908–917

Lambrot R, Xu C, Saint-Phar S, Chountalos G, Cohen T, Paquet M, Suderman M, Hallett M, Kimmins S (2013) Low paternal dietary folate alters the mouse sperm epigenome and is associated with negative pregnancy outcomes. Nat Commun 4:2889

Lee JT (2012) Epigenetic regulation by long noncoding RNAs. Science 338:1435–1439

Li E, Bestor TH, Jaenisch R (1992) Targeted mutation of the DNA methyltransferase gene results in embryonic lethality. Cell 69(6):915–926

Liu WM, Pang RTK, Chiu PCN, Wong BPC, Lao K, Lee KF, Yeung WSB (2012) Sperm-borne microRNA-34c is required for the first cleavage division in mouse. Proc Natl Acad Sci U S A 109:490–494

Liu J, Yue Y, Han D, Wang X, Fu Y, Zhang L, Jia G, Yu M, Lu Z, Deng X, Dai Q, Chen W, He C (2014) A METTL3-METTL14 complex mediates mammalian nuclear RNA N6-adenosine methylation. Nat Chem Biol 10:93–95

Liu Z, Oyola MG, Zhou S et al (2015) Knockout of the histone demethylase Kdm3b decreases spermatogenesis and impairs male sexual behaviors. Int J Biol Sci 11(12):1447–1457

Manikkam M, Tracey R, Guerrero-Bosagna C, Skinner MK (2012) Pesticide and insect repellent mixture (permethrin and DEET) induces epigenetic transgenerational inheritance of disease and sperm epimutations. Reprod Toxicol 34(4):708–719

Manning M, Lissens W, Liebaers I, Van Steirteghem A, Weidner W (2001) Imprinting analysis in spermatozoa prepared for intracytoplasmic sperm injection (ICSI). Int J Androl 24(2):87–94

Marques CJ, Costa P, Vaz B, Carvalho F, Fernandes S, Barros A, Sousa M (2008) Abnormal methylation of imprinted genes in human sperm is associated with oligozoospermia. Mol Hum Reprod 14(2):67–74

Marques CJ, Francisco T, Sousa S, Carvalho F, Barros A, Sousa M (2010) Methylation defects of imprinted genes in human testicular spermatozoa. Fertil Steril 94:585–594

Marques-Pinto A, Carvalho D (2013) Human infertility: are endocrine disruptors to blame? Endocr Connect 2:15–29

Mercer TR, Dinger ME, Mattick JS (2009) Long non-coding RNAs: insights into functions. Nat Rev Genet 10:155–159

Minor A, Chow V, Ma S (2011) Aberrant DNA methylation at imprinted genes in testicular sperm retrieved from men with obstructive azoospermia and undergoing vasectomy reversal. Reproduction 141(6):749–757

Moazed D (2009) Small RNAs in transcriptional gene silencing and genome defence. Nature 457:413–420

Mohamed el SA, Song WH, Oh SA, Park YJ, You YA, Lee S, Choi JY, Kim YJ, Jo I, Pang MG (2010) The transgenerational impact of benzopyrene on murine male fertility. Hum Reprod 25(10):2427–2433

Nettersheim D, Heukamp LC, Fronhoffs F, Grewe MJ, Haas N, Waha A, Honecker F, Waha A, Kristiansen G, Schorle H (2013) Analysis of TET expression/activity and 5mC oxidation during normal and malignant germ cell development. PLoS One 8:e82881

Ni K, Dansranjavin, Rogenhofer N, Oeztuerk N, Deuker J, Bergmann M, Schuppe HC, Wagenlehner F, Weidner W, Steger K, Schagdarsurengin U (2016) TET enzymes are successively expressed during human spermatogenesis and their expression level is pivotal for male fertility. Hum Reprod 31:1411–1424

Nottke A, Colaiácovo MP, Shi Y (2009) Developmental roles of the histone lysine demethylases. Development 136:879–889

Oakes CC, La Salle S, Smiraglia DJ, Robaire B, Trasler JM (2007) Developmental acquisition of genome-wide DNA methylation occurs prior to meiosis in male germ cells. Dev Biol 307:368–379

Okada Y, Scott G, Ray MK, Mishina Y, Zhang Y (2007) Histone demethylase JHDM2A is critical for Tnp1 and Prm1 transcription and spermatogenesis. Nature 450(7166):119–123

Okada Y, Tateishi K, Zhang Y (2010) Histone demethylase JHDM2A is involved in male infertility and obesity. J Androl 31(1):75–78

Pacheco SE, Houseman EA, Christensen BC, Marsit CJ, Kelsey KT, Sigman M, Boekelheide K (2011) Integrative DNA methylation and gene expression analyses identify DNA packaging and epigenetic regulatory genes associated with low motility sperm. PLoS One 6:e20280

Pathak S, Kedia-Mokashi N, Saxena M, D'Souza R, Maitra A, Parte P, Gill-Sharma MK, Balasinor N (2009) Effect of tamoxifen treatment on global and insulin-like growth factor 2-H19 locus specific DNA methylation in rat spermatozoa and its association with embryo loss. Fertil Steril 91(5 Suppl):2253–2263

Peng H, Shi J, Zhang Y, Zhang H, Liao S, Li W, Lei L, Han C, Ning L, Cao Y et al (2012) A novel class of tRNA-derived small RNAs extremely enriched in mature mouse sperm. Cell Res 22:1609–1612

Peters AH, O'Carroll D, Scherthan H, Mechtler K, Sauer S, Schöfer C, Weipoltshammer K, Pagani M, Lachner M, Kohlmaier A, Opravil S, Doyle M, Sibilia M, Jenuwein T (2001) Loss of the Suv39h histone methyltransferases impairs mammalian heterochromatin and genome stability. Cell 107(3):323–337

Poplinski A, Tüttelmann F, Kanber D, Horsthemke B, Gromoll J (2010) Idiopathic male infertility is strongly associated with aberrant methylation of MEST and IGF2/H19 ICR1. Int J Androl 33(4):642–649

Rajabi H, Mohseni-Kouchesfehani H, Mohammadi-Sangcheshmeh A, Farifteh-Nobijari F, Salehi M (2016) Pronuclear epigenetic modification of protamine deficient human sperm following injection into mouse oocytes. Syst Biol Reprod Med 62:125–132

Rassoulzadegan M, Grandjean V, Gounon P, Vincent S, Gillot I, Cuzin F (2006) RNA-mediated non-Mendelian inheritance of an epigenetic change in the mouse. Nature 441(7092):469–474

Reik W, Dean W, Walter J (2001) Epigenetic reprogramming in mammalian development. Science 293:1089–1093

Richardson ME, Bleiziffer A, Tüttelmann F, Gromoll J, Wilkinson MF (2014) Epigenetic regulation of the RHOX homeobox gene cluster and its association with human male infertility. Hum Mol Genet 23(1):12–23

Rinn JL, Chang HY (2012) Genome regulation by long noncoding RNAs. Annu Rev Biochem 81:145–166

Salas-Huetos A, Blanco J, Vidal F, Godo A, Grossmann M, Carme Pons M et al (2015) Spermatozoa from patients with seminal alterations exhibit a differential miRNA profile. Fertil Steril 104:591–601

Salian S, Doshi T, Vanage G (2009) Perinatal exposure of rats to bisphenol A affects the fertility of male offspring. Life Sci 85(21–22):742–752

Sato A, Otsu E, Negishi H, Utsunomiya T, Arima T (2007) Aberrant DNA methylation of imprinted loci in superovulated oocytes. Hum Reprod 22:26–35

Sato A, Hiura H, Okae H et al (2011) Assessing loss of imprint methylation in sperm from subfertile men using novel methylation polymerase chain reaction Luminex analysis. Fertil Steril 95:129–134

Sawan C, Herceg Z (2010) Histone modifications and cancer. Adv Genet 70:57–85
Sendler E, Johnson GD, Mao S, Goodrich RJ, Diamond MP, Hauser R, Krawetz SA (2013) Stability, delivery and functions of human sperm RNAs at fertilization. Nucleic Acids Res 41:4104–4117
Shen F, Huang W, Huang JT, Xiong J, Yang Y, Wu K, Jia GF, Chen J, Feng YQ, Yuan BF, Liu SM (2015) Decreased N(6)-methyladenosine in peripheral blood RNA from diabetic patients is associated with FTO expression rather than ALKBH5. J Clin Endocrinol Metab 100:E148–E154
Skinner MK (2014) Endocrine disruptor induction of epigenetic transgenerational inheritance of disease. Mol Cell Endocrinol 398(1–2):4–12
Skinner MK, Manikkam M, Tracey R, Guerrero-Bosagna C, Haque M, Nilsson EE (2013) Ancestral dichlorodiphenyltrichloroethane (DDT) exposure promotes epigenetic transgenerational inheritance of obesity. BMC Med 11:228
Stouder C, Paoloni-Giacobino A (2010) Transgenerational effects of the endocrine disruptor vinclozolin on the methylation pattern of imprinted genes in the mouse sperm. Reproduction 139(2):373–379
Stouder C, Paoloni-Giacobino A (2011) Specific transgenerational imprinting effects of the endocrine disruptor methoxychlor on male gametes. Reproduction 141(2):207–216
Stouder C, Somm E, Paoloni-Giacobino A (2011) Prenatal exposure to ethanol: a specific effect on the H19 gene in sperm. Reprod Toxicol 31:507–512
Stuppia L, Franzago M, Ballerini P, Gatta V, Antonucci I (2015) Epigenetics and male reproduction: the consequences of paternal lifestyle on fertility, embryo development, and children lifetime health. Clin Epigenetics 7:120
Sun WJ, Li JH, Liu S, Wu J, Zhou H, Qu LH, Yang JH (2016) RMBase: a resource for decoding the landscape of RNA modifications from high-throughput sequencing data. Nucleic Acids Res 44(D1):D259–D265
Tachibana M, Nozaki M, Takeda N, Shinkai Y (2007) Functional dynamics of H3K9 methylation during meiotic prophase progression. EMBO J 26:3346–3359
Tsai MC, Manor O, Wan Y, Mosammaparast N, Wang JK, Lan F, Shi Y, Segal E (2010) Chang HY long noncoding RNA as modular scaffold of histone modification complexes. Science 329(5992):689–693
Tseng Y, Liao H-F, Yu CY, Mo CF, Lin SP (2015) Epigenetic factors in the regulation of prospermatogonia and spermatogonial stem cells. Reproduction 150:R77–R91
Urdinguio RG, Bayón GF, Dmitrijeva M, Toraño EG, Bravo C, Fraga MF, Bassas L, Larriba S, Fernández AF (2015) Aberrant DNA methylation patterns of spermatozoa in men with unexplained infertility. Hum Reprod 30:1014–1028
Vassena R, Boue S, Gonzalez-Roca E, Aran B, Auer H, Veiga A, Izpisua Belmonte JC (2011) Waves of early transcriptional activation and pluripotency program initiation during human preimplantation development. Development 138:3699–3709
Walsh CP, Chaillet JR, Bestor TH (1998) Transcription of IAP endogenous retroviruses is constrained by cytosine methylation. Nat Genet 20:116–117
Wang C, Yang C, Chen X, Yao B, Yang C, Zhu C, Li L, Wang J, Li X, Shao Y et al (2011) Altered profile of seminal plasma microRNAs in the molecular diagnosis of male infertility. Clin Chem 57:1722–1731
Webster KE, O'Bryan MK, Fletcher S, Crewther PE, Aapola U, Craig J, Harrison DK, Aung H, Phutikanit N, Lyle R, Meachem SJ, Antonarakis SE, de Kretser DM, Hedger MP, Peterson P, Carroll BJ, Scott HS (2005) Meiotic and epigenetic defects in Dnmt3L-knockout mouse spermatogenesis. Proc Natl Acad Sci U S A 102(11):4068–4073
Wu Q, Song R, Ortogero N, Zheng H, Evanoff R, Small CL, Griswold MD, Namekawa SH, Royo H, Turner JM et al (2012) The RNase III enzyme DROSHA is essential for microRNA production and spermatogenesis. J Biol Chem 287:25173–25190
Yang Y, Huang W, Huang JT, Shen F, Xiong J, Yuan EF, Qin S, Zhang M, Feng YQ, Yuan BF, Liu SM (2016) Increased N6-methyladenosine in human sperm RNA as a risk factor for asthenozoospermia. Sci Rep 6:2434

Yue Y, Liu J, He C (2015) RNA N6-methyladenosine methylation in post-transcriptional gene expression regulation. Genes Dev 29:1343–1355

Zamudio NM, Chong S, O'Bryan MK (2008) Epigenetic regulation in male germ cells. Reproduction 136(2):131–146

Zhang X, Ho SM (2011) Epigenetics meets endocrinology. J Mol Endocrinol 46:R11–R32

Zhao X, Yang Y, Sun BF, Shi Y, Yang X, Xiao W, Hao YJ, Ping XL, Chen YS, Wang WJ, Jin KX, Wang X, Huang CM, Fu Y, Ge XM, Song SH, Jeong HS, Yanagisawa H, Niu Y, Jia GF, Wu W, Tong WM, Okamoto A, He C, Rendtlew Danielsen JM, Wang XJ, Yang YG (2014) FTO-dependent demethylation of N6-methyladenosine regulates mRNA splicing and is required for adipogenesis. Cell Res 24:1403–1419

Zheng G, Dahl JA, Niu Y, Fedorcsak P, Huang CM, Li CJ, Vågbø CB, Shi Y, Wang WL, Song SH, Lu Z, Bosmans RP, Dai Q, Hao YJ, Yang X, Zhao WM, Tong WM, Wang XJ, Bogdan F, Furu K, Fu Y, Jia G, Zhao X, Liu J, Krokan HE, Klungland A, Yang YG, He C (2013) ALKBH5 is a mammalian RNA demethylase that impacts RNA metabolism and mouse fertility. Mol Cell 49:18–29

Sperm Chromatin Compaction and Male Infertility

17

Aniket Patankar and Priyanka Parte

Abstract

Nucleosome, the fundamental unit of chromatin, is histone octamer composed of dimers of each histone H2A, H2B, H3, and H4. Histones are the key epigenetic players and regulate chromatin architecture. During later stages of spermatogenesis, extensive remodeling of chromatin takes place in which somatic histones get replaced by testis-specific histones, which in turn get replaced by transition proteins and finally by protamines. Disturbances that impair this highly orchestrated process may result in loose DNA packing, endangering its integrity. This reflects on sperm morphology and motility, resulting in teratozoospermia and asthenozoospermia and consequently infertility. These sperm are unable to reach the oocyte and, if they do, fail to fertilize. Assisted fertilization in the form of IVF or ICSI may help overcome this hindrance; however, the risk of failure at early embryonic developmental stages or preimplantation loss increases dramatically. This review provides an update on our current understanding of the role of sperm chromatin compaction in sperm function and the impact of its failure on male fertility.

Keywords

Male infertility • IVF/ICSI • Chromatin compaction • Histone modifications Protamine • Testis-specific histones

---

A. Patankar • P. Parte (✉)
Department of Gamete Immunobiology, National Institute for Research in Reproductive Health, Mumbai, Maharashtra, India
e-mail: partep@nirrh.res.in

R. Singh, K. Singh (eds.), *Male Infertility: Understanding, Causes and Treatment*, DOI 10.1007/978-981-10-4017-7_17

Key Points

- Chromatin packaging is an integral part of spermatogenesis, and sperm DNA is packed into almost crystalline status that is at least six times more condensed in comparison to mitotic chromosomes.
- During spermiogenesis, somatic histones in the haploid spermatid are replaced by testis-specific histones, which in turn are replaced by transition proteins and finally by protamines, leading to dense chromatin compaction in sperm.
- Sperm histones undergo several posttranslational modifications, predominantly methylation and acetylation, to repress the transcriptional activity in sperm.
- Compaction or proper packaging of chromatin is essential for shutting down the transcription activity in sperm and also for protecting its DNA from damage during its transit from testis through epididymis into the female reproductive tract.
- Defects in chromatin packaging affect the morphology of sperm and its transcriptional activity and are associated with infertility or the outcome of ARTs.
- Significantly higher histone-protamine ratios are observed in sperm from infertile men; a direct correlation exists between sperm protamine levels, DNA integrity, and sperm quality.

17.1 Introduction

WHO estimates as reported in 2012 indicate that about 50 million couples worldwide suffer from infertility (Mascarenhas et al. 2012). Male infertility accounts for almost 50% of the infertility. Asthenozoospermia, oligoasthenozoospermia, oligozoospermia, teratozoospermia, globozoospermia, azoospermia, and aspermia are the observed manifestations in male infertility. There is a considerable population of infertile individuals where none of these manifestations are observed and thus are referred to as idiopathic. Although research world over has been overwhelming with respect to female infertility, with respect to male infertility, it is limited probably because of the general perception that all problems of male infertility can be bypassed using assisted reproductive technologies (ARTs) such as IVF and ICSI. Increasing evidence is now available on the problems associated with ICSI (Wennerholm et al. 2000; Belva et al. 2007; Bonduelle et al. 2004; Morris et al. 2002). One of the biggest drawbacks of ICSI is that the genetic quality of sperm is overlooked leading to embryonic loss despite successful fertilization following ICSI. Genetic quality of sperm is determined by the integrity of its DNA and its compaction during spermiogenesis. A positive correlation has been observed between chromatin condensation and successful pregnancy in IUI and ICSI couples (Ioannou et al. 2016; Irez et al. 2015; Morris et al. 2002). In order to understand the impact of chromatin compaction on male fertility, it is imperative to understand the process of chromatin condensation.

17.2 DNA Packaging and Chromatin Compaction During Spermatogenesis

Spermatogenesis is a well-synchronized and tightly regulated process by which haploid male germ cells are formed. In the third and final stage of spermatogenesis, i.e., spermiogenesis, the haploid round spermatids undergo extensive morphological changes and nuclear remodeling to give rise to structurally distinct cell, the spermatozoa. Mature spermatozoon contains nucleus carrying haploid male genome, which is sixfold more compact as compared to any somatic cell of the body. Two major nucleoproteins involved in DNA compaction are nucleosomes and protamines. Nucleosome is the histone octamer composed of dimer of each histone H2A, H2B, H3, and H4. Histone H1 binds to the DNA in between two nucleosomes and is thought to be involved in higher-order chromatin structure formation. Each histone of nucleosome core particle (NCP) is divided into two parts: structured region made up of core and C-terminal region of histone and N-terminal unstructured tail which protrudes out of the nucleosome and interacts with DNA. Approximately 146 bp DNA is wrapped around each nucleosome. Posttranslational modifications (PTMs) like acetylation, phosphorylation, and ubiquitination occurring especially on N-terminus can influence chromatin structure either directly by adding negative or positive charge and altering histone-DNA interaction or indirectly by recruiting modification-specific chromatin remodeling factor (Pivot-Pajot et al. 2003).

Protamines are arginine-rich, small, basic, major nucleoproteins in sperm. They are synthesized in late-stage spermatid. Around 80–85% sperm DNA is compact due to protamination. In case of mammals, protamines are of two types protamine 1 (P1) and protamine 2 (P2). The presence of P1 in association with sperm DNA can be observed in nearly all vertebrates, whereas P2 is present only in primates, many rodents, and a subset of other placental mammals (Balhorn 2007). The number of protamine genes and copies present per haploid genome varies from species to species. Mammals have single-copy genes of P1 and P2, located on chromosome 16 (Reeves et al. 1989). P1and P2 are products of gene Prm1 and Prm2, respectively. The precursor protein of Prm2 undergoes proteolytic processing at its N-terminus to give rise to p2, p3, and p4. P2 family proteins, p2, p3, and p4, differ in 3–4 residues at N-terminus. The arginine-rich DNA-anchoring domains by which protamines bind with the negatively charged DNA and the multiple serine and threonine residues that can be used as phosphorylation sites form the structural elements of protamines (Balhorn 2007). The cysteine residues allow disulfide bond formation and thus link two adjacent protamines, which leads to further compaction of DNA.

During the process of spermiogenesis, nucleohistone to nucleoprotamine transition occurs. This transition is not direct but a gradual process, comprising of well-defined events, which involves first replacement of somatic histone by testis-specific histone variants and subsequently by transition proteins and then protamines.

At the round spermatid stage, the DNA compaction is the same as that of any somatic cell of the body. After the completion of second meiotic division, there is a surge of transcription observed characterized by two features not observed in somatic cells, namely, (a) the use of specialized transcriptional machinery and (b) the expression of large numbers of spermatogenic-specific genes which includes transcription of proteins like transition proteins, protamine, etc. required for spermiogenesis. At the same time, hyperacetylation of somatic histone H4 is observed. In vitro studies have indicated the role of hyperacetylated histones in nucleosome disassembly and replacement of histone by protamines (Oliva et al. 1987; Awe and Renkawitz-Pohl 2010). It has also been shown that bromodomain-containing protein (BRDT) binds with the hyperacetylated H4 and initiates nuclear remodeling (Pivot-Pajot et al. 2003; Moriniere et al. 2009).

Hyperacetylated histones then get replaced by testis-specific histone variants. Excepting histone H4, testis variants have been reported till date for core histones H2A, H2B, H3, and linker histone H1. During spermiogenesis, testis-specific histones get replaced by transition proteins (TP). Mammals, including mouse, rat, human, ram, and boar, predominantly have two types of transition proteins, viz., transition protein 1 (TP1) and transition protein 2 (TP2) (Akama et al. 1996; Chevaillier et al. 1998; Steger et al. 1998). Both TP1 and TP2 are encoded by single-copy genes, Tnp1 and Tnp2, respectively (Rathke et al. 2014). TP1 is a 6200 Da protein with about 20% arginine and 20% lysine, distributed uniformly, and no cysteine (Kistler et al. 1975). TP2 is a 13,000 Da protein with about 10% arginine, 10% lysine, and 5% cysteine (Grimes et al. 1975). It has a highly basic C-terminal domain and an N-terminal domain that forms zinc fingers (Meetei et al. 2000). TP1 is abundantly expressed (Heidaran et al. 1988), and its sequence is highly conserved in various mammals as compared to TP2 (Kremling et al. 1989). The role of TPs is not extensively studied. $TP1^{-/-}$ and $TP2^{-/-}$ knockout mice have been shown to be less fertile than normal mice and show abnormal chromatin condensation (Zhao et al. 2001). TP1 and TP2 double-knockout mice are sterile, and spermatogenesis is severely impaired suggesting their important role in spermiogenesis (Zhao et al. 2004).

Transition proteins remain associated with DNA for a short period of time and rapidly get replaced by protamines. Immediately after their synthesis, protamines get phosphorylated. Phosphorylation is thought to be essential for their nuclear transport as protamines can bind to their nuclear receptor and get transported only when phosphorylated (Mylonis et al. 2004). After binding of protamine to DNA, dephosphorylation takes place, and the disulfide bond formed between protamine further compacts the DNA. Chromodomain helicase DNA-binding protein 5 (Chd5) has a key role in the DNA compaction. It is involved in H4 hyperacetylation, histone variant expression, and removal and replacement of the histones with nucleoprotamines, and Chd5 deficiency in mice leads to defective sperm chromatin compaction and infertility (Li et al. 2014). Low expression of Chd5 has also been observed in the testis of infertile men by the same group.

17.3 Testis-Specific Histones

Replacement of histone by protamine is not 100%, and about 5–15% histones are still retained in mature human spermatozoa (Tanphaichitr et al. 1978; Gatewood et al. 1987; Zalensky et al. 2002). Retained histones have been found to be specifically enriched in the regulatory region of genes that are important for the earliest development stages postfertilization (Hammoud et al. 2009). Later it was shown by MNase sequencing that infertile males have random distribution of retained histones in spermatozoa (Hammoud et al. 2011). Testis-specific histone variants are thought to have specific biological function during spermiogenesis, as demonstrated by knockout studies with different variants. Table 17.1 summarizes the testis-specific histone variants known to date and their localization and influences on fertility.

Table 17.1 Testis-specific histone variants and their localization and influences on fertility

Histone variants	Chromosome location (human)	Characteristics/ localization	Knockout phenotype		References
			Male	Female	
H1					
H1T2 (testis-specific H1 histone)/ HANP1	12q13.11	In haploid male germ cells until the histone-to-protamine transition; in spermatid nuclei at the apical pole under the acrosome	Males, infertile; sperm morphology, abnormal; sperm nuclei, protamine 1 and 2 weakly detectable; sperm motility, altered; fertilizing ability, negative with IVF, but positive with ICSI; spermiogenesis, delayed nuclear condensation and aberrant elongation, acrosome detachment, and fragmented DNA	Females, fertile	Tanaka et al. (2005), Martianov et al. (2005), Catena et al. (2006)
Histone H1t	6p22.1	In pachytene spermatocytes and persists until chromatin reorganization in postmeiotic stages	Fertile, exhibit no spermatogenesis abnormalities, show enhanced gene expression of the canonical subtypes H1.1, H1.2, and H1.4	Not known	Drabent et al. (1996, 2003)
HILS1	17q21.33	Strongly expressed in nuclei of elongating and elongated spermatids	Not known	Not known	Iguchi et al. (2003, 2004), Yan et al. (2003)
H2A					
Histone H2A type 1-A (TH2A)	6p22.2	Pachytene spermatocytes	Double knockout for TH2A/TH2B, sterile; testis and epididymis weight reduced; sperm count, reduced; secondary spermatocytes at interkinesis, more abundant in mutant testis	Double knockout for TH2A/TH2B, effect seen on early embryonic development; oogenesis and folliculogenesis, normal; maternal TH2A/ TH2B involved in activation of paternal genome postfertilization	Trostle-Weige et al. (1982), Shinagawa et al. (2014, 2015)

(continued)

Table 17.1 (continued)

Histone variants	Chromosome location (human)	Characteristics/ localization	Knockout phenotype		References
			Male	Female	
Histone H2A-Bbd type 1	Xq28	Mouse homolog H2A. Lap1is targeted to the transcription start site of active genes expressed during specific stages of spermatogenesis; depleted cells demonstrate widespread changes in gene expression, a net downregulation of transcription, and disruption of normal mRNA splicing patterns	Not known	Not known	Nekrasov et al. (2013)
Histone H2A.X	11q23.3	γH2AX accumulates on unsynapsed sex chromosomes during the zygotene stage; H2Ax$^{-/-}$ mice lack γH2AX accumulation as well as meiotic sex chromosome inactivation (MSCI) initiation	Males, growth retarded, immune deficient, and infertile with hypogonadism, spermatogenesis arrested at pachytene stage	Females, fertile but litter size smaller than heterozygous or wild type	Celeste et al. (2002), Fernandez-Capetillo et al. (2003), Ichijima et al. (2012)
Histone H2A.Z	4q24	Not known	Required for early embryonic development; lack of functional H2A.Z leads to embryonic lethality	Not known	Faast et al. (2001)

H2B					
Histone H2B type 1-A (TH2B)	6p22.2	Expression starts in leptotene spermatocytes and then reaches an intense signal in round spermatids	TH2B knockout male mice are fertile; $Th2b^{+/tag}$ mice show arrest at condensing spermatid stage leading to lack of sperm in epididymis and consequently infertility; double knockout for TH2A/TH2B, sterile; testis and epididymis weight reduced; sperm count, reduced; secondary spermatocytes at interkinesis, more abundant in mutant testis	Double knockout for TH2A/TH2B in female affects early embryonic development; oogenesis and folliculogenesis were normal in mutant female mice	Shinagawa et al. (2014, 2015), Montellier et al. (2013), van Roijen et al. (1998)
Histone H2B type W-T	Xq22.2	Not known	No information on knock outs; polymorphisms (−9C>T and 368A>G) in the 5′ UTR associated with male infertility in Chinese and Korean population	Not known	Lee et al. (2009), Ying et al. (2012)
Spermatid-specific H2B ssH2B	Not known	Specifically synthesized and expressed in round spermatids and decrease before the bulk of histones becomes degraded	Not known	Not known	Unni et al. (1995)
H3					
Histone H3.1t	1q42.13	H3t synthesized in spermatogonia and remains detectable in spermatocytes and early spermatids	Not known	Not known	Trostle-Weige et al. (1984)

(continued)

Table 17.1 (continued)

Histone variants	Chromosome location (human)	Characteristics/ localization	Knockout phenotype		References
			Male	Female	
H3.3	H3F3A (1q42.12), H3F3B (17q25.1)	H3.3 protein seen in the nuclei of spermatogonia and leptotene spermatocytes; H3.3 levels peak during mid-stage in pachytene spermatocytes and persist throughout meiosis and most of spermiogenesis	H3f3b KO male, infertile; testis and sperm morphology, abnormal; histone PTMs and gene expression in testes affected; most prominent changes occurring at genes involved in spermatogenesis; defective chromatin reorganization and reduced protamine incorporation seen	Not known	Yuen et al. (2014)
H3.5/H3F3C	12p11.21	ChIP-seq analysis revealed that H3.5 accumulated around transcription start sites (TSSs) in testicular cells; H3.5 mRNA specifically expressed in seminiferous tubules (human testis)	Not known	Not known	Schenk et al. (2011), Urahama et al. (2016)

Among the retained histones, TH2B is the major testis-specific histone variant present in mature sperm. TH2B differs from H2B mainly at its N-terminus. N-terminus of H2B has been shown to be associated with chromosome condensation in meiotic cell (de la Barre et al. 2001). N-terminus of TH2B has additional three potential phosphorylation sites (Ser12, Thr23, and Thr34) and repositioning of two others (ser5 and ser60), which are not seen in H2B, resulting in a different phosphorylation map of the N-terminal tail for TH2B (Pradeepa and Rao 2007).

Presence of TH2A/TH2B in nucleosome has been shown to induce a more open chromatin structure (Padavattan et al. 2015). This open chromatin structure facilitates the removal of histones and their replacement by protamines, thus enabling further compaction of DNA (Montellier et al. 2013). We have earlier reported reduced TH2B in asthenozoospermic individuals (Parte et al. 2012). However, TH2B knockout male mice have been shown to be fertile as the absence of TH2B is compensated by overexpression of somatic H2B variants and modifications on other histones. But Th2b$^{+/tag}$ mice show arrest at condensing spermatid stage leading to lack of sperm in epididymis and consequently infertility (Montellier et al. 2013). However, double knockout for TH2A/TH2B causes defect in spermatogenesis in males. Histone replacement during spermiogenesis is also affected. The mice showed reduced testis and epididymis weight and are sterile. Secondary spermatocytes at interkinesis (the interphase between meiosis I and II) are more abundant in the mutant testis than in the wild type, suggesting extended interkinesis in mutant mice (Shinagawa et al. 2015). Interestingly, it is the TH2A and TH2B from oocyte that is involved in activation of paternal genome postfertilization (Shinagawa et al. 2014). The dynamic changes in chromatin structure during spermiogenesis, epididymal maturation, and up to early embryonic development are summarized in Fig. 17.1.

17.4 Histone Modifications in Sperm and Their Influence on Sperm Fertilizing Ability/Embryonic Development

Several posttranslational modifications (PTMs) have been observed in mouse and human sperm (Fig. 17.2). In mouse sperm, 26 PTMs have been reported in specific residues of core histones and linker histone and 11 PTMs on PRM1 and PRM2 (Brunner et al. 2014). Comprehensive assessment of the histone modifications in normal human sperm revealed 102 modifications (Schon et al. 2015). Modifications are observed on the linker histone H1, the canonical histones, as well as their variants. While modifications on H4 are conserved, those on H3 vary between individuals. The modifications are not altered on cryopreservation of the sperm. Some PTMs of histones are uniquely distributed in human sperm, and this distribution

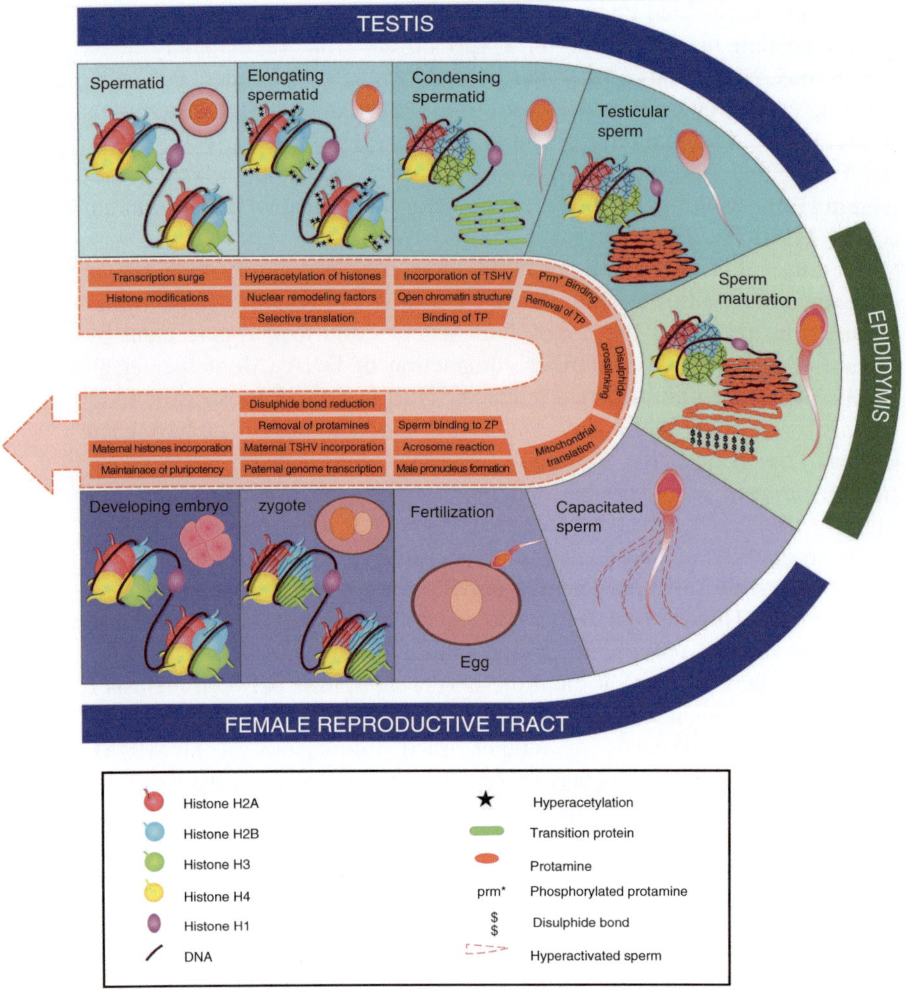

Fig. 17.1 Chromatin dynamics in the sperm during its journey from a round spermatid to the formation of an embryo. *Circles with no pattern* represent somatic histones, *circles with pattern* indicate respective testis-specific histones, while *circles with diagonal lines* indicate the respective maternally expressed testis-specific histones. *TSHV* testis-specific histone variant, *TP* transition protein

varies among individuals and also between the sperm of a single individual (Krejci et al. 2015). Variations among individuals have been observed in the levels of H3K9me1, H3K9me2, H3K27me3, H3K36me3, and H3K79me1 in the sperm-head fractions. Levels of acetylated (ac) histones H4 are relatively stable. Lower levels of H3K9ac, H3K9me1, H3K27me3, H3K36me3, and H3K79me1 are seen in sperm with P2 deficiency. H3K9me2 and levels of P2 show a strong correlation. While the localization of H3 lysine 4 methylation (H3K4me) or H3 lysine 27 methylation

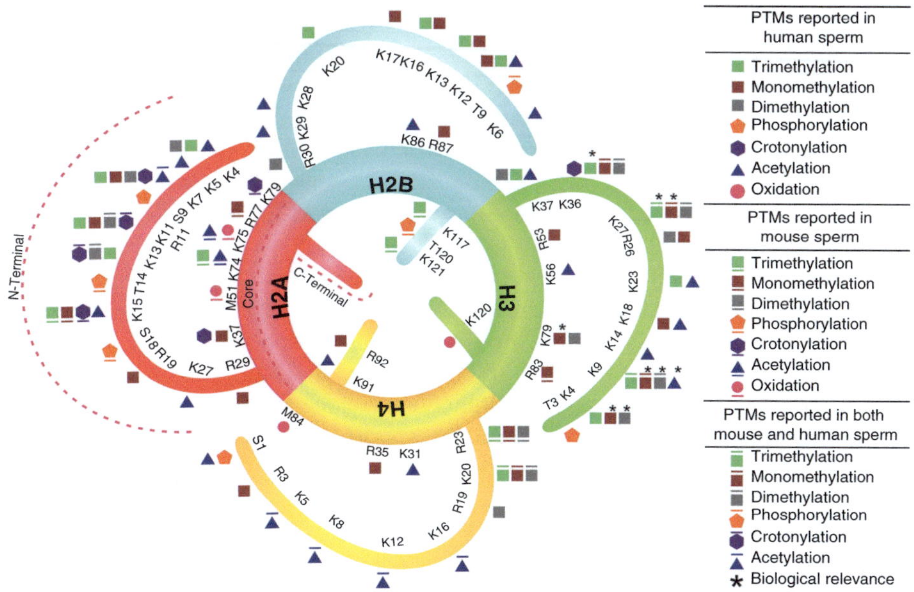

Fig. 17.2 Posttranslational modifications (PTMs) reported on the core histones in mouse and human spermatozoa. Colors *red*, *blue*, *green*, and *yellow* represent histones H2A, H2B, H3, and H4, respectively. The *extension outside the circle* represents the N-terminal region, and that *inside it* represents the C-terminal region. The *segments on the circle* represent core region of the respective histone. The PTMs reported on histones in spermatozoa of mouse, human, or both are as shown in the legend to the figure

(H3K27me) is highly similar in the gametes of infertile men compared with fertile men, a reduction in the amount of H3K4me or H3K27me retained at developmental transcription factors and certain imprinted genes has been noted. Also, the methylation status of certain developmental promoters and imprinted loci are altered in a subset of infertile men (Hammoud et al. 2011). Recently, histone PTMs and their relative abundance in distinct stages of mouse spermatogenesis and in human spermatozoa have been identified (Luense et al. 2016). They observed a strong conservation of histone PTMs for histone H3 and H4 between mouse and human sperm; however, H1, H2A, and H2B showed very little conservation (Luense et al. 2016).

In sperm, genes relevant to spermatogenesis are marked by H3K4me2, and the genes involved in developmental regulation are marked by H3K27me3 (Brykczynska et al. 2010). While H3K4me2 is an activating mark, H3K27me3 has been shown to be a repressor of genes. This means that prior to fertilization, the genes involved in early embryonic development are repressed by H3K27me3, while those involved in spermatogenesis are maintained in an active state by H3K4me2. Reduction in H3K4me2 induced by human KDM1A histone lysine 4 demethylase transgene overexpression during mouse spermatogenesis has been shown to severely impair development and survival of the offspring, a defect which is also seen in two subsequent generations (Siklenka et al. 2015). H3K4me2 was reduced at the CpG islands of genes involved in development. While this region is majorly marked by H3K27me3, work of Brykczynska has shown that some of the developmentally regulated genes are also marked by H3K4me2. H3K4me3 has also been demonstrated to be important for spermatogenesis; loss of H3K4 methyltransferase MII2 reduces H3K4me3 consequently rendering the male mice sterile (Glaser et al. 2009). H3K9me2/1-specific demethylase JHDM2A also known as JMJD1A is essential for spermatogenesis, and its loss causes infertility in male mice due to incomplete chromatin condensation (Okada et al. 2007).

Stage-specific modifications have been identified for TH2B with acetylated TH2B being most abundant in spermatogonia (28.9%) compared to spermatocytes (8.3) and spermatids (11.2%). At the C-terminus, phosphorylation at K116 and methylation at K117 were observed in combination in TH2B isolated from these stages (Lu et al. 2009). However, its functional relevance is not known. Various PTMs like acetylation, methylation, and phosphorylation have been identified on TH2B from tetraploid spermatocyte and haploid spermatid. LC–MS/MS analysis of TH2B from spermatocytes identified six acetylation, three monomethylation, and one phosphorylation site, while that of TH2B from round spermatids identified four acetylation and two monomethylation sites. In silico analysis showed altered histone-histone as well as histone-DNA interactions in TH2B-bearing nucleosome. Also acetylation on N-terminal tail of TH2B has been shown to weaken its interactions with the DNA (Pentakota et al. 2014). Its physiological relevance remains to be determined.

17.5 Protamines and Male Infertility

Protamines and histones are the two major nuclear proteins in many vertebrate species including mice, rat, human, etc. These proteins play major roles during chromatin condensation at spermiogenesis. Several reports indicate that P1 and P2 are expressed in nearly the same amount in fertile human sperm and alteration in P1/P2 ratio is associated with male infertility (Aoki et al. 2005a, 2006; Zhang et al. 2006; Hammoud et al. 2009).

The first documented report highlighting the importance of protamines revealed the absence of protamines in the spermatozoa of infertile (oligozoospermic) patients (Silvestroni et al. 1976). This was followed by a study on 7 infertile and 17 fertile individuals where increased P1/P2 ratio in six of the seven patients was observed (Balhorn et al. 1988). Thereafter, a good number of studies have indicated that fertile men express P1 and P2 in same amount, while alteration in this ratio correlates with male infertility; infertile men show either decreased or increased P1/P2 ratio (Balhorn 2007; Balhorn et al. 1988; Belokopytova et al. 1993; de Yebra et al. 1993; Mengual et al. 2003; Aoki et al. 2005a). Balhorn's group has shown that the percentage of protamines is different in the patients with abnormal seminal parameters compared to patients with normal parameters. Also within the heterogeneous population of spermatozoa, round-headed spermatozoa from patients contain less protamines and more histones and intermediate proteins than normal spermatozoa.

Further, protamine levels vary between individual sperm of infertile males and correlate with viability and DNA integrity (Aoki et al. 2006). Interestingly, studies employing Percoll separation for fractionating sperm have shown that P1/P2 ratio and total protamine from different Percoll fractions within the same sample were not significantly different. However, there were significant differences in P1/P2 ratios in the oligozoospermic and asthenozoospermic groups as compared to normozoospermic indicating that P1/P2 and amount of protamine retention were independent of morphology and motility of sperm cells (Mengual et al. 2003).

This alteration in P1/P2 ratio might be due to alteration in expression of either of two protamines or both. Several studies have indicated lower P2 and increased P2 precursor in infertile men, indicating abnormality in the processing of precursor protein (de Yebra et al. 1993, 1998; Carrell and Liu 2001; Torregrosa et al. 2006). Aoki et al. observed a P1/P2 ratio around 1 in fertile donors; in infertile group, the P1/P2 ratios were either less than 0.8, between 0.8 and 1.2, or greater than 1.2 in 13.6%, 46.7%, and 39.7% of the patients, respectively. P1 and P2 were both under-expressed in patients with a normal P1/P2 ratio. In patients with a high P1/P2 ratio, P1 was normally expressed and P2 was under-expressed. They also reported that patients with abnormal P1/P2 ratios displayed significantly reduced semen quality and sperm penetration ability (Aoki et al. 2005a).

Several studies have also reported the presence of protamine transcript in sperm. Significantly aberrant protamine mRNA ratio was found in infertile individuals, and it correlates with DNA fragmentation and IVF success (Steger et al. 2001; Rogenhofer et al. 2013; Ni et al. 2014a). Significantly higher PRM1 and PRM2 mRNA copy numbers have been observed in normozoospermic versus teratozoospermic samples (Savadi-Shiraz et al. 2015). In contrast, transition protein 2 (TNP2) transcript abundance was significantly higher in teratozoospermic samples and positively correlated with sperm-head defects.

17.6 Protamines, DNA Compaction, and Integrity

Protamines are essential for sperm-specific packing of DNA. Compaction of DNA shuts off transcription as the DNA is no more amenable to the transcription factors and RNA polymerase. It also protects the DNA from any damage thus maintaining its integrity. This ensures that postfertilization the paternal genome is delivered in a form that allows developing embryo to accurately express genetic information. DNA compaction during chromatin condensation changes a transcriptionally active chromatin into a transcriptionally silent chromatin, and all the genes that are required for spermatogenesis and sperm function are transcribed prior to this transition, i.e., until the round spermatid stage. Live imaging studies in *Drosophila* have shown that histone-to-protamine transition starts 50–60 h after completion of meiosis and lasts for 5–6 h (Awe and Renkawitz-Pohl 2010). In mice although there is no direct evidence such as live imaging, indirect evidences suggest that this transition starts approximately 156 h after completion of meiosis and it lasts for 120–126 h, i.e., from step 10 to step 15 of spermiogenesis (reviewed by Rathke et al. 2014).

This period is characterized by DNA breaks and repair which allows relief from the torsion stress and facilitates removal of the histones and replacement by transition proteins and subsequently protamines (Marcon and Boissonneault 2004). Thereafter, selective translation of the stored mRNA takes place as per the requirements of the sperm.

It is well established that about 5–15% histones are retained in normal human sperm. Elegant studies by Hammoud et al. have shown gene clusters important for embryonic development to be associated with the retained histones in sperm of fertile men (Hammoud et al. 2009). This implies that improper packaging due to higher histone retention as seen in sperm chromatin of infertile men may expose many more gene clusters. A subsequent study by the same group showed that in infertile men, histones retention was random genome-wide, unlike fertile men where the histone retention was seen only at specific gene clusters (Hammoud et al. 2011). The epigenetic marks H3K4me or H3K27me were also reduced on the retained histones in the infertile men. They speculate that these changes may be responsible for the poor reproductive outcome post ICSI/IVF in infertile men.

Any defects in chromatin packaging wreaks havoc with the sperm ability to fertilize or sire a viable offspring either by allowing the untimely transcription of certain genes, allowing certain modifications of histones that may switch the transcription on or off, or increasing the vulnerability of the DNA to drug-induced damage. Observations from the chromodomain helicase DNA-binding protein 5 (CHD5) KO mice are a testimony to the effect of improper condensation on sperm morphology and fertility of the male offspring (Zhuang et al. 2014). H4 hyperacetylation, which is vital for histone replacement during spermiogenesis, is reduced in these mice, and the sperm show deformed nuclei and abnormal head morphology. However, in these mice transcription of important genes, controlling spermatogenesis was not affected. Several groups have shown very lucidly the correlation between protamine compaction, DNA integrity, and sperm quality (Franken et al. 1999; García-Peiró et al. 2011; Manochantr et al. 2012; Utsuno et al. 2014). Chromatin packaging as studied by CMA_3 and acidic aniline blue staining negatively correlates with normal sperm morphology (Franken et al. 1999). Utsuno et al. observed abnormal protamination in significantly higher number of spermatozoa with abnormal head morphology compared to those with normal head morphology. DNA fragmentation was also higher in the protamine-deficient spermatozoa. Studies on DNA damage in men undergoing IVF treatment revealed a positive association between DNA damage and abnormal sperm morphology and motility and negative correlation with sperm concentration (Morris et al. 2002). Protamine 2-deficient mice sperm demonstrate a direct correlation between PRM2 haploinsufficiency and frequency of DNA damage as seen from comet assays and ultrastructural analysis (Cho et al. 2003). In studies with human sperm, a positive correlation has been shown between protamine deficiency and sperm DNA damage (De Iuliis et al. 2009; Nili et al. 2009; Tarozzi et al. 2009; Razavi et al. 2010; Manochantr et al. 2012; Utsuno et al. 2014). Several studies have correlated altered P1/P2 ratio with susceptibility to DNA damage (Aoki et al. 2005b, 2006).

17.7 DNA Integrity and ART Outcomes

DNA integrity also influences sperm penetration and fertilizing ability, IVF and embryo quality, and development in ICSI outcome (Khara et al. 1997; Carrell et al. 1999; Carrell and Liu 2001; Nasr-Esfahani et al. 2004; de Mateo et al. 2009). DNA fragmentation and CMA3 positivity indicative of protamine deficiency negatively correlate with the fertilization rate in ICSI patients; DNA methylation negatively correlated with DNA fragmentation (Tavalaee et al. 2009). However, Tarozzi et al. observed a close relationship between sperm protamination and fertilization and pregnancy only in IVF; in ICSI there was a correlation between DNA fragmentation and pregnancy (Tarozzi et al. 2009). In men enrolled for ICSI, a positive association was seen between sperm damage and impairment of postfertilization embryo cleavage (Morris et al. 2002). In another study of individuals referred for ICSI, CMA3 positivity showed a significant negative correlation with fertilization rate post ICSI (Iranpour 2014). An isolated study using cleavage-stage frozen-thawed embryos from cycles of IVF and ICSI has however observed no significant difference in the biochemical pregnancy, clinical pregnancy, and miscarriage rates between sperm showing DFI <30% and those >30% (Ni et al. 2014b). The group did find some association between DFI and blastocyst formation in the ICSI group. A recent study investigating the influence of sperm DNA fragmentation on the pregnancy outcome and pregnancy loss after ART in couples going for either autologous ICSI, ICSI using donor eggs, or IUI observed that while the pregnancy rates were not significantly different, pregnancy losses correlated positively with the DNA fragmentation which was measured as DNA fragmentation index (DFI). The study indicates that sperm samples showing DFI >27% are associated with an increased risk of early pregnancy loss (Rilcheva Violeta et al. 2016). A similar observation has been reported earlier (Jin et al. 2015). Additionally, this group observed that when the DFI exceeded 27.3%, the live birth and implantation rates were significantly reduced in women with reduced ovarian reserve vis-a-vis women with normal ovarian reserve.

Conclusions and Future Directions

DNA integrity and its proper compaction in sperm are vital to its fertilizing ability as well as for early embryonic development in the preimplantation stage. Poor DNA compaction in sperm severely hampers its fertilizing ability and further development that accounts for fertility loss in natural conception or poor success of IVF/ICSI procedures. While literature is replete with evidences on histone retention and protamine deficiency in infertile cases, our knowledge on impact of several histone modifications on the fertility of male is limited and needs attention. Further research in this direction may identify sperm chromatin tests that may predict the success of ARTs. At the same time, further studies are needed to understand the significance of the retained histones in sperm maturation and their contribution toward the fertilizing ability of sperm.

Acknowledgments We acknowledge with gratitude the research contributions of all the authors whose work has been cited in this review (IR/394/07-2016). We also thank Mr. Vivian Lobo, a graduate student from my laboratory, for his inputs with the figures and Mr. Vaibhav Shinde for his help with the graphic design. We thank the Department of Biotechnology, India for the Junior Research Fellowship to Mr Aniket Patankar.

References

Akama K, Sato H, Furihata-Yamauchi M, Komatsu Y, Tobita T, Nakano M (1996) Interaction of nucleosome core DNA with transition proteins 1 and 3 from boar late spermatid nuclei. J Biochem 119:448–455

Aoki VW, Liu L, Carrell DT (2005a) Identification and evaluation of a novel sperm protamine abnormality in a population of infertile males. Hum Reprod 20:1298–1306

Aoki VW, Moskovtsev SI, Willis J, Liu L, Mullen JB, Carrell DT (2005b) DNA integrity is compromised in protamine-deficient human sperm. J Androl 26:741–748

Aoki VW, Emery BR, Liu L, Carrell DT (2006) Protamine levels vary between individual sperm cells of infertile human males and correlate with viability and DNA integrity. J Androl 27:890–898

Awe S, Renkawitz-Pohl R (2010) Histone H4 acetylation is essential to proceed from a histone- to a protamine-based chromatin structure in spermatid nuclei of *Drosophila melanogaster*. Syst Biol Reprod Med 56:44–61

Balhorn R (2007) The protamine family of sperm nuclear proteins. Genome Biol 8:227

Balhorn R, Reed S, Tanphaichitr N (1988) Aberrant protamine 1/protamine 2 ratios in sperm of infertile human males. Experientia 44:52–55

Belokopytova IA, Kostyleva EI, Tomilin AN, Vorob'ev VI (1993) Human male infertility may be due to a decrease of the protamine P2 content in sperm chromatin. Mol Reprod Dev 34:53–57

Belva F, Henriet S, Liebaers I, Van Steirteghem A, Celestin-Westreich S, Bonduelle M (2007) Medical outcome of 8-year-old singleton ICSI children (born >or =32 weeks' gestation) and a spontaneously conceived comparison group. Hum Reprod 22:506–515

Bonduelle M, Bergh C, Niklasson A, Palermo GD, Wennerholm UB (2004) Medical follow-up study of 5-year-old ICSI children. Reprod Biomed Online 9:91–101

Brunner AM, Nanni P, Mansuy IM (2014) Epigenetic marking of sperm by post-translational modification of histones and protamines. Epigenetics Chromatin 7:2

Brykczynska U, Hisano M, Erkek S, Ramos L, Oakeley EJ, Roloff TC, Beisel C, Schubeler D, Stadler MB, Peters AH (2010) Repressive and active histone methylation mark distinct promoters in human and mouse spermatozoa. Nat Struct Mol Biol 17:679–687

Carrell DT, Liu L (2001) Altered protamine 2 expression is uncommon in donors of known fertility, but common among men with poor fertilizing capacity, and may reflect other abnormalities of spermiogenesis. J Androl 22:604–610

Carrell DT, Emery BR, Liu L (1999) Characterization of aneuploidy rates, protamine levels, ultrastructure, and functional ability of round-headed sperm from two siblings and implications for intracytoplasmic sperm injection. Fertil Steril 71:511–516

Catena R, Ronfani L, Sassone-Corsi P, Davidson I (2006) Changes in intranuclear chromatin architecture induce bipolar nuclear localization of histone variant H1T2 in male haploid spermatids. Dev Biol 296:231–238

Celeste A, Petersen S, Romanienko PJ, Fernandez-Capetillo O, Chen HT, Sedelnikova OA, Reina-San-Martin B, Coppola V, Meffre E, Difilippantonio MJ et al (2002) Genomic instability in mice lacking histone H2AX. Science 296:922–927

Chevaillier P, Chirat F, Sautiere P (1998) The amino acid sequence of the ram spermatidal protein 3--a transition protein TP3 or TP4? Eur J biochem 258:460–464

Cho C, Jung-Ha H, Willis WD, Goulding EH, Stein P, Xu Z, Schultz RM, Hecht NB, Eddy EM (2003) Protamine 2 deficiency leads to sperm DNA damage and embryo death in mice. Biol Reprod 69:211–217

De Iuliis GN, Thomson LK, Mitchell LA, Finnie JM, Koppers AJ, Hedges A, Nixon B, Aitken RJ (2009) DNA damage in human spermatozoa is highly correlated with the efficiency of chromatin remodeling and the formation of 8-hydroxy-2′-deoxyguanosine, a marker of oxidative stress. Biol Reprod 81:517–524

de la Barre AE, Angelov D, Molla A, Dimitrov S (2001) The N-terminus of histone H2B, but not that of histone H3 or its phosphorylation, is essential for chromosome condensation. EMBO J 20:6383–6393

de Mateo S, Gazquez C, Guimera M, Balasch J, Meistrich ML, Ballesca JL, Oliva R (2009) Protamine 2 precursors (Pre-P2), protamine 1 to protamine 2 ratio (P1/P2), and assisted reproduction outcome. Fertil Steril 91:715–722

de Yebra L, Ballesca JL, Vanrell JA, Bassas L, Oliva R (1993) Complete selective absence of protamine P2 in humans. J Biol Chem 268:10553–10557

de Yebra L, Ballesca JL, Vanrell JA, Corzett M, Balhorn R, Oliva R (1998) Detection of P2 precursors in the sperm cells of infertile patients who have reduced protamine P2 levels. Fertil Steril 69:755–759

Drabent B, Bode C, Bramlage B, Doenecke D (1996) Expression of the mouse testicular histone gene H1t during spermatogenesis. Histochem Cell Biol 106:247–251

Drabent B, Benavente R, Hoyer-Fender S (2003) Histone H1t is not replaced by H1.1 or H1.2 in pachytene spermatocytes or spermatids of H1t-deficient mice. Cytogenet Genome Res 103:307–313

Faast R, Thonglairoam V, Schulz TC, Beall J, Wells JR, Taylor H, Matthaei K, Rathjen PD, Tremethick DJ, Lyons I (2001) Histone variant H2A.Z is required for early mammalian development. Curr Biol 11:1183–1187

Fernandez-Capetillo O, Mahadevaiah SK, Celeste A, Romanienko PJ, Camerini-Otero RD, Bonner WM, Manova K, Burgoyne P, Nussenzweig A (2003) H2AX is required for chromatin remodeling and inactivation of sex chromosomes in male mouse meiosis. Dev Cell 4:497–508

Franken DR, Franken CJ, De La Guerre H, De Villiers A (1999) Normal sperm morphology and chromatin packaging: comparison between aniline blue and chromomycin A3 staining. Andrologia 31:361–366

García-Peiró A, Martínez-Heredia J, Oliver-Bonet M, Abad C, Amengual MJ, Navarro J, Jones C, Coward K, Gosálvez J, Benet J (2011) Protamine 1 to protamine 2 ratio correlates with dynamic aspects of DNA fragmentation in human sperm. Fertil Steril 95:105–109

Gatewood JM, Cook GR, Balhorn R, Bradbury EM, Schmid CW (1987) Sequence-specific packaging of DNA in human sperm chromatin. Science 236:962–964

Glaser S, Lubitz S, Loveland KL, Ohbo K, Robb L, Schwenk F, Seibler J, Roellig D, Kranz A, Anastassiadis K et al (2009) The histone 3 lysine 4 methyltransferase, Mll2, is only required briefly in development and spermatogenesis. Epigenetics Chromatin 2:5

Grimes SR Jr, Platz RD, Meistrich ML, Hnilica LS (1975) Partial characterization of a new basic nuclear protein from rat testis elongated spermatids. Biochem Biophys Res Commun 67:182–189

Hammoud SS, Nix DA, Zhang H, Purwar J, Carrell DT, Cairns BR (2009) Distinctive chromatin in human sperm packages genes for embryo development. Nature 460:473–478

Hammoud SS, Nix DA, Hammoud AO, Gibson M, Cairns BR, Carrell DT (2011) Genome-wide analysis identifies changes in histone retention and epigenetic modifications at developmental and imprinted gene loci in the sperm of infertile men. Hum Reprod 26:2558–2569

Heidaran MA, Showman RM, Kistler WS (1988) A cytochemical study of the transcriptional and translational regulation of nuclear transition protein 1 (TP1), a major chromosomal protein of mammalian spermatids. J Cell Biol 106:1427–1433

Ichijima Y, Sin HS, Namekawa SH (2012) Sex chromosome inactivation in germ cells: emerging roles of DNA damage response pathways. Cell Mol Life Sci 69:2559–2572

Iguchi N, Tanaka H, Yomogida K, Nishimune Y (2003) Isolation and characterization of a novel cDNA encoding a DNA-binding protein (Hils1) specifically expressed in testicular haploid germ cells. Int J Androl 26:354–365

Iguchi N, Tanaka H, Yamada S, Nishimura H, Nishimune Y (2004) Control of mouse hils1 gene expression during spermatogenesis: identification of regulatory element by transgenic mouse. Biol Reprod 70:1239–1245

Ioannou D, Miller D, Griffin DK, Tempest HG (2016) Impact of sperm DNA chromatin in the clinic. J Assist Reprod Genet 33:157–166

Iranpour FG (2014) Impact of sperm chromatin evaluation on fertilization rate in intracytoplasmic sperm injection. Adv Biomed Res 3:229

Irez T, Sahmay S, Ocal P, Goymen A, Senol H, Erol N, Kaleli S, Guralp O (2015) Investigation of the association between the outcomes of sperm chromatin condensation and decondensation tests, and assisted reproduction techniques. Andrologia 47:438–447

Jin J, Pan C, Fei Q, Ni W, Yang X, Zhang L, Huang X (2015) Effect of sperm DNA fragmentation on the clinical outcomes for in vitro fertilization and intracytoplasmic sperm injection in women with different ovarian reserves. Fertil Steril 103:910–916

Khara KK, Vlad M, Griffiths M, Kennedy CR (1997) Human protamines and male infertility. J Assist Reprod Genet 14:282–290

Kistler WS, Geroch ME, Williams-Ashman HG (1975) A highly basic small protein associated with spermatogenesis in the human testis. Investig Urol 12:346–350

Krejci J, Stixova L, Pagacova E, Legartova S, Kozubek S, Lochmanova G, Zdrahal Z, Sehnalova P, Dabravolski S, Hejatko J et al (2015) Post-translational modifications of histones in human sperm. J Cell Biochem 116:2195–2209

Kremling H, Luerssen H, Adham IM, Klemm U, Tsaousidou S, Engel W (1989) Nucleotide sequences and expression of cDNA clones for boar and bull transition protein 1 and its evolutionary conservation in mammals. Differentiation 40:184–190

Lee J, Park HS, Kim HH, Yun YJ, Lee DR, Lee S (2009) Functional polymorphism in H2BFWT-5′UTR is associated with susceptibility to male infertility. J Cell Mol Med 13:1942–1951

Li W, Wu J, Kim SY, Zhao M, Hearn SA, Zhang MQ, Meistrich ML, Mills AA (2014) Chd5 orchestrates chromatin remodelling during sperm development. Nat Commun 5:3812

Lu S, Xie YM, Li X, Luo J, Shi XQ, Hong X, Pan YH, Ma X (2009) Mass spectrometry analysis of dynamic post-translational modifications of TH2B during spermatogenesis. Mol Hum Reprod 15:373–378

Luense LJ, Wang X, Schon SB, Weller AH, Lin Shiao E, Bryant JM, Bartolomei MS, Coutifaris C, Garcia BA, Berger SL (2016) Comprehensive analysis of histone post-translational modifications in mouse and human male germ cells. Epigenetics Chromatin 9:24

Manochantr S, Chiamchanya C, Sobhon P (2012) Relationship between chromatin condensation, DNA integrity and quality of ejaculated spermatozoa from infertile men. Andrologia 44:187–199

Marcon L, Boissonneault G (2004) Transient DNA strand breaks during mouse and human spermiogenesis:new insights in stage specificity and link to chromatin remodeling. Biol Reprod 70:910–918

Martianov I, Brancorsini S, Catena R, Gansmuller A, Kotaja N, Parvinen M, Sassone-Corsi P, Davidson I (2005) Polar nuclear localization of H1T2, a histone H1 variant, required for spermatid elongation and DNA condensation during spermiogenesis. Proc Natl Acad Sci U S A 102:2808–2813

Mascarenhas MN, Flaxman SR, Boerma T, Vanderpoel S, Stevens GA (2012) National, regional, and global trends in infertility prevalence since 1990: a systematic analysis of 277 health surveys. PLoS Med 9:e1001356

Meetei AR, Ullas KS, Rao MR (2000) Identification of two novel zinc finger modules and nuclear localization signal in rat spermatidal protein TP2 by site-directed mutagenesis. J Biol Chem 275:38500–38507

Mengual L, Ballesca JL, Ascaso C, Oliva R (2003) Marked differences in protamine content and P1/P2 ratios in sperm cells from percoll fractions between patients and controls. J Androl 24:438–447

Montellier E, Boussouar F, Rousseaux S, Zhang K, Buchou T, Fenaille F, Shiota H, Debernardi A, Hery P, Curtet S et al (2013) Chromatin-to-nucleoprotamine transition is controlled by the histone H2B variant TH2B. Genes Dev 27:1680–1692

Moriniere J, Rousseaux S, Steuerwald U, Soler-Lopez M, Curtet S, Vitte AL, Govin J, Gaucher J, Sadoul K, Hart DJ et al (2009) Cooperative binding of two acetylation marks on a histone tail by a single bromodomain. Nature 461:664–668

Morris ID, Ilott S, Dixon L, Brison DR (2002) The spectrum of DNA damage in human sperm assessed by single cell gel electrophoresis (Comet assay) and its relationship to fertilization and embryo development. Hum Reprod 17:990–998

Mylonis I, Drosou V, Brancorsini S, Nikolakaki E, Sassone-Corsi P, Giannakouros T (2004) Temporal association of protamine 1 with the inner nuclear membrane protein lamin B receptor during spermiogenesis. J Biol Chem 279:11626–11631

Nasr-Esfahani MH, Salehi M, Razavi S, Mardani M, Bahramian H, Steger K, Oreizi F (2004) Effect of protamine-2 deficiency on ICSI outcome. Reprod Biomed Online 9:652–658

Nekrasov M, Soboleva TA, Jack C, Tremethick DJ (2013) Histone variant selectivity at the transcription start site: H2A.Z or H2A.Lap1. Nucleus 4:431–438

Ni K, Steger K, Yang H, Wang H, Hu K, Chen B (2014a) Sperm protamine mRNA ratio and DNA fragmentation index represent reliable clinical biomarkers for men with varicocele after microsurgical varicocele ligation. J Urol 192:170–176

Ni W, Xiao S, Qiu X, Jin J, Pan C, Li Y, Fei Q, Yang X, Zhang L, Huang X (2014b) Effect of sperm DNA fragmentation on clinical outcome of frozen-thawed embryo transfer and on blastocyst formation. PLoS One 9:e94956

Nili HA, Mozdarani H, Aleyasin A (2009) Correlation of sperm DNA damage with protamine deficiency in Iranian subfertile men. Reprod Biomed Online 18:479–485

Okada Y, Scott G, Ray MK, Mishina Y, Zhang Y (2007) Histone demethylase JHDM2A is critical for Tnp1 and Prm1 transcription and spermatogenesis. Nature 450:119–123

Oliva R, Bazett-Jones D, Mezquita C, Dixon GH (1987) Factors affecting nucleosome disassembly by protamines in vitro. Histone hyperacetylation and chromatin structure, time dependence, and the size of the sperm nuclear proteins. J Biol Chem 262:17016–17025

Padavattan S, Shinagawa T, Hasegawa K, Kumasaka T, Ishii S, Kumarevel T (2015) Structural and functional analyses of nucleosome complexes with mouse histone variants TH2a and TH2b, involved in reprogramming. Biochem Biophys Res Commun 464:929–935

Parte PP, Rao P, Redij S, Lobo V, D'Souza SJ, Gajbhiye R, Kulkarni V (2012) Sperm phosphoproteome profiling by ultra performance liquid chromatography followed by data independent analysis (LC-MS(E)) reveals altered proteomic signatures in asthenozoospermia. J Proteome 75:5861–5871

Pentakota SK, Sandhya SP, Sikarwar A, Chandra N, Satyanarayana Rao MR (2014) Mapping post-translational modifications of mammalian testicular specific histone variant TH2B in tetraploid and haploid germ cells and their implications on the dynamics of nucleosome structure. J Proteome Res 13:5603–5617

Pivot-Pajot C, Caron C, Govin J, Vion A, Rousseaux S, Khochbin S (2003) Acetylation-dependent chromatin reorganization by BRDT, a testis-specific bromodomain-containing protein. Mol Cell Biol 23:5354–5365

Pradeepa MM, Rao MR (2007) Chromatin remodeling during mammalian spermatogenesis: role of testis specific histone variants and transition proteins. Soc Reprod Fertil Suppl 63:1–10

Rathke C, Baarends WM, Awe S, Renkawitz-Pohl R (2014) Chromatin dynamics during spermiogenesis. Biochim Biophys Acta 1839:155–168

Razavi SH, Nasr-Esfahani MH, Deemeh MR, Shayesteh M, Tavalaee M (2010) Evaluation of zeta and HA-binding methods for selection of spermatozoa with normal morphology, protamine content and DNA integrity. Andrologia 42:13–19

Reeves RH, Gearhart JD, Hecht NB, Yelick P, Johnson P, O'Brien SJ (1989) Mapping of PRM1 to human chromosome 16 and tight linkage of Prm-1 and Prm-2 on mouse chromosome 16. J Hered 80:442–446

Rilcheva Violeta S, Ayvazova Nina P, Ilieva Lyubomira O, Ivanova Svetlana P, Konova EI (2016) Sperm DNA integrity test and assisted reproductive technology (art) outcome. J Biomed Clin Res 9(1):21–29

Rogenhofer N, Dansranjavin T, Schorsch M, Spiess A, Wang H, von Schonfeldt V, Cappallo-Obermann H, Baukloh V, Yang H, Paradowska A et al (2013) The sperm protamine mRNA ratio as a clinical parameter to estimate the fertilizing potential of men taking part in an ART programme. Hum Reprod 28:969–978

Savadi-Shiraz E, Edalatkhah H, Talebi S, Heidari-Vala H, Zandemami M, Pahlavan S, Modarressi MH, Akhondi MM, Paradowska-Dogan A, Sadeghi MR (2015) Quantification of sperm specific mRNA transcripts (PRM1, PRM2, and TNP2) in teratozoospermia and normozoospermia: new correlations between mRNA content and morphology of sperm. Mol Reprod Dev 82:26–35

Schenk R, Jenke A, Zilbauer M, Wirth S, Postberg J (2011) H3.5 is a novel hominid-specific histone H3 variant that is specifically expressed in the seminiferous tubules of human testes. Chromosoma 120:275–285

Schon SB, Luense LJ, Wang X, Donahue G, Garcia BA, Bartolomei MS, Berger SL (2015) A comprehensive assessment of histone modifications in human sperm. Fertil Steril 104:e296–e2e7

Shinagawa T, Takagi T, Tsukamoto D, Tomaru C, Huynh LM, Sivaraman P, Kumarevel T, Inoue K, Nakato R, Katou Y et al (2014) Histone variants enriched in oocytes enhance reprogramming to induced pluripotent stem cells. Cell Stem Cell 14:217–227

Shinagawa T, Huynh LM, Takagi T, Tsukamoto D, Tomaru C, Kwak HG, Dohmae N, Noguchi J, Ishii S (2015) Disruption of Th2a and Th2b genes causes defects in spermatogenesis. Development 142:1287–1292

Siklenka K, Erkek S, Godmann M, Lambrot R, McGraw S, Lafleur C, Cohen T, Xia J, Suderman M, Hallett M et al (2015) Disruption of histone methylation in developing sperm impairs offspring health transgenerationally. Science 350:aab2006

Silvestroni L, Frajese G, Fabrizio M (1976) Histones instead of protamines in terminal germ cells of infertile, oligospermic men. Fertil Steril 27:1428–1437

Steger K, Klonisch T, Gavenis K, Drabent B, Doenecke D, Bergmann M (1998) Expression of mRNA and protein of nucleoproteins during human spermiogenesis. Mol Hum Reprod 4:939–945

Steger K, Failing K, Klonisch T, Behre HM, Manning M, Weidner W, Hertle L, Bergmann M, Kliesch S (2001) Round spermatids from infertile men exhibit decreased protamine-1 and -2 mRNA. Hum Reprod 16:709–716

Tanaka H, Iguchi N, Isotani A, Kitamura K, Toyama Y, Matsuoka Y, Onishi M, Masai K, Maekawa M, Toshimori K et al (2005) HANP1/H1T2, a novel histone H1-like protein involved in nuclear formation and sperm fertility. Mol Cell Biol 25:7107–7119

Tanphaichitr N, Sobhon P, Taluppeth N, Chalermisarachai P (1978) Basic nuclear proteins in testicular cells and ejaculated spermatozoa in man. Exp Cell Res 117:347–356

Tarozzi N, Nadalini M, Stronati A, Bizzaro D, Dal Prato L, Coticchio G, Borini A (2009) Anomalies in sperm chromatin packaging: implications for assisted reproduction techniques. Reprod Biomed Online 18:486–495

Tavalaee M, Razavi S, Nasr-Esfahani MH (2009) Influence of sperm chromatin anomalies on assisted reproductive technology outcome. Fertil Steril 91:1119–1126

Torregrosa N, Dominguez-Fandos D, Camejo MI, Shirley CR, Meistrich ML, Ballesca JL, Oliva R (2006) Protamine 2 precursors, protamine 1/protamine 2 ratio, DNA integrity and other sperm parameters in infertile patients. Hum Reprod 21:2084–2089

Trostle-Weige PK, Meistrich ML, Brock WA, Nishioka K, Bremer JW (1982) Isolation and characterization of TH2A, a germ cell-specific variant of histone 2A in rat testis. J Biol Chem 257:5560–5567

Trostle-Weige PK, Meistrich ML, Brock WA, Nishioka K (1984) Isolation and characterization of TH3, a germ cell-specific variant of histone 3 in rat testis. J Biol Chem 259:8769–8776

Unni E, Zhang Y, Kangasniemi M, Saperstein W, Moss SB, Meistrich ML (1995) Stage-specific distribution of the spermatid-specific histone 2B in the rat testis. Biol Reprod 53:820–826

Urahama T, Harada A, Maehara K, Horikoshi N, Sato K, Sato Y, Shiraishi K, Sugino N, Osakabe A, Tachiwana H et al (2016) Histone H3.5 forms an unstable nucleosome and accumulates around transcription start sites in human testis. Epigenetics Chromatin 9:2

Utsuno H, Miyamoto T, Oka K, Shiozawa T (2014) Morphological alterations in protamine-deficient spermatozoa. Hum Reprod 29:2374–2381 van Roijen HJ, Ooms MP, Spaargaren MC, Baarends WM, Weber RF, Grootegoed JA, Vreeburg JT (1998) Immunoexpression of testis-specific histone 2B in human spermatozoa and testis tissue. Hum Reprod 13:1559–1566

Wennerholm UB, Bergh C, Hamberger L, Lundin K, Nilsson L, Wikland M, Kallen B (2000) Incidence of congenital malformations in children born after ICSI. Hum Reprod 15:944–948

Yan W, Ma L, Burns KH, Matzuk MM (2003) HILS1 is a spermatid-specific linker histone H1-like protein implicated in chromatin remodeling during mammalian spermiogenesis. Proc Natl Acad Sci U S A 100:10546–10551

Ying HQ, Scott MB, Zhou-cun A (2012) Relationship of SNP of H2BFWT gene to male infertility in a Chinese population with idiopathic spermatogenesis impairment. Biomarkers 17:402–406

Yuen BT, Bush KM, Barrilleaux BL, Cotterman R, Knoepfler PS (2014) Histone H3.3 regulates dynamic chromatin states during spermatogenesis. Development 141:3483–3494

Zalensky AO, Siino JS, Gineitis AA, Zalenskaya IA, Tomilin NV, Yau P, Bradbury EM (2002) Human testis/sperm-specific histone H2B (hTSH2B). Molecular cloning and characterization. J Biol Chem 277:43474–43480

Zhang X, San Gabriel M, Zini A (2006) Sperm nuclear histone to protamine ratio in fertile and infertile men: evidence of heterogeneous subpopulations of spermatozoa in the ejaculate. J Androl 27:414–420

Zhao M, Shirley CR, Yu YE, Mohapatra B, Zhang Y, Unni E, Deng JM, Arango NA, Terry NH, Weil MM et al (2001) Targeted disruption of the transition protein 2 gene affects sperm chromatin structure and reduces fertility in mice. Mol Cell Biol 21:7243–7255

Zhao M, Shirley CR, Mounsey S, Meistrich ML (2004) Nucleoprotein transitions during spermiogenesis in mice with transition nuclear protein Tnp1 and Tnp2 mutations. Biol Reprod 71:1016–1025

Zhuang T, Hess RA, Kolla V, Higashi M, Raabe TD, Brodeur GM (2014) CHD5 is required for spermiogenesis and chromatin condensation. Mech Dev 131:35–46

18 Proteomics of Male Infertility

Ashima Sinha and Savita Yadav

Abstract

Screening of male infertility cases via clinical semen analysis is not apt to diagnose the causes in 30–50% of infertility cases. Such idiopathic cases with no known cause are difficult to monitor and thus provide momentum toward enhanced and sensitive diagnostic tools for infertility examinations. "Omics," the system biology approach to study the biological system on a large scale, includes proteomics as a newcomer. The era of clinical proteomics in combination with bioinformatics has emerged as a new tool to identify novel molecular markers for pathology. The supremacy of proteomics technology to characterize the proteome content of a cell or tissue on a large scale has enabled it to explicate both global and targeted proteins. Both seminal plasma and sperm serve to have the potential to be a preliminary material for identifying protein signatures related to infertility. The current chapter illustrates the lacunae allied with the clinical semen analysis for infertility investigations and exemplifies the role of clinical semen proteomics in male infertility identification.

Keywords

Sperm proteome • Seminal plasma proteome • Differentially expressed proteins • Male infertility

A. Sinha • S. Yadav (✉)
Department of Biophysics, All India Institute of Medical Sciences, New Delhi, India
e-mail: ashimaashi19@gmail.com; savita11@gmail.com

R. Singh, K. Singh (eds.), *Male Infertility: Understanding, Causes and Treatment*, DOI 10.1007/978-981-10-4017-7_18

Key Points

- Semen is rich in proteins, making it important to study their functional roles.
- Proteomics provides information beyond gene expression as proteins are the actual functional molecules.
- Sperm proteins are known to undergo posttranslational modifications, making proteomic studies important.
- Semen proteomic studies hold the potential for providing better platform for diagnosis of idiopathic male infertility.

18.1 Introduction

Clinical proteomics in combination with bioinformatics has emerged as a new tool to identify novel molecular markers for pathology. From the DNA sequence, one cannot extract information about the level of protein expression. Thus, the proteome which is the protein compliment of the genome is defined as the sum of all the protein species occuring during the lifetime of an individual, isoforms of the protein and the posttranslational modifications (PTMs) (Jungblut et al. 2008). The proteome fluctuates in response to the internal and external stimuli and undergoes disease-specific changes. Study of the differentially expressed sperm proteins that regulate fertilization holds potential in unraveling the molecular signatures related to male infertility. Transcriptome profiling of the sperm holds less potential for the post testicular investigations as sperm is both transcriptionally and translationally silent, thus PTMs play important role in inducing physiological changes responsible for fertilization.

Presently, the concept of seminal plasma (SP), as a biological fluid and a noninvasive clinical sample for urogenital diagnosis and for biomarker discovery of male reproductive disorders, is gaining attention. SP is an affluent and easily available source of protein identification owing to its high protein content of 35–55 mg/mL. As SP is a collection of fluids secreted from testis, epididymis, and other male accessory glands, it serves to be a reservoir of proteins crucial for sperm capacitation, sperm-zona pellucida interaction, and sperm-egg fusion (Tomar et al. 2012). SP is a possible target for early detection of male reproductive cancers (prostate and testicular), since proteins representative for cancer emerge earlier in SP than in blood serum (Drabovich et al. 2014).

In the current chapter, we first briefly discuss the risk factors associated with male infertility followed by the clinical tests available for its assessment and the lacunae allied with them followed by the importance of proteomics in infertility diagnosis. In the second part of the chapter, we broadly review the proteomics data of SP and sperm with special reference to male infertility.

18.2 Functional Tests for Male Fertility

Despite of physical examination, hormone analysis, and semen analysis, the etiology of male infertility in a large number of cases remains idiopathic. A number of normozoospemic patients appear for assisted reproduction due to failure of natural conception for which the reasons are not obvious. Therefore, there is need to develop functional tests for male fertility, which can identify the lack of fertile (functional) sperm ejaculates with normal semen parameters according to the WHO criteria. It is not that we entirely lack the information about the potential functional analysis parameters; however, most of them suffer some or other limitations for bringing in regular clinical practice and are in need of further refinements. One such marker could be reactive oxygen species (ROS) level, which is well known to functionally impair spermatozoa. However, the short duration of activity of these molecules restricts the direct testing of ROS and its use as a method for infertility evaluation (Aitken et al. 1991). DNA fragmentation index (DFI) could be another such test; however, the lack of a standard DNA fragmentation test displaying universally agreed cutoff values limits its use as a diagnostic test for infertility assessment. The third could be the analysis of sperm/semen proteins for functional importance. The third aspect needs a lot of research on sperm proteomics, which has great potential in identifying the proteins of interest for this purpose.

18.3 Proteomics as a Diagnostic Tool for Evaluating Male Infertility

In the context of male infertility, clinical semen analysis providing a note for the concentration, motility, morphology of the sperm, and many other diagnostic tests still fail to screen the 30–40% infertility cases (Turek 2005). Thus, there is an unmet need for sensitive diagnostic tools for infertility investigations. With new scaling heights in molecular biology research, proteomics has expanded its horizon in sighting the pathological complexities of infertility and its causes. The protein biomarkers may help us toward better understanding of unknown causes of male infertility by dealing with the physiological functions of the proteins at tissue level that, in turn, can guide us to find better therapeutic solutions. It not only provides a platform to discover biomarkers of infertility but may also help in devising effective male contraceptives (Tomar et al. 2012; Upadhyay et al. 2013). By means of proteomic approaches, both global and targeted protein expression, regulation, and modification of proteins in various biological systems can be studied (Kolialexi et al. 2008). Several proteomic approaches are applied to study the proteome, its PTMs, and other different aspects that are directly linked with male fertility status. In the coming pages, a glimpse of the proteomic tools used to study the proteome will be discussed.

18.4 Proteomic Workflow for Sperm Characterization

The fundamental process of sperm proteome analysis comprises of purification of the sperm cells from the seminal fluid and making it free of contaminating cells (leukocytes, epithelial cells) and SP. Purity of the sperm cell preparation is a very decisive step and is the only step at the preliminary stage for the entry of minor contamination that could result in false-positive results. To circumvent this problem, density gradient centrifugation using Percoll or direct swim up method is used for isolating sperm cells in humans (de Mateo et al. 2013). Another parameter to be considered in sperm proteomic study is to scrutinize the component of sperm to be concerned, target the entire cell or subcellular fractionation, and explore specific cell compartments. An advantage of subcellular proteomics is that it may allow the detection of low abundance proteins that may well escape detection in whole cell approaches (de Mateo et al. 2011).

Subsequent to the purification of sperm cells or the subcellular fractions, solubilized proteins are digested for peptide generation which is identified by Mass Spectrometry (MS). Essentially, two alternatives are there for MS identification of proteins: (1) separation of proteins followed by protein digestion and peptide identification and (2) generation of peptides by direct digestion of the proteins present in the crude mixture followed by their identification. Conventionally, proteins were separated using two-dimensional electrophoresis (2DE) followed by LC-MS-based protein identification. Low abundant proteins are tricky to detect via 2DE pertaining to its low sensitivity. With the advancements in proteomics field, 2DE was replaced with a more specific and sensitive technique of differential gel electrophoresis (DIGE), which labels the proteins with Cy dye and gives information about the differential expression of proteins. With these 2DE approaches, small number of proteins were elucidated, whereas LC-MS/MS-based studies depicted large number of proteins. The mass spectra generated by MS are evaluated using computer-based algorithm to determine whether peptides found in protein databases could produce spectra that resemble those observed experimentally.

To gain access about the biological information of the identified proteins, they are then categorically distributed using Gene Ontology (GO) database into three domains, cellular component, molecular function, and biological process. By and large, the outcome of a sperm proteomic study depends on (1) sample preparation, (2) protein extraction, (3) reduction of sample complexity, (4) optimum protein separation by advanced gel electrophoresis and/or LC, (5) MS protein identification with sufficient mass resolution and mass accuracy, (6) advanced computational analysis of peptide and protein data, (7) the bioinformatics analysis for the establishment of potential protein interactions and the clustering of molecular functions of newly identified proteins, and (8) essential verification analysis of proteomic data, using immunoblotting, biochemical assays, confocal microscopy, and/or functional testing (Holland and Ohlendieck 2015).

18.5 Whole Sperm Proteomics

In the past, 2DE was the method of choice to investigate the proteome of the sperm. The number of proteins distinguished in 2D maps ranged from 10 to 200 approximately. For the first time, Naaby-Hansen established a 2D map of the neutral and acidic human spermatozoa proteins using 2DE (Table 18.1). The study found 260 proteins ranging from 20 to 200 kDa and pI 4.5 and 7.8 (Naaby-Hansen 1990). In subsequent study by the same author and the colleagues, near about 1397 vectorially labeled sperm surface proteins belonging to membrane protein fractions were

Table 18.1 Sperm proteome studies

Pathological condition	Technique used	Major outcomes of the study	Function	Reference
Whole sperm proteome				
Normozoospermic	2D-PAGE	A total of 260 proteins ranging from 20 to 200 kDa and pI 4.5 and 7.8 were identified	Established a 2D map of the neutral and acidic human spermatozoa proteins using 2DE	Naaby-Hansen (1990)
Normozoospermic	2D-PAGE with nonequilibrium pH gradient electrophoresis (NEPHGE)	1397 sperm membrane protein were identified with novel isoforms of actin, beta-tubulin, pH 20, and several phosphotyrosine-containing proteins	Sperm membrane proteome	Naaby-Hansen et al. (1997)
Normozoospermic	LC-MS/MS	A total of 1760 proteins, with 1350 proteins corresponding to soluble fractions, 719 for insoluble fractions, and 309 for both soluble and insoluble fractions were identified		Johnston et al. (2005)
Normozoospermic	2D-PAGE and MALDI-TOF	A total of 98 proteins were identified	Major role in transcription, translation, protein turnover (23%), energy production (23%), metabolism (6%), cell cycle, apoptosis, and oxidative stress (10%)	Martínez-Heredia et al. (2006)
Normozoospermic	LC-MS/MS	A total of 1056 proteins were identified in Triton X-100 soluble and insoluble sperm fractions	Analysis of the metabolic proteome	Baker et al. (2007)
Normozoospermic	LC-MS/MS	A total of 4675 unique sperm proteins were identified	A total of 227 testis-specific proteins with different functions were characterized	Wang et al. (2013)

(continued)

Table 18.1 (continued)

Pathological condition	Technique used	Major outcomes of the study	Function	Reference
Subcellular fractions				
Normozoospermic (nuclei)	LC-MS/MS	A total of 403 proteins were identified	Zinc fingers and transcription factors were deduced for the first time which may be responsible for epigenetic marking and embryonic development. Histone proteins were the most abundant family	de Mateo et al. (2011)
Normozoospermic (tail)	LC-MS/MS	A total of 1049 proteins were identified	Proteins identified were related to sperm tail structure and motility (11%) and metabolism and energy production (26%). Metabolic proteome (24%) comprised of enzymes involved in lipid metabolism, including enzymes for mitochondrial beta-oxidation. Peroxisomal proteins were also a part of tail proteome	Amaral et al. (2013)
Normozoospermic (head and tail)	LC-MS/MS	A total of 1429 proteins were identified with 721 tail proteins and 521 head proteins	Proteases were localized in the head region, and structural and motility-related proteins were localized in sperm tail	Baker et al. (2013)
Sperm fibrous sheath	MS/MS	Unique ADP/ATP carrier protein, glycolytic enzymes, and sorbitol dehydrogenase were identified	Provides a clue that ATP is regulated independent of mitochondrial oxidation via the principal piece of the flagellum	Kim et al. (2007)
Sperm membrane proteins	MS and Edman degradation	Heat shock proteins were identified	HSP chaperones are accessible for surface labeling on human sperm	Naaby-Hansen and Herr (2010)
Comparative proteomics				
Fertile donors vs. failed fertilization at IVF	LC-MS/MS and MALDI-TOF/MS	20 differential proteins were identified	Two proteins, viz., secretory actin-binding protein and outer dense fiber protein 2/2, were overexpressed in the patient	Pixton et al. (2004)
Patients with normal fertilization vs. failed fertilization	2D-PAGE, LC-MS/MS, and MALDI-TOF/MS	12 differential proteins were identified	Proteins associated with gamete interaction, viz., the laminin receptor LR67 and the L-xylulose reductase were identified	Frapsauce et al. (2009)

Sperm of infertile patients vs. healthy fertile sperm	MALDI-TOF/MS	24 differential proteins were identified	Proteins belonged to different functional groups: sexual reproduction, metabolic process, response to wounding, cell growth, and/or maintenance. Proteins involved in cell communication, proliferation, and differentiation pathways were also identified	Xu et al. (2012)
Asthenozoospermic vs. normozoospermic	2D, MALDI-TOF/MS	10 differential proteins were identified	Energy metabolism enzymes, viz., isocitrate dehydrogenase subunit, carbonic anhydrase, were downregulated in low motility sperm, whereas phosphoglycerate mutase 2, triosephosphate isomerase, and glutamate oxaloacetate transaminase-1 were upregulated	Zhao et al. (2007)
Asthenozoospermic vs. normozoospermic	2D-PAGE, MALDI-TOF/MS	17 differential proteins were identified	Cytoskeletal actin-B, annexin-A5, cytochrome C oxidase-6B, histone H2A, PIP and precursor, calcium binding protein-S100A9 (2 spots), clusterin precursor, dihydrolipoamide dehydrogenase precursor, fumarate hydratase precursor, heat shock protein-HSPA2, inositol-1 monophosphatase, 3-mercapto-pyruvate sulfurtransferase/dienoyl-CoA isomerase precursor, proteasome subunit-PSMB3, semenogelin 1 precursor, and testis-expressed sequence 12 were identified	Martínez-Heredia et al. (2008)
Asthenozoospermic vs. normozoospermic	2D-PAGE, MALDI-TOF/MS	12 differential proteins were identified	Phosphorylated forms of tubulin, reduced expression of gamma-tubulin were identified	Chan et al. (2009)
Asthenozoospermic vs. normozoospermic	2D-PAGE, MALDI-TOF/MS	8 differential proteins were identified	Proteins related to protein turnover, folding, and stress response proteins were identified	Siva et al. (2010)
Asthenozoospermic vs. normozoospermic	Nano UPLC-MS	66 differential phospho proteins were identified	Differentially expressed phosphorylated proteins were identified	Parte et al. (2012)

(continued)

Table 18.1 (continued)

Pathological condition	Technique used	Major outcomes of the study	Function	Reference
Asthenozoospermic vs. normozoospermic	2D-PAGE, MALDI-TOF/MS	16 differential proteins were identified	GRP78, lactoferrin, SPANXB, PGK2, flagellin, DJ-1, XPA binding protein 2, CAB2, GPX4, and GAPDH were the first to be identified as differentially expressed proteins in idiopathic asthenospermia patients	Shen et al. (2013)
Normozoospermic sperm samples with different IVF outcomes (pregnancy vs. no pregnancy)	TMT labeling, SDS-PAGE, LC-MS	66 differential proteins were identified	Proteins identified were involved in chromatin assembly and lipoprotein metabolism which have role in spermatogenesis	Azpiazu et al. (2014)
Asthenozoospermic vs. normozoospermic	TMT labeling, LC-MS/MS	80 differential proteins were identified	GO, cellular pathways, and clustering analyses indicated proteins associated with protein folding/degradation, vesicle trafficking, cytoskeleton, and energetic metabolism	Amaral et al. (2014)
Asthenozoospermic vs. normozoospermic	Label-free quantitative LC-MS/MS	127 differential proteins were identified	Functional category analyses indicated spermiogenesis, and motility-related proteins involved in that were related to metabolism, vesicle biogenesis cytoskeletal regulation, and protein degradation	Liu et al. (2015a)
Aged men and young asthenozoospermia patients	2D-PAGE, MALDI-TOF/MS	22 differential proteins were identified	Prostate and testis expressed 1 (PATE1) existed in both aged men and young asthenozoospermia patients. PATE1 was involved in sperm-egg penetration and sperm motility	Liu et al. (2015b)
Asthenozoospermic vs. normozoospermic (tail)	2D-PAGE, MALDI-TOF/MS	14 differential proteins were identified		Hashemitabar et al. (2015)

Globozoospermic vs. normozoospermic	2D-DIGE, MALDI-TOF/MS	35 differential proteins were identified	A total of nine proteins were found to be upregulated, and 26 proteins were found to be downregulated in round-headed spermatozoa. The differential proteins had important roles in spermatogenesis, cell skeleton, metabolism, and spermatozoa motility	Liao et al. (2009)
Oligoastheno zoospermic vs. normozoospermic	2D-DIGE, MALDI-TOF/MS	4 differential proteins were identified	Semenogelin II precursor and clusterin isoform 1 were not seen in the semen of infertile men	Thacker et al. (2011)
Functional proteomics				
Capacitated vs. non-capacitated normal sperm	2D-PAGE, MALDI-TOF/MS	Sperm phosphoproteome analysis was conducted for the first time revealing valosin-containing protein/p97 and two members of the A kinase-anchoring protein (AKAP) to be tyrosine phosphorylated during capacitation	Protein phosphorylation is the prerequisite for capacitation	Ficarro et al. (2003)
Capacitated vs. non-capacitated normal sperm	2D-PAGE, MALDI-TOF/MS	Proteins involved in energy metabolism (ATP synthase subunit alpha, L-asparaginase), flagellar organization (tubulin beta-2C chain, outer dense fiber protein, A-kinase anchor protein 4), protein turnover (heat shock-related 70 kDa protein 2) were downregulated after in vitro-induced capacitation seminal proteins such as clusterin and PIP were upregulated	Probably the motility apparatus of the capacitated sperm is deregulated, which may be induced by apoptosis-like mechanism	Secciani et al. (2009)

(continued)

Table 18.1 (continued)

Pathological condition	Technique used	Major outcomes of the study	Function	Reference
Capacitated normal sperm	Label-free phosphoproteomics	The activity of tyrosine phosphorylation kinase insulin growth factor 1 receptor (IGF1R) was augmented during sperm capacitation	Increased activity of tyrosine phosphorylation kinase IGF1R during sperm capacitation can be target for improvement in sperm functions in infertile men	Wang et al. (2015)
Impaired sperm-egg recognition vs. normal sperm	Label-free MS	Reduced expression of molecular chaperone, heat shock 70 kDa protein 2 (HSPA2)	Interaction analysis showed that HSPA2 was found in close association with two other proteins, sperm adhesion molecule 1 (SPAM1) and arylsulfatase A (ARSA)	Redgrove et al. (2012)
Acrosome reacted sperm	MS	Angiotensin-converting enzyme (ACE) and protein disulfide isomerase A6 (PDIA6) as the other new interacting partners of HSPA2	ACE and important component of the complex as its pharmacological inhibition significantly reduced the ability of human spermatozoa to undergo acrosome reaction	Bromfield et al. (2016)

catalogued using same 2DE technique (Naaby-Hansen et al. 1997). Additionally, the study revealed novel isoforms of actin, beta-tubulin, PH-20, and several phosphotyrosine-containing proteins in human sperm (Naaby-Hansen et al. 1997).

The low sensitivity issue associated with conventional 2DE approach hinders the detection of low abundant proteins. For this reason, researchers switched over to a more sensitive tool of MS. For the first time, extensive sperm proteome analysis of the soluble and insoluble sperm fractions was carried out using LC-MS/MS by Johnston and coauthors. Proteome analysis displayed about 1760 proteins in total, with 1350 proteins corresponding to soluble fractions, 719 for insoluble fractions, and 309 for both soluble and insoluble fractions. This sperm proteome characterization provides a physiologically relevant index of proteins (Johnston et al. 2005). In another study of its kind investigating the whole sperm proteome, 2DE separation was followed by matrix-assisted laser desorption/ionization time of flight-MS (MALDI-TOF/MS) identification of the proteins. The investigation revealed a total of 98 proteins with assigned functions. The proteins identified had major role in energy production (23%), transcription, translation, protein turnover (23%), cell cycle, apoptosis and oxidative stress (10%), and metabolism (6%). The functional details described in the present study paid impetus toward the better understanding of the sperm proteins (Martínez-Heredia et al. 2006).

Li et al. obtained a 2D reference map of 3872 proteins using narrow range pH strips and multiple 2D gels and identified the protein spots by MALDI-TOF/MS analysis, thus providing a comprehensive view of the sperm proteome, which can be useful in studying deregulations related to sperm infertility (Li et al. 2007). Similar to this, using comprehensive protocol of LC-MS/MS, the triton X-100 soluble and insoluble sperm fractions were analyzed, and 1056 different proteins were obtained (Baker et al. 2007). This is the first published list of identified proteins in human spermatozoa using LC-MS/MS analysis (Baker et al. 2007).

In a most extensive report, a total of 4675 unique proteins from human sperm have been successfully identified, of which 227 were testis specific. Furthermore, 500 proteins were annotated as drug targets, thus providing in-depth knowledge about the candidate targets for the development of male contraceptive drugs (Wang et al. 2013). In a recent study published by Amaral and her colleagues, the highest number of sperm proteins listed till date has been reported (Amaral et al. 2014). These proteins were reported to be involved in various functional pathways, such as metabolism, apoptosis, cell cycle, meiosis, and membrane trafficking. As discussed previously, functional annotations for the proteins identified are provided by using the GO catalogue.

18.6 Sperm Subcellular Proteomics

Subcellular proteomics helps in sorting different proteins from different compartments of the isolated sperm, viz., the head, tail, nucleus, and membrane proteins. As sperm is a cell with distinct sections having specific cellular roles, evaluation of the proteins from these compartments provides a clear view of the functioning of the sperm, events associated with fertilization and the proteins, which are responsible for male infertility. Another relevance of studying the different fractions of the cell is the identification of the specific cellular localization of each protein and also the less abundant proteins. Extending this view, de Mateo and coauthors elucidated the sperm nuclear proteome highlighting some interesting facts not discussed before. Nuclear proteins are potentially relevant for epigenetic marking, proper fertilization, and embryo development. The study revealed a total of 403 proteins from the sperm nuclei, with histones as the most abundant family, zinc fingers, and transcription factors were deduced for the first time and may be responsible for epigenetic marking and embryonic development (de Mateo et al. 2011).

In another study reviewing the sperm tail proteins, a number of proteins were identified by LC-MS/MS and were found to be involved in metabolism and energy production, motility, and structure of the tail (Amaral et al. 2013). Interestingly, some peroxisomal proteins were also exposed in the investigation, thus paying momentum to the fact that both mitochondrial and peroxisomal pathways are active in the sperm and are imperative for the motility of the sperm (Amaral et al. 2014). Moreover, Baker and colleagues isolated and analyzed the proteome of sperm head and tail jointly from the same sperm sample, clearly pointing the compartmentalized expression of the head and tail proteins (Baker et al. 2013). For example, energy-providing proteins were found to be present in the tail, whereas the proteases were localized in the head region (Baker et al. 2013).

Similarly, some investigators have isolated the human sperm fibrous sheath, a cytoskeletal element unique to spermiogenesis. The proteomic analysis identified unique ADP/ATP carrier protein, glycolytic enzymes (reported for the first time), and sorbitol dehydrogenase in the fibrous sheath of the sperm (Kim et al. 2007). The presence of these proteins in the fibrous sheath provide a clue that ATP is regulated independent of mitochondrial oxidation via the principal piece of the flagellum (Kim et al. 2007). Highly heterogeneous structures of the sperm such as the head, mid-piece, and tail are enveloped under the sperm surface membrane. Throughout the epididymal transit and during initial events of fertilization (capacitation, zona binding, acrosomal reaction), the sperm membrane proteins experience complex remodeling.

Sperm membrane proteins are presumably entailed in fertilization process, critically sperm-oocyte interaction, and capacitation. By means of discrete enrichment techniques, several authors have analyzed membrane calcium binding proteins (Naaby-Hansen et al. 2010), heat shock proteins (Naaby-Hansen and Herr 2010), and membrane proteins with an affinity for zona pellucida (Nixon et al. 2015). As the membrane proteins have crucial role in the fertilization events, most of the studies focused toward categorization of surface antigens, which are involved in infertility and their use as immune contraceptives (Shetty et al. 2001; Bohring et al. 1999).

18.7 Comparative Proteomics of Anomalous Behavior of Sperm Proteins

Abnormal semen parameters are the most common cause of male infertility as suggested by the World Health Organization. Treatment of infertility using intra cytoplsmic sperm injection (ICSI) is a very effectual and routinely used procedure.

Nearly, 5% of in vitro fertilization (IVF) attempts have an unpredictable failure fate, regardless of normal sperm parameters. However, in the literature, there are accumulating evidences in humans that sperm defects such as defective zona binding or the zona-induced acrosome reaction count for 56% of total fertilization failure in assisted conception (Liu and Baker 2000, 2003). Being a heterogeneous phenotype, it is not necessary that these defects are the only underlying cause of reproductive failure, and thus other factors involved need to be assessed. The molecular nature of these defects is also needed to be traced, and for this several proteomics studies have been reported in the literature toward the potential identification of the sperm protein defects that might be responsible for failed fertilization at the IVF.

Using proteomics strategy, Pixton et al. compared the sperm proteome profile of the fertile donors with that of the patient who experienced failed fertilization at IVF inspite of normal semen parameters and found 20 consistent protein differences in the patient proteome profile (Pixton et al. 2004). Similarly, in another report, proteins associated with gamete interaction, viz., the laminin receptor LR67 and the L-xylulose reductase, have been found (Frapsauce et al. 2009). Xu and his group also focused on the major proteins extracted from infertile patients with normal semen parameters but failed IVF. The study revealed a total of 24 altered proteins, which were involved in energy production, structure and movement, and cell signaling and regulation (Xu et al. 2012). Abnormal morphology (globozoospermia), reduced motility (asthenozoospermia), and reduced number of sperms (oligo-/azoospermia) are other probable causes of male infertility, and comparative studies related to these pathological conditions at the proteome level have displayed a whole lot of proteins that are differentially expressed.

There are proteomic studies assessing differential expression pertaining to asthenozoospermic patients and normal fertile donors. For sperm motility, ATP production is of prime importance, and glycolysis and oxidative phosphorylation are the accepted pathways for ATP production in the mammalian sperm mitochondria. Enzymes associated with energy metabolism such as isocitrate dehydrogenase subunit, carbonic anhydrase, and glycolytic enzymes have been identified in a study (Zhao et al. 2007). Other proteins related to low sperm motility include Rho GDP-dissociation inhibitor 1 and outer dense fiber protein (sperm structural proteins) (Zhao et al. 2007), phosphorylated forms of tubulin, reduced expression of gamma-tubulin (Chan et al. 2009), various HSPs, disturbed cAMP-mediated protein kinase A signaling, and abnormal actin regulation (Parte et al. 2012), protein turnover, and folding and stress response proteins (proteasome alpha 3 subunit and heat shock-related 70 kDa protein 2) (Siva et al. 2010), which can be used for establishing biomarker signature for low sperm motility thus improving its diagnosis.

Recently, Hashemitabar and colleagues have isolated and compared the proteome of sperm tail fractions of asthenozoospermic semen samples with that of normal fertile donors using 2DE and MALDI-TOF MS/MS. The authors found differentially expressed proteins related to turnover, folding and stress response (proteasome alpha 3 subunit and heat shock-related 70 kDa protein 2), energy metabolism, sperm movement, stress response, signaling and transport, antioxidant activity, and structural proteins (Hashemitabar et al. 2015). Globozoospermia (round-headed spermatozoa with an absent acrosome) diagnosed by the presence of 100% round-headed spermatozoa on semen analysis is an aberrant nuclear membrane and mid-piece defect making the patients with this condition absolutely infertile. Using DIGE coupled with MS over 61 protein spots were analyzed with nine proteins to be upregulated and 26 proteins to be downregulated in round-headed spermatozoa compared with normal spermatozoa. The differential proteins had important roles in a variety of cellular processes and structures, including spermatogenesis, cell skeleton, metabolism, and spermatozoa motility (Liao et al. 2009). Recently, Saraswat et al performed shotgun proteomic analysis (label free-LC- MS) of the sperm cells and seminal plasma proteins in normal and AS samples. The authors included 667 and 429 proteins for quantification in sperm and SP samples respectively. The investigators inferred that sperm motility pathway defects are reflected in sperm proteomic signatures and the seminal plasma data set does not imitate any of these defective pathways (Saraswat et al. 2017).

Oligoasthenozoospermia is a condition where both the sperm concentration and cellular motility are deranged posing the individual as infertile. Four unique proteins, semenogelin II precursor, prolactin-induced protein, clusterin isoform 1, and prostate-specific antigen (PSA) isoform 1 preproprotein were predominant in the semen of healthy men; however, semenogelin II precursor and clusterin isoform 1 were not seen in the semen of infertile men, suggesting unique differences in the spermatozoa protein profiles of fertile and infertile men (Thacker et al. 2011).

18.8 Functional Proteomics

The fate of fertilization is reliant on two hallmark events, viz., capacitation and acrosome reaction. Freshly ejaculated sperm goes through a number of functional modifications to accomplish fertilization proficiency. The process of acquiring the fertilizing potential starts by ejaculation and finally ends in the female reproductive tract. Austin and Chang in 1951 independently told that the sperm resides in the female tract to attain fertilizing capability and named it "capacitation" (Austin 1951, 1952; Chang 1951). Broadly, capacitation is defined as an ongoing process occurring during the sperm transport through female reproductive tract rendering sperm to undergo functional modifications, thus transforming it to competently fertile. The process is physiologically not complete until the spermatozoon reaches the oocyte (Bailey 2010). Once the sperm is competent, it binds to the zona pellucida, undergoes acrosome reaction followed by hyperactivated motility, and finally fuses with the oocyte (Bailey 2010).

Another major event succeeding capacitation is the acrosome reaction, which is calcium-dependent exocytosis triggered by the binding of the sperm to the oocytes zona pellucida (ZP) (Florman and Storey 1982). Outer acrosomal membrane fuses to the overlying plasma membrane at multiple points, thus liberating the entire contents, which pass through ZP and fuse with the oocyte plasma membrane. A prerequisite for this event is that the sperm should have undergone previous capacitation. Molecular mechanisms underlying capacitation and acrosome reaction are poorly understood. c-AMP-dependent tyrosine phosphorylation is a landmark for capacitation (Ficarro et al. 2003). Capacitated human sperm phosphoproteome analysis conducted for the first time revealed valosin-containing protein/p97 and two members of the A kinase-anchoring protein (AKAP) to be tyrosine phosphorylated during capacitation (Ficarro et al. 2003).

In another study of its kind, in vitro-induced capacitation proteome changes were illustrated. Altered proteome profile of normal versus capacitated sperm suggested that proteins involved in flagellar organization (tubulin beta-2C chain, outer dense fiber protein, A-kinase anchor protein 4), energy metabolism (ATP synthase subunit alpha, L-asparaginase), and protein turnover (heat shock-related 70 kDa protein 2) were downregulated after in vitro-induced capacitation (Secciani et al. 2009). Proteins that, instead, increased as a consequence of in vitro capacitation were seminal proteins such as clusterin and prolactin-inducible protein (PIP). The results indicate that the motility apparatus of the capacitated sperm is deregulated, which may probably be induced by apoptosis-like mechanism (Secciani et al. 2009). Label-free quantitative phosphoproteomics has been newly applied to investigate the overall phosphorylation events during sperm capacitation in humans and the phosphorylation sites involved. The results showed that the activity of insulin growth factor 1 receptor (IGF1R) tyrosine kinase is appreciably augmented during sperm capacitation posing it to be the target for improvement in sperm functions in infertile men (Wang et al. 2015).

The recognition and binding of spermatozoon to an ovulated oocyte is an imperative cellular event. Emerging evidences advocate for the concerted action of several sperm proteins for the accomplishment of sperm-egg fusion (Redgrove et al. 2011, 2012; Bromfield et al. 2016). Proteomic analysis of two such complexes using electrospray ionization mass spectrometry recognized the several components of the multimeric 20S proteasome and chaperonin-containing TCP-1 (CCT) complexes, with zona pellucida binding protein (ZPBP2) as a component of one of the complexes (Redgrove et al. 2011). Label-free MS-based comparative proteome analysis of sperm possessing an impaired capacity for sperm-egg recognition with normal cells revealed a reduced expression of the molecular chaperone and heat shock 70 kDa protein 2 (HSPA2) (Redgrove et al. 2012). Interaction analysis showed that HSPA2 was found in close association with two other proteins, sperm adhesion molecule 1 (SPAM1) and arylsulfatase A (ARSA), both of which have previously been implicated in sperm-egg interaction. The depletion of HSPA2 in the infertile patients posed impetus to the significance of this multimeric complex in arbitrating the sperm-egg contact thus paying attention to the male infertility causes (Redgrove et al. 2012).

Recently, Bromifield and colleagues have identified angiotensin-converting enzyme (ACE) and protein disulfide isomerase A6 (PDIA6) as the other new interacting partners of HSPA2, thus forming a multimeric protein complex participating in fertilization cascade. Moreover, the complex dwells in the membrane raft microdomains located in the peri-acrosomal region of the sperm head. Functional significance of the protein complex was assessed by inhibiting ACE, which significantly reduced the ability of human spermatozoa to undergo acrosome reaction (Bromfield et al. 2016).

18.9 Analysis of PTMs in Sperm Cells

In the rapidly changing environment persisting within the cell, the fragile homeostasis/balance is sustained by the proteins, which are the focal point of all the biological functions operative within the cell. The intricate process of transcription and translation (degradation), which govern the protein abundance, a composite network of intra- and intermolecular interactions, PTMs (affecting protein activity and function) aid in adjusting to the dynamic alterations in the cellular environment. Usual aging, disease onset, and many other biological processes are the consequence of slight changes within this network. Mature spermatozoa are almost transcriptionally and translationally silent, and to attain its fertile destiny, it relies on PTMs that play important roles in sperm functions. Phosphorylation being the most commonly studied PTMs has been detected on approximately 17,500 proteins, and roughly one-third of the proteins in eukaryotic cell are phosphorylated at any time (Mann et al. 2002). Other recurrent PTMs are ubiquitination (~8100 proteins), glycosylation (~4500), lysine acetylation (~6700 proteins), and lysine methylation (~2400 proteins) (Pagel et al. 2015).

The molecular mechanisms underlying capacitation and acrosome reaction are poorly understood, and phosphorylation and glycosylation are the most prominent PTMs during these two processes. c-AMP-dependent tyrosine phosphorylation is a landmark for capacitation (Ficarro et al. 2003). Capacitated human sperm phosphoproteome analysis conducted for the first time revealed valosin-containing protein/p97 and two members of the A kinase-anchoring protein (AKAP) to be tyrosine phosphorylated during capacitation (Ficarro et al. 2003). Label-free quantitative phosphoproteomics has been newly applied to investigate the overall phosphorylation events during sperm capacitation in humans and the phosphorylation sites involved. The results showed that the activity of insulin growth factor 1 receptor (IGF1R) tyrosine kinase is appreciably augmented during sperm capacitation posing it to be the target for improvement in sperm functions in infertile men (Wang et al. 2015).

Recent reports state that sperm motility is coupled with α-tubulin acetylation (Bhagwat et al. 2014) based on the finding that protein acetylation can modulate proteasomal degradation of core histones and axonemal microtubule construction (Yu et al. 2015). Using proteomics approaches, global lysine acetylation profiles of normal uncapacitated sperm were characterized reporting 973 lysine-acetylated sites that matched to 456 human sperm proteins, including 671 novel lysine acetylation sites and 205 novel lysine-acetylated proteins. Another imperative discovery of the study was novel acetylation of voltage-dependent anion channel 2 at Lys-74 in the asthenozoospermic sperm cells (Yu et al. 2015). A number of proteins have been found acetylated at lysine residues in human capacitated sperm with functions in motility, capacitation, acrosome reaction, and sperm-egg interaction, thus proving to be an evidence for the importance of lysine acetylation in the sperm (Sun et al. 2014). O-linked or N-linked glycosylation is another PTM reported in the developing spermatozoa during the epididymal descent. The role of glycosylation in cell-cell recognition, adhesion, and recognition is well established, and in the sperm, it helps in gamete binding.

Using high-throughput glyco-FASP technique for the enrichment of glycopeptides and then subjecting to tandem MS analysis, 554 N-glycosylation sites and 297 N-glycosylated proteins in human sperm were identified (Wang et al. 2013). About 91% of the N-glycoproteins were either lysosomal, extracellular, or membrane proteins, and via in vitro fertilization assay, it was evident that glutathione peroxidase 4 (GPX4), a membrane glycoprotein, was effectively involved in gamete interactions (Wang et al. 2013). A recent study has stated that excessive sumoylation is a marker of defective spermatozoa as some flagellar proteins, glycolytic and mitochondrial enzymes, and some heat shock proteins were found to be sumoylated at abnormally higher levels in nonmotile, two-tailed, microcephalic, and acephalic sperm (Vigodner et al. 2013). These sumoylated proteins were detected in the neck, flagella, and head regions as revealed by immunofluorescence and electron microscopy (Vigodner et al. 2013).

18.10 Seminal Plasma Proteomics in the Assessment of Male Fertility Status

As SP is a collective fluid derived from several organs and has protein constituents specific for the organ, therefore, differences in protein composition of SP might indicate an ongoing pathological process in a specific organ (Drabovich et al. 2014). For example, PSA, found at much higher concentrations in the semen than in the blood serum, is identified as a marker for prostatic diseases and is used for prostate cancer diagnosis. Proteome analysis of SP has raised the expectations for improved diagnosis and stratification of wide range of diseases (Davalieva et al. 2012). As SP is a collection of secretion of various tissue-specific proteins secreted by different male reproductive organs, it serves to be a potential source of protein biomarkers. SP proteome is subjected to alteration owing to male reproductive system disorders, thus leading to higher concentrations of organ-specific proteins, which can be quantified accurately by MS. Furthermore, it's proteome analysis could drive early diagnosis of testicular and prostate cancers as any cancer-specific protein appears much early in the SP than in the blood serum (Drabovich et al. 2014).

18.11 Seminal Plasma Proteome

The ejaculate is composed of 10% spermatozoa and 90% SP, with pH ranging from 7.2 to 8.0. SP serves to be the vehicle for the transport of spermatozoa during ejaculation from the male urethra thus escorting them to the female reproductive tract. Cell-free DNA, RNA, and microRNAs have also been identified in the SP, with microRNAs likely to be involved in spermatogenesis as their roles need to be further explored.

For the first time, SP proteins were electrophoretically separated in 1942 by a group of scientists (Gray and Huggins 1942; Ross et al. 1942), thus illustrating four protein components, α-globulin, β-globulin, γ-globulin, and albumin. Later advancements in the separation techniques resulted in the detection of nearly 40 proteins (Sensabaugh 1978). With the advent of new analytical paradigms in the field of electrophoretic protein separation through succeeding decades does the scientists were able to cut through the details of the SP proteome. In the early 1980s, Edwards and colleagues separated SP proteins by 2D-PAGE followed by blotting of proteins to nitrocellulose membrane and accordingly detected 200 proteins (Edwards et al. 1981).

The complexity of the SP was further attested by the introduction of high-throughput protein separation and identification tools, viz., soft ionization and MS.

In a study suggesting the role of SP proteins in impaired spermatogenesis, 750 proteins were identified including prostatic acid phosphatase (PAP), PSA, Zn-a-2-glycoprotein, glycodelin, and clusterin (Starita-Geribaldi et al. 2001). Furthermore, Fung et al. studied the SP proteome using LC-ESI thereby confirming the fact that the proteins were posttranslationally modified and that the multiple spots matching to the same parent protein were the isoforms of the same protein (Fung et al. 2004). Low molecular weight SP proteins of <30 kDa as truncated forms of semenogelin I and II, cystatin S, cystatin C, and variants of PIP were also identified.

In-depth analysis of SP proteome was conducted by Pilch and Mann in 2006, who catalogued a total of 932 proteins specific to each organ that has contributed to the formation of SP: seminal vesicle, prostate, epididymis, and Cowper's gland (Pilch and Mann 2006). Extracellular proteins secreted by the male sex glands, prostasomal proteins (originated from the epithelial lining of the prostate acini), and the proteins originated due to epithelial shredding were the three most prominent categories identified by the investigators. A large proportion of the proteome included proteins having functions in immunological reactions, providing metabolic sustainability and protection for the spermatozoa, involved in clot formation and liquefaction (Pilch and Mann 2006). Most recently, the investigators have used 2D liquid chromatography separations coupled to electrospray ionization and detection of mass spectra with Orbitrap™ and identified thousands of SP proteins. Largest library of SP proteins reported till date is of 3200 proteins in total as identified by Batruch et al. (2012).

Human SP contains a large array of proteins of clinical importance, and their characterization is imperative (Table 18.2). During fertilization process, the contact between the sperm and the egg is the decisive step for the future embryo to develop, and glycosaminoglycans (GAGs) have been reported to be vital for cell–cell interactions and communications. In the male reproductive biology, heparin, a GAG, is reported in processes, such as capacitation and acrosome reaction, and certain heparin-binding proteins (HBPs) interact with these GAGs present in the female reproductive tract, thus facilitating zona pellucida induction. Our group identified and characterized seven HBPs in the seminal fluid using affinity chromatography followed by MALDI-TOF/MS identification (Kumar et al. 2008). The major HBPs were semenogelin I fragment, semenogelin II, lactoferrin and its fragments, PSA, homolog of bovine SP proteins (BSP), zinc finger protein (Znf 169), and fibronectin fragments (Kumar et al. 2008).

As an extension to the abovementioned study, we also identified a group of concanavalin-A binding glycoproteins using affinity chromatography and subsequently identified them by MALDI-TOF/MS (Tomar et al. 2011). The major proteins identified in this study included aminopeptidase N, PSA, PAP, zinc-alpha-2-glycoprotein (ZAG), lactoferrin, Izumo sperm-egg fusion protein, progestogen-associated endometrial protein, and PIP (Tomar et al. 2011). Among the recent of all the studies done by our group, glycosylation sites, glycan compositions, and structures for 243 glycopeptides belonging to 73 N-glycosylation sites on 50 glycoproteins have been elucidated. Majority of the glycoproteins were complex type (83%) followed by high-mannose containing (10%) and hybrid type (7%), and most of the glycoproteins were either sialylated, fucosylated, or both (Saraswat et al. 2016).

Table 18.2 Seminal plasma proteome studies

Pathological condition	Technique used	Major outcomes of the study	Function	Reference
Whole SP proteome				
Normozoospermic	Fourier transform MS	A total of 932 proteins were identified	A large proportion of the proteome included proteins having functions in immunological reactions, providing metabolic sustainability and protection for the spermatozoa, involved in clot formation and liquefaction	Pilch and Mann (2006)
Normozoospermic	MudPIT-MS	A total of 3200 proteins were identified	Proteome included proteins having functions in immunological reactions, providing metabolic sustainability and protection for the spermatozoa, involved in clot formation and liquefaction	Batruch et al. (2012)
Normozoospermic	Affinity chromatography and MALDI-TOF	Major HBPs were semenogelin I fragment, semenogelin II, lactoferrin and its fragments, (PSA, homolog of bovine seminal plasma proteins (BSP), zinc finger protein (Znf 169), and fibronectin fragments	HBPs interact with the GAGs present in the female reproductive tract thus facilitating zona pellucida induction	Kumar et al. (2008)
SP proteome (concanavalin-A binding glycoproteins)	Affinity chromatography and MALDI-TOF		The major proteins identified in this study included aminopeptidase N, PSA, PAP, ZAG, lactoferrin, Izumo sperm-egg fusion protein, progestogen-associated endometrial protein, and PIP	Tomar et al. (2011)

(continued)

Table 18.2 (continued)

Pathological condition	Technique used	Major outcomes of the study	Function	Reference
SP glycoproteome	Affinity chromatography, MS/MS	Glycosylation sites, glycan compositions, and structures for 243 glycopeptides belonging to 73 N-glycosylation sites on 50 glycoproteins have been elucidated	Glycoproteins are involved in sperm and egg interaction	Saraswat et al. (2016)
Comparative proteomics				
Azoospermic vs. normozoospermic	MS	Nearly 700 proteins with acid phosphatase, PSA, Zn-a-2-glycoprotein, glycodelin, and clusterin were identified	Impaired spermatogenesis	Starita-Geribaldi et al. (2001)
Azoospermic vs. normozoospermic	2D-DIGE, MS/MS	STAB2, CP135, GNRP, and PIP as the potential markers		Yamakawa et al. (2007)
Nonobstructive Azoospermia vs. normozoospermic	MS/MS	28 differential proteins identified		Bai et al. (2007)
Asthenozoospermic vs. normozoospermic	LC-MS/MS	100 differential proteins identified	Epididymal secretory protein E1 and epididymal secretory protein E4 were increased in asthenozoospermic SP thus pointing toward the functional abnormalities in the prostate and epididymis contributing to abnormal sperm motility. Also downregulation of DJ-1 protein, involved in regulating oxidative stress concluded that increased levels of reactive oxygen species due to deregulated DJ-1	Wang et al. (2009)
Normal vs. asthenozoospermic, oligozoospermic, azoospermic SP	2D-DIGE, LC-MS/MS	8 proteins were differentially expressed in Azoospermia group	Fibronectin, PAP, proteasome subunit alpha type-3, beta-2-microglobulin, galectin-3-binding protein, PIP, and cytosolic nonspecific dipeptidase	Davalieva et al. (2012)

Normal vs. globozoospermic, oligozoospermic, SP	LC-MS/MS	20 proteins were differentially expressed	The functional analysis identified biological regulation as the major processes affected and determined that most of the identified proteins were of extracellular origin	Sharma et al. (2013a)
Infertile men vs. healthy donors to identify oxidative stress-related proteins	LC-MS/MS	14 differential proteins identified	PIP was found to be more abundantly present in men with increased levels of ROS. Gene ontology annotations showed extracellular distribution of proteins with a major role in antioxidative activity and regulatory processes	Sharma et al. (2013b)
Oligoasthenoteratozoospermia vs. normozoospermic	LC-MS/MS	Tubulin-folding cofactor B, alpha-1-antichymotrypsin, aldose reductase		Herwig et al. (2013)
Oligoasthenozoospermic vs. normal SP	2D-DIGE, LC-ESI/MS	4 differential proteins were identified	Epididymal secretory protein E1 and galectin-3-binding protein were under-expressed and lipocalin-1 and PIP	Giacomini et al. (2015)

The accuracy and proficiency of intracellular signaling pathways is under the influence of multiprotein complexes. Our group for the first time reported that ZAG is present as complex with PIP (Hassan et al. 2008). Human serum albumin (HSA), known to preserve in sperm motility, is proposed by our group as another plausible interacting partner of PIP (Kumar et al. 2012). Additionally, Tomar et al. identified the other interacting partners of PIP by co-immunoprecipitation followed by MS, suggesting semenogelin 1 fragments binding with PIP, thus providing its link in aiding spermatozoa to acquire motility (Tomar et al. 2013).

Later we also purified and characterized a zinc-binding high molecular weight multi-protein complex from human SP. The complex contained isoforms/fragments of different proteins with PSA, ZAG, PAP, and PIP as the major proteins of this complex (Yadav et al. 2011). Among the low molecular weight proteins, our group purified three cystatins (cysteine proteinase inhibitors), viz., cystatin 9, cystatin SN, and SAP-1 (N-terminal truncated form of cystatin S), and studied their enzyme kinetics (Yadav et al. 2013). Further interaction studies conducted on SAP-1 and heparin concluded that SAP-1 interacts with heparin and the binding is dependent on the chain length of heparin (Yadav et al. 2015).

Prostasomes, the membrane-enveloped vesicles secreted by the epithelial lining of the prostate acini, are rich source of intracellular proteins and are important for spermatozoa survival. Utleg et al. studied the composition of these prostasomes using LC-MS/MS and reported 139 proteins including enzymes, structural proteins, GTP binding proteins, and transport proteins. More importantly, majority of the proteins were secreted by the prostate (Utleg et al. 2003). Apart from focusing on the protein components of SP, researchers have also investigated the peptide constituents of the SP (Kausler and Spiteller 1992; Goverde et al. 1998; O'Mahony et al. 2000).

18.12 Comparative SP Proteomics with Clinical Objectives

From clinical viewpoint, the relevance of SP proteomics lays in the identification of male infertility-associated biomarkers. As SP constitutes the 90% of the total semen volume with the rest 10% engaged by spermatozoa and also higher concentration of tissue-specific proteins are present in it, it is a probable source of protein biomarkers. Extensive literature is there dealing with the identification of SP constituents; however, studies having inclination towards male infertility with extensive comparative analysis of SP proteome providing a correlation between SP proteins and male infertility are meager. In the first study of its kind, Starita-Geribaldi and colleagues studied the SP proteome in impaired spermatogenesis and compared the proteome from fertile men with vasectomized or azoospermic men and revealed nearly 700 proteins including acid phosphatase, PSA, ZAG, glycodelin, and clusterin (Starita-Geribaldi et al. 2001).

Most commonly, male infertility is diagnosed by a laboratory-based semen analysis for the presence of spermatozoa in the seminal fluid. Azoospermia, a condition with absence of sperm in the semen, is the most severe form of male infertility

(Jarow et al. 1989). Major attempt in this direction was made by Yamakawa et al. who analyzed the differential expression of proteins with respect to azoospermia condition and displayed stabilin 2 (STAB2), 135 kD, centrosomal protein (CP135), guanine nucleotide, releasing protein (GNRP), and PIP as the potential markers (Yamakawa et al. 2007). In another report, SP of nonobstructive azoospermia patients and healthy fertile males were compared showing 28 differentially expressed proteins (Bai et al. 2007). Our group purified PIP by immunoprecipitation and quantified its level in azoospermic SP samples using ELISA kit and found no significant change in its concentration in normozoospermia and oligozoospermia, while its expression was downregulated in azoospermia, thus paying impetus to the above findings of Yamakawa indicating PIP to be a plausible marker of azoospermia (Tomar et al. 2012).

Comparative proteomic analysis of normal and asthenozoospermic SP proteomes revealed 741 proteins, most of which were of epididymal and prostate origin. Moreover, epididymal secretory protein E1 and epididymal secretory protein E4 were increased in asthenozoospermic SP, thus pointing toward the functional abnormalities in the prostate and epididymis contributing to abnormal sperm motility (Wang et al. 2009). Another crucial finding of the study was the downregulation of DJ-1 protein, which is involved in regulating oxidative stress thus concluding that increased levels of reactive oxygen species due to deregulated DJ-1is an indicator of poor semen quality (Wang et al. 2009). High-resolution multidimensional protein identification technology (MudPIT) analyzed the SP proteome of normal and post-vasectomy (PV) data sets, thus reporting 32 proteins unique to controls and three unique to PV patients (Batruch et al. 2011).

The same authors recently catalogued more than 2000 proteins in nonobstructive azoospermia (NOA) subjects. Some of the proteins identified in this study, viz., LDHC, ELSPBP1, CES7, A2M, OVCH2, PTGDS, GPR64, and ALDH1A1, can possibly serve as markers differentiating NOA from obstructive azoospermia (Batruch et al. 2012). The only diagnostic protocol to differentiate between the obstructive azoospermia (OA) and NOA is testicular biopsy. In an attempt to identify markers distinguishing the two, Drabovich et al. identified two proteins, epididymis-expressed ECM1 and testis-expressed TEX101, which differentiated OA and NOA with high specificities and sensitivities (Drabovich et al. 2013). Using DIGE approach, differential protein expression was studied between normal, AS, oligozoospermic, and azoospermic men and found statistically significant increased expression of eight proteins in azoospermia compared with at least one of the other studied groups. The proteins were fibronectin, PAP, proteasome subunit alpha type-3, beta-2-microglobulin, galectin-3-binding protein, PIP, and cytosolic nonspecific dipeptidase, thus providing a deeper insight to the azoospermia condition (Davalieva et al. 2012).

In the search for a panel of common proteins in the fertile males that might be crucial for successful reproduction, a group of investigators performed high-throughput proteomic analysis. Of the 900 proteins resolved, 83 were common in set of five fertile men whose partners conceived 3 months before the study was initiated. Semenogelin I, semenogelin II, olfactory receptor 5R1, lactoferrin, hCAP18, spindling 1, and clusterin were the common proteins among them suggesting their vigor for reproduction (Milardi et al. 2012). In a recent investigation, differential protein expression of men with abnormal sperm count and sperm morphology was studied (Sharma et al. 2013a). Proteomics analysis revealed 20 differentially expressed proteins among the 4 groups with altered sperm count and abnormal morphology. Among the proteins identified, 3 were downregulated in the group with normal sperm count and abnormal morphology (NA), 1 in oligozoospermia and normal morphology (ON) group, 1 in oligozoospermia and abnormal morphology (OA) group while 2 were upregulated in the ON and OA groups. SP serves as an antioxidant reservoir, antioxidants remove the excess ROS generated in the body thus maintaining a balance and prohibiting oxidative stress. Imbalance in the levels of ROS is reported in SP of infertile men (Wang et al. 2009). Taking this into account, the same researchers studied the molecular mechanisms underlying oxidative stress and sperm dysfunction in infertile men by proteomic profiling. The oxidative stress parameters were assessed (ROS, antioxidant concentration, and DNA damage), and subjects were classified as ROS+ and ROS–. Proteomic analysis revealed 14 proteins in all, with seven proteins common in both the groups. Levels of PIP were elevated in men with increased ROS levels, and gene ontology annotation displayed the extracellular distribution of proteins with a major role in antioxidative activity and regulatory processes (Sharma et al. 2013b).

Similarly, Herwig et al. compared the SP of oligoasthenoteratozoospermia samples with normal fertile males and identified proteins related to oxidative stress, viz., tubulin-folding cofactor B, alpha-1-antichymotrypsin, and aldose reductase (Herwig et al. 2013). Recently, a comparative analysis of oligoasthenozoospermic and normal SP samples revealed that two proteins, namely, epididymal secretory protein E1 and galectin-3-binding protein, were under-expressed in oligoasthenozoospermia and two other proteins, lipocalin-1 and a PIP form, were overexpressed, thus suggesting their involvement in the pathology of idiopathic oligoasthenozoospermic condition (Giacomini et al. 2015).

Conclusion

As a concluding remark, research pertaining to SP proteomics for the search of biomarkers related to specific conditions of male infertility is still ongoing. Integrative approach of proteomic analysis and functional studies annotating the cellular pathways affected has paved the pathway for a deeper insight in mechanisms of male infertility-related pathology. Current research embraces the capability for the development of innovative and clinically relevant male infertility biomarkers using noninvasive procedures, which may provide a better platform for the patients undergoing treatment. Apart from this, success of ART in cases of infertility also needs to be explored. Diverse conditions of infertility have associated with the different sets of proteins. Nevertheless, the data obtained from these studies is heterogeneous as only a small subset of independent studies reporting a small fraction of proteins is found to be overlapping, reason being the use of different proteomic approaches and its combinations. However, the appearance of high-throughput MS-based techniques allows more detailed investigation of the proteomes of interest, among which is human seminal plasma proteome, and holds promise on more reproducible results in the future. Semen proteomics has the potential to provide information about the regulatory mechanisms of male infertility which is poorly understood till date. The identified proteins should be studied further in deep to find out their exact roles in male infertility. These studies may provide new approaches for management of male infertility.

References

Aitken RJ, Irvine DS, Wu FC (1991) Prospective analysis of sperm-oocyte fusion and reactive oxygen species generation as criteria for the diagnosis of infertility. Am J Obstet Gynecol 164(2):542–551

Amaral A, Castillo J, Estanyol JM, Ballescà JL, Ramalho-Santos J, Oliva R (2013) Human sperm tail proteome suggests new endogenous metabolic pathways. Mol Cell Proteomics 12(2):330–342

Amaral A, Castillo J, Ramalho-Santos J, Oliva R (2014) The combined human sperm proteome: cellular pathways and implications for basic and clinical science. Hum Reprod Update 20(1):40–62

Austin C (1951) Observations on the penetration of the sperm into the mammalian egg. Aust J Biol Sci 4(4):581–596

Austin CR (1952) The 'capacitation'of the mammalian sperm. Nature 170(4321):326

Azpiazu R, Amaral A, Castillo J, Estanyol JM, Guimerà M, Ballescà JL, Balasch J, Oliva R (2014) High-throughput sperm differential proteomics suggests that epigenetic alterations contribute to failed assisted reproduction. Hum Reprod 29(6):1225–1237

Bai J, Sun L, Chen SL, Zhang LW, Ma JL, Cong YL (2007) Comparative analysis of proteins in seminal plasma of non-obstructive azoospermia patients and healthy fertile males. Zhonghua Nan ke Xue 13(7):579–583

Bailey JL (2010) Factors regulating sperm capacitation. Syst Biol Reprod Med 56(5):334–348

Baker MA, Reeves G, Hetherington L, Müller J, Baur I, Aitken RJ (2007) Identification of gene products present in Triton X-100 soluble and insoluble fractions of human spermatozoa lysates using LC-MS/MS analysis. Proteomics Clin Appl 1(5):524–532

Baker MA, Naumovski N, Hetherington L, Weinberg A, Velkov T, Aitken RJ (2013) Head and flagella subcompartmental proteomic analysis of human spermatozoa. Proteomics 13(1):61–74

Batruch I, Lecker I, Kagedan D, Smith CR, Mullen BJ, Grober E, Lo KC, Diamandis EP, Jarvi KA (2011) Proteomic analysis of seminal plasma from normal volunteers and post-vasectomy patients identifies over 2000 proteins and candidate biomarkers of the urogenital system. J Proteome Res 10(3):941–953

Batruch I, Smith CR, Mullen BJ, Grober E, Lo KC, Diamandis EP, Jarvi KA (2012) Analysis of seminal plasma from patients with non-obstructive azoospermia and identification of candidate biomarkers of male infertility. J Proteome Res 11(3):1503–1511

Bohring C, Krause W (1999) The characterization of human spermatozoa membrane proteins-surface antigens and immunological infertility. Electrophoresis 20(4–5):971–976

Bromfield EG, McLaughlin EA, Aitken RJ, Nixon B (2016) Heat shock protein member A2 forms a stable complex with angiotensin converting enzyme and protein disulfide isomerase A6 in human spermatozoa. Mol Hum Reprod 22(2):93–109

Bhagwat S, Dalvi V, Chandrasekhar D, Matthew T, Acharya K, Gajbhiye R, Kulkarni V, Sonawane S, Ghosalkar M, Parte P (2014) Acetylated α-tubulin is reduced in individuals with poor sperm motility. Fertil Steril 101(1):95–104

Chan CC, Shui HA, Wu CH, Wang CY, Sun GH, Chen HM, Wu GJ (2009) Motility and protein phosphorylation in healthy and asthenozoospermic sperm. J Proteome Res 8(11):5382–5386
Chang MC (1951) Fertilizing capacity of spermatozoa deposited into the fallopian tubes. Nature 168(4277):697–698
Davalieva K, Kiprijanovska S, Noveski P, Plaseska-Karanfilska D, Plaseski T, Kocevska B (2012) Human seminal plasma proteome study: a search for male infertility biomarkers. Balkan J Med Genetics 15(Supplement):35–38
de Mateo S, Castillo J, Estanyol JM, Ballescà JL, Oliva R (2011) Proteomic characterization of the human sperm nucleus. Proteomics 11(13):2714–2726
de Mateo S, Estanyol JM, Oliva R (2013) Methods for the analysis of the sperm proteome. Methods Mol Biol 927:411–422
Drabovich AP, Dimitromanolakis A, Saraon P, Soosaipillai A, Batruch I, Mullen B, Jarvi K, Diamandis EP (2013) Differential diagnosis of azoospermia with proteomic biomarkers ECM1 and TEX101 quantified in seminal plasma. Sci Transl Med 5(212):1–10
Drabovich AP, Saraon P, Jarvi K, Diamandis EP (2014) Seminal plasma as a diagnostic fluid for male reproductive system disorders. Nat Rev Urol 11(5):278–288
Edwards JJ, Tollaksen SL, Anderson NG (1981) Proteins of human semen. I Two-dimensional mapping of human seminal fluid. Clin Chem 27(8):1335–1340
Ficarro S, Chertihin O, Westbrook VA, White F, Jayes F, Kalab P, Marto JA, Shabanowitz J, Herr JC, Hunt DF, Visconti PE (2003) Phosphoproteome analysis of capacitated human sperm evidence of tyrosine phosphorylation of a kinase-anchoring protein 3 and valosin-containing protein/p97 during capacitation. J Biol Chem 278(13):11579–11589
Florman HM, Storey BT (1982) Mouse gamete interactions: the zona pellucida is the site of the acrosome reaction leading to fertilization in vitro. Dev Biol 91(1):121–130
Frapsauce C, Pionneau C, Bouley J, de Larouziere V, Berthaut I, Ravel C, Antoine JM, Soubrier F, Mandelbaum J (2009) Unexpected in vitro fertilization failure in patients with normal sperm: a proteomic analysis. Gynecol Obstet Fertil 37(10):796–802
Fung KY, Glode LM, Green S, Duncan MW (2004) A comprehensive characterization of the peptide and protein constituents of human seminal fluid. Prostate 61(2):171–181
Giacomini E, Ura B, Giolo E, Luppi S, Martinelli M, Garcia RC, Ricci G (2015) Comparative analysis of the seminal plasma proteomes of oligoasthenozoospermic and normozoospermic men. Reprod Biomed Online 30(5):522–531
Goverde HJM, Bisseling JGA, Wetzels AMM, Braat DDM, Pesman GJ, Sweep FCGJ, Meuleman EJH (1998) A neuropeptide in human semen: oxytocin. Arch Androl 41(1):17–22
Gray S, Huggins C (1942) Electrophoretic analysis of human semen. Exp Biol Med 50(2):351–353
Hashemitabar M, Sabbagh S, Orazizadeh M, Ghadiri A, Bahmanzadeh M (2015) A proteomic analysis on human sperm tail: comparison between normozoospermia and asthenozoospermia. J Assist Reprod Genet 32(6):853–863
Hassan MI, Kumar V, Singh TP, Yadav S (2008) Purification and characterization of zinc alpha2-glycoprotein-prolactin inducible protein complex from human seminal plasma. J Sep Sci 31(12):2318–2324
Herwig R, Knoll C, Planyavsky M, Pourbiabany A, Greilberger J, Bennett KL (2013) Proteomic analysis of seminal plasma from infertile patients with oligoasthenoteratozoospermia due to oxidative stress and comparison with fertile volunteers. Fertil Steril 100(2):355–366
Holland A, Ohlendieck K (2015) Comparative profiling of the sperm proteome. Proteomics 15(4):632–648
Jarow JP, Espeland MA, Lipshultz LI (1989) Evaluation of the azoospermic patient. J Urol 142(1):62–65
Johnston DS, Wooters JOE, Kopf GS, Qiu Y, Roberts KP (2005) Analysis of the human sperm proteome. Ann N Y Acad Sci 1061(1):190–202
Jungblut PR, Holzhütter HG, Apweiler R, Schlüter H (2008) The speciation of the proteome. Chem Cent J 2(16):1–10
Kausler W, Spiteller G (1992) Analysis of peptides of human seminal plasma by mass spectrometry. Biol Mass Spectrom 21(11):567–575

Kim YH, Haidl G, Schaefer M, Egner U, Mandal A, Herr JC (2007) Compartmentalization of a unique ADP/ATP carrier protein SFEC (sperm flagellar energy carrier, AAC4) with glycolytic enzymes in the fibrous sheath of the human sperm flagellar principal piece. Dev Biol 302(2):463–476

Kolialexi A, Mavrou A, Spyrou G, Tsangaris GT (2008) Mass spectrometry-based proteomics in reproductive medicine. Mass Spectrom Rev 27(6):624–634

Kumar V, Hassan M, Kashav T, Singh TP, Yadav S (2008) Heparin-binding proteins of human seminal plasma: purification and characterization. Mol Reprod Dev 75(12):1767–1774

Kumar S, Tomar AK, Singh S, Saraswat M, Singh S, Singh TP, Yadav S (2012) Human serum albumin as a new interacting partner of prolactin inducible protein in human seminal plasma. Int J Biol Macromol 50(2):317–322

Li LW, Fan LQ, Zhu WB, Nie HC, Sun BL, Luo KL, Liao TT, Tang L, Lu GX (2007) Establishment of a high-resolution 2-D reference map of human spermatozoal proteins from 12 fertile sperm-bank donors. Asian J Androl 9(3):321–329

Liao TT, Xiang Z, Zhu WB, Fan LQ (2009) Proteome analysis of round-headed and normal spermatozoa by 2-D fluorescence difference gel electrophoresis and mass spectrometry. Asian J Androl 11(6):683–693

Liu DY, Baker HW (2000) Defective sperm–zona pellucida interaction: a major cause of failure of fertilization in clinical in-vitro fertilization. Hum Reprod 15(3):702–708

Liu DY, Baker HW (2003) Disordered zona pellucida–induced acrosome reaction and failure of in vitro fertilization in patients with unexplained infertility. Fertil Steril 79(1):74–80

Liu Y, Guo Y, Song N, Fan Y, Li K, Teng X, Guo Q, Ding Z (2015a) Proteomic pattern changes associated with obesity-induced asthenozoospermia. Andrology 3(2):247–259

Liu FJ, Liu X, Han JL, Wang YW, Jin SH, Liu XX, Liu J, Wang WT, Wang WJ (2015b) Aged men share the sperm protein PATE1 defect with young asthenozoospermia patients. Hum Reprod 30(4):861–869

Mann M, Ong SE, Grønborg M, Steen H, Jensen ON, Pandey A (2002) Analysis of protein phosphorylation using mass spectrometry: deciphering the phosphoproteome. Trends Biotechnol 20(6):261–268

Martínez-Heredia J, Estanyol JM, Ballescà JL, Oliva R (2006) Proteomic identification of human sperm proteins. Proteomics 6(15):4356–4369

Martínez-Heredia J, de Mateo S, Vidal-Taboada JM, Ballescà JL, Oliva R (2008) Identification of proteomic differences in asthenozoospermic sperm samples. Hum Reprod 23(4):783–791

Milardi D, Grande G, Vincenzoni F, Messana I, Pontecorvi A, De Marinis L, Castagnola M, Marana R (2012) Proteomic approach in the identification of fertility pattern in seminal plasma of fertile men. Fertil Steril 97(1):67–73

Naaby-Hansen S (1990) Electrophoretic map of acidic and neutral human spermatozoal proteins. J Reprod Immunol 17(3):167–185

Naaby-Hansen S, Herr JC (2010) Heat shock proteins on the human sperm surface. J Reprod Immunol 84(1):32–40

Naaby-Hansen S, Flickinger CJ, Herr JC (1997) Two-dimensional gel electrophoretic analysis of vectorially labeled surface proteins of human spermatozoa. Biol Reprod 56(3):771–787

Naaby-Hansen S, Diekman A, Shetty J, Flickinger CJ, Westbrook A, Herr JC (2010) Identification of calcium-binding proteins associated with the human sperm plasma membrane. Reprod Biol Endocrinol 8(6):1–12

Nixon B, Bromfield EG, Dun MD, Redgrove KA, McLaughlin EA, Aitken RJ (2015) The role of the molecular chaperone heat shock protein A2 (HSPA2) in regulating human sperm-egg recognition. Asian J Androl 17(4):568–573

O'Mahony OA, Djahanbahkch O, Mahmood T, Puddefoot JR, Vinson GP (2000) Angiotensin II in human seminal fluid. Hum Reprod 15(6):1345–1349

Pagel O, Loroch S, Sickmann A, Zahedi RP (2015) Current strategies and findings in clinically relevant post-translational modification-specific proteomics. Expert Rev Proteomics 12(3):235–253

Parte PP, Rao P, Redij S, Lobo V, D'Souza SJ, Gajbhiye R, Kulkarni V (2012) Sperm phosphoproteome profiling by ultra performance liquid chromatography followed by data independent analysis (LC–MS E) reveals altered proteomic signatures in asthenozoospermia. J Proteomics 75(18):5861–5871

Pilch B, Mann M (2006) Large-scale and high-confidence proteomic analysis of human seminal plasma. Genome Biol 7(5):R40 1–10

Pixton KL, Deeks ED, Flesch FM, Moseley FL, Björndahl L, Ashton PR, Barratt CL, Brewis IA (2004) Sperm proteome mapping of a patient who experienced failed fertilization at IVF reveals altered expression of at least 20 proteins compared with fertile donors: case report. Hum Reprod 19(6):1438–1447

Redgrove KA, Anderson AL, Dun MD, McLaughlin EA, O'Bryan MK, Aitken RJ, Nixon B (2011) Involvement of multimeric protein complexes in mediating the capacitation-dependent binding of human spermatozoa to homologous zonae pellucidae. Dev Biol 356(2):460–474

Redgrove KA, Nixon B, Baker MA, Hetherington L, Baker G, Liu DY, Aitken RJ (2012) The molecular chaperone HSPA2 plays a key role in regulating the expression of sperm surface receptors that mediate sperm-egg recognition. PLoS One 7(11):e50851

Ross V, Moore DH, Miller EG (1942) Proteins of human seminal plasma. J Biol Chem 144(3):667–677

Saraswat M, Joenvaara S, Tomar AK, Singh S, Yadav S, Renkonen R (2016) N-glycoproteomics of human seminal plasma glycoproteins. J Proteome Res 15(3):991–1001

Saraswat M, Joenväärä S, Jain T, Tomar AK, Sinha A, Singh S, Yadav S, Renkonen R (2017) Human spermatozoa quantitative proteomic signature classifies normo- and asthenozoospermia. Mol Cell Proteomics 16(1):57–72

Secciani F, Bianchi L, Ermini L, Cianti R, Armini A, La Sala GB, Focarelli R, Bini L, Rosati F (2009) Protein profile of capacitated versus ejaculated human sperm. J Proteome Res 8(7):3377–3389

Sensabaugh GF (1978) Isolation and characterization of a semen-specific protein from human seminal plasma: a potential new marker for semen identification. J Forensic Sci 23(1):106–115

Sharma R, Agarwal A, Mohanty G, Jesudasan R, Gopalan B, Willard B, Yadav SP, Sabanegh E (2013a) Functional proteomic analysis of seminal plasma proteins in men with various semen parameters. Reprod Biol Endocrinol 11(38):1–20

Sharma R, Agarwal A, Mohanty G, Du Plessis SS, Gopalan B, Willard B, Yadav SP, Sabanegh E (2013b) Proteomic analysis of seminal fluid from men exhibiting oxidative stress. Reprod Biol Endocrinol 11(85):1–15

Shen S, Wang J, Liang J, He D (2013) Comparative proteomic study between human normal motility sperm and idiopathic asthenozoospermia. World J Urol 31(6):1395–1401

Shetty J, Diekman AB, Jayes FC, Sherman NE, Naaby-Hansen S, Flickinger CJ, Herr JC (2001) Differential extraction and enrichment of human sperm surface proteins in a proteome: identification of immunocontraceptive candidates. Electrophoresis 22(14):3053–3066

Siva AB, Kameshwari DB, Singh V, Pavani K, Sundaram CS, Rangaraj N, Deenadayal M, Shivaji S (2010) Proteomics-based study on asthenozoospermia: differential expression of proteasome alpha complex. Mol Hum Reprod 16(7):452–462

Starita-Geribaldi M, Poggioli S, Zucchini M, Garin J, Chevallier D, Fenichel P, Pointis G (2001) Mapping of seminal plasma proteins by two-dimensional gel electrophoresis in men with normal and impaired spermatogenesis. Mol Hum Reprod 7(8):715–722

Sun G, Jiang M, Zhou T, Guo Y, Cui Y, Guo X, Sha J (2014) Insights into the lysine acetylproteome of human sperm. J Proteomics 109:199–211

Thacker S, Yadav SP, Sharma RK, Kashou A, Willard B, Zhang D, Agarwal A (2011) Evaluation of sperm proteins in infertile men: a proteomic approach. Fertil Steril 95(8):2745–2748

Tomar AK, Sooch BS, Raj I, Singh S, Singh TP, Yadav S (2011) Isolation and identification of Concanavalin A binding glycoproteins from human seminal plasma: a step towards identification of male infertility marker proteins. Dis Markers 31(6):379–386

Tomar AK, Sooch BS, Singh S, Yadav S (2012) Differential proteomics of human seminal plasma: a potential target for searching male infertility marker proteins. Proteomics Clin Appl 6(3–4):147–151

Tomar AK, Sooch BS, Raj I, Singh S, Yadav S (2013) Interaction analysis identifies semenogelin I fragments as new binding partners of PIP in human seminal plasma. Int J Biol Macromol 52:296–299

Turek PJ (2005) Practical approaches to the diagnosis and management of male infertility. Nat Clin Pract Urol 2(5):226–238

Upadhyay RD, Balasinor NH, Kumar AV, Sachdeva G, Parte P, Dumasia K (2013) Proteomics in reproductive biology: beacon for unraveling the molecular complexities. Biochim Biophys Acta 1834(1):8–15

Utleg AG, Yi EC, Xie T, Shannon P, White JT, Goodlett DR, Hood L, Lin B (2003) Proteomic analysis of human prostasomes. Prostate 56(2):150–161

Vigodner M, Shrivastava V, Gutstein LE, Schneider J, Nieves E, Goldstein M, Feliciano M, Callaway M (2013) Localization and identification of sumoylated proteins in human sperm: excessive sumoylation is a marker of defective spermatozoa. Hum Reprod 28(1):210–223

Wang J, Wang J, Zhang HR, Shi HJ, Ma D, Zhao HX, Lin B, Li RS (2009) Proteomic analysis of seminal plasma from asthenozoospermia patients reveals proteins that affect oxidative stress responses and semen quality. Asian J Androl 11(4):484–491

Wang G, Wu Y, Zhou T, Guo Y, Zheng B, Wang J, Bi Y, Liu F, Zhou Z, Guo X, Sha J (2013) Mapping of the N-linked glycoproteome of human spermatozoa. J Proteome Res 12(12):5750–5759

Wang J, Qi L, Huang S, Zhou T, Guo Y, Wang G, Guo X, Zhou Z, Sha J (2015) Quantitative phosphoproteomics analysis reveals a key role of insulin growth factor 1 receptor (IGF1R) tyrosine kinase in human sperm capacitation. Mol Cell Proteomics 14(4):1104–1112

Xu Wl, Hu H, Wang Z, Chen X, Yang F, Zhu Z, Fang P, Dai J, Wang L, Shi H, Li Z, Qiao Z (2012) Proteomic characteristics of spermatozoa in normozoospermic patients with infertility. J Proteomics 75(17):5426–5436

Yadav VK, Kumar V, Chhikara N, Kumar S, Manral P, Kashav T, Saini S, Srinivasan A, Singh S, Singh TP, Yadav S (2011) Purification and characterization of a native zinc-binding high molecular weight multiprotein complex from human seminal plasma. J Sep Sci 34(9):1076–1083

Yadav VK, Chhikara N, Gill K, Dey S, Singh S, Yadav S (2013) Three low molecular weight cysteine proteinase inhibitors of human seminal fluid: purification and enzyme kinetic properties. Biochimie 95(8):1552–1559

Yadav VK, Mandal RS, Puniya BL, Singh S, Yadav S (2015) Studies on the interactions of SAP-1 (an N-terminal truncated form of cystatin S) with its binding partners by CD-spectroscopic and molecular docking methods. J Biomol Struct Dyn 33(1):147–157

Yamakawa K, Yoshida K, Nishikawa H, Kato T, Iwamoto T (2007) Comparative analysis of interindividual variations in the seminal plasma proteome of fertile men with identification of potential markers for azoospermia in infertile patients. J Androl 28(6):858–865

Yu H, Diao H, Wang C, Lin Y, Yu F, Lu H, Xu W, Li Z, Shi H, Zhao S, Zhou Y (2015) Acetylproteomic analysis reveals functional implications of lysine acetylation in human spermatozoa (sperm). Mol Cell Proteomics 14(4):1009–1023

Zhao C, Huo R, Wang FQ, Lin M, Zhou ZM, Sha JH (2007) Identification of several proteins involved in regulation of sperm motility by proteomic analysis. Fertil Steril 87(2):436–438

Part III

Management of Male Infertility

Advancing Paternal Age: The Ticking Biological Clock

19

Rima Dada and Vidhu Dhawan

Abstract

Since spermatogenesis is a continuous process that occurs throughout life, paternal age often gets less consideration for childbearing. However, a number of studies have reported a significant impact of advanced paternal age on the time to pregnancy, adverse pregnancy outcomes and the birth of children with congenital deformities. This chapter provides an overview of the impact of advanced paternal age on the loss of fertility and increased likelihood of passing birth defects and genomic changes that can have significant impact on the coming generations.

Keywords

Paternal age • Age and infertility • Age and congenital abnormalities

Key Points

- Testosterone level declines by 0.4–2% per year, resulting in a decrease in libido and spermatogenesis.
- There is evidence of epigenetic changes with ageing that may affect the quality of gametes.
- DNA accumulates defects with ageing that are more likely to pass on to the next generation.
- Advanced paternal age is more likely to contribute to congenital defects.

R. Dada (✉) • V. Dhawan
Laboratory for Molecular Reproduction and Genetics,
Department of Anatomy, AIIMS, New Delhi, India
e-mail: rima_dada@rediffmail.com

R. Singh, K. Singh (eds.), *Male Infertility: Understanding, Causes and Treatment*, DOI 10.1007/978-981-10-4017-7_19

19.1 Introduction

Over the last few decade, the trend for advanced parental age has increased in couples due to professional and social commitments. There is a precipitous decline in female fecundity after 30–35 years of age with an increase in various adverse reproductive events from infertility, pregnancy complications to perinatal morbidity and mortality. The 'ticking biological clock' for women is well understood, but this clock ticks for men too and appears to tick faster (Crow 2000; Thacker 2004). The anecdotal reports about older fathers have always sounded fascinating, and the oldest paternity has been noted scientifically as 94 years (Seymour et al. 1935). However, the advanced paternal age adversely affects testicular functions, semen parameters, sperm DNA integrity and sperm telomere length and increases de novo mutation rate and chromosomal and epigenetic alterations. Accumulated chromosomal aberrations and mutations in male germ cells may lead to the increased risk of reduced fertility, poor implantation and pregnancy rates and an increased risk of birth defects and childhood disease burden.

A steep increase in the age of childbearing by men tends to invite a host of negative reproductive outcomes. According to CDC birth statistics, the birth rate for men 25–44 years is increasing with a decline seen in men <25 years (Hamilton et al. 2003). There is a strong association between paternal age and extension of sperm DNA strand breaks. With increased age, there is a concomitant increase in DNA damage with increase in the incidence of semen abnormalities (Singh et al. 2003; Moskovtsev et al. 2006, 2009). With age, the incidence of mutations in spermatozoa rises due to repeated premeiotic cell divisions, decreased antioxidant capacity and other diseases which are more likely to appear with ageing. This adversely affects embryogenesis because advanced paternal age is also associated with advanced maternal age and suboptimal quality of oocyte. This may lead to incomplete, inefficient aberrant repair of sperm DNA damage by oocyte postfertilization. The advanced age at the planning of the first child not only decreases the chances of conception but also increases the risk of DNA damage, gene deletions and chromosomal aneuploidies. A host of these factors are associated with pregnancy loss or the birth of children with congenital abnormalities. This chapter highlights the impact of advanced paternal age on fertility, its possible odd outcomes and the need to counsel couples for timely planning of family.

19.2 Testicular Morphology and Semen Parameters

Age-related morphological changes in testes affect spermatogenic efficiency, characterized by a predominance of multinucleated spermatogonia, megalospermatocytes, giant spermatids along with seminiferous tubular diverticula and thickening of the basal membrane (Johnson et al. 1988; Kuhnert and Nieschlag 2004; Dakouane et al. 2005). Testicular sclerosis occurs as a result of defective vascularization in senile testis and systemic arteriosclerosis (Sasano and Ichijo 1969). A decrease in testicular volume is attributed to a decrease in both Sertoli cells and Leydig cells. The Sertoli cells are seen to accumulate cytoplasmic lipid droplets, and Leydig cells may also be multinucleated (Johnson 1986; Holstein 1989). A pioneering study by Auger et al. (1995) found a 2.6% decline in sperm concentration, a 0.3% decline in motile sperm number and a 0.7% decline in the percentage of normal sperm morphology, with increased paternal age. There is a decrease in seminal volume and seminal fructose concentration with age, whereas zinc and α-glucosidase, secreted by prostate and epididymis, remain constant (Rolf et al. 1996).

19.3 Testicular Functions and Reproductive Hormones

Alterations in testicular functions develop gradually, which has been considered as a rough indicator of spermatogenesis. The testicular volume remains constant over quite a long time and is documented to decrease only in the eighth decade of life (Kuhnert and Nieschlag 2004). The changes in testicular volume are seen to be associated with the levels of follicle-stimulating hormone (FSH), inhibin B and testosterone. The resultant increased FSH levels associate with a decrease in the inhibin B/FSH ratio as well as a decrease in Sertoli cell mass and testosterone levels in the testis (Weiner-Megnazi et al. 2012). Testosterone levels peak around 20 years of age (Bhasin and Buckwalter 2001) and continue to decline by about 0.4–2% every year after 30 years of age (Feldman et al. 2002; Abram McBride et al. 2016). Decreasing testosterone levels were quantified by Massachusetts Male Ageing Study (MMAS), as a cross-sectional decline of 0.8%/year of age and a longitudinal decline of 1.6%/year, over a 10-year follow-up data (Morley et al. 1997).

The decrease in testosterone levels with age has led to the development of 'late-onset' hypogonadism (LOH) in contrast to hypogonadism, which is a more general term referring to the state with decreased testicular volume, impaired sperm production and low testosterone levels. Age-related below-normal testosterone levels and associated symptoms have also been addressed as 'andropause', 'symptomatic androgen deficiency', 'age-related hypogonadism' and 'testosterone deficiency (TD)'. This hypogonadism in ageing men is characterized by poor libido, fatigue and loss of cognitive functions (Sharma et al. 2015; Abram McBride et al. 2016). Testosterone production is regulated by the hypothalamic–pituitary–gonadal (HPG) axis via the production of luteinizing hormone (LH) (Abram McBride et al. 2016). Failure in this delicate balance can result in primary, secondary or mixed hypogonadism. The predominant form of testosterone deficiency in ageing men is mixed with primary and secondary hypogonadism components. The levels of luteinizing hormone may vary with age due to the decrease in Leydig cell number and subsequent decrease in sensitivity of the HPG axis to feedback inhibition and/or decreased LH pulse amplitude despite normal pulse frequency. The decrease in LH pulse amplitude may subsequently be related to decreased neuronal cell secretion of gonadotrophin-releasing hormone (Kaufman and Vermeulen 2005).

19.4 Spermatozoa Have Limited Repair Capacity

The aetiology of DNA damage in the spermatozoa is complex and is chiefly induced by oxidative stress and thus making it vital to understand the nature and extent of DNA damage. Oxidative stress is one of the major causes of defective sperm function, which disrupts the sperm DNA integrity, induces single double-strand breaks, shortens telomeres, alters sperm methylome, oxidizes DNA bases and interstand and intrastand crosslinking and also causes fragmentation of mt and nuclear DNA also (Shamsi et al. 2008; Mishra et al. 2014). It thus limits the fertilizing potential because of parallel damage to lipids and proteins in the sperm plasma membrane. Spermatozoa are particularly vulnerable to lipid peroxidation because they have high concentrations of unsaturated fatty acids, which further triggers the mitochondria for the generation of high levels of superoxide anion as a prelude to entering the intrinsic apoptotic cascade (Aitken and De Iullis 2010; Aitken et al. 2012, 2013).

Unfortunately, spermatozoa have very little capacity to respond to such an attack because they have a highly truncated base excision repair mechanism as they only possess the first enzyme in the base excision repair (BER) pathway, 8-oxoguanineglycosylase 1 (OGG1). The latter successfully creates an abasic site, but the spermatozoa cannot process the oxidative lesion further because of the lack of downstream proteins (APE1, XRCC1) needed to complete the repair process. These are repaired only by oocyte at the time of fertilization. However, ageing oocyte and the presence of extensive sperm DNA damage in ageing sperm may overwhelm the oocyte repair mechanism postfertilization with persistence of these mutagenic bases in the child, and that may be an alternative explanation for paternal age effects (Smith et al. 2013).

19.5 Paternal Age and Mutations in Sperm DNA

The female fertility reaches a natural limit marked by the occurrence of menopause, which leads to cessation of ovarian function due to inevitable loss of female gametes. The male reproductive functions have been observed to decline gradually with age and spermatogenesis throughout life. Each oocyte produced by the female undergoes 22 germ line divisions and two meiotic divisions, and regardless of her age, female oocytes do not undergo any further divisions after that. In males, as spermatogenesis is continuous throughout life, ageing increases the number of cell divisions, hence increased number of chromosomal replications with advanced age reaching a number of 840 replications by the age of 50 years. This increase in the number of cycles of DNA replication with advanced paternal age brings more copy-error mutations as per the Penrose's *copy-error hypothesis* (Penrose 1955).

Fathers bequeath more mutations with advanced age, and the germ line mutation rate is higher than in females, mainly because of many more germ-cell divisions. As compared to females, the number of cell divisions in males is seven times higher at the age of 20 and 25 times higher at the age of 40 (Crow 2000; Taylor et al. 2006). The mutation rate tends to further increase due to the decrease in sperm DNA integrity with ageing and accumulation of highly mutagenic oxidized DNA adducts like

8-hydroxy 2-deoxy guanosine. Oxidative DNA damage gets accumulated with advanced paternal age and also by the exposure to various exogenous and lifestyle factors like smoking, excess alcohol intake, psychological stress, increased BMI and exposure to xenobiotics and infections (Kumar et al. 2015). In a genome-wide association study by Kong et al. (2012), the average de novo mutation rate was found to be 1.20×10^{-8} per nucleotide per generation with an average father's age of 29.7 years. There is a striking effect of over two additional mutations every year or an exponential effect of doubling of paternal mutations every 16.5 years (Kong et al. 2012). The effect of hazardous environmental conditions and various demographic characteristics driven by forces of genetic drift, gene flow and natural selection cannot be negated.

The mutation rate for base substitutions is much higher in the ageing male. The critical phase for induction of de novo mutations is the post-meiotic events during spermiogenesis (Wyrobek et al. 2006; Crow 2006). NR5A1 nuclear receptor, also known as steroidogenic factor 1 mutations, has been reported in 46,XY disorders of sex development in 4% men with unexplained severe spermatogenic failure (Bashamboo et al. 2010). Men with non-obstructive azoospermia were reported to have de novo point mutations in Y-chromosomal gene USP9Y (Sun et al. 1999). On the contrary, small deletions or rearrangements do not show the paternal age effect. This has been observed in larger genes encoding for neurofibromatosis, Duchenne muscular dystrophy, Wilms' tumour or retinoblastoma (Crow 2000). Ongoing studies in our lab (Kumar et al. 2015) in fathers of children with nonfamilial sporadic heritable retinoblastoma (RB) showed higher levels of oxidative DNA adducts in blood of children with RB who were born to fathers who smoked or who were above 35 years of age. Advanced age in fathers and limited detection of DNA damage and repair in sperm and its dependence on oocyte to repair DNA damage may result in incomplete removal of DNA lesions due to suboptimal quality of oocyte (associated with advanced maternal age) and extensive DNA damage thus persists.

A variable incidence of different dominant mutations due to varied base substitutions and deletions has been observed in children as the age of the father increased. Advanced maternal age has been reported as the only well-documented non-genetic risk factor for trisomies in humans, but recent studies have found that trisomy 21 is primarily associated with advanced paternal age when the female partner is >35 years of age (Sartorius and Nieschlag 2010). Advanced paternal age has not been associated with trisomy 18 and even less likely with trisomy 13. No relationship of advanced paternal age has been observed with the birth of an offspring with anencephaly or encephalocele. No significant relationship between either maternal or paternal age has been observed in Klinefelter's syndrome as well. Nevertheless, an increase in the likelihood of these disorders with advancing paternal age cannot be denied.

19.6 Paternal Age Effect (PAE) Disorders

The first remarkable statement about the association of paternal age with birth disorders was given by Wilhelm Weinberg in 1912 when he noticed the sporadic cases of achondroplasia in the last-born children of sibship. This was further strengthened 40 years later by Penrose who gave the 'copy-error hypothesis' owing to more number of germ line mutations in men. Paternal age effect (PAE) disorders are small group of such type of rare disorders with an increased risk for spontaneous congenital disorders and common complex diseases (some cancers, schizophrenia, autism, bipolar disorder) (Goriely and Wilkie 2012; Goriely et al. 2013). Replication error is not the only underlying mechanism in such disorders. The common factor in these disorders lies in dysregulation of spermatogonial cell behaviour with an effect mediated by specific mutations in genes encoding components of the tyrosine kinase receptor/RAS/MAPK signalling pathways (Maher et al. 2014). These are random mutations occurring during mitotic divisions of spermatogonial stem cells (SSCs) that confer a selective/growth advantage on mutant SSCs, leading to a clonal expansion of mutant cells significantly above the background mutation rate. The clonal expansion takes place in the testes of all men, leading to the relative enrichment of mutant sperm over time. This phenomenon is known as *Selfish Spermatogonial Selection* and skews the mutational profile of sperm as men age, enriching the de novo mutations in offsprings of older fathers (Goriely et al. 2013).

Nine autosomal-dominant disorders (Apert, Crouzon, Pfeiffer and Muenke syndromes, achondroplasia, Costello and Noonan syndromes and multiple endocrine neoplasia types 2A and 2B), corresponding to specific point mutations within five genes (FGFR2, FGFR3, HRAS, PTPN11and RET), have been ascribed to the PAE disorders. 99% of individuals with Apert syndrome carry either of the two transversions (c.755C>G or c.758C>G), encoding substitutions in two adjacent amino acids (p. Ser252Trp or p. Pro253Arg, respectively) located within the extracellular region of the receptor tyrosine kinase protein fibroblast growth factor receptor-2 (FGFR2) (Wilkie et al. 1995). It is characterized by craniosynostosis (premature fusion of the cranial sutures) and severe syndactyly of both hands and feet. Single nucleotide substitution mutation (encoding a p. Gly380Arg mutant protein) in FGFR3 causes more than 95% of achondroplasia cases (Rousseau et al. 1994), which is the most common cause of short-limbed dwarfism.

Crouzon and Pfeiffer syndromes are witnessed to overlap clinically and are caused by any of more than 50 specific activating point mutations in fibroblast growth factor receptor 2 (FGFR2) gene. Craniosynostosis is seen to occur in Apert syndrome, but limb abnormalities are milder (Kan et al. 2002). Muenke's syndrome develops because of a single c.749C>G transversion in FGFR3 (resulting in a point substitution Pro250Arg equivalent to the FGFR2 Apert-causing Pro253Arg) and is the most common genetic cause of coronal craniosynostosis (Vajo et al. 2000). Costello and Noonan syndromes are a part of neuro–cardio–facial cutaneous syndromes or RASopathies and present with variable combinations of distinctive craniofacial features, short stature, failure to thrive, developmental delay and skin, cardiac and skeletal abnormalities (Aoki et al. 2008). 90% of Costello syndrome patients have the c.34G>A transition in HRAS (Gly12Ser) at a well-known mutation hotspot in tumorigenesis, while ~50% of Noonan syndrome mutations are detected within the PTPN11 gene (encoding SHP2-containing tyrosine phosphatase). The last two PAE disorders, multiple endocrine neoplasia types 2A (Men2A) and 2B (Men2B), are caused by allelic mutations within the RET receptor tyrosine kinase (Aoki et al. 2008; Tartaglia et al. 2010).

A noticeable phenotypic overlap is observed between different PAE syndromes though they are clearly distinct and have well-defined complex pathological entity. These features highlight the pleiotropic role played by the PAE genes during development, whereas the clinical overlaps of these features point out to the fact that these genes are required in common cellular contexts in shared molecular pathways.

19.7 Paternal Age, Sperm DNA Integrity and Reproductive Outcomes

Moskovtsev et al. (2006) reported that there was twice an increase in DNA fragmentation index (DFI) from less than 30 years of age (15.2%) to ≥45 years of age (32.0). In an ongoing study in our department, we observed that an increase in DNA damage is associated with decreased probability of conception with increase in time to pregnancy. Decreased sperm DNA integrity is associated with an increased risk of recurrent miscarriages, congenital birth defects and childhood carcinomas. Advanced paternal age has been seen to affect the rates of fertilization, implantation, pregnancy and miscarriage. The impact of paternal age on the seminal oxidative stress and DNA integrity in our laboratory showed an increase in seminal ROS from 58.3 to 115.7 relative light units (RLU)/s/million sperm and an increase in DFI from 32.6 to 42.3% from 2 to 40 years of age. Damage to the germ cells entering meiosis will precipitate an increase in apoptosis, thus making the sperm cell susceptible to accumulate damage to its genome and epigenome right from the time they are formed till conception (Tremellon 2008; Dada et al. 2012; Aitken et al. 2012, 2013).

19.8 Increased Telomere Length in Offsprings of Old Fathers

The response of the germ cells to an increase in stress is up-regulation of telomerase activity and increase in telomere length of the spermatozoa. Milder level of oxidative stress may thus compensate, and this is one of the effects which may favour the survival of the offspring by increasing the telomere length. The telomere length is a paternally inherited trait so the offsprings of ageing fathers will have longer telomeres and this can be explained as a biological resistance to ageing process (Unryn et al. 2005). A strong and positive correlation has been showed between increasing paternal age and telomere length (Aston et al. 2012). By contrast if the germ cells are exposed to oxidative stress post-meiotically as in cases of infertility patients undergoing ART, the telomerase can no longer increase and the telomere length will be abnormally short, posing serious health hazards for the offspring.

With age the telomeres of leukocytes tend to decrease, while that of sperm tend to increase in length (Aston et al. 2012). Consistent with increase in sperm telomere length, a correlation between paternal age at birth and leukocyte telomere length of the individuals has been reported (Prescott et al. 2012). The paternal age contribution to offspring leukocyte telomere length is stronger than the maternal contribution (Broer et al. 2013), and the paternal age shows cumulative effect across generations (Eisenberg et al. 2012). A recent study analysed leukocyte and sperm telomere length in the same individual in relation to spermatogenic activity and parents' age at birth by recruiting 18–19 years old high school students. Sperm and leukocyte telomere length showed correlation, but sperm telomere length was significantly longer. Also, a positive correlation between sperm telomere length and total sperm number was observed. This increase telomere length may have health implications, though they do not seem to affect the risk of cancer (Chang 2012).

19.9 Age-Related Changes in Sperm Epigenome

Epigenetics is a stable heritable modification on histone tails but not the DNA sequence that leads to altered gene expression. The sperm cell has a highly differentiated and specialized morphology, and the epigenome of human sperm matters for embryogenesis. Epigenetic factors suggest that sperm play diverse and critical roles in embryonic development. Methylation of cytosine residues, typically found at cytosine phosphate guanine dinucleotides (CpGs), in the DNA by DNMTs (DMA methyl transferases) is the most important mechanism regulating the process of gene expression and capable of regulatory control over gene activation or silencing. DNA hypomethylation is associated with gene transcriptional activity, whereas hypermethylation is associated with gene silencing activity (Carrell and Hammoud 2010; Carrell 2012). Epigenetic patterns are shown to be silenced/disrupted by various environmental and endogenous factors such as age, diet and lifestyle factors, including smoking or drug intake (Sharma et al. 2015). These epigenetic events may impair or inhibit key steps of fertilization, implantation and/or embryo development.

Epigenetic modifications have been shown to not only affect normal cellular function but also to be involved in ageing and cancer and as a mechanism where environmental influences come into play. Jenkins et al. studied the impact of ageing on DNA methylation in 17 fertile donors by methylation array approach. They identified 139 regions in sperm DNA that were hypomethylated and eight regions that were significantly hypermethylated with age. They reported that 117 genes were associated with these regions and a portion of age-related changes in sperm DNA methylation were located at genes associated with schizophrenia and bipolar disorder (Jenkins et al. 2014). Epigenetic marks within sperm are specifically associated with genes that regulate transcription and developmental processes. As the embryo grows, these imprints are maintained in somatic tissues, but erased in primordial germ cells so that imprints can be re-established during gametogenesis. The various methylation changes during development make the epigenome vulnerable to interference from environmental exposure (Kumar et al. 2015). Epigenetic programming plays an important role in an organism's response to environmental stress during critical developmental periods. Understanding the epigenetics of sperm can be a potential mean to decipher the mechanisms of pluripotency, which has broad implications for potential therapies.

Conclusion

In the last decade, an alarming increase in delayed marriages and delayed parenthood has been observed. This is due to several reasons like increased contraceptive use and professional pressures. Though the effect of advanced maternal age on fertility and health of the offspring is well documented, the impact of advanced paternal age on fertility is less well investigated. Hence, there is little awareness about the impact of father's age on fertility and the health of offspring. The use of sperm from old men for ART may also lead to pre- and post-implantation losses and congenital malformations. In a number of cases of miscarriage, the role of paternal age may be significant; however, investigation of this requires well-planned studies on aborted foetuses. Thus, there is a need to increase awareness to not delay childbearing as ageing affects the quality of gametes, which is usually associated with adverse pregnancy outcomes. Nevertheless, the rate of testicular ageing can be slowed by adopting a healthy lifestyle and practice of meditation and yoga.

References

Abram McBride J, Carson CC III, Coward RM (2016) Testosterone deficiency in the aging male. Ther Adv Urol 8(1):47–60

Aitken RJ, De Iuliis GN (2010) On the possible origins of DNA damage in human spermatozoa. Mol Hum Reprod 16:3–13

Aitken RJ, Jones KT, Robertson SA (2012) Reactive oxygen species and sperm function- in sickness and in health. J Androl 33(6):1096–1102

Aitken RJ, Smith TB, Lord T, Kuczera L, Koppers AJ, Naumovski N, Connaughton H, Baker MA, De Iuliis GN (2013) On methods for the detection of reactive oxygen species generation by human spermatozoa: analysis of the cellular responses to catechol oestrogen, lipid aldehyde, menadione and arachidonic acid. Andrology 1:192–205

Aitken RJ, Smith TB, Jobling MS, Baker MA, DeIuliis GN (2014) Oxidative stress and male reproductive health. Asian J Androl 16:31–38

Aoki Y, Niihori T, Narumi Y, Kure S, Matsubara Y (2008) The RAS/MAPK syndromes: novel roles of the RAS pathway in human genetic disorders. Hum Mutat 29:992–1006

Aston KI, Hunt SC, Susser E, Kimura M, Factor-Litvak P et al (2012) Divergence of sperm and leukocyte age-dependent telomere dynamics: implications for male-driven evolution of telomere length in humans. Mol Hum Reprod 18:517–522

Auger J, Kunstmann JM, Czyglik F, Jouannet P (1995) Decline in semen quality among fertile men in Paris during the past 20 years. N Engl J Med 332:281–285

Bashamboo A, Ferraz-de-Souza B, Lourenço D, Lin L, Sebire NJ, Montjean D et al (2010) Human male infertility associated with mutations in NR5A1 encoding steroidogenic factor 1. Am J Hum Genet 87:505–512

Bhasin S, Buckwalter TG (2001) Testosterone supplementation in older men: a rational idea whose time has not yet come. J Andol 22(5):718–731

Broer L, Codd V, Nyholt DR, Deelen J, Mangino M, Willemsen G et al (2013) Meta-analysis of telomere length in 19713 subjects reveals high heritability, stronger maternal inheritance and a paternal age effect. Eur J Hum Genet 21:1163–1168

Carrell DT (2012) Epigenetics of the male gamete. Fertil Steril 97(2):267–274
Carrell DT, Hammoud SS (2010) The human sperm epigenome and its potential role in embryo development. Mol Hum Reprod 16(1):37–47
Chang M (2012) Long telomeres: too much of good thing. Biomol Concepts 3(4):387–393
Crow JF (2000) The origins, patterns and implications of human spontaneous mutation. Nat Rev Genet 1:40–47
Crow JF (2006) Age and sex effects on human mutation rates: an old problem with new complexities. J Radiat Res (Tokyo) 47(Suppl B):B75–B82
Dada R et al (2012) Epigenetics and its role in male infertility. J Assist Reprod Genet 29:213–223
Dakouane M, Bicchieray L, Bergere M, Albert M, Vialard F, Selva J (2005) A histomorphometric and cytogenetic study of testis from men 29–102 years old. Fertil Steril 83:923–928
Eisenberg DT, Hayes MG, Kuzawa CW (2012) Delayed paternal age of reproduction in humans is associated with longer telomeres across two generations of descendants. Proc Natl Acad Sci U S A 109(26):10251–10256
Feldman HA, Longscope C, Derby CA et al (2002) Age trends in the level of serum testosterone and other hormones in middle-aged men: longitudinal results from the Massachusetts male aging study. J Clin Endocrinol Metab 87(2):589–598
Goriely A, Wilkie AOM (2012) Paternal age effect mutations and selfish spermatogonial selection: causes and consequences for human disease. Am J Hum Genet 90(2):175–200
Goriely A, McGrath JJ, Hultman CM, Wilkie AOM, Malaspina D (2013) "Selfish spermatogonial selection": a novel mechanism for the association between advance paternal age and neurodevelopmental disorders. Am J Psychiatry 170:599–608
Hamilton BE, Martin JA, Sutton PD (2003) Births: preliminary data for 2002. Natl Vital Stat Rep 51(11):1–20
Holstein AF (1989) Morphological evidence for the involution of spermatogenesis during senescence. In: Holstein AF (ed) Reproductive biology and medicine. Diesbach, Berlin, pp 66–77
Jenkins TG, Aston KI, Pflueger C, Cairns BR, Carrell DT (2014) Age-associated sperm DNA methylation alteration: possible implications in offspring disease susceptibility. PLoS Genet 10(7):e1004458
Johnson L (1986) Spermatogenesis and aging in the human. J Androl 7:331–354
Johnson L, Abdo JG, Petty CS, Neaves WB (1988) Effect of age on the composition of seminiferous tubular boundary tissue and on the volume of each component in humans. Fertil Steril 49:1045–1051
Kan SH, Elanko N, Johnson D, Cornejo-Roldan L, Cook J, Reich EW et al (2002) Genomic screening of fibroblast growth-factor receptor 2 reveals a wide spectrum of mutations in patients with syndromic craniosynostosis. Am J Hum Genet 70:472–486
Kaufman J, Vermeulen A (2005) The decline of androgen levels in elderly men and its clinical and therapeutic implications. Endocr Rev 26:833–876
Kong A, Frigge ML, Masson G, Besenbacher S, Sulem P, Magnusson G et al (2012) Rate of de novo mutations, father's age, and disease risk. Nature 488(7412):471–475
Kuhnert B, Nieschlag E (2004) Reproductive functions of the ageing male. Hum Reprod Update 10:327–339
Kumar S, Chawla B, Bisht S, Yadav R, Dada R (2015) Tobacco use increases oxidative DNA damage in spermpossible etiology of childhood cancer. Asian Pac J Cancer Prev 16(16): 6967–6972
Maher GJI, Goriely A, Wilkie AOM (2014) Cellular evidence for selfish spermatogonial selection in aged human testes. Andrology 2(3):304–314
Mishra S, Kranthi V, Kumar R, Malhotra N, Mohanty K, Pathak V, Dada R (2014) Oxidative damage to sperm DNA: clinical implications. Andrology 3(1):116
Morley JE, Kaiser FE, Perry HM 3rd et al (1997) Longitudinal changes in testosterone, luteinizing hormone, and follicle-stimulating hormone in healthy older men. Metabolism 46(4): 410–413

Moskovtsev SI, Willis J, Mullen JB (2006) Age-related decline in sperm deoxyribonucleic acid integrity in patients evaluated for male infertility. Fertil Steril 85:496–499

Moskovtsev SI, Willis J, White J, Mullen JB (2009) Sperm DNA damage: correlation to severity of semen abnormalities. Urology 74:789–793

Penrose LS (1955) Parental age and mutation. Lancet 269:312–313

Prescott J, Du M, Wong JY, Han J (2012) De vivo I paternal age at birth is associated with offspring leukocyte telomere length in the nurses' health study. Hum Reprod 27(12):3622–3631

Rolf C, Behre HM, Nieschlag E (1996) Reproductive parameters of older couples compared to younger men of infertile couples. Int J Androl 19:135–142

Rousseau F, Bonaventure J, Legeai-Mallet L, Pelet A, Rozet JM, Maroteaux P, Le Merrer M, Munnich A (1994) Mutations in the gene encoding fibroblast growth factor receptor-3 in achondroplasia. Nature 371:252–254

Sartorius GA, Nieschlag E (2010) Paternal age and reproduction. Hum Reprod Update 16:65–79

Sasano N, Ichijo S (1969) Vascular patterns of the human testis with special reference to its senile changes. Tohoku J Exp Med 99:269–280

Seymour F, Duffy C, Korner A (1935) A case of authentic fertility in a man of 94. JAMA 105:1423–1424

Shamsi MB, Kumar R, Dada R (2008) Evaluation of nuclear DNA damage in human spermatozoa in men opting for assisted reproduction. Indian J Med Res 127(2):115–123

Sharma R, Agarwal A, Rohra VK, Assidi M, Abu-Elmagd M, Turki RF (2015) Effects of increased paternal age on sperm quality, reproductive outcome and associated epigenetic risks to offspring. Reprod Biol Endocrinol 13:35

Singh NP, Muller CH, Berger RE (2003) Effects of age on DNA double-strand breaksand apoptosis in human sperm. Fertil Steril 80:1420–1430

Smith TB, Dun MD, Smith ND, Curry BJ, Connaughton HS et al (2013) The presence of a truncated base excision repair pathway in human spermatozoa that is mediated by OGG1. J Cell Sci 126:1488–1497

Sun C, Skaletsky H, Birren B, Devon K, Tang Z, Silber S et al (1999) An azoospermic man with a de novo point mutation in the Y-chromosomal gene USP9Y. Nat Genet 23:429–432

Tartaglia M, Zampino G, Gelb BD (2010) Noonan syndrome: clinical aspects and molecular pathogenesis. Mol Syndromol 1:2–26

Taylor J, Tyekucheva S, Zody M et al (2006) Strong and weak male mutation bias at different sites in primate genomes: insights from the human-chimpanzee comparison. Mol Biol Evol 23(3):565–573

Thacker PD (2004) Biological clock ticks for men, too: genetic defects linked to sperm of older fathers. JAMA 291(14):1683–1685

Tremellen K (2008) Oxidative stress and male infertility-a clinical perspective. Hum Reprod Update 14:243–258

Unryn BM, Cook LS, Riabowol KT (2005) Paternal age is positively linked to telomere length of children. Aging Cell 4:97–101

Vajo Z, Francomano CA, Wilkin DJ (2000) The molecular and genetic basis of fibroblast growth factor receptor 3 disorders: the achondroplasia family of skeletal dysplasias, Muenkecraniosynostosis, and Crouzon syndrome with acanthosis nigricans. Endocr Rev 21:23–39

Weinberg W (1912) Zur Verebung des zwergwuches. Arch Rassen-u Gesel Biol 9:710–718

Weiner-Megnazi Z, Auslender R, Dirnfeld M (2012) Advanced paternal age and reproductive outcome. Asian J Androl 14(1):69–76

Wilkie AM, Slaney SF, Oldridge M, Poole MD, Ashworth GJ, Hockley AD, Hayward RD, David DJ, Pulleyn LJ, Rutland P et al (1995) Apert syndrome results from localized mutations of FGFR2 and is allelic with Crouzon syndrome. Nat Genet 9:165–172

Wyrobek AJ, Eskenazi B, Young S, Arnheim N, Tiemann-Boege I, Jabs EW et al (2006) Advancing age has differential effects on DNA damage, chromatin integrity, gene mutations, and aneuploidies in sperm. Proc Natl Acad Sci U S A 103:9601–9606

Food, Nutrition, and Male Fertility 20

Sudha Bhagwati and Rajender Singh

"The food you eat can be either the safest and most powerful form of medicine or the slowest form of poison"
—Ann Wigmore.

Abstract

In a number of lower organisms and seasonal breeders, availability of food is a key determinant in shaping the time for reproduction and fertility. Therefore, food and nutrition strongly affect fertility, even in nonseasonal breeders. Eating a balanced diet is the key to good overall health. Food habits and their inherent components vary greatly across the globe. Making nutritious food a part of the regular diet can ameliorate health and upkeep fertility. Deficiency of nutrients and antioxidants can decrease fertility as various reports have supported the role of antioxidants in fertility. This chapter provides a comprehensive coverage of dietary elements that provide essential nutrients, cofactors, and antioxidants for the maintenance of good reproductive potential and fertility and improve prophylaxis against infertility.

Keywords

Food and fertility • Vitamins and fertility • Antioxidants and spermatogenesis • Soy food • Coenzyme Q10 • Vitamins A, C, and E

Key Points

- Food and nutrients play vital roles in male reproduction and fertility.
- Food contains antioxidants, N-acetylcysteine, vitamin C, vitamin E, CoQ10, selenium, and zinc that significantly improve sperm health by reducing oxidative stress.

S. Bhagwati • R. Singh (✉)
Male Reproductive Biology Laboratory, CSIR-Central Drug Research Institute, Lucknow, Uttar Pradesh, India
e-mail: rajender_singh@cdri.res.in

R. Singh, K. Singh (eds.), *Male Infertility: Understanding, Causes and Treatment*, DOI 10.1007/978-981-10-4017-7_20

- Soy foods, fatty acids, and obesogens (endocrine-disrupting chemicals) may compromise spermatogenesis and thus decline male fertility.
- Eating healthy and nutritious diet can reduce chances of male infertility.
- Taking nutritious and balanced diet, avoiding obesity, and including required vitamin supplements can be a good prophylactic measure against male infertility.

20.1 Introduction

Food and reproduction are the basic needs, which mark the very basis of survival and perpetuation of all species. Food and reproduction are closely linked aspects and every species tries to ensure both these essentials for survival. In a number of avian species, food availability is the principal factor that has shaped the timing of breeding season (Davies and Deviche 2014). Lack's theory (1950, 1968) postulated that the breeding timing has a genetic basis and seasonal variations in food supply select genotypes of birds laying eggs such that the nestling stage coincides with the peak in food availability. Similarly, in a number of other species such as wild boar (*Sus scrofa*), the availability of high quality of acorns and olives correlated with higher body weight, more breeding females and a larger litter size than in the years of poor production of these foods (Massei et al. 1996). Further, the age at puberty in beef cows is inversely proportional to the availability of nutrition (Schillo et al. 1992). These evidences clearly indicate an important impact of food and nutrition on fertility.

Humans are not seasonal breeders, but studies on the other animals suggest that food can have a significant impact on fecundity in humans. Worldwide, people eat various kinds of food. Some eat plant products like fruits, vegetable, cereals, pulses, etc., while others eat animal products like red meat, egg, fish, etc. In India, people mostly eat plant-based diet, while in the western countries, people are more dependent on animal-based food products. Some of these products affect male fertility positively while others have negative effects. Vitamins, minerals, and fatty acids are essential parts of the diet and are well known to affect male fertility. The levels of these nutrients vary greatly across the foods described above. Interestingly, similar to the variation in food habits across the globe, semen parameters vary greatly across major populations (Vujkovic et al. 2009). A large fraction of these variations may be explained by differences in food habits. In fact, there is research showing that there are certain fertility foods a man can include to his diet to help increase the odds of conceiving a baby, and other food items can actually impair men's fertility.

Nutritional status and lifestyle factors are considered as crucial determinants of normal healthy reproductive function. Nutrition has a significant impact on sperm health. What men eat reflects in their fertility. Research shows that having a poor diet and regular alcohol consumption, for instance, can compromise the quality and quantity of sperm and make conception more difficult. Food and nutrition are known to affect fertility, and nutrients are accredited to affect molecular mechanisms and balance in physiological functions. The use of nutrients to treat infertility is documented, but there is no specific heed to food and nutritional recommendations for infertility in classical medicine reference books. Therefore, there is a need to write an account of food and nutrition in relation to male fertility. This chapter provides a summary of food and nutrients that improve male fertility.

20.2 Dietary Essentials

20.2.1 Carbohydrate

Very less is known about how carbohydrates influence male reproductive health. An observational study explains that consumption of cereal and fruit was positively associated with semen quality (Braga et al. 2012). Additionally, one case-control study conducted on 30 men suffering from poor semen quality and 31 normal healthy controls reported that the control group had comparatively higher intake of raw or cooked vegetables (lettuce and tomato) and fruits (apricots and peaches), whereas intake of potato was higher and that of fruits and vegetables was lower in the case group (Mendiola et al. 2009). In yet another study, Eslamian et al. (2012) stated that men reporting higher consumption of fruit and vegetables showed lower risk of asthenozoospermia. Subgroup analysis on the basis of fruit and vegetable consumption showed that orange intake was negatively related to the risk of asthenozoospermia. Among vegetables, the intake of dark green vegetables and tomatoes was linked with a lower risk of asthenozoospermia (Eslamian et al. 2012).

20.2.2 Protein

A few studies analyzed the association between different dietary sources of protein with male reproductive health. Swan et al. (2007) described that maternal beef intake as well as anabolic steroids in beef result in an alteration in the male fetus development in utero and have adverse effects on his reproductive capacity. Sperm concentration of son was negatively correlated with mother's weekly beef consumption. In sons of high beef consumers (>7 beef meals per week), sperm concentration decreased by 24.3% in comparison with men whose mothers consumed less beef (Swan et al. 2007). Similarly, an observational study conducted on 250 male patients undergoing intracytoplasmic sperm injection (ICSI) therapy reported that meat consumption was significantly higher in infertile cases as compared to healthy individuals (Braga et al. 2012). A case-control study on 72 asthenozoospermic and 169 normozoospermic patients also reported identical findings. The study showed that the odds of asthenozoospermia were 2.03 times higher in the topmost tertile processed meat consumers. Furthermore, the odds of asthenozoospermia were 0.47 lower for those in the highest tertile of poultry product consumers (Eslamian et al. 2012).

Although soy food is regarded as a vegetable source of protein, some studies indicated that it adversely affects sperm parameters due to its high content of isoflavone. Soybean is a member of family Fabaceae. It is a legume and native of East Asia. It has a significant role in the treatment of some cancers, such as colon, prostate, and breast. Soy and soy-derived products contain isoflavones which mimic the actions of estrogens and may exert adverse effects on male fertility. The intake of 15 soy-based food by 99 males of subfertile couples for a period of 3 months showed that high intake of soy foods and soy isoflavones associated with lower sperm concentration (Chavarro et al. 2008; Modaresi et al. 2011) showed that the 20, 30, and 50% soy diet had a negative effect on male reproductive system in mice with a decline in primary spermatocytes and sperm count (Modaresi et al. 2011). The adverse effects of soy food on male fertility are due to the presence of isoflavones. Isoflavones are the type of naturally occurring isoflavonoids, which act as phytoestrogens and adversely affect sperm health and the male reproductive system.

20.2.3 Fat

More than 33% of the daily caloric intake of the human diet in most parts of the world contains fats and oils together (Bialostosky et al. 2002). Evidence from literature suggests that dietary fatty acids (FAs) may have substantial effects on male fertility. Bongalhardo et al. (2009) showed that birds fed fish and corn showed the highest and lowest n-3 polyunsaturated fatty acids (PUFA), respectively, in sperm. Diet comprising few distinct lipid sources differentially alters the lipid content of sperm head and body membrane, with minor effects on sperm characteristics (Bongalhardo et al. 2009). The fat composition of sperm membrane may affect sperm maturation in epididymis as the epididymal maturation is known to bring significant changes in sperm plasma membrane by extracting certain lipids (Rana et al. 1991).

Three types of natural fatty acids include saturated, monounsaturated, and polyunsaturated. Polyunsaturated fatty acids (PUFAs) are needed for various processes including growth, reproduction, vision, and brain development. Since they cannot be synthesized by the human body, they are regarded as essential fatty acids (Mazza et al. 2007). A clinical study conducted for analyzing the level of PUFA and saturated fatty acids in semen suggested that spermatozoa of asthenozoospermic patients have lower levels of PUFA compared with saturated fatty acids and this may contribute to the poor motility of sperm in these men (Tavilani et al. 2006).

20.3 Dietary Pattern

The dietary pattern has a significant impact on semen quality and hence on male reproductive health. A cross-sectional and observational study was conducted on 188 young men, dependent on two different dietary patterns (Prudent and Western) in the years 2009–2010 at the University of Rochester. Prudent diet included high intake of fish, chicken, fruit, vegetables, legumes, and whole grains, while Western diet included high intake of red and processed meat, refined grains, pizza, snacks, high-energy drinks, and sweets. Semen samples were collected and analyzed for sperm count, motility, and sperm morphology and compared between the two dietary patterns. The consumption of a prudent dietary pattern was found to be significantly associated with higher progressive sperm motility and not associated to sperm concentration and morphology. On the other hand, the consumption of Western diet showed neither positive nor negative association with conventional semen parameters. Therefore, it can be concluded that prudent diet or inclusion of at least a few of prudent components in the diet may help upkeep sperm motility (Gaskins et al. 2012).

20.4 Food Spices

Nigella sativa is a medicinal spice, which is also known as black cumin. It has a potent bioactive compound known as thymoquinone that is used to treat epilepsy and allergies and boost the immune system. Seeds and alcoholic extract of *Nigella sativa* are found to improve spermatogenesis and hence male fertility potential (Mohammad et al. 2009). *Nigella sativa* is used as a food spice in some countries.

Seeds are the major source of the active components of this plant and used in the traditional medicine as a natural therapy for a variety of disorders and manifestations such as headache, dizziness, bronchial asthma, nasal congestion, fever, diarrhea, inflammation, cough, influenza, eczema, toothache, hypertension, diabetes, kidney and liver dysfunctions, lung diseases, rheumatism, parasitic infections, hypercholesterolemia, gastrointestinal disorders, and overall general well-being, for more than twenty centuries (Ahmad et al. 2013).

There are many other plant products like *Asparagus racemosus*, *Chlorophytum borivilianum*, *Crocus sativus*, *Curculigo orchioides*, *Mucuna pruriens*, *Tribulus terrestris*, *Trichopus zeylanicus*, *Withania somnifera*, *Zingiber officinale*, etc. that have potential pro-fertility activities. These plants may not be used as food items, but their human use for overcoming male infertility and sexual debilities has been documented. Some of these products and their specific uses have been described in detail in Chapter 21.

20.5 Nutrients and Vitamins

Humans have evolved with a sophisticated and complex antioxidant protection system to protect cells and organs of the body from reactive oxygen species. It involves a number of components, both endogenous and exogenous in origin, that function interactively and synergistically to neutralize the free radicals (Percival 1998). These components include (1) nutrient-derived antioxidants like vitamin C (ascorbic acid), vitamin E (tocopherols and tocotrienols), carotenoids, and other low molecular weight compounds such as lipoic acid, glutathione, etc.; (2) antioxidant enzymes, like superoxide dismutase (SOD), glutathione peroxidase, and glutathione reductase, which catalyze the quenching reactions of free radicals; and (3) metal-binding proteins, for example, ferritin, lactoferrin, ceruloplasmin, and albumin that confiscate free iron and copper ions that are able to catalyze oxidative reactions. There are many other antioxidant phytonutrients present in an extensive variety of plant foods (Ford

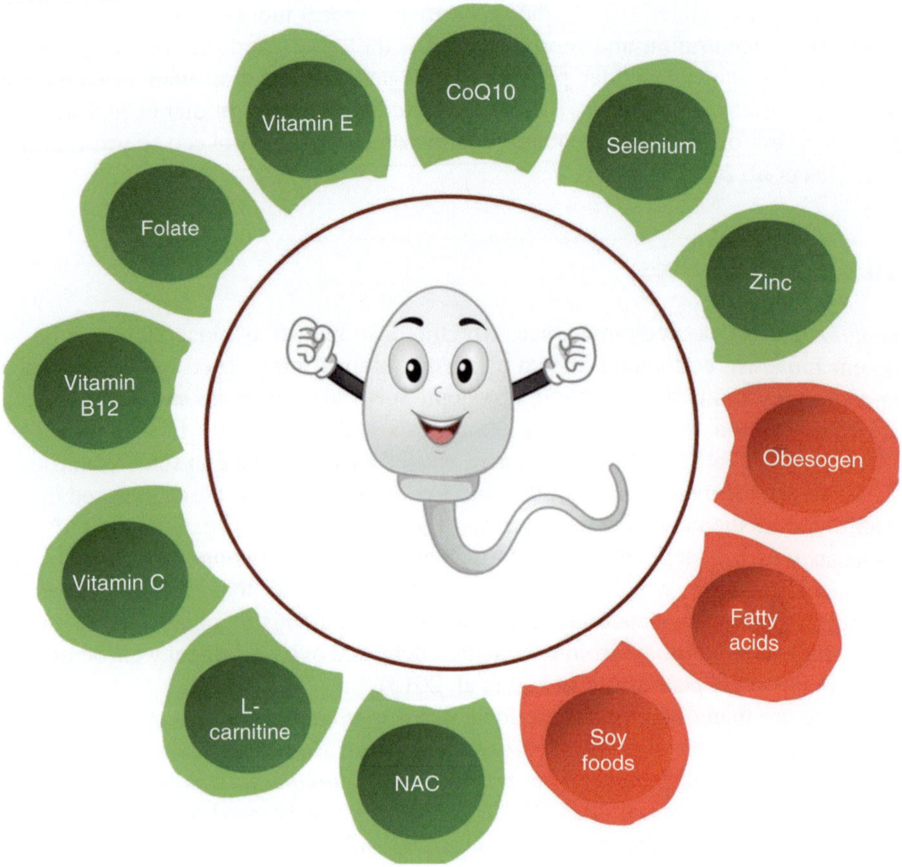

Fig. 20.1 Food/nutrition/vitamins that affect spermatogenesis. The items shown in green have positive effect and those shown in red have negative effect

et al. 1999). The eating habits of various species are set to provide the best reproductive fitness and fecundity. However, due to a number of environmental and stress factors, the regular eating habits may not always remain adequate to maintain the level of various nutrients and extra antioxidants required under these conditions. A detailed description of food and nutrients which improve various parameters of male reproductive health is given below (Fig. 20.1).

20.5.1 Zinc

Zinc is an important trace mineral present in the cells throughout the body. It is required for body's defensive (immune) system to work properly. It plays a crucial role in cell division, cell growth, wound healing, and the breakdown of carbohydrates. It is also required for the sense of smell and taste, during pregnancy and infancy.

Zinc is regarded as one of the most significant trace minerals for male reproductive health; increasing zinc levels in infertile men has been shown to increase sperm count and improve the morphology, form, function, and quality of sperm, thus improving male fertility. Hunt et al. (1992) showed that there is a significant decrease in seminal volume, serum testosterone concentration, and sperm morphology in young men due to zinc depletion in diet (Hunt et al. 1992) According to the World Health Organization (WHO) guidelines (5th edition), the lower reference limit for seminal zinc is ≥2.4 μmol/ejaculate, and level below this range may be a risk for infertility.

Oysters, calf liver, sesame seeds, beef, lamb, pumpkin seeds, yogurt, turkey, peas, venison, and shrimps are the major food sources of zinc. Zinc can be degraded by cooking; therefore, it is important to eat some foods in their raw forms, which are high in zinc.

20.5.2 Selenium

Selenium is an essential trace mineral, which our body requires in small amount. Selenium is necessary for the production of special proteins called antioxidant selenoproteins for protection against oxidative stress caused by the reactive oxygen species (ROS) and reactive nitrogen species (NOS) (Tinggi 2008). It exerts its biological functions through selenoproteins that contain amino acid selenocysteine. There are 25 selenoproteins encoded by the human genome.

Selenium is a requisite element for the production of sperm. A study showed that hyperlipidemia has significant adverse effects on male fertility, which can be ameliorated by diet supplemented with probiotics, inorganic selenium, or selenium-enriched probiotics (Ibrahim et al. 2012). Malondialdehyde (MDA) is a lipid peroxidation marker and the level of MDA is high in semen of infertile men. A vitamin intervention study reported that oral supplementation of vitamin E and selenium caused a significant decrease in MDA concentration in sperm and an improvement in sperm motility (Keskes-Ammar et al. 2003). In another study, a combination therapy with selenium and vitamin E was found to be effective for the treatment of asthenospermia and asthenoteratospermia and the induction of spontaneous pregnancy (Moslemi and Tavanbakhsh 2011). It has also been shown that oral supplementation of selenium and N-acetylcysteine improved all semen parameters, such as sperm count and motility (Safarinejad and Safarinejad 2009). Specific effects of nutrients and vitamins on semen parameters are detailed in Fig. 20.2.

Brazil nuts, mushrooms, cereals, egg, liver, cod, sardines, halibut, tuna, salmon, shrimp, snapper, and turkey are the food products that provide selenium.

20.5.3 CoQ10

Coenzyme Q10 is a naturally occurring quinone, present ubiquitously in the animal body. It is also known as ubiquinone and ubidecarenone. A critical role of CoQ10 is

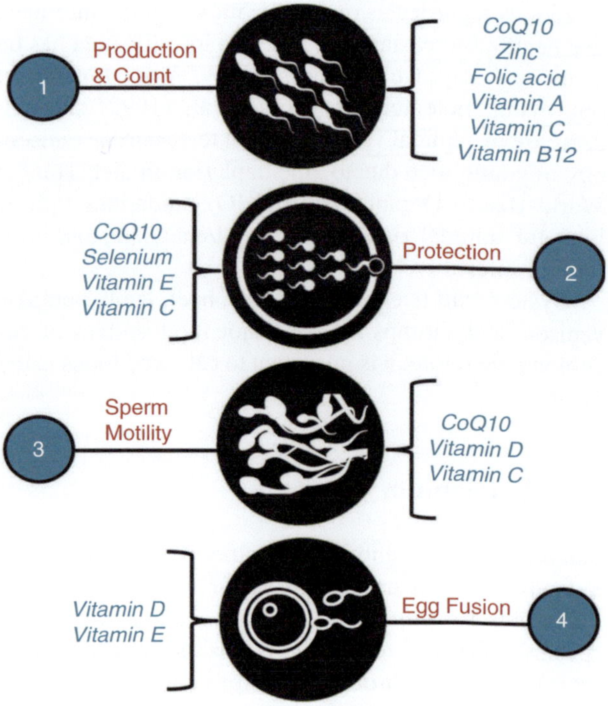

Fig. 20.2 Effect of nutrients/vitamins on specific semen parameters and fertility

as an electron carrier in the mitochondrial respiratory chain complex. It is one of the most important lipophilic antioxidants, protecting the production of free radicals as well as oxidation of proteins, lipids, and DNA. Decreased levels of CoQ10 in humans are observed in many pathological conditions, for example, cardiac disorders, neurodegenerative diseases, AIDS, cancer, infertility, etc. (Bentinger et al. 2007). In these cases, treatment involves pharmaceutical supplementation or increased consumption of CoQ10 with meals as well as treatment with suitable chemical compounds like folic acid or vitamin B group, which significantly increase ubiquinone biosynthesis in the body.

CoQ10 is a vital antioxidant that helps in preventing cellular damage caused by free radicals, thus protecting DNA. CoQ10 is necessary for sperm motility and is, therefore, a crucial nutrient which affects male fertility. In vitro and in vivo studies done by Lewin and Lavon (1997) reported a significant increase in sperm cell motility after treatment with coenzyme Q10 in humans (Lewin and Lavon 1997). Additionally, some other recent studies also have shown that CoQ10 can increase sperm health, particularly sperm motility (Balercia et al. 2004; Balercia et al. 2009; Safarinejad 2009; Mancini and Balercia 2011).

CoQ10 is abundantly present in seafood and organ meats, though it is very difficult to obtain through the diet, especially for vegetarians and vegans. CoQ10 ubiquinol supplementation is the best way to obtain CoQ10.

20.5.4 Vitamin E

Vitamin E is an important antioxidant, which protects body tissues from damage caused by the free radicals. It functions as an essential lipid soluble antioxidant or radical scavenger. It limits the production of free radicals in tissues by reacting with them to form a tocopheryl radical, which will then be reduced by a hydrogen donor (such as vitamin C) and thus return to its reduced state (Traber and Stevens 2011). Vitamin E is also being used as a commercial antioxidant in ultrahigh molecular weight polyethylene used in hip and knee implants to replace defective joints to help resist oxidation (UHMWPE Biomaterials Handbook). Vitamin E also plays a role in neurological functions (Muller 2010) and inhibition of platelet coagulation (Dowd and Zheng 1995). Vitamin E also protects lipids and prevents the oxidation of polyunsaturated fatty acids.

Vitamin E has been shown in various studies to improve sperm health and motility in men. Vitamin E is found to play a significant role in improving the in vitro functions of spermatozoa, which is evaluated by zona-binding test (Kessopoulou et al. 1995). In other studies, vitamin E and selenium supplementation in combination were found to improve sperm parameters like motility (Keskes-Ammar et al. 2003; Moslemi and Tavanbakhsh 2011). It is also known as "tocopherol" that literally means to bear young. It is an important antioxidant that helps to protect DNA damage and maintains the DNA integrity of sperm and egg cells (Kessopoulou et al. 1995).

Vitamin E is abundantly present in spinach, sunflower seeds, olives, papaya, almonds, and dark green leafy vegetables.

20.5.5 Folic Acid/Folate/Vitamin B9

Folate is an important factor required for the production and maintenance of new cells, DNA and RNA synthesis, preventing DNA damage, and thus preventing cancer (Kamen 1997). This is also involved in the biosynthesis of nitrogen bases, nucleic acids, and some amino acids like creatine, methionine, and serine. Folic acid prevents spina bifida and neural tube defects. Its ability to lower the level of homocysteine suggests that it might have a positive influence on cardiovascular diseases. The role of folic acid in maintaining good health may extend beyond these ailments to encompass other birth defects, several kinds of cancer, dementia, Down syndrome, and serious conditions affecting pregnancy outcome (Lucock 2000).

Research suggests that folic acid can potentially improve sperm health. A double-blind, placebo-controlled interventional study showed an increase of about 74% in total normal sperm count in previously subfertile and normal fertile men taking 66 mg/day of zinc with 5000 mcd/day of folic acid (Wong et al. 2002). Folate is required for DNA synthesis pathways and repair. Folate deficiency hinders DNA synthesis, cell division, and reproduction (Forges et al. 2007). Men with low levels of seminal plasma folate have increased risks of low sperm density and count

(Wallock et al. 2001). In an observational study, 70 fertile and 63 subfertile men undergoing in vitro fertilization or intracytoplasmic sperm injection treatment were assessed for semen parameters and tHcy (total homocysteine), folate, cobalamin, and pyridoxine concentrations in seminal plasma and blood. In case of fertile men, seminal plasma folate level was inversely correlated with the DNA fragmentation index (Boxmeer et al. 2009). Fertilization of egg with an abnormal sperm may lead to birth defects such as Down syndrome or an increased chance of miscarriage, making folate pathway crucial for reproductive health.

According to the European Association of Urology, antioxidant treatment (folic acid, vitamin E, zinc, selenium) has a positive effect on semen quality. Lentils, spinach, pinto beans, asparagus, navy beans, black beans, garbanzo beans, kidney beans, and collard greens are the food sources of folic acid.

20.5.6 Vitamin B12

Also known as cobalamin, B12 helps the body convert food (carbohydrate) into fuel (glucose), which is used to produce energy. Vitamin B12 is a critically important vitamin for maintaining healthy nerve cells and helping synthesis and repair of DNA and RNA. Vitamin B12 works closely with folate or folic acid in helping the formation of red blood cells. Folate and vitamin B12 work together to produce S-adenosylmethionine (SAM), a derived amino acid involved in immune function and mood.

It is required for cellular replication and studies suggest that cobalamin deficiency can cause reduced sperm count and motility. In an observational study, 70 fertile and 63 subfertile men undergoing in vitro fertilization or intracytoplasmic sperm injection treatment were assessed for semen parameters and tHcy (total homocysteine), folate, cobalamin, and pyridoxine concentrations in seminal plasma. In the fertile control men, cobalamin was found to positively correlate with sperm count, but inversely correlate with ejaculate volume (Boxmeer et al. 2009). In another study on male albino rats, vitamin B12-deficient diet was given to animals for three different periods, (1) whole period (gestation to mature), (2) gestation period (gestation to weaning), and (3) immature period (3–12 weeks postnatal). This study suggested that dietary vitamin B12 deficiency during pregnancy may induce damage to germ cells of the embryo and affect maturation of spermatozoa (Watanabe et al. 2003).

Food sources rich in vitamin B12 include clams, oysters, muscles, liver, lamb, caviar (fish eggs), lobster, fish, crab beef, cheese, and eggs.

20.5.7 Vitamin C/Ascorbic Acid

Vitamin C is necessary for the development and maintenance of connective tissues and plays a crucial role in wound healing, bone formation, and the maintenance of healthy gums. It also plays an important role in a variety of metabolic functions such as activation of vitamin B and folic acid and conversion of cholesterol to bile acid and tryptophan (amino acid) to serotonin (the neurotransmitter). It is an antioxidant that protects the body from free radical-induced damage.

Infertile men possess considerably more sperm DNA damage than the normal fertile men, and vitamin C is found to improve sperm quality and protect sperm from DNA damage. It also helps in reducing the chance of miscarriage and chromosomal problems. A study conducted on males working in a battery manufacturing industry at Hyderabad (India) showed a significant increase in total sperm count and sperm motility and a significant decrease in abnormal sperm morphology after vitamin C prophylaxis (Vani et al. 2012). Greco et al. (2005) showed that sperm DNA damage can be efficiently treated with oral administration of antioxidants (Greco et al. 2005). Vitamin C also appears to keep sperm from clumping together, making them more motile. In a study that analyzed various semen parameters in oligospermic infertile men, before and after oral supplementation of vitamin C, it was concluded that vitamin C supplementation in infertile men might improve sperm count, motility, and morphology (Akmal et al. 2006).

It is also known as ascorbic acid and is abundantly present in plants and fruits, including red peppers, potatoes, broccoli, cranberries, tomatoes, cabbage, and citrus fruits.

20.5.8 L-Carnitine

This compound is synthesized in the liver, kidney, and brain and is composed of two amino acids, lysine and methionine. It performs a crucial role in the energy supply for tissues during fetal life and in the neonatal stage by regulating the influx of fatty acids into mitochondria. L-Carnitine regulates the level of acylo-CoA and CoA in the mitochondria and provides acetyl moieties for the biosynthesis of acetylcholine (Rospond and Chłopicka 2012). L-Carnitine also plays a vital role in the metabolism of lipids and by transporting long-chain fatty acids into mitochondria for beta-oxidation. L-Carnitine further functions as an antioxidant, favoring fatty acid replacement within previously oxidatively damaged membrane phospholipids. Availability of L-carnitine is compulsory in the developing fetus for various processes underlying fetal maturation.

Carnitine is a vital nutrient for sperm cells to function normally. Sperm requires high concentrations of carnitine for energy metabolism. A study showed a direct correlation between the level of free carnitine in seminal fluid and sperm count and motility (Johansen and Bohmer 1979). In a clinical study, it was found that L-carnitine is a potential factor, which significantly improves sperm motility and increases the rate of pregnancy. It is also a safe therapeutic for the treatment of asthenozoospermia (Wang et al. 2010). Carnitine, acetyl carnitine, L-arginine, and ginseng combined therapy significantly improved progressive sperm motility in men with asthenospermia (Morgante et al. 2010). Balercia et al. (2005) showed that supplementing with L-carnitine helps in improving sperm health and increasing sperm count and motility in the patients with low count and motility. L-Carnitine and L-acetyl carnitine in combination are found to be effective in increasing sperm kinetic properties in idiopathic asthenozoospermia patients and improves the total oxyradical scavenging capacity of seminal fluid (Balercia et al. 2005).

Red meat and dairy products are the major food sources of L-carnitine, but other foods include nuts, seeds, asparagus, brussels sprouts, collard greens, garlic, mustard greens, okra, kale, broccoli, apricots, bananas, bee pollen, artichokes, brewer's yeast, parsley, buckwheat, corn, oatmeal, rice bran, rye, and whole wheat.

20.5.9 *N*-Acetylcysteine (NAC)

N-Acetylcysteine or acetylcysteine (NAC) is a mucolytic agent used to loosen the thick mucus in the disorders such as cystic fibrosis or chronic obstructive pulmonary disease. It is also used for the treatment of numerous disorders, such as doxorubicin cardiotoxicity, ischemia-reperfusion cardiac injury, acute respiratory distress syndrome, bronchitis, chemotherapy-induced toxicity, HIV/AIDS, heavy metal toxicity, and psychiatric disorders. It acts as a crucial antioxidant as it reacts with OH, NO_2,• and CO_3– (Samuni et al. 2013).

NAC is a modified amino acid, which has potent antioxidant properties. It significantly reduces the destructive reactive oxygen species in human semen and improves impaired sperm function (Oeda et al. 1997). Ciftci et al. (2009) concluded from a research that in the NAC-treated group, the total antioxidant capacity of serum was greater and total peroxide and oxidative stress indices were lower as compared to the control group (Ciftci et al. 2009). Few more studies suggested that NAC is an important antioxidant, which can ameliorate sperm health by combating the reactive oxygen species (Safarinejad 2009; Reddy et al. 2011).

Granola, oat flakes, and vegetables like broccoli, red pepper, and onion are major sources of cysteine. Other plant sources include bananas, garlic, linseed, and wheat germ.

Other than derived amino acids, L-carnitine and *N*-acetylcysteine (NAC), a recent study also found some basic amino acids to have a positive effect on male infertility by improving production and quality of sperm. Supplementation with amino acids (lysine/methionine/threonine/tryptophan/valine) in a particular ratio (100:27:73:19:69) in boar diet improved sperm quality and subsequently increased the fertilization capacity and the number of live piglets (Dong et al. 2016).

20.6 Obesogens

Obesogens are the foreign chemicals, which disrupt normal development and balance of lipid metabolism. This may lead to obesity in some cases. There are various potential obesogens found everywhere, and people come in contact with them every day, intentionally or unintentionally, for example, bisphenol-A (BPA), high fructose corn syrup (HFCS), nicotine, arsenic, pesticides, organotins (tributyltin and triphenyltin), and perflurooctanoic acid (PFOA). All endocrine-disrupting chemicals

(EDC) are defined as obesogen and are well known to be associated with early puberty, reproductive dysfunctions, and infertility later in life of humans and animals (Diamanti-Kandarakis et al. 2009; Skakkebaek et al. 2001).

EDCs have transgenerational effects; they affect not only the exposed individual but also the subsequent generations. Ephemeral exposure of a gestating female rat during the gonadal sex determination to the EDCs such as vinclozolin (an antiandrogenic) or methoxychlor (an estrogenic compound) produced an adult phenotype in the F1 generation with decreased spermatogenesis (cell number and viability) and increased incidence of male infertility (Anway et al. 2005). This effect may be transmitted not due to the mutations in the DNA sequence, but through modifications in the factors such as DNA methylation and histone acetylation, which regulate gene expressions. Causes of obesity and its association with male infertility are described in detail in Chapter 11.

20.7 Nutrigenomics

Nutritional genomics or nutrigenomics is the study of how individual genetic differences can affect the way we respond to nutrients and other natural compounds we eat or how nutrients exert health effects by affecting gene expression (The NCMHD Center of Excellence for Nutritional Genomics, University of California, Davis). Nutrigenomics is an approach to understand the relationship between diet and health with integration to individual differences in the genetic makeup. Some of the changes thus introduced may be inherited from generation to generation, resulting in transgenerational effects. Munshi and Duvvuri (2008) explained how nutrients influence gene expression, i.e., mRNA synthesis (transcriptomics), protein synthesis (proteomics), and production of metabolites (metabolomics), by giving the example of genetic polymorphism (SNPs) which may be responsible for variations in individual's response to bioactive food components (Munshi and Duvvuri 2008).

The effect of dietary components on gene expression has not been well explored except a few commonly studied pathways. One such pathway relates folate and homocysteine cycle with the genes participating in one carbon metabolism pathway. MTHFR is the most commonly studied gene from this pathway. A number of studies have suggested a significant impact of MTHFR 677C > T polymorphism with various disorders, including male infertility. At the same time, folate is known to be important for spermatogenesis. Interventional studies have shown that folate supplementation improved sperm concentration in infertile men. A simple explanation for this observation may be adequate functioning of the folate pathway upon supplementation. Interestingly, Aarabi et al. (2015) in a recent study showed that apart from improvement in the blood folate levels, significant changes in the methylation level of differentially methylated regions of several imprinted loci (H19, DLK1/GTL2, MEST, SNRPN, PLAGL1, KCNQ1OT1) in sperm DNA were seen upon supplementation. Interestingly, a recent study has shown significant differences in the MTHFR promoter methylation between infertile individuals and controls (Aarabi et al. 2015). Karaca et al. (2016) showed that the percentage of MTHFR promoter methylation in infertile normozoospermic men was significantly higher in comparison to healthy controls (Karaca et al. 2016).

In another interesting double-blind, placebo-controlled interventional study, Ebisch et al. (2003) analyzed 677 C > T polymorphisms in 13 fertile versus 77 subfertile individuals and studied MTHFR-dependent response to sperm concentration upon folic acid/zinc sulfate supplementation. Daily capsules of folic acid (5 mg) and/or zinc sulfate (66 mg) versus placebo were recommended for 26 weeks. The authors found that the genotype frequencies between the two groups were comparable. Interestingly, sperm concentration increased significantly in wild types, but heterozygous and homozygotes did not show significant improvements. The study concluded that MTHFR genotype had a significant impact on the response to folic acid/zinc supplementation in subfertile individuals (Ebisch et al. 2003). Similar studies in other disorders have supported the role of gene polymorphisms in affecting the response to diet or nutrient supplementation. For example, a study on dietary folate intake showed an inverse association with promoter methylation in colorectal adenomas that was dependent on the MTHFR genotype (van den Donk et al. 2007). Similarly, the influence of a number of dietary or nutritional factors via their effects on promoter methylation and gene expression is likely in addition to their simple availability to act as enzyme cofactors.

20.8 Discussion and Future Directions

Apart from congenital disorders and genetic causes, the major reason for infertility is the hormonal imbalance and/or oxidative stress. Disturbance in the homeostasis of hormone levels and the antioxidant defense may result in increased production of reactive oxygen species, leading to slowing down or arrest of spermatogenesis. Hormonal imbalance may result in compromised spermatogenesis, which can be further decelerated by oxidative stress. Apart from affecting sperm production, oxidative stress can cause DNA damage. DNA of both parents is the future blueprint for the child. Impaired DNA is known to cause miscarriages, birth defects, and developmental problems in the offsprings. Studies have also shown a strong correlation between oxidative stress caused by free radicals and male infertility. If the physiology is otherwise perfect or close to that, food can have a significant impact on sperm production and fertility. Therefore, a general precaution and adequate attention to nutrition can have a sound effect on fertility.

The food products discussed in this book chapter have nutritional value and may not be adequate in regular diet. "Fertilica Choice Antioxidants" contains most of the important antioxidant nutrients in a capsule formulation. This blend is useful for both men and women, but especially for men with low sperm count and poor sperm health. The transgenerational effects of food and nutrition in relation to male fertility cannot be denied. Obesogens (such as endocrine-disrupting chemicals) have transgenerational effects on male fertility and decline the reproductive status of the future male progeny (Schug et al. 2011). Short-term exposure of a gestating female rat during the gonadal sex determination, to the EDCs like vinclozolin or methoxychlor, induces an adult phenotype in the F1 generation of decreased spermatogenic capacity and increased incidence of male infertility. These effects are transferred through the male germ line to the males of F1 to F4 generations. The transgenerational effects may not be transmitted due to mutation in the DNA sequence, but through modifications in the factors like DNA methylation and histone acetylation, which regulate gene expressions.

The capability of an environmental factor (such as endocrine-disrupting chemicals) to reprogram the germ line and promote a transgenerational condition has a remarkable association between evolutionary biology and disease etiology (Anway et al. 2005). It must be noted that fertility, reproductive potential, environment, and evolution are strongly correlated. There is a proof for the role of sperm-derived RNAs in arbitrating paternal transgenerational effects, with several categories of RNA recently discovered in sperm that are amenable to changes in diet, behavior, and stress (Sharma and Rando 2014). The research on the transgenerational impact of environmental factors, endocrine disruptors, and food is still in the infantile phase; further research in this area would bring forth the effects of food and nutrition unseen so far.

Conclusion

People with idiopathic male infertility can treat their disorder by supplementation of zinc, selenium, CoQ10, folic acid, vitamin B12, vitamin E, vitamin C, L-carnitine, and antioxidant-rich diet. Normal fertile individuals can prolong their fertility period by supplementation of a nutritious diet in their regular food habits and by avoiding food items like soy food, fatty acids, and obesogens that are potentially detrimental to spermatogenesis and fertility. We should prefer food therapy over drug therapy to increase the quality and quantity of semen as a measure to avoid or treat male infertility. Nutrigenomics research may open the way to personalized nutrition. Food may be seen as a requirement for regular course of life, but interestingly, food and nutrition play roles well beyond that as they have been shown to affect the DNA that we pass on to our subsequent generations. The impact of environmental toxicants, endocrine disruptors, stress, and other factors that affect fertility can be minimized or reversed by paying a little attention to daily nutritional requirements. Therefore, one must keep an eye on self-nutrition to upkeep fertility and pass the same to the coming generations.

Acknowledgments The authors would like to thank the Indian Council of Medical Research (Ref. No. 5/10/FR/5/2012-RHN) and the Council of Scientific and Industrial Research (CSIR), Govt. of India, for financial help.

References

Aarabi M, San Gabriel MC, Chan D, Behan NA, Caron M, Pastinen T, Bourque G, MacFarlane AJ, Zini A, Trasler J (2015) High-dose folic acid supplementation alters the human sperm methylome and is influenced by the MTHFR C677T polymorphism. Hum Mol Genet 24(22): 6301–6313

Ahmad A, Husain A, Mujeeb M, Khan SA, Najmi AK, Siddique NA, Damanhouri ZA, Anwar F (2013) A review on therapeutic potential of *Nigella sativa*: a miracle herb. Asian Pacific J Trop Biomed 3(5):337–352

Akmal M, Qadri JQ, Al-Waili NS, Thangal S, Haq A, Saloom KY (2006) Improvement in human semen quality after oral supplementation of vitamin C. J Med Food 9(3):440–442

Anway MD, Cupp AS, Uzumcu M, Skinner MK (2005) Epigenetic transgenerational actions of endocrine disruptors and male fertility. Science 308(5727):1466–1469

Balercia G, Mosca F, Mantero F, Boscaro M, Mancini A, Ricciardo-Lamonica G, Littarru G (2004) Coenzyme q 10 supplementation in infertile men with idiopathic asthenozoospermia: an open, uncontrolled pilot study. Fertil Steril 81(1):93–98

Balercia G, Regoli F, Armeni T, Koverech A, Mantero F, Boscaro M (2005) Placebo-controlled double-blind randomized trial on the use of L-carnitine, L-acetylcarnitine, or combined L-carnitine and L-acetylcarnitine in men with idiopathic asthenozoospermia. Fertil Steril 84(3):662–671

Balercia G, Buldreghini E, Vignini A, Tiano L, Paggi F, Amoroso S, Ricciardo-Lamonica G, Boscaro M, Lenzi A, Littarru G (2009) Coenzyme Q 10 treatment in infertile men with idiopathic asthenozoospermia: a placebo-controlled, double-blind randomized trial. Fertil Steril 91(5):1785–1792

Bentinger M, Brismar K, Dallner G (2007) The antioxidant role of coenzyme Q. Mitochondrion 7:S41–S50

Bialostosky K, Wright JD, Kennedy-Stephenson J, McDowell M, Johnson CL (2002) Dietary intake of macronutrients, micronutrients, and other dietary constituents: United States 1988–94. Vital Health Stat 245:1–158

Bongalhardo DC, Leeson S, Buhr MM (2009) Dietary lipids differentially affect membranes from different areas of rooster sperm. Poult Sci 88(5):1060–1069

Boxmeer JC, Smit M, Utomo E, Romijn JC, Eijkemans MJ, Lindemans J, Laven JS, Macklon NS, Steegers EA, Steegers-Theunissen RP (2009) Low folate in seminal plasma is associated with increased sperm DNA damage. Fertil Steril 92(2):548–556

Braga DPDAF, Halpern G, Rita de Cássia SF, Setti AS, Iaconelli A, Borges E (2012) Food intake and social habits in male patients and its relationship to intracytoplasmic sperm injection outcomes. Fertil Steril 97(1):53–59

Chavarro JE, Toth TL, Sadio SM, Hauser R (2008) Soy food and isoflavone intake in relation to semen quality parameters among men from an infertility clinic. Hum Reprod 23(11):2584–2590

Ciftci H, Verit A, Savas M, Yeni E, Erel O (2009) Effects of N-acetylcysteine on semen parameters and oxidative/antioxidant status. Urology 74(1):73–76

Davies S, Deviche P (2014) At the crossroads of physiology and ecology: food supply and the timing of avian reproduction. Horm Behav 66(1):41–55

Diamanti-Kandarakis E, Bourguignon JP, Giudice LC, Hauser R, Prins GS, Soto AM, Zoeller RT, Gore AC (2009) Endocrine-disrupting chemicals: an Endocrine Society scientific statement. Endocr Rev 30(4):293–342

Dong HJ, Wu D, Xu SY, Li Q, Fang ZF, Che LQ, Wu CM, Xu XY, Lin Y (2016) Effect of dietary supplementation with amino acids on boar sperm quality and fertility. Anim Reprod Sci 172:182–189

Dowd P, Zheng ZB (1995) On the mechanism of the anticlotting action of vitamin E quinone. Proc Natl Acad Sci 92(18):8171–8175

Ebisch IM, van Heerde WL, Thomas CM, van der Put N, Wong WY, Steegers-Theunissen RP (2003) C677T methylenetetrahydrofolate reductase polymorphism interferes with the effects of folic acid and zinc sulfate on sperm concentration. Fertil Steril 80(5):1190–1194

Eslamian G, Amirjannati N, Rashidkhani B, Sadeghi MR, Hekmatdoost A (2012) Intake of food groups and idiopathic asthenozoospermia: a case–control study. Hum Reprod 27(11):3328–3336

Ford ES, Will JC, Bowman BA, Narayan KV (1999) Diabetes mellitus and serum carotenoids: findings from the third National Health and Nutrition Examination Survey. Am J Epidemiol 149(2):168–176

Forges T, Monnier-Barbarino P, Alberto JM, Gueant-Rodriguez RM, Daval JL, Gueant JL (2007) Impact of folate and homocysteine metabolism on human reproductive health. Hum Reprod Update 13(3):225–238

Gaskins AJ, Colaci DS, Mendiola J, Swan SH, Chavarro JE (2012) Dietary patterns and semen quality in young men. Hum Reprod 27(10):2899–2907

Greco E, Iacobelli M, Rienzi L, Ubaldi F, Ferrero S, Tesarik J (2005) Reduction of the incidence of sperm DNA fragmentation by oral antioxidant treatment. J Androl 26(3):349–353

Hunt CD, Johnson PE, Herbel J, Mullen LK (1992) Effects of dietary zinc depletion on seminal volume and zinc loss, serum testosterone concentrations, and sperm morphology in young men. Am J Clin Nutr 56(1):148–157

Ibrahim HA, Zhu Y, Wu C, Lu C, Ezekwe MO, Liao SF, Haung K (2012) Selenium-enriched probiotics improves murine male fertility compromised by high fat diet. Biol Trace Elem Res 147(1–3):251–260

Johansen L, Bohmer T (1979) Motility related to the presence of carnitine/acetyl-carnitine in human spermatozoa. Int J Androl 2(1–6):202–210

Kamen B (1997) Folate and antifolate pharmacology. Semin Oncol 24(5 Suppl 18):S18–S30

Karaca, MZ, Konac E, Yurteri B, Bozdag G, Sogutdelen E, Bilen CY (2016) Association between methylenetetrahydrofolate reductase (MTHFR) gene promoter hypermethylation and the risk of idiopathic male infertility. Andrologia. 1–6

Keskes-Ammar L, Feki-Chakroun N, Rebai T, Sahnoun Z, Ghozzi H, Hammami S, Zghal K, Fki H, Damak J, Bahloul A (2003) Sperm oxidative stress and the effect of an oral vitamin E and selenium supplement on semen quality in infertile men. Arch Androl 49(2):83–94

Kessopoulou E, Powers HJ, Sharma KK, Pearson MJ, Russell JM, Cooke ID, Barratt CL (1995) A double-blind randomized placebo cross-over controlled trial using the antioxidant vitamin E to treat reactive oxygen species associated male infertility. Fertil Steril 64(4):825–831

Lewin A, Lavon H (1997) The effect of coenzyme Q 10 on sperm motility and function. Mol Asp Med 18:213–219

Lucock M (2000) Folic acid: nutritional biochemistry, molecular biology, and role in disease processes. Mol Genet Metab 71(1):121–138

Mancini A, Balercia G (2011) Coenzyme Q10 in male infertility: physiopathology and therapy. Biofactors 37(5):374–380

Massei G, Genov PV, Staines BW (1996) Diet, food availability and reproduction of wild boar in a Mediterranean coastal area. Acta Theriol 41(3):307–320

Mazza M, Pomponi M, Janiri L, Bria P, Mazza S (2007) Omega-3 fatty acids and antioxidants in neurological and psychiatric diseases: an overview. Prog Neuro-Psychopharmacol Biol Psychiatry 31(1):12–26

Mendiola J, Torres-Cantero AM, Moreno-Grau JM, Ten J, Roca M, Moreno-Grau S, Bernabeu R (2009) Food intake and its relationship with semen quality: a case-control study. Fertil Steril 91(3):812–818

Modaresi, M., Messripour, M. and Khorami, H., 2011. Effect of soybean on male reproductive physiology in mice. In: International Conference on Life Science and Technology, vol 3. IPCBEE, pp 15–18.

Mohammad MA, Mohamad MM, Dradka H (2009) Effects of black seeds (*Nigella sativa*) on spermatogenesis and fertility of male albino rats. Res J Med Med Sci 4(2):386–390

Morgante G, Scolaro V, Tosti C, Di Sabatino A, Piomboni P, De Leo V (2010) Treatment with carnitine, acetyl carnitine, L-arginine and ginseng improves sperm motility and sexual health in men with asthenopermia. Minerva Urol Nefrol 62(3):213–218

Moslemi MK, Tavanbakhsh S (2011) Selenium–vitamin E supplementation in infertile men: effects on semen parameters and pregnancy rate. Int J Gen Med 4:99

Muller DP (2010) Vitamin E and neurological function. Mol Nutr Food Res 54(5):710–718

Munshi A, Duvvuri VS (2008) Nutrigenomics: looking to DNA for nutrition advice. Indian J Biotechnol 7(1):32

Oeda T, Henkel R, Ohmori H, Schill WB (1997) Scavenging effect of N-acetyl-L-cysteine against reactive oxygen species in human semen: a possible therapeutic modality for male factor infertility? Andrologia 29(3):125–131

Percival, M., 1998. Antioxidants. Clinical Nutrition Insights. 1–4. Available at: https://acudoc.com/Antioxidants.PDF

Rana AP, Majumder GC, Misra S, Ghosh A (1991) Lipid changes of goat sperm plasma membrane during epididymal maturation. Biochim Biophys Acta Biomembr 1061(2):185–196

Reddy PS, Rani GP, Sainath SB, Meena R, Supriya CH (2011) Protective effects of N-acetylcysteine against arsenic-induced oxidative stress and reprotoxicity in male mice. J Trace Elem Med Biol 25(4):247–253

Rospond B, Chłopicka J (2012) The biological function of L-carnitine and its content in the particular food examples. Przegl Lek 70(2):85–91

Safarinejad MR (2009) Efficacy of coenzyme Q10 on semen parameters, sperm function and reproductive hormones in infertile men. J Urol 182(1):237–248

Safarinejad MR, Safarinejad S (2009) Efficacy of selenium and/or N-acetyl-cysteine for improving semen parameters in infertile men: a double-blind, placebo controlled, randomized study. J Urol 181(2):741–751

Samuni Y, Goldstein S, Dean OM, Berk M (2013) The chemistry and biological activities of N-acetylcysteine. Biochim Biophys Acta Gen Subj 1830(8):4117–4129

Schillo KK, Hall JB, Hileman SM (1992) Effects of nutrition and season on the onset of puberty in the beef heifer. J Anim Sci 70(12):3994–4005

Schug TT, Janesick A, Blumberg B, Heindel JJ (2011) Endocrine disrupting chemicals and disease susceptibility. J Steroid Biochem Mol Biol 127(3):204–215

Sharma U, Rando OJ (2014) Father-son chats: inheriting stress through sperm RNA. Cell Metab 19(6):894–895

Skakkebaek NE, Rajpert-De Meyts E, Main KM (2001) Testicular dysgenesis syndrome: an increasingly common developmental disorder with environmental aspects: opinion. Hum Reprod 16(5):972–978

Swan SH, Liu F, Overstreet JW, Brazil C, Skakkebaek NE (2007) Semen quality of fertile US males in relation to their mothers' beef consumption during pregnancy. Hum Reprod 22(6):1497–1502

Tavilani H, Doosti M, Abdi K, Vaisiraygani A, Joshaghani HR (2006) Decreased polyunsaturated and increased saturated fatty acid concentration in spermatozoa from asthenozoospermic males as compared with normozoospermic males. Andrologia 38(5):173–178

Tinggi U (2008) Selenium: its role as antioxidant in human health. Environ Health Prev Med 13(2):102–108

Traber MG, Stevens JF (2011) Vitamins C and E: beneficial effects from a mechanistic perspective. Free Radic Biol Med 51(5):1000–1013

van den Donk M, van Engeland M, Pellis L, Witteman BJ, Kok FJ, Keijer J, Kampman E (2007) Dietary folate intake in combination with MTHFR C677T genotype and promoter methylation of tumor suppressor and DNA repair genes in sporadic colorectal adenomas. Cancer Epidemiol Biomark Prev 16(2):327–333

Vani K, Kurakula M, Syed R, Alharbi K (2012) Clinical relevance of vitamin C among lead-exposed infertile men. Genet Test Mol Biomarkers 16(9):1001–1006

Vujkovic M, de Vries JH, Dohle GR, Bonsel GJ, Lindemans J, Macklon NS, van der Spek PJ, Steegers EAP, Steegers-Theunissen RPM (2009) Associations between dietary patterns and semen quality in men undergoing IVF/ICSI treatment. Hum Reprod 24(6):1304–1312

Wallock LM, Tamura T, Mayr CA, Johnston KE, Ames BN, Jacob RA (2001) Low seminal plasma folate concentrations are associated with low sperm density and count in male smokers and nonsmokers. Fertil Steril 75(2):252–259

Wang YX, Yang SW, Qu CB, Huo HX, Li W, Li JD, Chang XL, Cai GZ (2010) L-carnitine: safe and effective for asthenozoospermia. Natl J Androl 16(5):420–422

Watanabe T, Ohkawa K, Kasai S, Ebara S, Nakano Y, Watanabe Y (2003) The effects of dietary vitamin B_ < 12> deficiency on sperm maturation in developing and growing male rats. Congenit Anom 43(1):57–64

Wong WY, Merkus HM, Thomas CM, Menkveld R, Zielhuis GA, Steegers-Theunissen RP (2002) Effects of folic acid and zinc sulfate on male factor subfertility: a double-blind, randomized, placebo-controlled trial. Fertil Steril 77(3):491–498

Plant Products in the Management of Male Infertility

21

Sudha Bhagwati and Rajender Singh

"The management of fertility is one of the most important functions of adulthood"
—Germaine Greer.

Abstract

Male infertility is a disorder with an undefined etiology in about half of the cases. It has devastating effects on personal and social life of a couple. In the past few years, the modern medical sciences have prospered a lot; however, in spite of great advancements in synthetic products, the herbal products are still a preferred choice in terms of safety, affordability, and higher efficacy. Assisted reproductive technologies (ART) such as IUI, IVF, and ICSI promise to treat a few but a long-lasting curative effect, easy availability, natural way of healing, and fewer side effects make herbal plant products an attractive alternate for a larger section of society. As recommended by the Ayurvedic, Unani, and Siddha medicinal systems, plant products are gaining a substantial importance for infertility management. In this chapter, we have focused on some well-known selected plants with respect to the scientific evidence of their effects on male fertility and their availability in the Indian market.

Keywords

Male infertility treatment • Plants in male infertility • Plant extracts • *Mucuna pruriens* • *Withania somnifera* • *Asparagus racemosus* • *Tribulus terrestris*

S. Bhagwati • R. Singh (✉)
Male Reproductive Biology Laboratory, CSIR-Central Drug Research Institute, Lucknow, Uttar Pradesh, India
e-mail: rajender_singh@cdri.res.in

R. Singh, K. Singh (eds.), *Male Infertility: Understanding, Causes and Treatment*, DOI 10.1007/978-981-10-4017-7_21

Key Points

- According to the World Health Organization (WHO) estimates, due to poverty and lack of access to modern medicine, about 65–80% of the world's population living in the developing countries depends essentially on plants for primary health care.
- In Ayurveda "Vajikarana Rasayana" has been suggested as an effective treatment for male sexual debilities and infertility.
- Plant products have aphrodisiac, adaptogenic, and antioxidant properties and offer holistic benefits to overcome male infertility.
- Medicinal plant products contain steroidal saponins, flavonoids, and alkaloids as active components, which may account for their pro-fertility effects.
- Plant products such as *Asparagus racemosus*, *Chlorophytum borivilianum*, *Curculigo orchioides*, *Mucuna pruriens*, *Tribulus terrestris*, *Withania somnifera*, etc., improve the level of testosterone and increase libido.

21.1 Introduction

Infertility is defined as the inability of a couple to achieve pregnancy after 1 year or more of regular and unprotected intercourse. Around the world, 1/6 couples trying to conceive have difficulties. Male reproductive capability is compromised in about 50% of the infertile couples. It is certain that stressful lifestyle has increased the number of cases suffering from sexual dysfunction and infertility. Oligozoospermia, sexual, and ejaculatory dysfunctions are also important factors for the decrease in conception other than congenital and immunological factors. Male infertility can have a great impact on a man's life affecting his self-esteem, confidence, and his sense of manhood. It can devastate a young couple's life by influencing their sexual, procreative, and marital needs and usually results in unconsummated marriages and divorce. To be childless generates problems for many couples regarding the lack of future support in old age and awful social suffering. To save a couple from these undesirable consequences due to all these reasons, it is desirable and highly worthwhile to treat infertility.

According to *Charak Samhita*, the healthy life has three main pillars—balanced diet, proper sleep, and healthy sexual and marital life. Ayurvedic remedies have long been used to address the complications of infertility. Vajikarana Rasayana was a very popular and effective treatment of male infertility in the ancient Indian medicinal system, which is still a method of choice alone or in combination with other methods of infertility management. In Sanskrit "Vaji" defines horse, the symbol of potency and performance; therefore, Vajikarana means producing a horse's vigor, particularly the animal's great potential for sexual activity. There are more than a 100 formulations used for Vajikarana, which include plant parts such as root, leaves, seeds, etc., of various medicinal plants. These formulations are used for good physique, potency, sexual exhilaration, and strength. The plants most commonly used in Vajikarana Rasayana are *Mucuna pruriens*, *Asparagus racemosus*, *Madhuca indica*, *Withania somnifera*, *Pueraria tuberosa*, *Saccharum officinarum*, *Tribulus terrestris*, *Bambusa erundinacea*, etc. (Dalal et al. 2013).

Nevertheless, specific antioxidants, revivers, and nutrients such as vitamins are available in the contemporary medicinal systems; however, none of them provides as multifarious constituents as may be required to withstand the multifaceted problems of male infertility. Plant products are the ultimate substitutes when general and varied effects are desired. Among the methods used to treat male infertility, medicinal plants have been used empirically as pure herbs, extracts, or semi-purified compounds. These herbal products are used for the treatment of erectile dysfunction, decreased libido, and sperm disorders. Many in vitro, in vivo, and clinical studies manifest the empirical use of plant products in the improvement of male fertility parameters. More than 50% of the world's population trusts herbal products for medicinal purposes.

Since the lack of libido is one of the causes of male infertility, the plants improving sexual drive help in treating male infertility. Some of the plants have already been tested on humans and found to be very potent in treating sexual debility and male infertility. Some plants are scientifically proven for reviving sexual desire, such as ethanolic extract of *Trichopus zeylanicus*; ethanolic extract of *Vanda tessellata* flowers; lipidic extract from *Lepidium meyenii*; extract of *Turnera diffusa*, *Pfaffia paniculata*, and *Tribulus terrestris*; and roots of *Panax ginseng*, or *Panax quinquefolius*. Infertile couples use both traditional medicine and modern therapies as treatment, but drugs derived from natural plant sources have always been considered safe and promising. The research on natural products as a source of potential drug has been of resurgent interest in developing as well as the developed countries due to the various reasons, namely, conventional medicine can be ineffective and abusive and wrong use of a synthetic drug may cause adverse effects.

A number of plant products claimed for pro-fertility effects have now been established to possess these properties by modern scientific experimentation, while a few still lack such evidence. A few plants with strong pro-fertility properties have been subjected to preparation of specific extracts and fractionation of active constituents. Mechanistic studies on a number of plants have shown the presence of antioxidant properties, which is a major factor in the treatment of male infertility. However, other properties specific to a number of plants have also been reported (Table 21.1). We have provided a detailed description of each plant and the status of the experimental evidence in support of their claimed effects.

Table 21.1 List of some acclaimed plants in practice for treatment of male infertility

Plants	Probable mechanism of action	Used part	Major constituent	Scientific evidence
Asparagus racemosus	Regeneration of seminiferous tubules	Root powder, hydroalcoholic, and aqueous extract	Saponins, shatavarin I–IV	Wani et al. (2011); Devi et al. (2004)
Asteracantha longifolia	Increases level of testosterone and spermatogenesis	Ethanolic extract of seed	Cardiac glycoside, phenol, aponins	Chauhan et al. (2011); Doss (2009)
Chlorophytum borivilianum	Antioxidant, aphrodisiac	Aqueous extract of root	Steroidal saponins, Borivillianosides	Rath and Panja (2013); Thakur et al. (2009); Kenjale et al. (2008); Visavadiya and Narasimhacharya (2007)
Crocus sativus	Antioxidant, aphrodisiac	Aqueous extract of stigma	Crocin	Hosseinzadeh et al. (2008); Heidary et al. (2008)
Curculigo orchioides	Aphrodisiac	Root extract	Saponins, glycosides Curculigosides A–D	Agrahari et al. (2010)
Dioscorea bulbifera	Reducing reactive oxygen species	Tuber	Diosgenin, alkaloids, flavonoids, vit C and B	Son et al. (2007); Okwu and Ndu (2006)
Morinda citrifolia	Protective effects against oxidative injuries and reduce lipid peroxidation in sperm membranc	Fruit, leaves, roots	Alkaloids, polysaccharides, lignans, anthraquinones	Wang et al. (2011); Kamiya et al. (2004); Hirazumi and Furusawa (1999)
Mucuna pruriens	Reducing ROS and GnRH signaling	Seed powder	L-Dopa, sterols, alkaloids	Singh et al. (2013); Lieu et al. (2010)
Tribulus terrestris	Increases level of T, DHT, and DHEAS	Dried fruit powder	Protodioscin	Sellandi et al. (2012)

Table 21.1 (continued)

Plants	Probable mechanism of action	Used part	Major constituent	Scientific evidence
Trichopus zeylanicus	Increases mating performance	Leaf ethanolic extract	Flavonoid glycosides, glycolipids	Tharakan and Manyam 2005; Singh et al. (2001); Subramoniam et al. (1997)
Withania somnifera	Increases T and LH and reduce FSH and PRL	Lyophilized aqueous extract of root	Withaferin, withanolides	Ahmad et al. (2010); Singh et al. (2001); Abdel-Magied et al. (2001)
Zingiber officinale	Increases level of testosterone	Rhizome aqueous and ethanol extract	Polyphenol, vit C, β-carotene, and flavonoids	Oyeyipo et al. 2014; Morakinyo et al. (2008)

21.2 Plants with Pro-fertility/Aphrodisiac Properties

21.2.1 *Asparagus racemosus*

Asparagus racemosus is a medicinal plant of family Asparagaceae that grows in tropical and subtropical India, Nepal, and Sri Lanka. It is commonly known as Shatavari or Shatmull. It has been a part of Vajikarana Rasayana in the traditional Indian medicinal system and was very popular for its aphrodisiac role. Shatavari roots contain majorly four steroid saponins, namely, shatavarin I–IV. The root powder of *A. racemosus* has already been tested for treatment of male infertility in rats (Fig. 21.1). It has been given to men in combination with other medicinal plants such as Gokshura and Ashwagandha (Devi et al. 2004). Hydro-ethanolic and aqueous extracts of the roots of *A. racemosus* were examined for male infertility treatment and found to be potent (Wani et al. 2011). We demonstrated that *A. racemosus* root powder restores spermatogenesis by reducing the level of reactive oxygen species and restoration of hormonal imbalance at the central endocrine axis (our unpublished study). Pure herb powder or syrup of Shatavari is available in the Indian market from different manufacturers with the brand names like Himalaya Shatavari (capsule and syrup), Patanjali Shatavari Churna (powder), Organic India Shatavari (capsule), and many others. Only a few side effects of *Asparagus* have been noted. It might cause weight gain in some people, and allergic reaction to herb may lead to coughing, runny nose, inflammation of skin, occupational asthma, etc. (Alok et al. 2013). Patients with edema because of kidney disorder or impaired heart function should not use Shatavari.

Fig. 21.1 Figure showing plants with pro-spermatogenesis/fertility activity

Fig. 21.1 (continued)

Plant Fruit Dried fruit

Tribulus terrestris

Plant Leaves and flower

Trichopus zeylanicus

Plant Root Dried root

Withania somnifera

Plant Flower Rhizome

Zingiber officinale

Fig. 21.1 (continued)

21.2.2 *Asteracantha longifolia*

A. longifolia is an annual herb of family Acanthaceae, which has been used in the ancient Indian medicinal system for management of several disorders. It is a native of tropical and subtropical regions and is abundantly present in India. It is commonly known as Kokilaksha and Ikshura. Roots, seeds, and leaves have been used in Indian medicinal system for treatment of various diseases, but seeds are conventionally used to treat sexual frailty, erectile dysfunction, premature ejaculation, and oligozoospermia (Chauhan et al. 2011). Ethanolic extract of seeds was administered to male rats for 28 days, and a significant increase in sperm count and fructose level of seminal vesicles was seen (Chauhan et al. 2011). Cardiac glycosides, phenols, steroids, saponins, and tannins have been proven as major chemical constituents present in *A. longifolia* by phytochemical screening (Doss 2009). "Speman" manufactured by Himalaya is a drug, which comprises of *A. longifolia* in combination with some other herbs (*M. pruriens*, *T. terrestris*, *W. somnifera*) for the treatment of erectile dysfunction and infertility. No human trial has been done on the powder of seed, leaf, or roots of plant, but "Speman" has been tested on humans. Speman has been tested on 30 idiopathic oligospermia patients with administration of two tablets twice daily for 6 months. A significant increase in sperm count and motility was observed (Agrawal and Kulkarni 2003). No side effect has been reported till date.

21.2.3 *Chlorophytum borivilianum*

Chlorophytum is a medicinal plant of family Papilionaceae with manifold therapeutic values. It is a native of tropical forest of peninsular India and commonly known as Safed Musli. Seeds and roots of Safed Musli are in use for treating male infertility since ages. It possesses adaptogenic and immunomodulatory properties, which treat impotency and sterility and enhance male potency. Aqueous extract of tuberous roots of *C. borivilianum* (CB) has already been evaluated for its aphrodisiac and spermatogenic potential on rats (Visavadiya and Narasimhacharya 2007; Kenjale et al. 2008; Thakur et al. 2009). Aqueous extract of CB has also been tested on infertile human patients and found to be effective in improving male sexual health (Rath and Panja 2013). Roots of *C. borivilianum* contain major chemical constituents such as steroidal saponins, namely, neotigogenin, neohecogenin, stigmasterol, tokorogenin, sapogenins, fructans, magnesium, etc. (Thakur et al. 2009). We could not find any drug based on *C. borivilianum*. There are no known side effects of Safed Musli if taken under prescribed doses. Higher doses may however lead to gastrointestinal problems.

21.2.4 *Crocus sativus*

Crocus sativus is a member of family Iridaceae, and dried stigma is known as saffron or kesar or kumkuma that is used for medicinal purposes since ages. This plant is a native of Greece and Southwest Asia and was first cultivated in Greece. Dried stigmas (threadlike parts of the flower) are used for medicinal purposes. A study designed to evaluate the antioxidant properties of saffron in humans used 50 mg with drinking milk administered three times a week and reported a significant increase in sperm count and motility (Heidary et al. 2008). Another recent study has been done on rats exposed to cadmium that showed improved sperm parameters in saffron-treated animals (Asadi et al. 2013). Aqueous extract of *C. sativus* and its active chemical constituent "crocin and safranal" were assessed for their aphrodisiac activity in male rats, and crocin was reported to be a potent aphrodisiac at all doses (100, 200, and 400 mg/kg body weight) and aqueous extract specially at doses 160 and 320 mg/kg body weight; however, safranal did not show aphrodisiac effects (Hosseinzadeh et al. 2008). This is popular as strong aphrodisiac and contains major chemical constituent such as crocin and picrocrocin. *C. sativus* is present in combination with some other medicinal plants in marketed drug "Speman forte Vet" manufactured by Himalaya for improvement of spermatogenesis and sperm quality. Side effects of *C. sativus* include dry mouth, anxiety, dizziness, nausea, and headache. Allergic reaction may be seen in some people. A large dose of saffron by mouth is unsafe as it can cause poisoning; bloody diarrhea; bleeding from nose, lips, and eyelids; and other serious side effects (Wüthrich et al. 1997). Doses of 12–20 g can cause death (Wüthrich et al. 1997).

21.2.5 *Curculigo orchioides*

It is often known as "Kali Musli" and "Golden Eye Grass" that belongs to the family Hypoxidaceae and is well known for its aphrodisiac character in the ancient Indian and Chinese medicinal systems. It is a native of China, Japan, Indian subcontinent, Papuasia, and Micronesia. Rhizome or roots of the plant are commonly used for medicinal purposes. The major chemical constituents present in *C. orchioides* are glycosides and polysaccharides such as starch, tannins, resin, hemicelluloses, mucilage, etc. Fresh rhizome contains sapogenin (yuccagenin) and alkaloids (lycorin) (Irshad et al. 2006). The rhizome of *C. orchioides* is described in Ayurveda as a Vajikarana Rasayana. Ethanolic extract of rhizomes was evaluated for sexual behavior of male rats and found to be potent for enhancing sexual performance (Chauhan et al. 2007). It also improves sperm count in heat-exposed rats as compared to heat-exposed positive control groups (Chauhan et al. 2007). No drug is available in the market that contains Kali Musli for male infertility management. It has not been tested for pro-male fertility effects in humans despite being a potential aphrodisiac. Therefore, further exploration of this product is warranted. There is no report of side effects of *C. orchioides* till date.

21.2.6 *Dioscorea bulbifera*

D. bulbifera is a native of the Indian subcontinent and Africa and is commonly known as "Varahi or air potato." It is a tropical and subtropical plant of family Dioscoreaceae. Tuber of *D. bulbifera* is observed as pungent, tonic, aphrodisiac, stomachic, and anthelmintic in Ayurveda. Antioxidative and hypolipidemic effects of *D. bulbifera* have been proven on high-cholesterol-diet rats in which diosgenin (steroidal saponins), a major chemical constituent of *Dioscorea* spp., showed increased level of superoxide dismutase (SOD) in the plasma and liver and catalase in erythrocytes and the liver (Son et al. 2007). It also contains bioactive compounds comprising saponins, alkaloids, flavonoids, tannins, phenols, vitamin C (ascorbic acid), and vitamin B complex (niacin, riboflavin, thiamin) (Okwu and Ndu 2006). This plant has not been investigated in detail for pro-fertility activity in humans. As it has antioxidant properties and vitamin B, it must be potent in treatment of male infertility and requires further attention of scientists in this field. We could not find any drug based on *D. bulbifera* for male infertility treatment. The overdose of air potato causes gastrointestinal reactions like vomiting, diarrhea, abdominal pain, etc. It may also cause certain damage to the liver and kidneys.

21.2.7 *Morinda citrifolia*

Morinda citrifolia is an evergreen tree of family Rubiaceae, which is commonly known as "Noni or Indian mulberry." It is a native of Southeast Asia and Australasia. Fruits, leaves, and roots are used to enhance the sexual strength in men. Juice of fruit is used as an alternative medicine for many ailments such as inflammatory, infections, and cancers. It has been explored for its effective adaptogenic property and relieves from stress in ICR mice (Wang et al. 2011). No study has been undertaken on humans for testing its pro-fertility effects. It contains majorly alkaloids, polysaccharides, lignans, etc. (Kamiya et al. 2004). The pure herb is available, but no formulated drug for treatment of male infertility in market. No information is available regarding its adverse effects.

21.2.8 *Mucuna pruriens*

Mucuna pruriens is a member of family Papilionaceae and has been used for the treatment for male sexual debility since ages. It is commonly known as "velvet bean, Yokohama velvet bean, cowage, and lacuna bean." The seeds of this plant are mainly used for treating male infertility in Indian medicinal system. L-Dopa is a major constituent present in *M. pruriens*. Along with L-dopa, this plant is an abundant source of alkaloids such as prurienine, prurieninine, prurienidine, etc. The seed powder of *M. pruriens* improves sperm count and motility by reducing the oxidative stress and DNA damage in Sprague-Dawley rats and is a potent product for management of free radical mediated disorders (Singh et al. 2013).

The defensive mechanisms of *M. pruriens* on rats as well as humans have already been investigated. Seed powder was orally administered to infertile men that significantly ameliorated physiological stress and seminal plasma lipid peroxide level, resulting in improved sperm count and motility (Shukla et al. 2010). Crude powder of seed and ethanolic extract of *M. pruriens* have been assessed for treatment of male infertility. The ethanolic extract of *M. pruriens* was administered to male Wistar albino rats for different doses and time periods. This increased mounting frequency, intromission frequency, and ejaculation latency and decreased the mounting latency, intromission latency, post-ejaculatory interval, and inter-intromission interval at a particular dose of 200 mg/kg body weight (Suresh et al. 2009). Aqueous extract could be used for the same purpose because extract in water was found to be active in the improvement of Parkinsonism (Lieu et al. 2010). We demonstrated that *M. pruriens* and its major chemical constituent L-dopa significantly recuperate the spermatogenic loss caused by ethinyl estradiol administration in SD male rats by lowering the reactive oxygen species level, reestablishing mitochondrial membrane potential and regulating the level of apoptosis (Singh et al. 2013). There are many drugs present in the Indian market, which contain *Mucuna* such as "Confido," "Speman," and "Tentex forte" by Himalaya wellness. It is safe with very few side effects like headache and pounding heartbeat and symptoms of psychosis including confusion, agitation, hallucination, and delusions (Infante et al. 1990).

21.2.9 *Tribulus terrestris*

It is a flowering plant of family Zygophyllaceae, a native of temperate and tropical regions of the Old World. The common names are Bullhead, Burra Gokharu, and Bindii. Roots and seed pods have been used in Ayurveda, Unani, and Siddha medicines and also in Chinese medicine since long back. Protodioscin is the major chemical constituent present in *T. terrestris*, which is known for its aphrodisiac properties. It restores libido by increasing the level of testosterone, dihydrotestosterone, and dehydroepiandrosterone sulfate (Gauthaman and Ganesan 2008). The experimental studies on *T. terrestris* have already been done on primates, rabbits, castrated rats (Gauthaman and Ganesan 2008), and humans (Sellandi et al. 2012). "Libilov" by Nutrica, Inc. is a marketed drug, which is composed of purified extracts of *T. terrestris*, *Ginkgo biloba*, and natural amino acid (L-arginine). It may cause stomach upset and also may affect the blood glucose level.

21.2.10 *Trichopus zeylanicus*

It is a rare berry plant commonly known as Arogyapacha or Kerala Ginseng that is native to India. It has been historically used by Kani tribe in India as an antifatigue medicine (Pushpangadan et al. 1988). It is a medicinal herb of family Dioscoreaceae with many pharmacological activities, such as antihepatotoxic, antifatigue, and antiulcer (Tharakan and Manyam 2005). The ethanolic extract of *T. zeylanicus* leaves has been proven active in the enhancement of mount and mating performance in male mice (Subramoniam et al. 1997). It contains NADH, polyphenols, and sulfhydryl compounds, which have the ability to lower ROS level. The antioxidant activity may be an important mechanism of action of *T. zeylanicus* to withstand fatigue. This plant also has adaptogenic (Singh et al. 2001) and aphrodisiac (Subramoniam et al. 1997) properties apart from antioxidant and antifatigue properties. Studies on humans could be undertaken using ethanolic extract of *T. zeylanicus* as it remains unexplored. We could not identify any drug based on this plant in the market that promises to improve sexual health or fertility. There are no reports on the side effects of *T. zeylanicus*.

21.2.11 *Withania somnifera*

Withania somnifera is commonly known as winter cherry and is a member of the family Solanaceae. It is a native of India, North Africa, and the Middle East. The practitioners of the traditional medicinal system in India account *W. somnifera* as "Indian Ginseng" (Grandhi et al. 1994). Tuberous roots, berries, and leaves are the plant parts, which were generally used in the traditional medicine system. The lyophilized aqueous extract of *Withania* was reported to increase testicular weight, diameter of seminiferous tubules, seminiferous epithelial cell layers, and serum levels of ICSH (interstitial cell stimulating hormone) with a synchronous reduction in serum testosterone and FSH level in immature Wistar rats (Abdel-Magied et al. 2001). We demonstrated that *W. somnifera* and its major constituent withaferin A improve spermatogenesis due to their restorative efficacy on reproductive hormones and reduction in oxidative stress (our unpublished study). We also carried out a trial on humans to evaluate the value of *W. somnifera* root extract in treating idiopathic male infertility (Mahdi et al. 2011). Administration of *W. somnifera* was found to improve semen quality by significantly reducing oxidative stress and level of serum FSH and prolactin, increasing serum testosterone, and luteinizing hormone levels (Ahmad et al. 2010). The major chemical constituents of *W. somnifera* are alkaloids such as withanine, somniferine, somniferinine, withananine, and somnine. "Himalaya Wellness" has manufactured some drugs in combination with *W. somnifera* with some other medicinal plants. These drugs are available in the market with the brand names such as "Confido," "Himplasia," "Speman," and "Tentex Forte." Although it is safe when taken by mouth for short time periods, overdoses of Ashwagandha may cause stomach upset, diarrhea, and vomiting.

21.2.12 *Zingiber officinale*

It is a flowering plant of the family Zingiberaceae whose rhizome has been used in the traditional medicine to treat various ailments like nausea, diarrhea, and arthritis for ages. It is indigenous to South China and eventually spread to other parts of Asia and West Africa. The Rhizome of plant has been used in traditional medicine for treatment of various disorders. Many studies have been done on pro-fertility properties of *Z. officinale.* The beneficial effects of *Z. officinale* on male reproductive function by increased sperm count, motility, and testosterone and decreased malondialdehyde level have been proven (Morakinyo et al. 2008). In a very recent study, aqueous extract of the rhizome of *Z. officinale* was found to have preventive effect on nicotine-induced infertility (Oyeyipo et al. 2014). Different animal models such as cyclophosphamide, aspartame-induced male rats, and male broilers have also been used, but no human trial has been done till date. Antioxidant components such as flavonoids, polyphenols, and tannins are the major chemical constituents present in different extracts (water, methanol, and ethanol) of *Z. officinale* (Prakash 2010). There is no drug available in the market for male infertility treatment. Burning feeling in mouth, abdominal pain, and diarrhea may occur due to overdose.

21.3 Discussion

Sexual potentiality is a principal element of quality of life and subject matter for well-being in humans. Several factors such as obesity, stress, alcohol, tobacco consumption, and excessive use of synthetic medicines have elevated the risk of erectile dysfunction and sexual impairment, a leading cause of male infertility. The number of erectile dysfunction and infertility cases has risen significantly over the last few decades. Several new factors such as pollution and lack of physical exercise contribute to the increasing number of infertility cases, a majority of which fail to show any defined cause. The semen quality has seen an appreciable decline over the last few decades (Auger et al. 1995; Le Moal et al. 2014; Borges Jr. et al. 2015; Romero-Otero et al. 2015). The rising trend of male infertility suggests far higher requirement of new medicinal options to tackle male infertility. Therefore, further efforts to explore medicinal options for infertility treatment must continue.

Most of the plants listed in this article provide various benefits that culminate into better overall and sexual health. Some of the common mechanisms of action include their effects on the central nervous system, which results in alleviated stress and better regulation of the central control of hormone release. Scientific studies have supported their action on the central nervous system as a number of these products improve the levels of peripheral hormones that ultimately result in higher level of testosterone and increased libido. Further, these products provide a number of flavonoids, alkaloids, and other ingredients for the synthesis of steroids and messengers that result in activation of signaling cascades leading to better blood flow to the reproductive organs, thereby resulting in a significant improvement in sexual function and health (Fig. 21.2). Some of the plants are well-known aphrodisiacs, which enhance sexual power, thus raising the confidence in the sexual act. Elevated sexual confidence makes sexual life pleasurable and increases the chances of regular coitus. The normalization of the circulating hormone levels and testosterone translates into a higher sperm count and motility, thereby increasing the chances of conception and fertility. Strong adaptogenic properties increase the level of nitric oxide and the level of reactive oxygen species, leading to a significant improvement in the qualitative parameters of sperm function. Apart from all the above, the nutritional elements present in these plants need further investigation, which could be one of the several reasons for their effects. The actual effect of a number of plant products on sperm count and motility has been demonstrated in animal models and human trials.

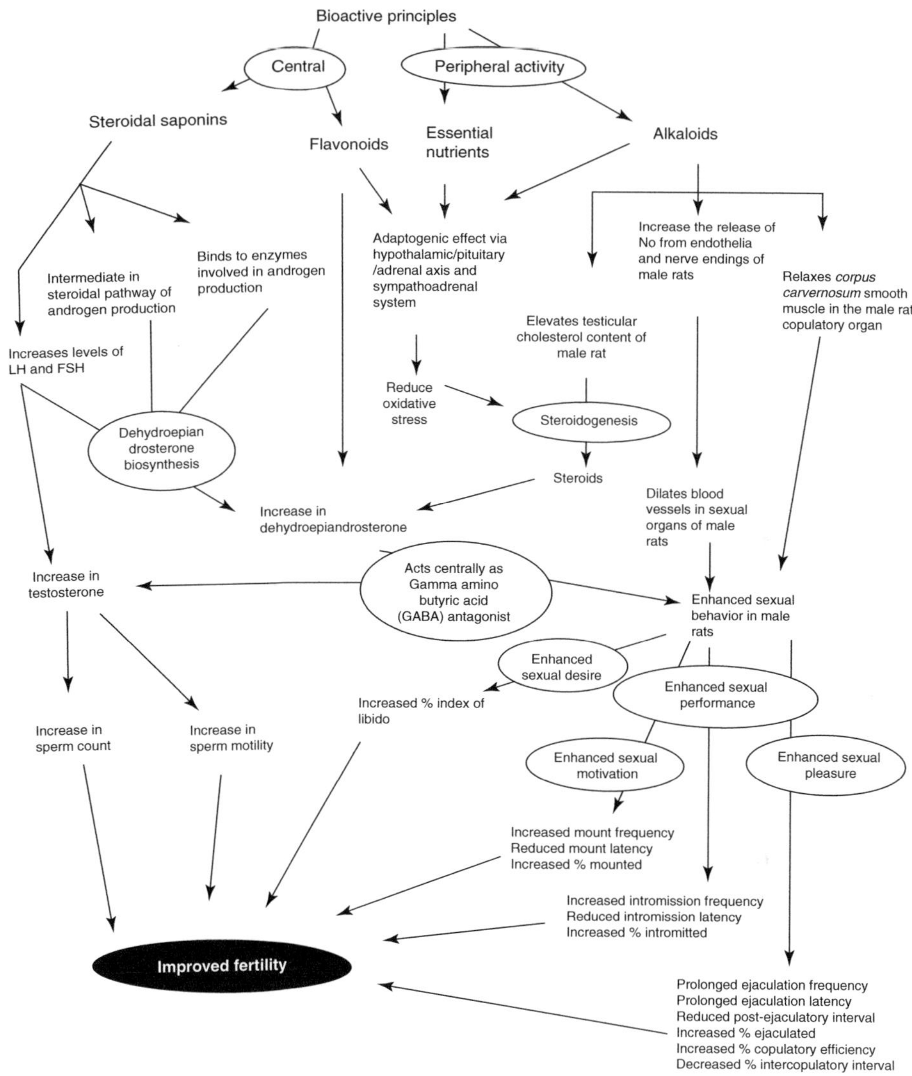

Fig. 21.2 Possible mode of action of pro-fertility plants (Adapted from Yakubu and Akanji 2011)

The plant products are not only used as medications for the patients but may also be used for prophylactic purposes to provide strength and immunity against fertility-related issues. In fact, the references in Ayurveda and other systems of medicine suggest their use for strength building. Since antioxidant mechanisms quench reactive oxygen species and free radicals to protect against damage to spermatozoa, antioxidant properties of medicinal plants play significant and diverse roles in reproduction and could possibly be used as preventive medication against sexual debility and infertility. Most of the scientific studies till date have not explored the available products from prophylactic point of view; therefore, further studies are required to identify their potential use as preventive medicine. Nevertheless, the products suggested to be good for fertility might be consumed for these purposes as they are grossly nontoxic and have been subjected to experimentation by common men over the last several years. Appropriate use of promising products could help not only in improving fertility but also in maintaining good reproductive and sexual health for a longer duration. Most of the plants used as aphrodisiac agents have not been rigorously experimentally studied on humans.

According to the World Health Organization (WHO) estimates, due to poverty and lack of access to modern medicine, about 65–80% of the world's population living in the developing countries depends essentially on plants for primary health care. The importance of medicinal plants in the management of male infertility in India as well as in other countries is indubitable. The plant-based products are gaining preference over other modern forms of medicine. As far as infertility is concerned, the scope of plant products is on its way to rise. The need of the hour is an exhaustive scientific research to generate support in evidence of their claimed benefits and assess toxic effects, if any. Toxicity evaluation should be made a part of the scientific studies. Therefore, exhaustive research into the chemical and biological properties and toxicological aspects of less explored medicinal plants is still needed to determine their aphrodisiac and pro-fertility properties. Most of the plant products are regarded as safe and are unlikely to cause side effects; however, the overdose may lead to some mild side effects. As mentioned above, the often-noted side effects are gastrointestinal problems due to the presence of various polyphenols in plant products and extracts.

21.4 Conclusion and Future Prospects

Successful treatment of sexual impairment may improve not only a sexual relationship but also the overall quality of life. The medicinal plants are popular among people since ages due to their affordability, safety, and higher efficacy. Curiosity in the traditional medicine has led to the expeditious research and studies of various herbal medications employed for sexual impairment. Many herbs have been used by people from different cultures to manage various conditions of male infertility such as erectile dysfunction, lack of libido, abnormal sperm, etc. The lack of clinical efficacy data and safety issues may be a concern for some of the potential users of these medications. Therefore, there is a crucial need to conduct clinical studies to endorse the traditional claims. All of the plants listed in this article are effective in the treatment of male infertility, but it remains to be explored if a combination would be more effective. For example, "Confido," "Himplasia," "Speman," etc., are some products available in the market, which contain more than one product in combination. Desire of natural aphrodisiacs necessitates rigorous studies to comprehend their effects on humans and address safety concerns.

The lack of human trials is a major concern in the promotion of traditional products for their pro-fertility effects. Human trials need to be undertaken for a large number of plants alone and in combination. As detailed above, only a few plants have been tried on human patients, and evidence for many others is based on literature or animal experiments. As plant products are safe without serious adverse effects, they can be tested on humans to accelerate drug discovery in this field. Most of the studies till date have been undertaken on animal models, particularly rodents. Since the effect may vary in the higher individuals, human trials on the selected products must be undertaken to gain further insights into their effects. Molecular mechanisms could be studied in the animal models. Apart from the analysis of the common markers of apoptosis and antioxidant mechanisms, it is required to expand research to study the effects on the central nervous system, hormone levels, and overall well-being.

One of the most challenging aspects of the future of plant products is regarding their translation into drugs by means of laboratory synthesis. The activity-guided fractionation of potential plants may lead to the identification of the active ingredients; however, this needs to be subjected to laboratory synthesis to make it economical and reduce the burden on natural sources. Identification of the active compounds would also accelerate drug development by means of preparation of substitutes with better absorption, efficacy, and reduced side effects. Most of the active plant compounds are quite complex in nature, which are not easy to synthesize. The use of some of these products in their natural form may be justified as far as the plant parts used do not affect the existence of the species and are easily reproduced. The ultimate aim of activity-guided fractionation is to identify the plant chemical for synthesis in the laboratory. Therefore, further research into this field needs a lot of scientific analysis to decide the depth of investigations and synthesis of the identified chemical constituents. Further advancements in the field of chemistry are equally important to get breakthroughs in laboratory synthesis of complex compounds, which are most abundant in plants and other natural resources.

Acknowledgment The authors would like to thank the Indian Council of Medical Research (Ref. No. 5/10/FR/5/2012-RHN) and the Council of Scientific and Industrial Research (CSIR), Govt. of India for the financial help.

References

Abdel-Magied EM, Abdel-Rahman HA, Harraz FM (2001) The effect of aqueous extracts of *Cynomorium coccineum* and *Withania somnifera* on testicular development in immature Wistar rats. J Ethnopharmacol 75(1):1–4

Agrahari AK, Panda SK, Meher A, Padhan AR, Khaliquzzama M (2010) Phytochemical screening of *Curculigo orchioides* Gaertn. Root tubers. J Chem Pharm Res 2(2):107–111

Agrawal HSK, Kulkarni S (2003) Efficacy and safety of Speman in patients with oligospermia: an open clinical study. Indian J Clin Pract 14:29–31

Ahmad MK, Mahdi AA, Shukla KK, Islam N, Rajender S, Madhukar D, Shankhwar SN, Ahmad S (2010) *Withania somnifera* improves semen quality by regulating reproductive hormone levels and oxidative stress in seminal plasma of infertile males. Fertil Steril 94(3):989–996

Alok S, Jain SK, Verma A, Kumar M, Mahor A, Sabharwal M (2013) Plant profile, phytochemistry and pharmacology of *Asparagus racemosus* (Shatavari): a review. Asian Pac J Trop Dis 3(3):242–251

Asadi MH, Zafari F, Sarveazad A, Abbasi M, Safa M, Koruji M, Yari A, Miran RA (2013) Saffron improves epididymal sperm parameters in rats exposed to cadmium. Nephrourol Mon 6(1):e12125

Auger J, Kunstmann JM, Czyglik F, Jouannet P (1995) Decline in semen quality among fertile men in Paris during the past 20 years. N Engl J Med 332(5):281–285

Borges E Jr, Setti AS, Braga DPDAF, Figueira RDCS, Iaconelli A Jr (2015) Decline in semen quality among infertile men in Brazil during the past 10 years. Int Braz J Urol 41(4):757–763

Chauhan NS, Rao CV, Dixit VK (2007) Effect of *Curculigo orchioides* rhizomes on sexual behaviour of male rats. Fitoterapia 78(7):530–534

Chauhan NS, Sharma V, Dixit VK (2011) Effect of *Asteracantha longifolia* seeds on the sexual behaviour of male rats. Nat Prod Res 25(15):1423–1431

Dalal PK, Tripathi A, Gupta SK (2013) Vajikarana: treatment of sexual dysfunctions based on Indian concepts. Indian J Psychiatry 55(Suppl 2):S273

Devi PR, Laxmi V, Charulata C, Rajyalakshmi A (2004) "Alternative medicine"—a right choice for male infertility management. In: International congress series, vol 1271. Elsevier, Amsterdam, pp 67–70

Doss A (2009) Preliminary phytochemical screening of some Indian Medicinal Plants. Ancient Science of Life 29(2):12–16

Gauthaman K, Ganesan AP (2008) The hormonal effects of *Tribulus terrestris* and its role in the management of male erectile dysfunction–an evaluation using primates, rabbit and rat. Phytomedicine 15(1):44–54

Grandhi A, Mujumdar AM, Patwardhan B (1994) A comparative pharmacological investigation of Ashwagandha and ginseng. J Ethnopharmacol 44(3):131–135

Heidary M, Vahhabi S, Reza Nejadi J, Delfan B, Birjandi M, Kaviani H, Givrad S (2008) Effect of saffron on semen parameters of infertile men. Urol J 5(4):255–259

Hirazumi A, Furusawa E (1999) An immunomodulatory polysaccharide-rich substance from the fruit juice of *Morinda citrifolia*(noni) with antitumour activity. Phytother Res 13(5):380–387

Hosseinzadeh H, Ziaee T, Sadeghi A (2008) The effect of saffron, *Crocus sativus* stigma, extract and its constituents, safranal and crocin on sexual behaviors in normal male rats. Phytomedicine 15(6):491–495

Infante M, Perez A, Simao M, Manda F, Baquete E, Fernandes A, Cliff J (1990) Outbreak of acute toxic psychosis attributed to *Mucuna pruriens*. Lancet 336(8723):1129

Irshad S, Singh J, Jain SP, Khanuja SPS (2006) Curculigo orchioides Gaertn.(Kali Musali): an endangered medicinal plant of commercial value. Nat Prod Rad 5(5):369–372

Kamiya K, Tanaka Y, Endang H, Umar M, Satake T (2004) Chemical constituents of *Morinda citrifolia* fruits inhibit copper-induced low-density lipoprotein oxidation. J Agric Food Chem 52(19):5843–5848

Kenjale R, Shah R, Sathaye S (2008) Effects of *Chlorophytum borivilianum* on sexual behaviour and sperm count in male rats. Phytother Res 22(6):796–801

Le Moal J, Rolland M, Goria S, Wagner V, De Crouy-Chanel P, Rigou A, De Mouzon J, Royère D (2014) Semen quality trends in French regions are consistent with a global change in environmental exposure. Reproduction 147(4):567–574

Lieu CA, Kunselman AR, Manyam BV, Venkiteswaran K, Subramanian T (2010) A water extract of *Mucuna pruriens* provides long-term amelioration of parkinsonism with reduced risk for dyskinesias. Parkinsonism Relat Disord 16(7):458–465

Mahdi AA, Shukla KK, Ahmad MK, Rajender S, Shankhwar SN, Singh V, Dalela D (2011) Withania somnifera improves semen quality in stress-related male fertility. Evid Based Complement Alternat Med 2011.

Morakinyo AO, Adeniyi OS, Arikawe AP (2008) Effects of *Zingiber officinale* on reproductive functions in the male rat. Afr J Biomed Res 11(3)

Okwu DE, Ndu CU (2006) Evaluation of the phytonutrients, mineral and vitamin contents of some varieties of yam (Dioscorea sp.) Int J Mol Med Adv Sci 2(2):199–203

Oyeyipo IP, Obembe OO, Oladokun OO, Raji Y (2014) Sperm function and fertility profile following nicotine administration in male rats: protective potentials of *Zingiber officinale*. Int J Green Pharm 8(2)

Prakash J (2010) Chemical composition and antioxidant properties of ginger root (*Zingiber officinale*). J Med Plant Res 4(24):2674–2679

Pushpangadan P, Rajasekharan S, Ratheshkumar PK, Jawahar CR, Nair VV, Lakshmi N, Amma LS (1988) 'Aogyappacha' (*Trichopus zeylanicus* gaerin), the 'Ginseng'of Kani tribes of Agashyar hills (Kerala) for ever green health and vitality. Anc Sci Life 8(1):13

Rath SK, Panja AK (2013) Clinical evaluation of root tubers of Shweta Musali (*Chlorophytum borivilianum* L.) and its effect on semen and testosterone. *Ayu* 34(3):273

Romero-Otero J, Medina-Polo J, García-Gómez B, Lora-Pablos D, Duarte-Ojeda JM, García-González L, García-Cruz E, Rodríguez-Antolín A (2015) Semen quality assessment in fertile men in Madrid during the last 3 decades. Urology 85(6):1333–1338

Sellandi TM, Thakar AB, Baghel MS (2012) Clinical study of *Tribulus terrestris* Linn. in Oligozoospermia: a double blind study. *Ayu* 33(3):356

Shukla KK, Mahdi AA, Ahmad MK, Jaiswar SP, Shankwar SN, Tiwari SC (2010) *Mucuna pruriens* reduces stress and improves the quality of semen in infertile men. Evid Based Complement Alternat Med 7(1):137–144

Singh B, Gupta DK, Chandan BK (2001) Adaptogenic activity of a glyco-peptido-lipid fraction from the alcoholic extract of Trichopus zeylanicus Gaertn. Phytomedicine 8(4):283–291

Singh AP, Sarkar S, Tripathi M, Rajender S (2013) *Mucuna pruriens* and its major constituent L-DOPA recover spermatogenic loss by combating ROS, loss of mitochondrial membrane potential and apoptosis. PLoS One 8(1):e54655

Son IS, Kim JH, Sohn HY, Son KH, Kim JS, Kwon CS (2007) Antioxidative and hypolipidemic effects of diosgenin, a steroidal saponin of yam (Dioscorea spp.), on high-cholesterol fed rats. Biosci Biotechnol Biochem 71(12):3063–3071

Subramoniam A, Madhavachandran V, Rajasekharan S, Pushpangadan P (1997) Aphrodisiac property of *Trichopus zeylanicus* extract in male mice. J Ethnopharmacol 57(1):21–27

Suresh S, Prithiviraj E, Prakash S (2009) Dose-and time-dependent effects of ethanolic extract of *Mucuna pruriens* Linn. Seed on sexual behaviour of normal male rats. J Ethnopharmacol 122(3):497–501

Thakur GS, Bag M, Sanodiya BS, Debnath M, Zacharia A, Bhadauriya P, Prasad GBKS, Bisen PS (2009) Chlorophytum borivilianum: a white gold for biopharmaceuticals and neutraceuticals. Curr Pharm Biotechnol 10(7):650–666

Tharakan B, Manyam BV (2005) Botanical therapies in sexual dysfunction. Phytother Res 19(6):457–463

Visavadiya NP, Narasimhacharya AVRL (2007) Ameliorative effect of *Chlorophytum borivilianum* root on lipid metabolism in hyperlipaemic rats. Clin Exp Pharmacol Physiol 34(3):244–249

Wang MY, Hurn J, Peng L, Nowicki D, Anderson G (2011) A multigeneration reproductive and developmental safety evaluation of authentic *Morinda citrifolia* (noni) juice. J Toxicol Sci 36(1):81–85

Wani JA, Achur RN, Nema RK (2011) Phytochemical screening and aphrodisiac activity of *Asparagus racemosus*. *Studies* 8:9

Wüthrich B, Schmid-Grendelmeyer P, Lundberg M (1997) Anaphylaxis to saffron. Allergy 52(4):476–477

Yakubu MT, Akanji MA (2011) Effect of aqueous extract of *Massularia acuminata* stem on sexual behaviour of male Wistar rats. Evid Based Complement Alternat Med 2011:738103

Lifestyle, Fertility, and Infertility Management

22

Rajender Singh

Abstract

With technological advancements, we have evolved into a species that is surrounded by a number of potential hazards that may endanger our own survival. One of the important aspects of the effect of modern lifestyle is increased incidence of infertility. Renewal being the most important requirement for a species, fertility loss can have radical consequences. Since fertility does not need to be earned, it is taken to be immune to changes in lifestyle and surroundings. While a number of poor lifestyle practices are adopted, we fall prey to other hazards inadvertently. We recognized the effect of lifestyle on fertility a little late, but fortunately it is neither too late nor too difficult to confront the lifestyle factors that may take a heavy toll on fertility. In this chapter, I have provided the most comprehensive review of a number of lifestyle factors that matter to fertility and simple ways to overcome their potential perils. Following a great lifestyle ardently may be a difficult task, but can have great therapeutic rewards.

Keywords

Lifestyle • Mobile phone • Laptop use • Wi-Fi radiation • Testicular heating • Smoking • Alcohol • Stress • Sleep disorders

R. Singh
Male Reproductive Biology Laboratory, CSIR-Central Drug Research Institute, Lucknow, Uttar Pradesh, India
e-mail: rajender_singh@cdri.res.in

R. Singh, K. Singh (eds.), *Male Infertility: Understanding, Causes and Treatment*, DOI 10.1007/978-981-10-4017-7_22

Key Points

- Laptop use, Wi-Fi radiations, television watching, occupational exposure, tight undergarments, and obesity can cause testicular heating.
- Paternal smoking is now well known to cause altered meiosis in testes, quantitative loss of semen parameters, contribute to pregnancy failure, and increase the risk of childhood cancer.
- Paternal alcohol use (moderate to heavy) is known to contribute to fetal alcohol syndrome, which encompasses a number of disorders seen in the children of alcoholics.
- Stress in various forms is a killer of fertility; failure of infertility treatment raises stress even further, making it more difficult to treat these individuals.
- Mobile phone radiations in heavy users may contribute to the loss of sperm production and quality, which may be compounded by Wi-Fi radiations.
- Disturbance of day-night cycle, abnormal/prolonged working hours, late-night television watching/laptop usage, etc. can contribute to sleep disturbance and decreased testosterone production and fertility.
- Exercise in general is good for fertility, but heavy exercise, for example, heavy cycling, may contribute to reduced semen quality and infertility.

22.1 Introduction

Prevention is better than cure is an old and apt proverb. It is applicable in almost every domain of life, in healthcare, the most. The adaptability of the biological system resists adverse changes in the body; therefore, development of an ailment takes place over a period of several years. Diagnosis of a disorder is generally done at a stage when significant damage to the body has already been caused. Therefore, treatment may show only partial recovery and may restore the original condition, and some of the changes are altogether irreversible. There are near and far connections in various molecular and organ systems of human body. Therefore, the damage caused by a clinical disorder is generally not restricted to the affected organ, making the management further difficult. Therefore, prophylaxis is much better than cure, especially in those that are modifiable by human causes.

The term lifestyle, which has for the last about a century described a central concept of Adlerian psychology, has recently gained importance in general and has become everyday vocabulary. For Adler, lifestyle represented the organismic ideas of the individual as an actor rather than a reactor of the purposiveness, goal-directedness, unity, self-consistency, and uniqueness of the individual and of the ultimately subjective determination of his actions (Adler 1992). The use of the term lifestyle has increased in the last few decades owing to a number of illnesses related to it as the causal or compounding factor. We are equipped with machines and technologies that make us more sedentary, expose to heat and radiations, potentially increasing the risk of a number of lifestyle-related disorders. Apart from technological advances, individual account of food habits, smoking, drinking, and other activities influences a number of health aspects. Due to sophistication in the recent times, the number of lifestyle disorders has gone up alarmingly. Cardiovascular disorders, atherosclerosis, stroke, hypertension, diabetes, and obesity are some of the disorders that have already attained top ranks among lifestyle disorders.

The incidence of infertility has increased in the last few decades. A number of studies have shown that the overall semen quality has declined significantly over the last few decades, and this correlates with increased incidence of infertility and other reproductive health disorders. The decline in semen quality is thought to be the result of a compound of factors, including changes in lifestyle. Unfortunately, the impact of lifestyle factors on reproductive health and fertility potential was recognized relatively late. It has come to light that a number of lifestyle factors such as the use of mobile phone, laptop computers, Wi-Fi and other radiations, smoking, drinking, hot baths, excessive exercise, stress, and sleep disorders may affect spermatogenesis and semen quality. Well-planned scientific studies have now established a number of these factors to negatively correlate with semen quality and fertility. Fortunately, lifestyle is the most easily modifiable factor. The current chapter presents a collective account of the lifestyle factors that have a significant impact on reproductive potential and fertility of men.

22.2 Activities Causing Testicular Heating

In humans and most of the mammals, spermatogenesis is temperature dependent and takes place at 2–8 °C below the body temperature (Ivell 2007). The difference in the temperature between the testicles and body was first reported in 1945 by Badenoch that was subsequently confirmed by others (Harrison and Weiner 1949; Kitayama 1965). Maintenance of this temperature difference is critical to spermatogenesis, and two mechanisms are in place to regulate the same. The scrotum is capable of expanding the skin surface for heat exchange and the other is regulation at the level of spermatic cord where heat is exchanged with the incoming blood. Testicular temperature is strongly dependent on the scrotal position and body posture. As in the case of most mammals, the scrotum in humans is supposed to hang, which keeps its temperature at the optimum (Zorgniotti and Macleod 1973). The best heat dissipation is possible in hanging and uncovered scrotum, and scrotal temperature is lower in walking posture in comparison to sitting posture. Ambulatory posture keeps testes away from body, and sitting or lying posture brings the testis between or on the thighs, raising their temperature by few degrees depending upon position, posture, and duration (Rock and Robinson 1965).

Experimental studies have demonstrated that testicular hyperthermia suppresses spermatogenesis in rats (Guo et al. 2007), mice (Li et al. 2013), monkeys (Zhang et al. 2006), and humans (Rao et al. 2015). Increasing the temperature of testicles by 1–3 °C can cease spermatogenesis altogether. The importance and impact of temperature on spermatogenesis can be understood by the fact that testicular heating has been tried as a method of contraception (Mieusset and B'ujan 1994). In one such technique, testes were fixed nonsurgically close to the inguinal canal by passing penis and the empty scrotum through a hole made in a close-fitting underwear.

Fig. 22.1 Lifestyle and other factors that adversely affect spermatogenesis and male fertility

In another method, immobilization was achieved by adding a ring of soft material surrounding the hole in the underwear. Both these techniques demonstrated that a daily mild increase in testicular temperature can act as a significant contraceptive (Mieusset and B'ujan 1994). Scientific studies have identified numerous external factors such as seating posture, clothing, lifestyle, use of laptop, television watching, playing video games, etc. that can result in testicular heating and impaired fertility. A number of these and other factors that affect spermatogenesis and fertility are detailed in Fig. 22.1.

22.2.1 Use of Laptop

Laptop computers have now become an indispensable part of the contemporary lifestyle. Apart from using the computers and electronics at office, we remain clung to these devices at home, even if for different reasons. From office work to household chores and entertainment, laptop computers are indispensable. Laptops are known to reach high internal temperature during continuous usage. Sheynkin et al. (2005) analyzed the effect of computer heat on scrotal temperature in 29 healthy volunteers in two separate sessions of 60 min. The study compared the testicular temperature with laptop in its position with approximated thighs without laptop. Rise in scrotal temperature with working laptop in lap position (2.88 °C) was significantly higher in comparison to the increase in the comparison group (2.18 °C) (Sheynkin et al. 2005). Since spermatogenesis is known to be very sensitive to temperature, a simple increase in testicular temperature by one degree can hamper spermatogenesis significantly.

Later, Sheynkin et al. (2011) undertook an observational study on 29 volunteers to find out the ways to protect elevation of scrotal temperature by means of using a lap pad and legs-apart sitting position in laptop users. The authors found that a lap pad is not an effective method of avoiding testicular heating; however, sitting in legs apart position may help in reducing scrotal hyperthermia in combination with a shorter duration of laptop use (Sheynkin et al. 2011). The use of laptop computers is increasing and serves many purposes. Therefore, simple precautions during laptop use can help evade the ill effects of heat on spermatogenesis. Laptop, as the name indicates, is generally kept in lap, particularly in a household setting. Laptop should be kept on a tabletop for work, and one should try to maintain the maximum possible distance between the testes and laptop. In case a table is not available, laptop should be kept away from the groin area. It could be kept on other surface such as a thick and heat-resistant lap pad. In any of these cases, the user should try to be in legs-apart position for effective testicular cooling. It is important to take short breaks during continued laptop use. For example, a 10-min break every 1 h of laptop use should help in protecting against testicular heating. Such breaks should preferably be used to take a short walk for better air circulation to the scrotal area.

22.2.2 Wi-Fi Radiations

In every home or office, we are surrounded by Wi-Fi signals that are used to communicate between various wireless applications that connect with each other or internet. With the number of wireless devices increasing, the use of 2.4 GHz EMRs has become very common, thus increasing exposure to these radiations. Electromagnetic radiations from various sources such as satellite links, wireless communication, microwave oven, frequency modulation radio, and television transmitters/antennas are some of the common indoor/outdoor EMR spectrum that we are exposed to most often. Widespread use of scientific, medical, industrial, military, and domestic applications using 2.45 GHz radio-frequency radiations is inevitable and a low-cost technology. Due to widespread and unlicensed use of this spectrum of radiations, their leakage into the environment and exposure are common. It has already been shown that 2.45 GHz microwave exposure causes an increase in caspase-3 and creatine kinase activities and a decrease in plasma level of testosterone and melatonin in exposed rats (Avendano et al. 2012). It is believed that exposure to Wi-Fi can increase the production of reactive oxygen species (Naziroğlu and Gümral 2009). This has been shown to lead to an increase in lipid peroxidation and a detrimental effect on reproductive tissues (Shokri et al. 2015).

In a study to evaluate the effect of Wi-Fi radiations on sperm parameters, Avendano et al. (2012) exposed semen samples of 29 healthy donors to wireless internet-connected laptop for 4 h and compared the results with a control sample incubated under the same conditions without Wi-Fi exposure. The authors found a significant decrease in progressive sperm motility and increase in sperm DNA fragmentation (Avendano et al. 2012). de Gannes et al. (2013) evaluated the effect of exposure to 2.45 GHz Wi-Fi signal on the reproductive system of male and female Wistar rats. However, the authors did not find any deleterious effect of Wi-Fi exposure on reproductive organs and fertility (de Gannes et al. 2013). In a recent study to assess the impact of Wi-Fi radiations on sperm parameters and testicular histomorphometry, Shokri et al. (2015) exposed rats to 2.45 GHz radiation. The animals were exposed to Wi-Fi radiations to different degrees for 2 months, finding that Wi-Fi exposure resulted in a deleterious effect in a time-dependent manner (Shokri et al. 2015).

The use of Wi-Fi radiations is unavoidable and growing. The scientific literature is divided on the actual impact of these radiations on reproductive organs. Nevertheless, some negative effect on sperm motility and DNA integrity is possible. Further well-planned studies must uncover any potential effect of these radiations on fertility. Nevertheless, general precautions in the use of Wi-Fi should help avoid their effect on spermatogenesis. The Wi-Fi should be switched off when not in use, for example, during nighttime, and Wi-Fi should preferably not be installed close to the area where you spend most of your time. The Wi-Fi should be of adequate strength according to the requirement, and the use of extra strong Wi-Fi despite a small coverage area should be avoided. Since these radiations are invisible, one has to be conscious about their presence, and simple precautions in their use can easily cut down exposure to and unwanted effects of these radiations.

22.2.3 Hot Bath and Spring

There are a number of activities, which may increase scrotal temperature temporarily. The modern lifestyle has introduced a number of such activities, for example, hot tubs, hot baths, steam baths, Jacuzzi, and a general habit of taking bath in hot water. Exposure to wet heat has been shown to be more detrimental in comparison to dry heat. Shefi et al. (2007) analyzed semen parameters in a cohort of men who had a remarkable history of wet heat exposure in the form of hot baths. The study compared the semen parameters before and after discontinuation of such activities and found that sperm count and motility showed significant improvements upon discontinuation of exposure to wet heat. The study concluded that exposure to wet heat has adverse but reversible effect on semen quality (Shefi et al. 2007). Hot spring is very popular in daily life, particularly in China. A number of hotels and resorts provide the facility of hot springs. Hot bath lovers may have this facility built at home. In a hot spring, the temperature is typically between 37 and 45 °C, which is much higher than scrotal temperature (Rao et al. 2016). This exposure is anticipated to be detrimental to spermatogenesis.

An experimental study provided a proof that hot baths associated with testicular heating can damage spermatogenesis and compromise semen quality. A recent prospective randomized clinical study was undertaken on 20 normozoospermic subjects who were divided into two groups for different schedules of hot baths at 43 °C for ten times of 30 min each (Rao et al. 2016). One of the groups underwent testicular warming for ten consecutive days and those in second group once every 3 days. Both the groups showed significant and reversible changes in disrupted mitochondrial membrane potential, sperm apoptosis, high DNA stainability, and changes in the expression of a number of proteins involved in heat stress and mitochondrial function. The study demonstrated that transient and frequent scrotal hyperthermia causes severe and reversible damage to spermatogenesis and that consecutive exposure had serious damage in comparison to intermittent exposure (Rao et al. 2016). As mentioned above, the impact of wet heat is much more in comparison to dry heat. Therefore, the use of hot baths, hot tubes, steam bath, sauna, hot springs, and any other activity that can cause testicular heating should be minimized, particularly in a repeated and regular manner.

22.2.4 Television Watching/Video Games

We could identify only two studies on the impact of television watching on semen parameters. In a study to evaluate the impact of physical activity and television watching, Gaskins et al. (2013) analyzed 189 men aged 18–22 for hours and type of physical activity and hours of TV watching over a period of 3 months (Gaskins et al. 2013). Comparison of data with semen parameters found that sperm count was directly related with moderate to vigorous activity such that men in the highest quartile of moderate to vigorous exercise (>15 h/week) had 73% higher sperm concentration than men in the lowest quartile (<5 h/week). TV watching was found to inversely correlate with sperm concentration and total sperm count. Men in the highest quartile of TV watching (>20 h/week) had 44% lower sperm concentration than men in the lowest quartile (0 h/week) (Gaskins et al. 2013; British J of Sports Medicine). A recent cross-sectional study on 1,210 healthy young Danish men undergoing fitness test for military services found that time spent watching television was associated with a poor sperm count (Priskorn et al. 2016). Quantitatively, men who watched television for more than 5 h per day had an average adjusted sperm count of 37 million/ml in comparison to 52 million/ml in the control group (Priskorn et al. 2016).

Some of the possible factors that may affect semen parameters as a result of television watching are sedentary time, testicular heating, and lack of physical movement. Depending upon the position, television watching may bring the scrotum in between or on the thighs, resulting in testicular heating. In case of cross-leg sitting in front of television, heat dissipation from the testis may get compromised. Further, people tend to eat while watching television, which may contribute to weight gain and increased BMI. Long hours of television watching correlate with poor physical activity. Increased BMI and poor physical activity are well known to have a significant negative correlation with sperm production and fertility. In order to avoid adverse impact, television watching should be minimized and interrupted by short breaks. Further, watching television should not be admixed with untimely eating. During television watching, one must be conscious about the position of testes in order to avoid prolonged testicular heating. The body posture and choice of clothes should be decided to facilitate adequate testicular cooling.

22.2.5 Occupational Exposure

Epidemiological studies have emphasized that occupational exposure to high temperatures such as in the case of electric welders, drivers, etc. can have deleterious side effects on spermatogenesis because of scrotal heat stress (Thonneau et al. 1997; Hjollund et al. 2000). Sitting on chair for long hours may result in poor scrotal cooling, which could be further complicated by cushioned chairs and tight underwear and jeans as well as trousers. In an interesting study on the impact of sitting hours and the type of chair used for this purpose, Koskelo et al. (2005) found that a statistically highly significant increase of up to 3 °C was recorded when subjects were sitting on commonly used cushioned chairs in comparison to the subjects sitting on saddle chair (Koskelo et al. 2005). The study concluded that the chairs, which increase scrotal temperature, may have contributed to a decline in semen quality in the sedentary society (Koskelo et al. 2005). The studies evaluating the impact of sitting hours have been rare, and further well-planned studies taking into consideration a number of parameters with respect to outer clothing, undergarments and number of sitting hours, and frequency and duration of intermittent breaks for underscoring the actual impact of prolonged seating on semen quality parameters are required.

With the recent advancements, a number of people are into sedentary jobs that may require continuous seating for several hours. As most of the working class spends about one third of their time in offices, studying the impact of seating hours on semen quality is extremely important. A number of occupations have been studied, but a job requiring sedentary office life needs to be investigated further for correlation with semen parameters. Indirect evidence suggests that sedentary sitting in chair can cause testicular heating; therefore, as a general precaution to avoid adverse effects on fertility, the office chair should be selected with care. For example, the chair should be less cushioned so as to allow maximum airflow around the groin area. Cross-leg seating for long duration and sitting in a sophisticated cushioned chair may compromise testicular cooling. The material of chair should allow adequate airflow, and the use of leather or any other synthetic and impermeable material should be avoided. Short breaks at work should be combined with a short walk so as to facilitate testicular cooling.

22.2.6 Choice of Clothes/Underwear

Layers of clothing impede heat exchange and can cause testicular heating. Clothing has been found to elevate scrotal temperature by 1.5–2 °C in comparison to unclothed state (Zorgniotti et al. 1982). Therefore, it is important to choose clothes carefully, particularly underwear and pants, in order to maximize airflow around testicles. A number of observational studies have been undertaken to assess the impact of loose or tight underwear on semen quality in individuals seeking infertility evaluation (Oldereid et al. 1991; Parazzini et al. 1995; Jung et al. 2001; Povey et al. 2012; Pacey et al. 2014). Some of these studies have reported an association of loose underwear with better semen quality (Parazzini et al. 1995; Jung et al. 2001; Povey et al. 2012), though others have shown no association (Oldereid et al. 1991; Pacey et al. 2014). Most of the interventional studies have shown that using underwear type device to hold the testes close to the body resulted in reduced semen quality parameters (Shafik 1992; Mieusset and B'ujan 1994; Ahmad et al. 2012). At least one study reported complete azoospermia after several months of the use of tight underfitting (Shafik 1992), another reported higher high DNA stainability and high DNA fragmentation index (Ahmad et al. 2012), and yet another reported no effect (Wang et al. 1997).

Some other studies randomized men to wear briefs and boxers for several months and found that using loose underwear associated with good semen quality, though the sample size was a limitation with these studies (Sanger and Friman 1990; Tiemessen et al. 1996). In an interesting recent study on the impact of the type of underwear on male fecundity, Sapra et al. (2016) collected data from a prospective preconception cohort conducted in 16 counties in Michigan and Texas, USA. Five hundred one couples were enrolled and followed for 12 months, during which semen analysis was undertaken and the men were classified into six categories on the basis of the type of underwear worn during daytime and bedtime. The classes were (1) briefs day/night, (2) boxer briefs day/night, (3) boxers day/night, (4) briefs day and boxers/none at night, (5) boxer briefs day and boxers/none at night, and (6) boxers day and none at night. Interestingly, it was found that men switching from their usual daytime underwear to boxers/none for bed showed the most prominent evidence of differences in semen quality endpoints in comparison to those sticking to briefs day/night. The study concluded that the choice of underwear during day/bed is associated with differences in semen parameters, though it did not correlate with time to pregnancy (Sapra et al. 2016).

It is conceivable from the above discussion that the choice of clothes, particular underwear, may affect testicular cooling and hence spermatogenesis. Buying clothes may be a matter of significant choice, but with evidence from the recent studies, buying under covers also requires your careful attention with respect to size and material. Cotton garments are likely to facilitate better air exchange in comparison to other materials. Choosing a loose underwear should help in effective testicular cooling. The brief-style underwear, which presses the scrotum close to the pelvic floor, should be avoided. The use of boxer-style underwear should be preferred, as their use is known to correlate with better semen quality. Further, when off work, the use of tight underwear should be avoided. During bedtime, switching to boxers without a pant/trouser over it is a good idea. The use of shorts during holidays and off-work hours should help in better air circulation around the scrotal area. General sensitization to the testicular heating should help one choose appropriate under- and overgarments for office and off-work hours.

22.2.7 Obesity and Increased BMI

Obesity is characterized by high BMI and excess body fat or white adipose tissue. An individual with BMI in 25–30 kg/m^2 is classified as overweight, and in excess of 30 kg/m^2 is classified as obese, though there are other more specific measures of obesity such as waist-to-hip ratio. Dietary factors in complex with sedentary lifestyle have given rise to the problem of obesity. According to WHO, about 16 billion adults were classified as overweight and 400 million as obese in 2005 (WHO 2009). Obesity is now rampant even among young school-going children. Fatty tissue converts testosterone to estrogen by increased aromatase activity (Roth et al. 2008). Dysregulated ratio of estrogen to testosterone affects spermatogenesis and other aspects of male reproduction. Among other problems in obesity that contribute to infertility are decreased libido and increased incidence of erectile dysfunction (Cheng and Ng 2007). Obesity not only leads to increased mass in thighs that presses testicles closer to them but also may result in fat deposition in testes (du Plessis et al. 2010). Increased testicular heating and dysregulation of hormone production and balance leads to impaired spermatogenesis (Hammoud et al. 2008).

Obesity is the root or associated cause of a number of disorders. With the advent of obesity starts a phase of poor overall health, which slowly starts affecting every organ of the body. Obese and diabetic males are frequently diagnosed with sleep apnea, characterized by fragmented sleep course due to repeated episodes of airway obstruction and hypoxia. Patients with sleep apnea may develop disturbance of pituitary-gonadal axis that affects night rise in the level of testosterone. A study on the male partners of 471 infertile couples from Korea measured testicular heating in relation with body mass index (BMI). The study reported that the temperature difference between the thigh and testicles was the highest (up to 1.5 °C) in the underweight and normal groups and the least in the obese group (Jo and Kim 2016). Testicular heating in the obese individuals has been linked with poor semen quality by a number of studies (du Plessis et al. 2010). The impact of obesity on male factor infertility has been covered in detail in Chap. 11.

Obesity and high BMI are well known causes of a number of disorders, which contribute significantly to morbidity and mortality. Obesity should be avoided as far as possible by choosing better lifestyle in the form of eating habits and exercise regimen. Once obesity signs in, it is very hard to get rid of its adverse health implications. Increased weight makes it difficult to adhere to a strict exercise schedule, thus, making it very difficult to regain fitness. Therefore, keeping away from obesity and high BMI is the best option. Regular exercise to maintain BMI in the normal range should be one of the most important objectives of daily timetable. Fortunately, it requires only as little as 4% (45 min) of a day or 2% (45 min × 4 days) of a week's total time for adequate workout.

22.2.8 Sports and Exercise

A number of sports activities such as soccer, hockey, cricket, and others require securing the testicles in a pouch for safety purposes. During intense activities in sports, a lot of heat generated in and around the groin area is not dissipated optimally. This can result in temporary hyperthermia in testes. There is evidence from indirect and related studies that sports and intense exercise can result in testicular hyperthermia (Vaamonde et al. 2016a, b). Similar innerwear devices have been demonstrated to impede spermatogenesis and have contraceptive effect (Mieusset and B'ujan 1994). In one such technique, testes were fixed nonsurgically close to the inguinal canal by passing penis and the empty scrotum through a hole made in a close-fitting underwear. In another method, immobilization was achieved by adding a ring of soft material surrounding the hole in the underwear. Both these techniques demonstrated that a daily mild increase in testicular temperature could act as a significant contraceptive (Mieusset and B'ujan 1994).

For all sports or exercise regimens, it must be kept in mind that the activities that require wearing tight undergarments or wrapping the testicles against the abdomen should be avoided in order to preserve fertility. In general, individuals participating in intense exercise as an obligation should avoid testicular tethering against the pelvic floor in order to avoid adverse impact on fertility. Testicles are meant to hang and let them follow their natural course as far as possible. The effects of testicular heating are reversible, and hence recovery is easily possible if such events are rare and infrequent. Nevertheless, repeated episodes, particularly on a daily basis as a preparation for competitive activities, may not allow sufficient time for recovery, resulting in testicular insult and loss of fertility.

22.3 Use of Mobile Phone

The use of mobile phone is inevitable. The impact of the mobile phone on human life has been perhaps the most revolutionary. Mobile phone came with calling facility, but engulfed a number of other useful features to become a many-in-one device that has replaced a number of other gadgets of daily use. Essentially, that has made us more dependent on this device leading to a close proximity between the mobile phone and us. The use of electromagnetic radiations (EMRs) in mobile phone communication and a number of other applications is increasing. The EMRs used in mobile phone are non-ionizing radiations; nevertheless, they can cause heating effect, which is detrimental to spermatogenesis. Therefore, several investigators have asked if mobile phone radiations in fact affect sperm production, motility, and fertility.

22.3.1 Animal Studies

One of the methods to evaluate the effect of mobile phone radiations is the study on animal models. Among the oldest studies, Dasdag et al. (1999) investigated the adverse effects of mobile phone radiations in male rats. Seminiferous tubule diameter was lower, and the rectal temperature was higher in the exposed groups than the control group. The authors, however, did not observe differences in the epididymal sperm count and the count of normal and abnormal sperm. In a relatively recent study on exposure of rats to EMRs using mobile phones, the authors found that the exposed group had a lower value of mean total sperm count and an increased percentage of apoptotic cells (Kesari et al. 2010). Other studies on animals exposed to mobile phone radiations have shown ill effects leading to reduction in testicular size (Desai et al. 2009) and degeneration of the seminiferous epithelium (Saunders and Kowalczuk 1981; reviewed in Agarwal et al. 2011).

22.3.2 Direct Exposure of Sperm

Another commonly used method of analyzing the effect of mobile phone radiations is in vitro exposure of human ejaculate to mobile radiations. In one such study, Erogul et al. (2006) exposed semen samples from 27 donors to EMR of 900 MHz cellular phone and found a significant decline in rapid progressive motility, slow progressive motility, and an increase in no-motility category of sperm movement; however, sperm concentration was not different between the two comparison groups. In another in vitro study, Falzone et al. (2008) exposed human spermatozoa to pulsed 900 MHz GSM mobile phone radiations and found that the exposure led to a significant decrease in two kinetic parameters, i.e., straight-line velocity and beat cross frequency, without change in mitochondrial membrane potential (Falzone et al. 2008). In an in vitro pilot study, Agarwal et al. (2009) exposed sperm from 23 healthy donors to mobile phone radiations and found that exposure led to a significant decrease in sperm motility and viability, increase in ROS level, and decrease in ROS-TAC score. In another in vitro study, Ahmad and Baig (2011) exposed semen samples of 22 individuals aged 20–35 to mobile phone radio frequency-electromagnetic waves (RF-EMW) for 1 h and found a significant decrease in sperm motility (Ahmad and Baig 2011).

22.3.3 Usage-Based Studies

The most direct method of evaluating the impact of mobile phone radiations on male fertility is usage-based correlation with semen quality. In a usage-based study, Fejes et al. (2005) collected mobile phone usage data from 371 individuals attending infertility clinic. The authors reported a significant negative correlation of the duration of possession and daily transmission time with the proportion of rapid progressive motile sperm and the proportion of slow progressively motile sperm. In the high transmitter group, the proportion of rapid progressive motile sperm was low in comparison to low transmitter group (Fejes et al. 2005). In another usage data-based study, Wdowiak et al. (2007) analyzed the effect of cell phone usage in individuals who appeared for marital infertility therapy. Comparison of the semen parameters among nonusers, moderate users, and heavy users showed an increase in the percentage of sperm with abnormal morphology with the duration of exposure to GSM phone (Wdowiak et al. 2007). Agarwal et al. (2008) collected mobile phone usage data from 361 participants and observed that the mean sperm motility, viability, and normal morphology were significantly different according to the level of exposure. The study concluded that decline in semen parameters was correlated with duration of daily exposure and was independent of initial semen quality (Agarwal et al. 2008).

22.3.4 Meta-Analysis on the Effect of Mobile Phone

A recent meta-analysis on the effects of mobile phone usage included data from 11 studies and found that mobile phone usage deteriorated a number of semen parameters including sperm concentration, sperm morphology, sperm motility, number of nonprogressive motile sperm, and the proportion of progressive motile sperm (Dama and Bhat 2013). The above study also undertook a pooled analysis on *in vitro* studies and found that exposure of spermatozoa to mobile phone radiations led to a significant decline in sperm quality. In-depth analysis showed that mobile radiation particularly deteriorated straight-line velocity, fast progressive motility, hypoosmotic swelling test score, major axis, minor axis, total sperm motility, and acrosome-reacted spermatozoa.

In another meta-analysis on the effect of electromagnetic wave (EMW) exposure in vitro, Fakhri et al. (2016) analyzed data from ten studies. Sperm motility in the unexposed and exposed samples were 17.70 ± 10.9% to 87.20 ± 7.32% and 18.40 ± 11.90% to 87.5 ± 8.57%, respectively. The mean differences for sperm motility and heterogeneity were REM:−4.57;CI (−7.11 to −2.03) and I2 = 69.38%; ρ heterogeneity <0.001, respectively. The percentage range of sperm viability in the unexposed and exposed samples were 50.78 ± 5.98% to 90.9 ± 3.7% and 48.43 ± 13.99 to 90.4 ± 4.1%, respectively, and for sperm viability, the mean differences for sperm motility and heterogeneity were REM-1.19; CI (−2.04 to −0.34) and I2 = 96.9%; ρ heterogeneity <0.001, respectively. The study concluded that exposure to EMW of mobile phone decreased sperm motility significantly; however, the decrease in sperm viability was not significant (Fakhri et al. 2016).

22.3.5 Precautionary Measures

The studies on the impact of mobile phones are very complex and difficult to design in a confounder-free manner. It is difficult to modify the schedule of mobile usage for a prospective study in the users. Individual variations in recalling the usage, the number of active usage hours, sleeping hours, proximity of mobile phone with testes, and the type of mobile device are some of the complicating factors in dissecting the actual impact of mobile phone radiations on semen parameters. Studies on human ejaculate may suffer from a number of limitations, such as direct exposure, which is not the case with actual usage. None of the studies on mobile phone radiations in humans collected data on most significant confounders affecting exposure to mobile phone radiations. At least two literature-based reviews warrant further analysis before reaching conclusions regarding the impact and the magnitude of the effect of mobile phone radiations on sperm parameters (Agarwal et al. 2011; Merhi 2012). Another review on general health effects of radio frequency suggests conducting well-planned and long-term studies, including those on exposure in young age (Ahlbom et al. 2004). Therefore, the quantum of the impact of mobile phone radiations remains dubious.

Irrespective of the inconclusive scientific and experimental evidence, there is a plethora of data suggesting adverse effects of mobile radiations on spermatogenesis and sperm function. The effect, even if mild and incapable of introducing infertility, may raise the risk of infertility in the presence of other comorbidities/factors. Therefore, we must be cautious in using this indispensable device. For example, keeping the mobile phone in a place other than the trousers' pocket would help keep the radiations away from testes. Another method of avoiding excessive exposure to mobile phone radiations is keeping the phone on table rather than pocket while sedentary. In ambulatory positions, prefer to keep mobile phone in the pocket of shirt, which offers the maximum distance between mobile phone and testes in comparison to carrying it in trouser pocket or hand. Among other measures to contain the exposure would be switching off mobile phone in night or keeping it away from bed. This would not only keep radiations away but also minimize sleep disturbances induced by mobile phones, which is another reason for a number of disorders, including sleep apnea and poor semen parameters. General lifestyle modifications that can help in upkeeping spermatogenesis and male fertility are shown in Fig. 22.2.

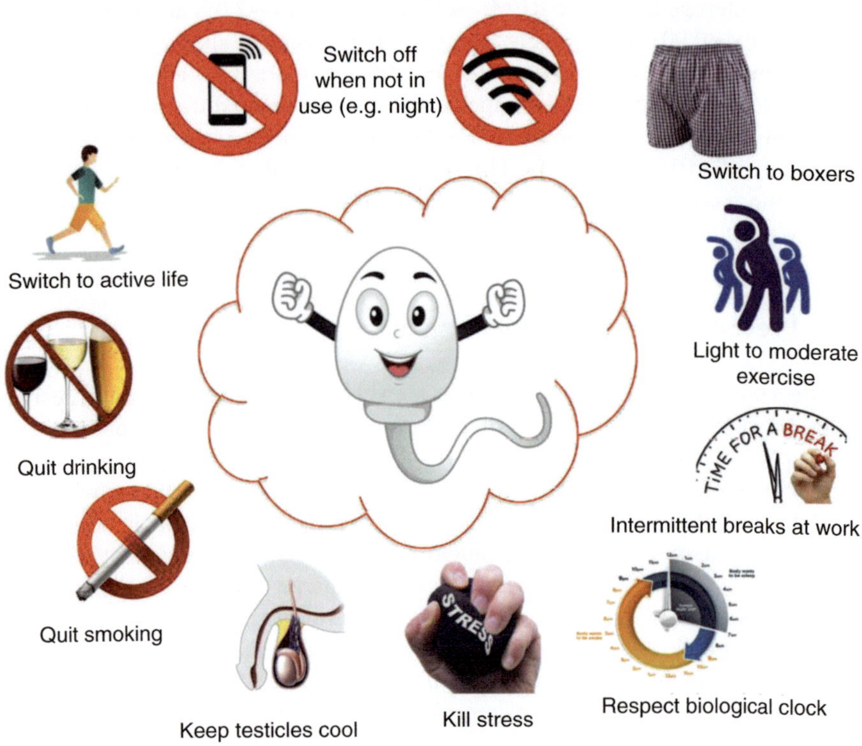

Fig. 22.2 Lifestyle modifications that can help upkeep spermatogenesis and fertility

22.4 Smoking

Approximately one third of male adults worldwide use tobacco, mainly in the form of cigarettes (WHO report 2015). Cigarette smoke consists of gases, vaporized liquids, and particles, and tobacco combustion produces more than 7000 compounds, with most of them being toxic in general including the reproductive health (Rodgman and Perfetti 2013). Cigarette smoke contains well-recognized mutagens that include cadmium, dimethylbenzanthracene, naphthalene, dimethylnitrosamine, methnaphthalene, and radioactive polonium. Most of these are well-known carcinogens, and their presence correlates with the increased risk of a number of cancers in smokers. Initially, a number of studies reported controversial findings regarding the impact of smoking on semen parameters; however, it is now largely agreed that smoking has an adverse impact on semen quality.

Smoking has been associated with lower sperm concentration, impaired sperm motility and morphology, increased DNA damage, and reduced cell viability (Cui et al. 2016). Cigarette smoke is known to lower the antioxidant capacity of human body, thereby lowering protection against any potential insult to the reproductive system. Benzo(a)pyrene, which is a highly mutagenic carcinogen, is found bound to DNA in higher quantities in smokers in comparison to non-smokers (Alexandrov et al. 2006). Cigarette smoking is thought to affect meiosis in ovaries and testes, semen parameters in a dose-dependent manner, number of retrieved oocytes leading to early menopause, inhibit embryo development post-fertilization, and raise the risk of childhood cancer (Zenzes 2000). A number of studies have reported significant adverse effects of smoking on sperm count and motility. In a recent study, Asare-Anane et al. (2016) compared semen parameters between smokers and non-smokers and found that the former have significantly lower semen volume, sperm count and motility, viability, and normal morphology (Asare-Anane et al. 2016).

22.4.1 Smoking Correlates with Infertility

In a cross-sectional on a rural Chinese population, Yang et al. (2016) analyzed the relation of male smoking with couple's infertility. The study included data from 7,025 couples and found that after adjusting for the confounding factors, the couples were more likely to suffer from infertility if the husband smoked before the first pregnancy. In-depth analysis showed that the risk started after a longer duration of 5–10 years of smoking. Further, a stronger association was observed in the groups with more than 10 years of smoking. Similar quantitative relationship was found for the number of cigarettes per day and the total number of cigarettes smoked (Yang et al. 2016). Among studies on large number of smokers, Meri et al. (2013) analyzed 396 smokers and 546 non-smokers and found that smokers had poor sperm motility and a higher proportion of abnormal sperm and leukocytes. Sperm count on the other hand was not affected. Further analysis found that the relationship between smoking and semen quality loss was directly proportional. Heavy smokers had poor semen parameters in comparison to non-heavy smokers (Meri et al. 2013).

In a retrospective cohort of 1,512 infertile patients, Zhang et al. (2013) included the highest number of smokers ($n = 737$) and a comparable control group ($n = 775$). The study found that smokers had a significant decrease in semen volume, rapid progressive motility, and sperm viability in comparison to non-smokers. Further, smokers had a significant increase in the number of immotile sperm and semen leukocytes. Almost all sperm motility parameters were lower in the smokers group. Eventually, the percentage of normal morphology sperm was significantly decreased in smokers, and the loss of normal sperm morphology showed a quantitative decrease with increased degree of smoking (Zhang et al. 2013). Most of the above studies agree that smoking causes a reduction in semen quality and quitting smoking may benefit the individuals with marginal loss of fertility. Smoking has been found to have multitude of adverse effects on male fertility by affecting almost everything that matters for competent sperm production.

22.4.2 Smoking Affects Hypothalamic-Pituitary-Gonadal Axis

Smoking has also been found to affect the functioning of the hypothalamic-pituitary-gonadal axis and disturbs the release of a variety of hormones, which inhibit luteinizing hormone and prolactin. A number of endocrine disruptors are known to increase infertility risk by the same means. In a study on the effect of tobacco, Ochedalski et al. (1994) reported that smokers had a relatively higher level of estradiol, LH, and FSH with decreased level of prolactin, though the level of testosterone and dihydrotestosterone was not significantly different from non-smokers (Ochedalski et al. 1994). Another study reported a positive dose-dependent relationship between smoking and testosterone and LH and that elements of tobacco smoke might interrupt the HPG axis leading to Leydig cell failure in smokers (Ramlau-Hansen et al. 2007). Nevertheless, another study on 889 fertile men divided into mild, moderate, and heavy smokers found no significant differences in FSH, LH, or serum total testosterone levels (Pasqualotto et al. 2006).

22.4.3 Smoking Affects Sperm Maturation and Varicocele

In addition to its action on hormone levels, smoking has also been shown to affect sperm maturation in the epididymis (Dacheux and Dacheux 2014). Smoking has been found to increase the loss of sperm count in the individuals with varicocele. It is evident that a combination of smoking with varicocele correlates with oligozoospermia with a tenfold greater incidence than non-smoking men with varicocele and five times more than the incidence in men who smoked but did not have varicocele (Klaiber et al. 1987). Smoking has been found to increase oxidative stress, and the latter is a well-known risk factor for loss of fertility. Therefore, smoking may cause infertility by promoting varicocele and hyperthermia of the scrotal region (Pasqualotto et al. 2006). It is well known that the vascular blood supply in the testicular cord is relatively insufficient. Smoking is known to further decrease oxygen tension in blood, which may compromise spermatogenesis by creating hypoxia in the testis. Smoking also correlates with reduced semen volume (Pasqualotto et al. 2006), which may affect the life and fertility of sperm after ejaculation.

22.4.4 Smoking Contributes to Erectile Dysfunction

Smoking may also affect the physiology of erection and the competence of semen to support sperm fertility. Erectile dysfunction is one of the major causes of infertility as discussed elsewhere in this book. Smoking has been conclusively found to be a risk factor for erectile dysfunction. A systematic review that included four prospective cohort studies and four case-control studies reported that smoking increases the risk of erectile dysfunction significantly (Cao et al. 2013) and cessation of smoking significantly improved physiological and sexual health in male smokers (Pourmand et al. 2004; Maiorino et al. 2015). A number of accessory glands such as the seminal vesicle, prostate, and bulbourethral pour their secretions in the ejaculate. In smokers, vesicular and prostatic parameters showed a decline (Pasqualotto et al. 2006; Harlev et al. 2015). Experimental proof has been provided to show that the secretions in the smokers are not competent enough for sperm fertility. Spermatozoa from non-smokers when exposed to seminal plasma of smokers have reduced sperm motility and acrosome reaction significantly (Arabi and Moshtaghi 2005). Conversely, incubation of spermatozoa from smokers with seminal plasma of non-smokers leads to a non-significant improvement in functional parameters of sperm (Mehran 2005).

22.4.5 Molecular Mechanism of Action of Smoking

Cigarette smoke condensate (CSC) contains several carcinogenic and teratogenic components, which result in accelerated germ cell death via the cytoplasmic transcription factor, aryl hydrocarbon receptor (AHR). A study on germ cell line, GC2, showed that CSC activates AHR and adversely affects the development of germ cells by disturbing the expression of a battery of genes participating in cell proliferation, cell cycle, apoptosis, and antioxidant mechanisms (Esakky and Moley 2016). It is worth noting that AHR is important during testicular sperm production and post-testicular sperm maturation. This report suggested that cigarette smoke exerted negative effects both genomically and non-genomically, contributing to the loss of sperm count and germinal epithelium (Esakky and Moley 2016).

22.4.6 Meta-Analysis Supports Adverse Effect of Smoking

A recent meta-analysis analyzed data for 5,865 participants from 20 studies, suggesting that cigarette smoking was associated with reduced sperm count, motility, and morphology. Further analysis revealed that the effect size was bigger in infertile individuals than the control group, and there was a quantitative effect of smoking as heavy smokers faced a greater decrease in comparison to mild smokers. Therefore, there is now sufficient and conclusive evidence that smoking impairs semen quality and fertility by affecting hormone levels, exposure to carcinogenic compounds, and contribution to erectile dysfunction, ultimately leading to reduced sperm motility, viability, and DNA integrity. Scientific studies have also uncovered the possible mechanism of actions involved in the adverse effects of smoking and tobacco chewing on semen quality. Smoking has also been reported to affect the epigenome in male infertility (Dong et al. 2016). The latter could have transgenerational effects, which needs further investigation.

22.4.7 Precautionary Measures

Smoking does not require any specific suggestion for its ill effects. The adverse health effects of smoking are well known and reviewed. Smoking in addition to general deterioration of health is known to cause cancer. Therefore, smoking should be avoided or reduced in quantity. Most of the studies on smoking have reported adverse effects of adult smokers; however, its effects on younger population around puberty needs further detailed investigations for effects on sexual and pubertal development. In the infertility patients who are heavy smokers, quitting smoking should help not only directly but also by improving the action of other therapeutic measures in place. In addition to active smoking, non-smokers should avoid exposure to passive smoking. Other forms of tobacco such as chewing tobacco also need to be curbed for betterment of semen parameters. Significant transgenerational effects of smoking may expose the coming generations to the risk of reproductive disorders, the full spectrum of which is yet to be studied.

22.5 Alcohol

Alcohol is used as food, entertainment, and recreation at personal and societal level. A large fraction of population drinks alcohol on a regular basis in a variety of forms. According to the World Health Organization (WHO) report, worldwide per capita consumption of alcoholic beverages in 2005 equaled 6.13 L of pure alcohol consumed by every person aged 15 years or older. WHO report also stated that the alcohol consumption worldwide is underscored in comparison to the actual usage. The consumption of alcohol by people of younger age group (13–15 years) has also increased significantly over a period of time. According to the WHO Global Survey of Alcohol and Health (2008), the 5-year trend of drinking showed an increase in the underage drinking in about 71% of the responding countries (WHO report 2008). Low usage of alcohol has been shown to have no detrimental effect, particular in short terms. Therefore, the adverse effects of alcohol are often discussed in relation to excessive use of alcohol.

In a classical study, Lloyd and Williams found that 72% of men with advanced alcoholic cirrhosis exhibited decreased libido and sexual potency (Lloyd and

Williams 1948). Further analysis in this study reported that out of 22 men with cirrhosis, 18 were unable to ejaculate, two produced no sperm at all, one had reduced sperm counts, and only one had normal sperm count. Lloyd and Williams also reported that 87% of men with advanced alcoholic cirrhosis exhibited reduced axillary hair, showing decreased testosterone action. Men with low consumption (10–40 g or approximately 1–3.5 drinks per day) showed no abnormal sperm forms (Pajarinen et al. 1996). Sperm abnormalities are generally seen in individuals consuming alcohol in moderate or high quantities. Moderate alcohol consumption (40–80 g or 3.5–7 drinks per day) was found to lead to slight alterations in sperm maturation. A history of heavy alcohol consumption (>80 g or 7 drinks per day) led to arrest of sperm development in about 20% of the cases (Pajarinen et al. 1996). Alcoholics who had not yet developed severe liver damage were found to have reduced sperm count in 40% of the cases, abnormal sperm shapes in 45%, and altered sperm motility in 50% (Villalta et al. 1997).

22.5.1 Alcohol Affects the Hypothalamic-Pituitary-Gonadal Axis

Experimental evidence suggests that alcohol can affect the production of LH by the pituitary gland. For LH production from the pituitary, GnRH released from the hypothalamus must interact with specific receptor on the surface of the pituitary cells. Upon GnRH binding, an enzyme protein kinase C must move from LH-producing cell to their surface. Alcohol has been shown to affect the movement of this protein to the cell surface (Steiner et al. 1997). In addition to its action on the testis and pituitary gland, alcohol is also known to directly affect the hypothalamus. A study on male rats found that alcohol administration significantly lowered GnRH level in the blood vessels connecting the hypothalamus to the pituitary gland (Ching et al. 1988). Apart from affecting GnRH secretion, alcohol also appears to affect the production of active GnRH molecules. Therefore, alcohol affects sperm production by acting at several points on the hypothalamus-pituitary-gonadal (HPG) axis. The effects could be compounded by small but additive effects on each of the main elements of the HPG axis, ultimately resulting in compromised sperm production or sperm maturity.

22.5.2 Alcohol Affects the Function of Leydig and Sertoli Cells

An interesting study on the action of alcohol on Leydig cell function undertaken on young healthy male volunteers having normal liver function who received alcohol over a period of 4 weeks found that testosterone in these individuals decreased as early as 5 days and continued falling during the entire duration of the study (Gordon et al. 1976). Alcohol is known to stimulate aromatase, which results in increased conversion of testosterone to estrogen in fat and liver (Gordon et al. 1979). High levels of estradiol are known to be detrimental to spermatogenesis. Numerous studies have suggested that alcohol abuse can result in shrinkage of the testis and impaired testosterone production (Adler 1992). These alterations can result in impotence and infertility. It has also been reported that alcohol may damage some of the Sertoli cell proteins that are required for sperm production (Zhu et al. 1997). Muthusami and Chinnaswamy (2005) in a study on alcoholics free from smoking found that heavy alcohol consumption resulted in increased levels of LH, FSH, and E2 and decreased levels of testosterone and prolactin. This was accompanied by a significant decrease in semen volume, sperm count, motility, and other morphological parameters of sperm quality (Muthusami and Chinnaswamy 2005). Another study on alcoholics found that moderate to high alcohol consumption resulted in a significant increase in morphologically abnormal nuclei and plasma membranes in sperm (Joo et al. 2012).

22.5.3 Paternal Alcoholism Contributes to Fetal Alcohol Syndrome

Excessive maternal alcohol consumption during pregnancy is well known to result in a number of disorders in the offsprings, collectively known as fetal alcohol spectrum disorders (FASD). A number of these disorders are known to inherit by epigenetic modifications. Fetal alcohol syndrome (FAS) is one of the worst outcomes, which is characterized by growth retardation, craniofacial abnormalities, and mental retardation. However, paternal contribution to FAS was shown by studies where mother had not consumed alcohol during pregnancy, but fathers were alcoholics (Lemoine et al. 1968). FAS was seen in about 75% of the children whose fathers were alcoholics (Abel 1983). Human and rodent studies have shown that preconception paternal alcohol intake was related with growth retardation, low birth weight, and congenital abnormalities (Friedler 1996; Passaro et al. 1998). It has now been confirmed that FASD may be the result of contribution of paternal and maternal exposure to alcohol preconception or maternal exposure during pregnancy (Abel 2004). Paternal alcohol exposure has been found to result in reduced global DNA methylation in the developing mouse fetus (Knezovich and Ramsay 2012).

22.5.4 Paternal Alcoholism Affects Fetal Development

Paternal alcohol consumption has also been shown to affect not only fertility but also the development of the fetus. Alcohol administration at moderate to high doses before mating in rats previously unexposed to alcohol resulted in the production of low birth weight in pups and also reduced litter size (Cicero et al. 1994). Recently, alcohol consumption has also been shown to have transgenerational effects by affecting the epigenome of father (Knezovich and Ramsay 2012). A recent study examining the effect of preconception alcohol exposure found that prenatal alcohol exposure of male mice resulted in significant changes in the paternally methylated imprinting control regions (H19 and Rasgrf1) in the sperm of exposed males and somatic DNA of the sired offspring. The study also reported a significant reduction in methylation at H19 CTCF and CTCF2 binding sites in the offsprings that correlated with a reduced weight at postnatal days 35–42 (Knezovich and Ramsay 2012).

22.5.5 Precautionary Measures

As mentioned above, low-level consumption of alcohol may not have severe effects on reproductive health. Nevertheless, it is generally seen that alcohol addiction increases with time and a number of people have genetic tendency to become alcoholics. Therefore, negligible effect of low consumption does not make one immune to adverse effects of alcohol. It is the low consumption which needs to be curbed to avoid development of alcoholism that has a number of health-deteriorating effects including poor reproductive health and sexual life. Alcoholics seeking infertility treatment should be strongly encouraged to quit drinking. If required, the patients may seek medical advice and be referred to habitation centers for quitting alcohol.

22.6 Stress

Stress can be defined as an internal state different from one's normal state of rest, which may be caused by an external or internal stressor (Selye 1955). Our body reacts to a number of situations by engaging in automatic responses in order to aid the individual cope up adequately with the noxious stimuli by temporarily changing the physiology (Cannon 1994). Popularly known as the "fight or flight response," the physiological changes during acute stress are thought to involve the activation of the sympathetic nervous system and inhibition of the parasympathetic system in order to help the individual tackle the potentially harmful situation (Cannon 1994). The physiological response initiates at the level of the hypothalamus, which upon recognizing the state of stress, stimulates the secretion of the neurotransmitters epinephrine (adrenalin) and norepinephrine (noradrenalin) in the blood stream by the adrenal medulla of the autonomic nervous system (McCorry 2007). These neurotransmitters stimulate the sympathetic nervous system that helps the body to better deal with the stressful situation. Upon activation of the hypothalamic-pituitary-adrenal axis in the stressful situation, the hypothalamus stimulates the pituitary gland to secrete a number of hormones including adrenocorticotropic hormone. The latter stimulates the adrenal glands to produce cortisol, which inhibits the sympathetic nervous system, aiding the return of homeostasis.

Persistent stress may result in continued activation of the sympathetic nervous system, resulting in a number of changes in the hypothalamic-pituitary-gonadal axis that affects spermatogenesis. Because of close association between the central nervous system and the gonadal function, disturbances at the level of CNS translate into changes in gametogenesis rather rapidly. Stimulation of the HPA axis may inhibit the HPG axis, thus altering spermatogenesis (Whirledge and Cidlowski 2013). While occasional stress can be handled with ease, prolonged stress can disturb the HPG axis, thereby affecting fertility potential. The effect of stress on fertility was initially seen as a form of female infertility that resulted in increased risk of dysmenorrhea, anovulation, infertility, and loss of pregnancy (Mishra et al. 2000). However, now there is sufficient and conclusive evidence that stress increases the risk of male factor infertility as well. It has been reported and well reviewed that stress results in increased production of reactive oxygen species that causes damage to the germ cells, resulting in increased production of abnormal sperm with poor motility and fertility potential (Lawson 2016).

22.6.1 Life and Work Stress

In one of the oldest studies assessing the impact of stress and work environment, Bigelow et al. (1998) analyzed semen parameters of 845 infertile men and found a reduction in the percentage of progressive sperm and an increase in the percentage of coiled tail sperm defects in welders, compared with unexposed subjects. The study also reported a significant dose-dependent relationship between perceived job stress and percentage of progressive sperm, total motile count, morphology, abnormal heads, and coiled tail defects (Bigelow et al. 1998). In a cross-sectional study, Auger et al. 2001 analyzed sperm morphological defects due to lifestyle and environmental factors of 1,001 male partners of pregnant women from four European cities. The authors found that significant variations of several sperm defects were related to stress, weekly working time, occupational posture, and metal welding (Auger et al. 2001).

In a study on evaluating the impact of life and work-related stress, Janevic et al. (2014) evaluated 193 men and measured the stress level including job strain, perceived stress, and stressful life events in relation with semen parameters. The authors found an inverse correlation between perceived stress score and sperm concentration, motility, and morphology. Men who experienced two or more stressful life events in the past year had lower percentage of motile sperm and a lower percentage of morphologically normal sperm in comparison to those having no stressful events (Janevic et al. 2014). A recent study analyzed the impact of stress and everyday factors on sperm DNA damage. The study collected data from 286 men attending infertility clinic who had a normal semen concentration or slight oligozoospermia and found that high occupational stress and age increased DNA fragmentation index. Since DNA integrity is an extremely important parameter indicative of fertility potential, the study highlights the impact of everyday stress on loss of sperm fertility (Radwan et al. 2016).

22.6.2 Stress-Related Life Events

A number of studies on stress-related to life events have reported a decline in sperm count, motility, or morphology (Hjollund et al. 2004; Gollenberg et al. 2010; Janevic et al. 2014). Among the large studies on life event-related stress, Gollenberg et al.

(2010) examined the association between stressful life events and semen parameters in 744 fertile men (Gollenberg et al. 2010). After adjusting for confounders, the authors found that men reporting 2+ recent stressful events had an increased risk of being classified below normal WHO values of sperm count, motility, and morphology criteria. Further, men with 2+ stressful life events had lower sperm concentration and lower percent sperm motility; however, morphology was less affected. Similarly, a number of studies on occupational stress showed detrimental impact of stress on sperm count, motility, and morphology (reviewed in Nordkap et al. 2016).

22.6.3 Psychological Stress

Among one of the recent studies evaluating the impact of psychological stress on semen parameters, Nouri et al. (2014) recruited 70 male partners to study the effect of psychological stress using hospital anxiety and depression score (HADS) questionnaire. The study reported that sperm count, motility, and morphologically normal sperm were lower in men having abnormal HADS. The study concluded that psychological stress primarily lowers total testosterone with rise in serum LH and FSH and reduces semen quality (Nouri et al. 2014). A recent cross-sectional study on the impact of psychological stress on male fertility evaluated 1,215 Danish men by getting a questionnaire on health and lifestyle, including self-related stress filled by general participants. It was found that the individuals with self-reported stress scores above an intermediate level had poorer semen quality in a dose-dependent manner, though no differences in the hormone levels were found (Nordkap et al. 2016).

Psychological stress may also affect the outcome of assisted reproductive techniques. Among the oldest studies on this aspect, Harrison et al. (1987) analyzed the impact of psychological stress on semen samples used in IVF setting. The investigators analyzed semen samples of 500 men such that one of the samples was collected before the IVF work-up and the other after ovum aspiration. The study found that the second sample had significantly lower sperm density, sperm count, and qualitative and quantitative motility (Harrison et al. 1987). Therefore, stress not only increases the risk of infertility but also hampers infertility treatment using natural or assisted methods. Specific and occasional stressful situations such as war and examinations also impact semen parameters negatively (Eskiocak et al. 2005; Abu-Musa et al. 2007; Lampiao 2009).

22.6.4 Infertility Itself Causes Stress

Stress makes a reciprocal relationship with infertility. A number of studies on the impact of fertility treatment-related psychological distress have framed the analysis in line with Lazarus and Folkman's (1984) landmark theory of the relationship between stress, appraisal, and coping. According to this, the patients who practice inadequate coping strategies for failure of infertility treatment result in a significantly higher distress, further negatively affecting the outcome of their infertility management (Lawson et al. 2014). It is known that most of the infertility patients suffer from some level of moderate to significant distress, which may increase tremendously upon initial failure of treatment trials. The inability to conceive may raise personal, familial, and societal stress, which reduces the fecundity further. Research on infertility patients undergoing treatment has found that 40–50% have mild to moderate depression, 2% have severe symptoms that worsen over time, and more than 50% suffer from anxiety (Cousineau and Domar 2007). These symptoms if present increase further upon initial failure of the treatment trials. Therefore, stress and infertility form a loop, one giving rise to the other, and this ultimately increases the level of both, distress and infertility, which makes it further difficult to treat these patients.

A number of other studies have now firmly established that stress due to a variety of factors including stressful life events, work stress, self-perceived stress, and psychological stress due to infertility cause a significant decline in semen quality by altering sperm count, motility, and the number of morphologically normal sperm (Nordkap et al. 2016). Most of the stress factors discussed above contribute to poor semen quality and infertility by increasing oxidative stress. In addition to the loss of sperm count and motility, oxidative stress also results in functional loss of sperm fertility and DNA integrity (Sawyer et al. 2003). Apart from damage to the nuclear DNA, mitochondrial DNA is particularly susceptible to oxidative stress-induced DNA damage. All these lead to reduced semen quality, significantly contributing to male infertility.

22.6.5 Precautionary Measures

As discussed above, stress comes in various forms, such as work pressure, job stress, household stress, and a combination of these culminating into psychological stress. Stress has a strong correlation with both male and female infertility. Stress and trauma early in life is well known to have transgenerational effects transmitted via sperm (Gapp et al. 2014). Stress management by medication may not help as far as reproductive fitness and fertility are concerned. A number of stress medications are known to be detrimental to fertility. Therefore, management of stress by other means such as exercise, yoga, and meditation is advisable. Better management and advance planning of daily and official chores would be an ideal strategy to avoid stress. Such management should preferably be used as a prophylaxis measure than a therapy; however, in the cases with stress, therapy in this form may be helpful too. Nevertheless, exercise in an appropriate dose would help alleviating stress and also improve the effect of other therapeutic measures employed to curb infertility.

22.7 Sleep Apnea

The physiology of human body and most of the animals is designed to follow a circadian rhythm, and a number of functions have been fine-tuned to take place optimally under light or dark conditions. In a classical study, Boyar et al. (1974)

found that testosterone level increases in relation to luteinizing hormone production during sleep in around puberty (Boyar et al. 1974). Axelsson et al. (2005) in a study on day sleep and night sleep found that testosterone levels increased during sleep periods and fell during walking and in awake position. In a study on sleep and sleep-related disorders, Schiavi et al. (1992) found that sleep disturbance correlates with decreased testosterone levels, and this may be one of the mechanisms for decrease in testosterone level with age (Schiavi et al. 1992). Similarly, another study showed that increase in testosterone was related to sleep and not the circadian rhythm (Luboshitzky et al. 2001). All these studies suggested that appropriate and disturbance-free sleep is critical to maintain adequate testosterone level for spermatogenesis and other reproductive health effects. In another study, Luboshitzky et al. (2005) reported that men with sleep apnea had lower LH and testosterone levels, which may be attributed to sleep disturbance and fragmented sleep.

22.7.1 Sleep Apnea Correlates with Low Testosterone

A number of studies in obese individuals have shown the presence of sleep apnea and low testosterone levels; however, it has been reported that the effect of sleep disorders on testosterone is independent of BMI (Gambineri et al. 2003). The relationship between sleep apnea and testosterone level is now well known (Hammoud et al. 2012). The reduction in testosterone level affects a number of fertility aspects, including spermatogenesis, sexual behavior, libido, and erectile function. The relationship between sleep apnea and sexual dysfunction has been suspected because of association between testosterone and sleep-related erection (Granata et al. 1997). Therefore, an altered sleep pattern can affect erectile function as a result of its effect on testosterone level. Budweiser et al. (2009) in a study on 401 men showed that sleep apnea was correlated with erectile dysfunction. In another study on a large set of men, Andersen et al. (2010) suggested a relationship between sleep apnea and erectile dysfunction after analysis on 467 men. A number of other studies have confirmed the relationship between sleep disturbances, erections, and quality of sexual life (Hammoud et al. 2008). Alvarenga et al. (2015) have demonstrated that rats exposed to 96 h of paradoxic sleep deprivation displayed poor sexual behavior and reduced performance (Alvarenga et al. 2009). Animal studies provide further proof of the effects of sleep disturbance on male sexual behavior, leading to changes in functional parameters and performance (Alvarenga et al. 2015).

22.7.2 Sleep Apnea May Affect Fertility

At least one study reported an adverse impact of sleep apnea on sperm count in rats along with a change in the sexual behavior and performance (Alvarenga et al. 2015). In this study, sexually experienced rats were subjected to paradoxic sleep deprivation or sleep restriction, and testosterone level, sperm count, and gene expression related to spermatogenesis and fertility were analyzed. The authors found that paradoxic sleep deprivation results in decreased level of testosterone, testicular gene expression, and a decrease in the number of live sperm. Further analysis showed that iNOS and hydroxysteroid 11b-dehydrogenase genes showed decreased expression in comparison to the control group. The study concluded that sleep disturbance can causes reduction in testosterone, sperm production, and male reproduction function by affecting testicular nitric oxide pathway (Alvarenga et al. 2015).

To the best of our knowledge, there has been no study demonstrating directly a relationship between sleep apnea and male infertility. In case of humans, it is difficult to undertake such studies. In a study on obstructive sleep apnea characterized by intermittent hypoxia and oxidative stress, Torres et al. (2014) subjected male mice to periodic hypoxia mimicking sleep apnea. The mice were tested for effective fertility by mating experiments. The authors observed that progressive sperm motility was significantly reduced in the test group. Further, the proportion of pregnant females and the number of fetuses per mating were significantly lower in the intermittent hypoxia group (Torres et al. 2014).

22.7.3 Precautionary Measures

The studies on sleep apnea and fertility are in the infancy and have to go a long way before a direct correlation can be established. Nevertheless, animal studies on fertility and indirect evidence in the form of low testosterone upon sleep disturbance in humans are good reasons to foresee an adverse impact of sleep disturbance on semen parameters and fertility. Amid, preliminary evidence, management of disturbance-free sleep hours seems to be very important. Sleep disturbance not only changes hormone levels but also may increase stress level, which is an independent factor for impaired fertility. Therefore, late hours work, night shift job, watching television late night, and other activities that may cause sleep disturbance should better be kept at a distance. Night-shift female workers have already been shown to have disturbed menstrual cycle and ovulation (Gamble et al. 2013). Infertile patients with the presence of disturbed sleep pattern or night-shift job should be encouraged to follow a natural work cycle to better manage infertility. Further studies on sleep apnea and its contribution to male infertility would be of significant interest. It would be very interesting to study the effect of lifestyle changes and other therapeutic interventions in alleviating the adverse effects of sleep apnea on testosterone level and fertility.

22.8 Exercise

Sedentary lifestyle has been linked with a number of disorders, such as diabetes, cardiovascular, muscle pain, poor bone health, and other lifestyle disorders. Given the multitude of adverse effects of sedentary lifestyle, it is conceivable that sedentary lifestyle may also affect spermatogenesis. A recent cross-sectional study on 1,210 healthy young Danish men undergoing fitness test for military services found that time spent watching television (sedentary time) was associated with a poor sperm count. Quantitatively, men who watched television for more than 5 h per day had an average adjusted sperm count of 37 million/ml in comparison to 52 million/per in the control group (Priskorn et al. 2016). Sedentary lifestyle can increase the risk of poor semen quality and infertility by a number of means including testicular heating, increased BMI and weight, and poor energy metabolism.

22.8.1 Obese Individuals Benefit from Exercise

Obese individuals looking for infertility solution benefit from losing weight, irrespective of their gender. In case of obese females, it has been found that losing weight improves the chances of natural or induced ovulation and pregnancy. Similarly, in case of males, it has been found that physically active subjects have better semen parameters and hormone levels in comparison to the sedentary males. In general, exercise helps maintain good overall health and immunity, thereby reducing the risk of a number of disorders such as cardiovascular, blood pressure, diabetes, etc. It is conceivable that better overall health and energy metabolism would provide a conducive environment for optimal spermatogenesis and fecundity. In a comparison of semen parameters, Vaamonde et al. (2012) compared physically active and sedentary subjects. Statistically significant difference in total progressive motility and morphology was seen. The study concluded that physically active subjects had better semen and hormonal parameters in comparison to the sedentary subjects (Vaamonde et al. 2012).

The above suggests a positive impact of exercise on overall health and fertility; however, exercise may be good or bad from fertility point of view depending upon the intensity, type, and objective. Many studies have emphasized the negative effects of exercise on human fertility, particularly in the female athletes. The adverse effects of exercise are clearer in case of females in the form of delayed menarche, oligomenorrhea, amenorrhea, inadequate luteal phase, and anovulatory cycles (Vaamonde et al. 2016a, b). A number of studies have mentioned adverse effects of intense exercise on female fertility in case of runners, cyclists, swimmers, gymnasts, figure skaters, and ballet dancers (Vaamonde et al. 2016a, b). Therefore, it is very important to distinguish between beneficial and harmful exercise from fertility point of view.

Interestingly, the impact of exercise on the male reproductive system should follow the same trend as in females; however, this is not easy to investigate in males due to wide variations in the level of semen parameters. In normal human males, the level of sperm production varies a lot across individuals and populations, making it difficult to identify subtle changes. Most of the studies on the impact of exercise on male fertility have been undertaken on athletes and not on the individuals of moderate fitness or those who follow a sedentary lifestyle. Therefore, it is very difficult to conclude about the scientific standpoint of the effect of exercise on male fertility. However, significant evidence suggests that exercise can have good or bad effects depending on the type, volume, and objective of exercise (Vaamonde et al. 2016a, b).

22.8.2 Exercise Affects Hypothalamic-Pituitary-Gonadal Axis

Acute and prolonged exercise regimen may lead to changes in the HPG axis, thus altering sperm production and maturation. Most of the literature on the impact of strenuous exercise in males is on athletes. Studies on cyclists have shown a detrimental effect of heavy cycling on spermatogenesis. It has been reported that cyclists who participate in competitions had lower testosterone levels than others (Lucia et al. 2001). In case of cyclists, most of the studies agree on a detrimental effect of intense cycling on HPG axis and sperm production. This could be due to continued friction between the saddle of the bicycle and the reproductive system (Grunbaum and Carrier 2002). Hypogonadal states have been reported in the individuals undergoing endurance training in the form of running or cycling for 10–20 h/week (Hackney et al. 2005). Similarly, in case of rowers, heavy training resulted in decreased free testosterone/cortisol ratio, when compared to basal levels (Vervoorn et al. 1991). In case of swimmers, there is controversy regarding the impact of exercise. Other studies reported changes in testosterone and other hormones, which returned to normal levels during recovery (Vaamonde et al. 2016a, b). In other sports activities, such as soccer and basketball, it is not clear if there is any detrimental or beneficial effect on male reproductive parameters.

22.8.3 Exercise May Affect Testosterone Level and Erectile Function

It is difficult to assess if exercise leads to adverse effects on the size of testicles and other accessory reproductive glands. However, it has been found that an adverse effect on hormone levels may result in altered testicular size in athletes with intense exercise schedule. This may be complexed with reduced activity of the accessory reproductive glands, thus compromising semen quality. For example, the saddle of bicycle could cause direct damage to the accessory sex glands by repeated rubbing, resulting in inadequate production of their secretions, thus affecting semen production. In case of strenuous exercise, tiredness and stress may lead to poor erections not enough to sustain the sexual intercourse. This may be further complexed by trauma to the reproductive organs during strenuous exercise. A majority of the studies on cyclists have reported erectile dysfunction (reviewed in Grunbaum and Carrier 2002). A number of conservative measures including altering the seat position, increasing the width of the saddle, increasing saddle cushioning, and designs that reduce perineal pressure have been suggested as measures to avoid the development of erectile dysfunctions in cyclists (Grunbaum and Carrier 2002).

22.8.4 Precautionary Measures

In conclusion, while light to moderate exercise may help us lose weight and improve fertility in comparison to a sedentary lifestyle, heavy exercise, particularly in the form of endurance, competition, and athletics, may lead to a decline in the semen quality and fertility. This may be insignificant in the initial years of training and exercise; however, continued pressure on the body and reproductive organs may lead to adverse effects in the long term that may cause a decline in fertility. Therefore, a correlation between high-load exercise and a negative impact on fertility is concluded. It is very important to define the low, normal, and intense levels of exercise. A regimen of workout of 45 min with light to moderate exercise four times a week should be considered optimal for overall health benefits including reproductive health. When exercise is solely introduced with an aim to improve fertility, it should not include cycling or other similar exercises, which could create testicular heating or cause trauma to the reproductive organs.

As a corrective measure to the problem of infertility, whether or not exercise should be advised depends on a number of parameters that include physical state of the individual, past history of exercise and trauma, weight and BMI, and age. In case of the individual with high BMI or those suffering from obesity, light to moderate exercise to reduce weight would be beneficial in improving fertility. Similarly, in case of individuals with normal weight but sedentary lifestyle, light to moderate exercise in the form of running, walking, or aerobics may help improve fertility. Heavy exercise should not be advised to the individuals undergoing infertility treatment. In the case of athletes, where intense exercise seems to be the only probable cause of infertility, limiting the exercise or stopping it altogether should be advised in order to improve fertility.

22.9 Discussion and Future Directions

In the recent years, there has been emphasis on a significant decline in semen quality over the last few decades. Appreciable decline in semen parameters has been reported; however, the contributing factors remain unidentified. Among a number of plausible factors behind this decline, lifestyle factors discussed above may be a few significant contributors. Fortunately, a number of these factors are easily modifiable to curb their ill effects on general and reproductive health. A high-quality lifestyle including a good regimen of exercise, minimal or cautious use of electronic gadgets and other appliances using radiations, avoiding testicular heating by taking precautionary measures at work or home, staying away from alcohol and smoking, and minimizing stress by appropriate and advanced management of life chores should help upkeep good reproductive health and fertility.

As discussed in details above, scientific research on the impact of mobile phone usage has provided initial details with most of the studies supporting an adverse impact of mobile phone radiations on seminal parameters. Nevertheless, a few studies have issued caution regarding the impact of these radiations on fertility; therefore, further well-planned studies can provide necessary data to take the debate to conclusion. Mobile phone usage has a number of confounding parameters, such as the number of hours of usage, proximity of phone with testes, type of device, and exposure during nighttime, all of which can affect the degree of their ill effects. Therefore, further studies on mobile phone usage should take these variables into consideration in order to work out the exact quantum of adverse effects this indispensable device may have on semen parameters.

Activities that could affect testicular heating, such as laptop usage, need further well-planned studies to identify their quantitative effects on semen parameters. We could identify only one study that analyzed the impact of laptop usage on testicular heating and one study that explored the ways to avoid testicular heating in laptop users. Retrospective and prospective studies taking into consideration the number of hours of laptop use at office and home, sitting position, body posture, position of laptop, frequency and duration of intermittent breaks, and other variables should be planned to firmly establish a relationship between laptop use and testicular heating and its potential impact on semen parameters. The studies discussed in this article have addressed changes in semen parameters, but research on the impact of testicular heating on fertility remains to be explored. Since the use of laptop starts early in life, studying its effects on pubertal development and semen parameters in adulthood holds an important place in fertility research.

Research on the impact of laptop and mobile phone use suffers from a number of limitations in addition to the above-mentioned confounding factors. A host of other variables, such as exercise and fitness regimen, nutritional status, and other stressful factors, may affect their effects on fertility. Since the radiations and heat emitted by these devices are invisible, we end up using them extravagantly. It must be kept in mind that these indispensable devices are getting smarter with time and they keep on performing a number of functions in the background even when not in use. Therefore, the devices such as mobile phone, Wi-Fi, and laptop should be switched off when not in use. Further, we must ensure a safe distance from them, and when unavoidable, we must ensure maximum possible distance between these devices and testes. For example, mobile phone should not be kept in the trousers' pocket when ambulatory and should be placed on a nearby table when static. These simple measures should be able to curb most of the adverse impacts these devices may silently have on fertility parameters.

The studies on alcohol, smoking, and stress have advanced to conclusively establish their negative effects on seminal parameters. Further studies on these aspects may focus on molecular investigations to identify fine molecular changes contributing to their ill effects on fertility. One of the important aspects to investigate in relation to these habits is their impact on the sexual development, puberty, and fertility in young people. It has been reported that smoking and alcohol are tightening grip in the young generation. Exposure to these in young age may interfere with sexual development and attainment of maturity. Therefore, further studies should investigate the effect of early exposure to these habits as a cause of infertility. Since both alcohol and smoking have been reported to have potential transgenerational effects, further research in this direction would provide significant clues to understand if the coming generations would have to pay for the poor lifestyle of the reproducing generation. Similarly, stress and trauma early in life have been reported to have transgenerational effects; therefore, there is utmost need to further investigate transgenerational impact of smoking, alcohol, and a stressful life in relation to semen parameters.

Exercise in general has good overall effects on health; however, engaging into intense exercise and endurance building should be carefully considered to avoid adverse effects on fertility. Nevertheless, light to moderate exercise with appropriate combination of aerobics, muscle strength building, and fitness regimen should be included in daily schedule of activities. Some of the good practices to improve lifestyle can have dramatic effects on semen parameters and reproductive health, particularly if adopted early in life. The list of lifestyle diseases and the overall health benefits of engaging into good lifestyle is growing, and I have presented their impact on an important aspect of species survival. A number of recent studies have emphasized the transgenerational impact of poor lifestyle activities and suggested that the coming generation may have to pay for the "sins" of father; however, the knowledge in this area is still in infancy. Therefore, it is imperative to pay sufficient attention to this important but easily manageable aspect to ensure good overall and reproductive health in the present and future generations.

Conclusion

Lifestyle factors such as smoking, alcohol, mobile and laptop use, exposure to heat and radiations, stress, and other activities adversely affect testosterone level, spermatogenesis, erectile function, and fertility. Fortunately, lifestyle is one of the easiest modifiable factors. General consciousness to health effects of poor lifestyle and numerous simple modifications are sufficient to wake us to adopt simple and sensible measures for better fertility. Smoking, alcohol, and stress have already been shown to have transgenerational impact, which can imprint poor sperm production and fertility in the DNA that we inherit to the coming generations. Therefore, engaging into good and fitness-oriented lifestyle should be one of our top priorities for the sake of our fertility and that of the generations to come.

Acknowledgments I would like to thank the Council of Scientific and Industrial Research (CSIR), Govt. of India for financial help under the network project, PROGRAM (BSC0101).

References

Abel EL (1983) Marihuana, tobacco, alcohol, and reproduction. CRC, Boca Raton, FL

Abel E (2004) Paternal contribution to fetal alcohol syndrome. Addict Biol 9(2):127–133

Abu-Musa AA, Nassar AH, Hannoun AB, Usta IM (2007) Effect of the Lebanese civil war on sperm parameters. Fertil Steril 88(6):1579–1582

Adler RA (1992) Clinical review 33: clinically important effects of alcohol on endocrine function. J Clin Endocrinol Metabol 74(5):957–960

Agarwal A, Deepinder F, Sharma RK, Ranga G, Li J (2008) Effect of cell phone usage on semen analysis in men attending infertility clinic: an observational study. Fertil Steril 89(1):124–128

Agarwal A, Desai NR, Makker K, Varghese A, Mouradi R, Sabanegh E, Sharma R (2009) Effects of radiofrequency electromagnetic waves (RF-EMW) from cellular phones on human ejaculated semen: an in vitro pilot study. Fertil Steril 92(4):1318–1325

Agarwal A, Singh A, Hamada A, Kesari K (2011) Cell phones and male infertility: a review of recent innovations in technology and consequences. Int Braz J Urol 37(4):432–454

Ahlbom A, Green A, Kheifets L, Savitz D, Swerdlow A (2004) ICNIRP (International Committee for Non-Ionizing Radiation Protection) Standing Committee on Epidemiology. Epidemiology of health effects of radiofrequency exposure. Environ Health Perspect 112:1741–1754

Ahmad L, Baig NM (2011) Mobile phone RF-EMW exposure to human spermatozoa: an in vitro study. Pak J Zool 43:1147–1154

Ahmad G, Moinard N, Esquerré-Lamare C, Mieusset R, Bujan L (2012) Mild induced testicular and epididymal hyperthermia alters sperm chromatin integrity in men. Fertil Steril 97(3): 546–553

Alexandrov K, Rojas M, Rolando C (2006) DNA damage by benzo (a) pyrene in human cells is increased by cigarette smoke and decreased by a filter containing rosemary extract, which lowers free radicals. Cancer Res 66(24):11938–11945

Alvarenga TA, Andersen ML, Velázquez-Moctezuma J, Tufik S (2009) Food restriction or sleep deprivation: which exerts a greater influence on the sexual behaviour of male rats? Behav Brain Res 202(2):266–271

Alvarenga TA, Hirotsu C, Mazaro-Costa R, Tufik S, Andersen ML (2015) Impairment of male reproductive function after sleep deprivation. Fertil Steril 103(5):1355–1362

Andersen ML, Santos-Silva R, Bittencourt LR, Tufik S (2010) Prevalence of erectile dysfunction complaints associated with sleep disturbances in Sao Paulo, Brazil: a population-based survey. Sleep Med 11(10):1019–1024

Arabi M, Moshtaghi H (2005) Influence of cigarette smoking on spermatozoa via seminal plasma. Andrologia 37(4):119–124

Asare-Anane H, Bannison SB, Ofori EK, Ateko RO, Bawah AT, Amanquah SD, Oppong SY, Gandau BBN, Ziem JB (2016) Tobacco smoking is associated with decreased semen quality. *Reprod Health 13*(1):90

Auger J, Eustache F, Andersen AG, Irvine DS, Jørgensen N, Skakkebaek NE, Suominen J, Toppari J, Vierula M, Jouannet P (2001) Sperm morphological defects related to environment, lifestyle and medical history of 1001 male partners of pregnant women from four European cities. Hum Reprod 16(12):2710–2717

Avendano C, Mata A, Sarmiento CAS, Doncel GF (2012) Use of laptop computers connected to internet through Wi-Fi decreases human sperm motility and increases sperm DNA fragmentation. Fertil Steril 97(1):39–45

Axelsson J, Ingre M, Åkerstedt T, Holmbäck U (2005) Effects of acutely displaced sleep on testosterone. J Clin Endocrinol Metabol 90(8):4530–4535

Bigelow PL, Jarrell J, Young MR, Keefe TJ, Love EJ (1998) Association of semen quality and occupational factors: comparison of case-control analysis and analysis of continuous variables. Fertil Steril 69(1):11–18

Boyar RM, Rosenfeld RS, Kapen S, Finkelstein JW, Roffwarg HP, Weitzman ED, Hellman L (1974) Human puberty simultaneous augmented secretion of luteinizing hormone and testosterone during sleep. *J Clin Invest 54*(3):609

Budweiser S, Enderlein S, Jörres RA, Hitzl AP, Wieland WF, Pfeifer M, Arzt M (2009) Sleep apnea is an independent correlate of erectile and sexual dysfunction. J Sex Med 6(11): 3147–3157

Cannon B (1994) Walter Bradford Cannon: reflections on the man and his contributions. Int J Stress Manag 1(2):145–158

Cao S, Yin X, Wang Y, Zhou H, Song F, Lu Z (2013) Smoking and risk of erectile dysfunction: systematic review of observational studies with meta-analysis. *PLoS One 8*(4):e60443

Cheng JYW, Ng EML (2007) Body mass index, physical activity and erectile dysfunction: an U-shaped relationship from population-based study. Int J Obes 31(10):1571–1578

Ching M, Valenca M, Negro-Vilar A (1988) Acute ethanol treatment lowers hypophyseal portal plasma luteinizing hormone-releasing hormone (LH-RH) and systemic plasma LH levels in orchidectomized rats. Brain Res 443(1):325–328

Childhood obesity prevention. WHO (2009) Website:http://www.who.int/dietphysicalactivity/childhood/WHO_new_childhoodobesity_PREVENTION_27nov_H R_PRINT_OK.pdf

Cicero TJ, Nock B, O'Connor LH, Sewing BN, Adams ML, Meyer ER (1994) Acute paternal alcohol exposure impairs fertility and fetal outcome. Life Sci 55(2):PL33–PL36

Cousineau TM, Domar AD (2007) Psychological impact of infertility. Best Pract Res Clin Obstet Gynaecol 21(2):293–308

Cui X, Jing X, Wu X, Wang Z, Li Q (2016) Potential effect of smoking on semen quality through DNA damage and the downregulation of Chk1 in sperm. Mol Med Rep 14(1):753–761

Dacheux JL, Dacheux F (2014) New insights into epididymal function in relation to sperm maturation. Reproduction 147(2):R27–R42

Dama MS, Bhat MN (2013) Mobile phones affect multiple sperm quality traits: a meta-analysis. F1000Res 2:40

Dasdag S, Ketani MA, Akdag Z, Ersay AR, Sari I, Demirtas ÖC, Celik MS (1999) Whole-body microwave exposure emitted by cellular phones and testicular function of rats. Urol Res 27(3):219–223

de Gannes FP, Billaudel B, Haro E, Taxile M, Le Montagner L, Hurtier A, Aissa SA, Masuda H, Percherancier Y, Ruffié G, Dufour P (2013) Rat fertility and embryo fetal development: influence of exposure to the Wi-Fi signal. Reprod Toxicol 36:1–5

Desai NR, Kesari KK, Agarwal A (2009) Pathophysiology of cell phone radiation: oxidative stress and carcinogenesis with focus on male reproductive system. *Reprod Biol Endocrinol* 7(1):1

Dong H, Wang Y, Zou Z, Chen L, Shen C, Xu S, Zhang J, Zhao F, Ge S, Gao Q, Hu H (2016) Abnormal methylation of imprinted genes and cigarette smoking assessment of their association with the risk of male infertility. Reprod Sci pii:1933719116650755

Du Plessis SS, Cabler S, McAlister DA, Sabanegh E, Agarwal A (2010) The effect of obesity on sperm disorders and male infertility. Nat Rev Urol 7(3):153–161

Erogul O, Oztas E, Yildirim I, Kir T, Aydur E, Komesli G, Irkilata HC, Irmak MK, Peker AF (2006) Effects of electromagnetic radiation from a cellular phone on human sperm motility: an in vitro study. Arch Med Res 37(7):840–843

Esakky P, Moley KH (2016) Paternal smoking and germ cell death: a mechanistic link to the effects of cigarette smoke on spermatogenesis and possible long-term sequelae in offspring. Mol Cell Endocrinol 435:85–93

Eskiocak S, Gozen AS, Yapar SB, Tavas F, Kilic AS, Eskiocak M (2005) Glutathione and free sulphydryl content of seminal plasma in healthy medical students during and after exam stress. Hum Reprod 20(9):2595–2600

Fakhri Y, Khoei HH, Sistanian F, Moloudizargari M, Adel M, Keramati H, Bay A, Avazpour M, Zandsalimi Y, Moradi B, Amirhajeloo LR (2016) Effects of electromagnetic wave from mobile phones on human sperm motility and viability: a systematic review and meta-analysis. Int J Med Res Health Sci 5(6):172–182

Falzone N, Huyser C, Fourie F, Toivo T, Leszczynski D, Franken D (2008) In vitro effect of pulsed 900 MHz GSM radiation on mitochondrial membrane potential and motility of human spermatozoa. Bioelectromagnetics 29(4):268–276

Fejes I, Závaczki Z, Szöllősi J, Koloszár S, Daru J, Kovacs L, Pal A (2005) Is there a relationship between cell phone use and semen quality? Arch Androl 51(5):385–393

Friedler G (1996) Paternal exposures: impact on reproductive and developmental outcome. An overview. Pharmacol Biochem Behav 55(4):691–700

Gambineri A, Pelusi CA, Pasquali R (2003) Testosterone levels in obese male patients with obstructive sleep apnea syndrome: relation to oxygen desaturation, body weight, fat distribution and the metabolic parameters. J Endocrinol Investig 26(6):493–498

Gamble KL, Resuehr D, Johnson C (2013) Shift work and circadian dysregulation of reproduction. Front Endocrinol 4:92

Gapp K, Jawaid A, Sarkies P, Bohacek J, Pelczar P, Prados J, Farinelli L, Miska E, Mansuy IM (2014) Implication of sperm RNAs in transgenerational inheritance of the effects of early trauma in mice. Nat Neurosci 17(5):667–669

Gaskins AJ, Mendiola J, Afeiche M, Jørgensen N, Swan SH, Chavarro JE (2013) Physical activity and television watching in relation to semen quality in young men. *Br J Sports Med* 49(4):265–270

Giessing J (2003) Choosing the most effective level of intensity for cardiovascular exercise. NSCA's Perform Train J 2(3):11–14
Global status report on alcohol and health. WHO (2008). Website:http://www.who.int/substance_abuse/publications/global_alcohol_report/msbgsruprofiles.pdf
Gollenberg AL, Liu F, Brazil C, Drobnis EZ, Guzick D, Overstreet JW, Redmon JB, Sparks A, Wang C, Swan SH (2010) Semen quality in fertile men in relation to psychosocial stress. Fertil Steril 93(4):1104–1111
Gordon GG, Altman K, Southren AL, Rubin E, Lieber CS (1976) Effect of alcohol (ethanol) administration on sex-hormone metabolism in normal men. N Engl J Med 295(15):793–797
Gordon GG, Southren AL, Vittek J, Lieber CS (1979) The effect of alcohol ingestion on hepatic aromatase activity and plasma steroid hormones in the rat. Metabolism 28(1):20–24
Granata AR, Rochira V, Lerchl A, Marrama P, Carani C (1997) Relationship between sleep-related erections and testosterone levels in men. J Androl 18(5):522–527
Grunbaum A, Khaleeq-ur-Rehman, Carrier S (2002) Bicycling and erectile dysfunction: a review of the literature. J Sex Reprod Med 2(2):75–79
Guo J, Tao SX, Chen M, Shi YQ, Zhang ZQ, Li YC, Zhang XS, Hu ZY, Liu YX (2007) Heat treatment induces liver receptor homolog-1 expression in monkey and rat Sertoli cells. Endocrinology 148(3):1255–1265
Hackney AC, Moore AW, Brownlee KK (2005) Testosterone and endurance exercise: development of the "exercise-hypogonadal male condition". Acta Physiol Hung 92(2):121–137
Hammoud AO, Wilde N, Gibson M, Parks A, Carrell DT, Meikle AW (2008) Male obesity and alteration in sperm parameters. Fertil Steril 90(6):2222–2225
Hammoud AO, Carrell DT, Gibson M, Peterson CM, Meikle AW (2012) Updates on the relation of weight excess and reproductive function in men: sleep apnea as a new area of interest. Asian J Androl 14(1):77–81. doi:10.1038/aja.2011.64
Harlev A, Agarwal A, Gunes SO, Shetty A, du Plessis SS (2015) Smoking and male infertility: an evidence-based review. World J Mens Health 33(3):143–160
Harrison RG, Weiner JS (1949) Vascular patterns of the mammalian testis and their functional significance. J Exp Biol 26(3):304–316
Harrison KL, Callan VJ, Hennessey JF (1987) Stress and semen quality in an in vitro fertilization program. Fertil Steril 48(4):633–636
Hjollund NHI, Bonde JPE, Jensen TK, Olsen J (2000) Diurnal scrotal skin temperature and semen quality. Int J Androl 23(5):309–318
Hjollund NHI, Bonde JPE, Henriksen TB, Giwercman A, Olsen J, Danish First Pregnancy Planner Study Team (2004) Reproductive effects of male psychologic stress. Epidemiology 15(1):21–27
Ivell R (2007) Lifestyle impact and the biology of the human scrotum. *Reprod Biol Endocrinol 5*(1):1
Janevic T, Kahn LG, Landsbergis P, Cirillo PM, Cohn BA, Liu X, Factor-Litvak P (2014) Effects of work and life stress on semen quality. Fertil Steril 102(2):530–538
Jo J, Kim H (2016) The relationship between body mass index and scrotal temperature among malc partners of subfertile couples. J Therm Biol 56:55–58
Joo KJ, Kwon YW, Myung SC, Kim TH (2012) The effects of smoking and alcohol intake on sperm quality: light and transmission electron microscopy findings. J Int Med Res 40(6):2327–2335
Jung A, Eberl M, Schill WB (2001) Improvement of semen quality by nocturnal scrotal cooling and moderate behavioural change to reduce genital heat stress in men with oligoasthenoteratozoospermia. Reproduction 121(4):595–603
Kesari KK, Kumar S, Behari J (2010) Mobile phone usage and male infertility in Wistar rats. Indian J Exp Biol 48(10):987–992
Kitayama T (1965) Study on testicular temperature in men. *Hinyokika Kiyo. Acta Urologica Japonica 11*(6):435
Klaiber EL, Broverman DM, Pokoly TB, Albert AJ, Howard PJ, Sherer JF (1987) Interrelationships of cigarette smoking, testicular varicoceles, and seminal fluid indexes. Fertil Steril 47(3):481–486

Knezovich JG, Ramsay M (2012) The effect of preconception paternal alcohol exposure on epigenetic remodeling of the h19 and rasgrf1 imprinting control regions in mouse offspring. Front Genet 3:10

Koskelo R, Zaproudina N, Vuorikari K (2005) High scrotal temperatures and chairs in the pathophysiology of poor semen quality. Pathophysiology 11(4):221–224

Lampiao, F., 2009. Variation of semen parameters in healthy medical students due to exam stress. Malawi Med J, 21(4) 166-167.

Lawson AK (2016) Psychological stress and infertility. In: Stevenson JS (ed) Fertility and assisted reproductive technology (ART): theory, research. Springer, New York

Lawson AK, Klock SC, Pavone ME, Hirshfeld-Cytron J, Smith KN, Kazer RR (2014) Prospective study of depression and anxiety in female fertility preservation and infertility patients. Fertil Steril 102(5):1377–1384

Lazarus RS, Folkman S (1984) Stress, appraisal, and coping. Springer Publishing Company, New York

Lemoine P, Harousseau H, Borteyru JP, Menuet JC (1968) Les enfants de parents alcooliques: Anomalies observees a propos de 127 cas. *Ouest Méd 21*(476–482):2

Li XX, Chen SR, Shen B, Yang JL, Ji SY, Wen Q, Zheng QS, Li L, Zhang J, Hu ZY, Huang XX (2013) The heat-induced reversible change in the blood-testis barrier (BTB) is regulated by the androgen receptor (AR) via the partitioning-defective protein (Par) polarity complex in the mouse. *Biol Reprod 89*(1):12

Lloyd CW, Williams RH (1948) Endocrine changes associated with Laennec's cirrhosis of the liver. Am J Med 4(3):315–330

Luboshitzky R, Zabari Z, Shen-Orr Z, Herer P, Lavie P (2001) Disruption of the nocturnal testosterone rhythm by sleep fragmentation in normal men. J Clin Endocrinol Metabol 86(3):1134–1139

Luboshitzky R, Lavie L, Shen-Orr Z, Herer P (2005) Altered luteinizing hormone and testosterone secretion in middle-aged obese men with obstructive sleep apnea. Obes Res 13(4):780–786

Lucia A, Diaz B, Hoyos J, Fernandez C, Villa G, Bandres F, Chicharro JL (2001) Hormone levels of world class cyclists during the Tour of Spain stage race. Br J Sports Med 35(6):424–430

Maiorino MI, Bellastella G, Esposito K (2015) Lifestyle modifications and erectile dysfunction: what can be expected? *Asian J Androl 17*(1):5

McCorry LK (2007) Physiology of the autonomic nervous system. Am J Pharm Educ 71(4):78

Mehran A (2005) The toxic effect of seminal plasma from smokers on sperm function in non-smokers. Zhonghua Nan Ke Xue 11(9):647–651

Merhi ZO (2012) Challenging cell phone impact on reproduction: a review. J Assist Reprod Genet 29(4):293–297

Meri ZB, Irshid IB, Migdadi M, Irshid AB, Mhanna SA (2013) Does cigarette smoking affect seminal fluid parameters? A comparative study. Oman Med J 28(1):12–16

Mieusset R, B'ujan LOUIS (1994) The potential of mild testicular heating as a safe, effective and reversible contraceptive method for men. Int J Androl 17(4):186–191

Mishra GD, Dobson AJ, Schofield MJ (2000) Cigarette smoking, menstrual symptoms and miscarriage among young women. Aust N Z J Public Health 24(4):413–420

Muthusami KR, Chinnaswamy P (2005) Effect of chronic alcoholism on male fertility hormones and semen quality. Fertil Steril 84(4):919–924

Naziroğlu M, Gümral N (2009) Modulator effects of L-carnitine and selenium on wireless devices (2.45 GHz)-induced oxidative stress and electroencephalography records in brain of rat. Int J Radiat Biol 85(8):680–689

Nordkap L, Jensen TK, Hansen ÅM, Lassen TH, Bang AK, Joensen UN, Jensen MB, Skakkebæk NE, Jørgensen N (2016) Psychological stress and testicular function: a cross-sectional study of 1,215 Danish men. Fertil Steril 105(1):174–187

Nouri K, Litschauer B, Sator M, Tiringer D, Ott J, Walch K, Hefler LA, Tempfer CB (2014) Decline of semen quality during IVF is not associated with subjective male stress. *Asian J Androl 16*(4):597

Ochedalski T, Lachowicz-Ochedalska A, Dec W, Czechowski B (1994) Examining the effects of tobacco smoking on levels of certain hormones in serum of young men. Ginekol Pol 65(2):87–93

Oldereid NB, Rui H, Purvis K (1991) Life styles of men in barren couples and their relationship to sperm quality. Int J Fertil 37(6):343–349

Pacey AA, Povey AC, Clyma JA, McNamee R, Moore HD, Baillie H, Cherry NM (2014) Modifiable and non-modifiable risk factors for poor sperm morphology. *Hum Reprod* 29(8):1629–1636. doi:10.1093/humrep/deu116

Pajarinen J, Karhunen PJ, Savolainen V, Lalu K, Penttilä A, Laippala P (1996) Moderate alcohol consumption and disorders of human spermatogenesis. Alcohol Clin Exp Res 20(2):332–337

Parazzini F, Marchini M, Luchini L, Tozzi L, Mezzopane R, Fedele L (1995) Tight underpants and trousers and risk of dyspermia. Int J Androl 18(3):137–140

Pasqualotto FF, Sobreiro BP, Hallak J, Pasqualotto EB, Lucon AM (2006) Cigarette smoking is related to a decrease in semen volume in a population of fertile men. BJU Int 97(2):324–326

Passaro KT, Little RE, Savitz DA, Noss J (1998) Effect of paternal alcohol consumption before conception on infant birth weight. *Teratology 57*(6):294–301

Pourmand G, Alidaee MR, Rasuli S, Maleki A, Mehrsai A (2004) Do cigarette smokers with erectile dysfunction benefit from stopping?: a prospective study. BJU Int 94(9):1310–1313

Povey AC, Clyma JA, McNamee R, Moore HD, Baillie H, Pacey AA, Cherry NM (2012) Modifiable and non-modifiable risk factors for poor semen quality: a case-referent study. *Hum Reprod* 27(9):2799–2806. doi:10.1093/humrep/des183

Priskorn L, Jensen TK, Bang AK, Nordkap L, Joensen UN, Lassen TH, Olesen IA, Swan SH, Skakkebaek NE, Jørgensen N (2016) Is sedentary lifestyle associated with testicular function? A cross-sectional study of 1,210 men. *Am J Epidemiol* 184(4):284–294. doi:10.1093/aje/kwv338

Radwan M, Jurewicz J, Merecz-Kot D, Sobala W, Radwan P, Bochenek M, Hanke W (2016) Sperm DNA damage—the effect of stress and everyday life factors. *Int J Impot Res* 28(4):148–154

Ramlau-Hansen CH, Thulstrup AM, Aggerholm AS, Jensen MS, Toft G, Bonde JP (2007) Is smoking a risk factor for decreased semen quality? A cross-sectional analysis. Hum Reprod 22(1):188–196

Rao M, Zhao XL, Yang J, Hu SF, Lei H, Xia W, Zhu CH (2015) Effect of transient scrotal hyperthermia on sperm parameters, seminal plasma biochemical markers, and oxidative stress in men. *Asian J Androl 17*(4):668

Rao M, Xia W, Yang J, Hu LX, Hu SF, Lei H, Wu YQ, Zhu CH (2016) Transient scrotal hyperthermia affects human sperm DNA integrity, sperm apoptosis, and sperm protein expression. Andrology. doi:10.1111/andr.12228. [Epub ahead of print] PubMed PMID: 27410176

Rock J, Robinson D (1965) Effect of induced intrascrotal hyperthermia on testicular function in man. *Am J Obstet Gynecol 93*(6):793–801

Rodgman A, Perfetti TA (2013) The chemical components of tobacco and tobacco smoke. CRC Press, Boca Raton, FL

Roth MY, Amory JK, Page ST (2008) Treatment of male infertility secondary to morbid obesity. Nat Clin Pract Endocrinol Metab 4(7):415–419

Sanger WG, Friman PC (1990) Fit of underwear and male spermatogenesis: a pilot investigation. Reprod Toxicol 4(3):229–232

Sapra KJ, Eisenberg ML, Kim S, Chen Z, Buck Louis GM (2016) Choice of underwear and male fecundity in a preconception cohort of couples. Andrology 4(3):500–508

Saunders RD, Kowalczuk CI (1981) Effects of 2 45 GHz microwave radiation and heat on mouse spermatogenic epithelium. Int J Radiat Biol Relat Stud Phys Chem Med 40(6):623–632

Sawyer DE, Mercer BG, Wiklendt AM, Aitken RJ (2003) Quantitative analysis of gene-specific DNA damage in human spermatozoa. *Mutat Res 529*(1):21–34

Schiavi RC, White D, Mandeli J (1992) Pituitary-gonadal function during sleep in healthy aging men. Psychoneuroendocrinolgy 17(6):599–609

Selye H (1955) Stress and disease. Laryngoscope 65(7):500–514

Shafik A (1992) Contraceptive efficacy of polyester-induced azoospermia in normal men. Contraception 45(5):439–451

Shefi S, Tarapore PE, Walsh TJ, Croughan M, Turek PJ (2007) Wet heat exposure: a potentially reversible cause of low semen quality in infertile men. Int Braz J Urol 33(1):50–57

Sheynkin Y, Jung M, Yoo P, Schulsinger D, Komaroff E (2005) Increase in scrotal temperature in laptop computer users. Hum Reprod 20(2):452–455

Sheynkin Y, Welliver R, Winer A, Hajimirzaee F, Ahn H, Lee K (2011) Protection from scrotal hyperthermia in laptop computer users. Fertil Steril 95(2):647–651

Shokri S, Soltani A, Kazemi M, Sardari D, Mofrad FB (2015) Effects of Wi-Fi (2.45 GHz) exposure on apoptosis, sperm parameters and testicular histomorphometry in rats: a time course study. *Cell J (Yakhteh) 17*(2):22

Steiner J, Kirsteins L, LaPaglia N, Lawrence A, Williams D, Emanuele N, Emanuele M (1997) The effect of acute ethanol (EtOH) exposure on protein kinase C (PKC) activity in anterior pituitary. Alcohol 14(3):209–211

Thonneau P, Ducot B, Bujan L, Mieusset R, Spira A (1997) Effect of male occupational heat exposure on time to pregnancy. Int J Androl 20(5):274–278

Tiemessen CJ, Evers JH, Bots RG (1996) Tight-fitting underwear and sperm quality. Lancet 347(9018):1844–1845

Torres M, Laguna-Barraza R, Dalmases M, Calle A, Pericuesta E, Montserrat JM, Navajas D, Gutierrez-Adan A, Farré R (2014) Male fertility is reduced by chronic intermittent hypoxia mimicking sleep apnea in mice. Sleep 37(11):1757–1765

Vaamonde D, Da Silva-Grigoletto ME, García-Manso JM, Barrera N, Vaamonde-Lemos R (2012) Physically active men show better semen parameters and hormone values than sedentary men. Eur J Appl Physiol 112(9):3267–3273

Vaamonde D, Agarwal A, du Plessis SS, Algar-Santacruz C, Kruger TF (2016a) Impact of physical activity and exercise on male reproductive potential: semen alterations. In: Vaamonde D, du Plessis SS, Agarwal A (eds) Exercise and human reproduction (pp. 101–124). Springer New York

Vaamonde D, du Plessis SS, Agarwal A (eds) (2016b) Exercise and human reproduction - induced fertility disorders and possible therapies. Springer, New York

Vervoorn C, Quist AM, Vermulst LJM, Erich WBM, De Vries WR, Thijssen JHH (1991) The behaviour of the plasma free testosterone/cortisol ratio during a season of elite rowing training. *Int J Sports Med 12*(03):257–263

Villalta J, Ballesca JL, Nicolas JM, Martinez de Osaba MJ, Antunez E, Pimentel C (1997) Testicular function in asymptomatic chronic alcoholics: relation to ethanol intake. Alcohol Clin Exp Res 21(1):128–134

Wang C, McDonald V, Leung A, Superlano L, Berman N, Hull L, Swerdloff RS (1997) Effect of increased scrotal temperature on sperm production in normal men. Fertil Steril 68(2):334–339

Wdowiak A, Wdowiak L, Wiktor H (2007) Evaluation of the effect of using mobile phones on male fertility. Ann Agric Environ Med 14(1):169–172

Whirledge S, Cidlowski JA (2013) A role for glucocorticoids in stress-impaired reproduction: beyond the hypothalamus and pituitary. *Endocrinology 154*(12):4450–4468

World Health Organization (2015) WHO global report on trends in prevalence of tobacco smoking 2015. World Health Organization. http://apps.who.int/iris/bitstream/10665/156262/1/9789241564922_eng.pdf

Yang F, Li L, Chen JP, Liu XQ, Zhong CL, Yang Y, Ren YF, Yuan W, Liang H, Miao MH (2016) Couple's infertility in relation to male smoking in a Chinese rural area. doi:10.4103/1008-682X.168685*Asian J Androl*

Zenzes MT (2000) Smoking and reproduction: gene damage to human gametes and embryos. Hum Reprod Update 6(2):122–131

Zhang XS, Zhang ZH, Jin X, Wei P, Hu XQ, Chen M, Lu CL, Lue YH, Hu ZY, Sinha Hikim AP, Swerdloff RS (2006) Dedifferentiation of adult monkey Sertoli cells through activation of extracellularly regulated kinase 1/2 induced by heat treatment. Endocrinology 147(3):1237–1245

Zhang ZH, Li LL, Yu Y, Zhang HG, Liu RZ (2013) Decline of semen quality and increase of leukocytes with cigarette smoking in infertile men. Int J Reprod BioMed 11(7):589–596

Zhu Q, Van Thiel DH, Gavaler JS (1997) Effects of ethanol on rat Sertoli cell function: studies in vitro and in vivo. Alcohol Clin Exp Res 21(8):1409–1417

Zorgniotti AW, Macleod J (1973) Studies in temperature, human semen quality, and varicocele. Fertil Steril 24(11):854–863

Zorgniotti A, Reiss H, Toth A, Ealfon AS (1982) Effect of clothing on scrotal temperature in normal men and patients with poor semen. *Urology 19*(2):176–178

Preserving Male Fertility

23

Rajender Singh

Abstract

Do you love your fertility? Obviously yes; but the most important question is how often do you think of preserving your fertility? Not so often or never. We tend to believe that fertility doesn't require any care and in some societies; this subject is secluded from family matters. Fertility loss may start right from the attainment of puberty and may get accelerated if not paid heed. Interestingly, the loss of fertility may have no associated symptoms in a large number of cases. Finding yourself infertile once you wish to have children may have distressing consequences, which could have been easily averted with simple nontherapeutic ways aimed at preserving fertility. Taking cues from the published literature, I have prescribed lifestyle interventions and other simple measures that may keep you way from prescriptions for infertility management.

Keywords

Preserve fertility • Upkeep fertility • Lifestyle • Smoking • Alcohol • Stress • Exercise and diet • Semen analysis

Key Points

- Avoid exposure to radiations in the form of mobile phone, Wi-Fi, and other equipments used in household or office settings.
- Minimize testicular heating by taking precautions in laptop use and television watching and by keeping away from hot baths, sauna, Jacuzzi, and other habits that may raise scrotal temperature.

R. Singh
Male Reproductive Biology Laboratory, CSIR-Central Drug Research Institute, Lucknow, Uttar Pradesh, India
e-mail: rajender_singh@cdri.res.in

R. Singh, K. Singh (eds.), *Male Infertility: Understanding, Causes and Treatment*, DOI 10.1007/978-981-10-4017-7_23

- Learn to tackle stress and anxiety due to occupational or psychological stress to minimize their effects on testosterone production, spermatogenesis, and fertility.
- Respect the biological clock and follow a balanced work-relax schedule to avoid sleep apnea and development of stress.
- Light to moderate exercise along with a balanced and nutritious diet is the key to better semen quality, but avoid regular heavy exercise.
- Semen analysis early in life can help preserve fertility in susceptible cases.
- Timely planning of family helps not only with good fertility but also with minimum risk of congenital abnormalities in the next generation.

23.1 Introduction

The nature has provided every species with the best tactics to survive even in the weirdest conditions. Security for food and reproduction are the two most important aspects intricately wired into all living beings. The logic of great investment in these two is easily understandable as they are indispensable for species survival. Food is the first priority for any living organism, and once food is secured, the hunt for reproduction comes to the fore. Superimposed on this, every species is provided with methods to improve its race by employing various methods of crossover and selection of the best partners for procreation. The biological system invests heavily in the process of reproduction. The gamete production (spermatogenesis in the context of this chapter) is a heavy energy demanding process, which deserves sufficient attention to earn returns on the investment.

Most of us are endowed the gift of procreation; however, we generally do not ponder about our fertility, until we think of starting a family. Unfortunately, for a number of individuals, it is too late by then. Men present a continuous spectrum of no, poor, moderate, and high fertility. To ensure species survival, nature has ensured multiple lines of defense to preserve fertility and perhaps that is why most of us can do without worrying much about fertility. However, the reproductive fitness varies greatly across individuals. The declining reproductive fitness is evident by reverse trends in semen quality and increased frequency of infertility cases in the recent years. The etiology in a large number of infertility cases remains unknown, complicating the treatment. Nevertheless, prevention is always better than cure.

Poor fertility early in life tends to fade sooner than rich fertility. This can be more dreadful in the cases where fertility loss is discovered too late. A number of infertility cases could have been fertile and had some general, inexpensive, common, and habitual precautions been in place. One must care for fertility from early in life and even after having done with the family planning. In a number of situations, people choose to have children even after family planning owing to a number of reasons from personal to accidental. From the initiation of puberty and spermatogenesis till the descent of spermatogenesis with andropause or age, caring for self-fertility can be rewarding. Irrespective of the marital status, availability of a partner or age, you must be conscious about your sperm and fertility so that you can initiate a pregnancy whenever required. Therefore, caring for fertility all through life should be one of our priorities.

Over the past several decades of research on fertility and infertility, we have learned a number of factors that contribute to infertility. Therefore, we have reached the crossroads where we should take a call and educate ourselves about the ways of fertility preservation. This would not only help in the maintenance of fertility for a longer duration but also help in contributing the best quality gametes for species propagation. In this chapter, I have attempted to put a number of evidence-based suggestions that can help preserve fertility for a longer duration, delay the onset of infertility, and avert it altogether in some cases. This guide is intended to help people understand and preserve fertility, the loss of which can have devastating impact later in life.

23.2 Avoid Exposure to Radiations

Radiations are inevitable in today's life owing to a number of applications we cannot do without. Therefore, suggestions to get rid of them would be impractical; nevertheless, the possibility of minimizing the use to "just essential" is always there. A bunch of studies has shown potential ill effects of mobile phone radiations (Agarwal et al. 2008), Wi-Fi, and other radiations used in household, clinic, or office settings (Fakhri et al. 2016). Some of the studies even claim that the radiations may not cause significant harm; nevertheless, it is sensible to avoid extra exposure. In a number of situations, we end up using them extravagantly. The carelessness may not cost us extra in monetary terms but can be very expensive in terms of their adverse impact on a number of health conditions, including reproductive health and fertility. We must pay sufficient attention to feel the presence of these invisible radiations and try to minimize exposure by switching off mobile phone, Wi-Fi, and other equipments, when not in use, in order to avoid their potential hazards on spermatogenesis and fertility.

23.3 Avoid Testicular Heating

A number of conditions such as using laptop in lap position, watching television, wearing tight underwear and tight jean pants, and other activities such as a number of sports activities that require tying up the scrotum against the pelvis, cross-leg seating, cushioned chair at work or home can result in testicular heating. Simple awareness about the activities, habits, or lifestyle factors that can result in testicular heating is sufficient to understand the necessary modifications that can help us avoid testicular heating. Testicular heating has been shown to result in significant adverse impact of spermatogenesis. A simple rise of 1 °C can lower spermatogenesis; significantly a rise of 2–3 °C can cease spermatogenesis altogether (Durairajanayagam et al. 2015). While there are biological mechanisms to protect against testicular heating, repeated episodes of heating cannot be tackled sufficiently to nullify all effects of testicular heating. Therefore, testicular heating should best be avoided to let spermatogenesis flourish. Testes are meant to hang and let them follow their natural course.

23.4 Learn to Kill Stress

Stress is a well-known detrimental factor for spermatogenesis. Stress has a strong impact on the hypothalamic-pituitary-gonadal axis, disturbances in which can leave spermatogenesis significantly impaired (Nordkap et al. 2016). A number of endocrinopathies are known to affect fertility in both the genders; therefore, keeping the HPG axis close to normal is the key to quality sperm production. Stress comes in avoidable and unavoidable forms. Life events, which are beyond our control cannot be evaded; nevertheless, their ill effects can be controlled by engaging in a number of activities that are well-known stress busters. Engaging in sports, exercise, and other hobbies relieves stress, and a small change can have a big effect on the endocrine system and stress management. Mismanagement and poor planning is a significant reason behind stress due to job, household chores, and other professional or domestic activities (Michie 2002). Anxiety in a variety of forms can result in stress in the lack of efficient management. Therefore, advance planning, efficient and smart work style, defining objectives, and engaging into strategy building can alleviate stress due to the above factors. Despite all these measures, some forms of stress would always strike the doors. Identifying ways to tackle stress by training your mind not to overreact can have a tremendous effect on stress management. The biological system is highly flexible and a fast learner. Therefore, attempts to train your brain in stress management can yield unprecedented results.

23.5 Quit Smoking and Alcohol

Smoking has been proven to be the cause of cancer and increases the risk of a number of cancers tremendously. Still, a large population of the world smokes. Moderate and heavy smoking has been shown to reduce sperm count and motility (Sharma et al. 2016). Smoking leaves a number of carcinogenic and xenobiotic compounds in our body, which need to be removed by the detoxification system. In heavy smoking, a large battery of the detoxification enzymes gets engaged in repeated cleaning of the smoking compounds inhaled. Under these conditions, the biological system remains under stress to keep the defense system on its toes by engaging in continuous clearance of the xenobiotics. Such a system is not capable of tackling any emergent stress in the form of testicular heating, poor nutrition, heavy load, job pressure, and psychological stress. Under adverse conditions, the biological system drained out as a result of smoking would surrender, leading to adverse impact on spermatogenesis. A small increase in the number of abnormal sperm in the ejaculate can compromise the fertility quite significantly despite the production of a good number of normal sperm. Further, smoking and drinking early in life can hamper attainment of puberty and may result in inadequate pubertal developmental. Such individuals are more prone to fertility loss, and the fertile period in these individuals may also be shorter. Therefore, one must stay away from these habits and try to quit if already in their grip.

23.6 Engage in Light to Moderate Exercise

Exercise in general is good for health. Exercise brings in a lot of vital improvements in the circulatory, excretory, musculoskeletal, neurosensory, and endocrine and immune systems, thus attaining a state of good overall health. Every organ system is connected to others, and therefore, good health in one aspect favors all others positively. Engaging in daily exercise is the way to good health; however, exercise can be good or bad depending upon exertion, intensity, type, and overall objective. Engaging in heavy exercises such as heavy gym, weight lifting, cycling, and other endurance- and resistant-building exercises may compromise spermatogenesis due to direct or indirect effects (Vaamonde et al. 2016). For example, in heavy cycling, the continuous pressure of saddle on reproductive organs can result in trauma, affecting spermatogenesis and semen production. In a number of other sports activities, tethering of scrotum in a tight pouch against the pelvic floor may be required, which can result in testicular heating. In general, engaging in exercise is good, but one must avoid repeated, particularly daily or several times every week, schedule of heavy exercise in order to reduce the possibility of adverse impact on spermatogenesis. For good part of exercise, spermatogenesis can be improved by engaging in light to moderate exercises, such as sports, walk, aerobics, running, light gym and other activities, which do not take a heavy toll on the system. One of the primary aims of exercise should be to keep body mass index (BMI) within normal limits (below 25) in order to keep good overall health.

23.7 Follow the Biological Clock

The biological clock in animals is set for a harmony with the environment and nature. The theory of evolution supports the origin of animals according to the environmental conditions that were existent and the cycle of light and darkness. Even today, our body brings necessary modifications in our system in a constant endeavor to better adapt with the changing environment. Therefore, harmony with the nature and environment is the best way to keep good health. As detailed above, the biological system is capable of adjusting for subtle and occasional deviations from the regular course of life and systematics; however, drastic, repeated, and haphazard deviations can have a heavy impact on the general health and fertility. A number of hormones including melatonin and testosterone are secreted during night's sleep, though the biological reasons behind this choice are unknown. However, that essentially means, following the biological clock is the key to good health and fertility is no exception. Significant alterations in the biological clock can be detrimental to fertility. Studies on sleep disturbance in animal models and human studies on sleep deprivation and sleep apnea have already shown significant alterations in testosterone and other hormone levels (Alvarenga et al. 2015). These changes have been predicted to translate into erectile dysfunction and poor semen parameters. Similarly, night shift women workers have been shown to have oligomenorrhea, anovulation, and dysmenorrhea (Gamble et al. 2013). Therefore, it is advisable to follow the natural course of sleep and work as far as possible, and significant deviations from this should be avoided in order to keep good fertility.

23.8 Eat a Balanced and Nutritious Diet

A balanced and nutritious diet has been emphasized since ages. A number of food elements have undergone significant debates with respect to the health benefits they offer. Food elements known to have beneficial overall health effects are generally good for fertility too. High-fat diet, excessive eating, poor glucose metabolism and lipid profile, and a low level of antioxidants affect spermatogenesis and fertility adversely. Therefore, taking a well-balanced and nutritious diet is the key to adequate spermatogenesis and fertility. A number of food items, such as a prudent diet and the relative levels of components of plant or animal origin, are known to affect spermatogenesis and fertility, which should be chosen carefully for inclusion in the regular diet. Adequate levels of essential vitamins/antioxidants such as B complex, C, E, L-carnitine, and CoQ10 must be taken to maintain the mineral balance required for spermatogenesis and good overall health. The choice of food and nutrition elements to keep good fertility should be based on the dietary habits of the individual. For example, vegetarians may have a deficiency of vitamin B12; hence, the diet and nutrition should be designed keeping in mind the requirement of B12. Certain phytoestrogens such as soy have been shown to have detrimental effects on spermatogenesis; hence, high consumption of these food items should be avoided. The eating regimen and diet composition should be decided in accordance with the lifestyle such that malnutrition, obesity, high blood pressure, etc., are kept at bay by necessary dietary modifications. Taking nutrients and vitamins regularly may not increase your fertility, but they may counteract the adverse changes under stressful conditions. The vitamins that are water soluble (B group and C) can be taken without the fear of toxicity; however, the lipid soluble vitamins (A, D, E, K) should be taken only when a test suggests their deficiency.

23.9 Have Regular Coitus

Sex is not only a way to procreate but also a way to keep good health. Sex has been shown to affect a number of signaling mechanisms mediated/induced by the release of oxytocin, vasopressin, dopamine, serotonergic signaling, endorphin, and endogenous morphinergic mechanisms, coupled to nitric oxide autoregulatory pathways (Esch and Stefano 2005). A number of these changes are known to alleviate stress levels. Therefore, having regular sex is one of the ways to keep good overall and sexual health. That is not all; regular sex may have other important implications. It has been shown in scientific studies that a second ejaculate showed better sperm parameters in comparison to the first ejaculate after 2–7 days of abstinence (Bahadur et al. 2016). It is known that sperm stored in the epididymis have to undergo apoptosis if not ejaculated. This may result in a significant number of sperm undergoing apoptosis in epididymis after a long period of abstinence, resulting in a significantly high proportion of such sperm in the first ejaculate after prolonged abstinence (Sunanda et al. 2014). Therefore, regular ejaculations can dispense sperm before they are directed to undergo apoptosis and thus improve sperm quality (Cannon 2013; Wilton et al. 1988). The frequency of sexual intercourse varies a lot across the globe, but from a fertility standpoint of view, it has been suggested that intercourse every alternate day is a good way to achieve pregnancy in comparison to daily or infrequent intercourse (Imer and Willbanks 2010). Therefore, regular sex has benefits beyond pleasure in stress management and keeping a store of better sperm. Some studies have even reported that regular ejaculations reduce the risk of prostate cancer. Prostate cancer, if present, would require therapeutic or surgical intervention for management, both of which are detrimental to the process of spermatogenesis and fertility due to their effects of testis, testosterone action, spermatogenesis, and semen formation. Therefore, engaging in regular sex may add more than pleasure in life and a method of up keeping fertility and reproductive health.

23.10 Semen Analysis Early in Life

Spermatogenesis and fertility is set once puberty is achieved. However, we do not care about this essential process until we think of starting a family. In the cases with adequate spermatogenesis, waiting till the initiation of family may not be a matter of thought. However, a significant number of individuals are poor producers of sperm or have poor semen parameters from the beginning. Being infertile can have serious personal, psychological, and societal impact; therefore, waiting to discover low fertility or infertility as late as while starting a family may leave one with no corrective measure available. We have come across a few patients who had sperm count in normal range (>20 and <50 million/mL) in the initial years of infertility investigations, which dropped to less than 15 million in subsequent years and in some cases to even azoospermia over a period of 5 years. Early semen analysis could be very helpful in such cases. Finding of abnormally low values of sperm count and motility in an individual can help by taking appropriate precautionary measures to maintain fertility and by keeping an eye on semen parameters over a period of time, just like organ surveillance and screening tests recommended in individuals susceptible to certain familial diseases. Detecting poor semen parameters early in life may also help in appropriate and timely family planning or to opt for cryopreservation in case a threat to fertility looks plausible. The age of 20 years may be suitable for undergoing the first semen analysis, and a subsequent analysis every 2–3 years should be a sensible choice to keep an eye on fertility.

23.11 Timely Planning of Family

The age at marriage or the age of planning the first child is increasing. Fertility in males starts around the age of 17 and in females around the age of 14 once they achieve puberty. However, the age at marriage or the first child is generally very late in comparison to this age. There are numerous reasons for pushing the marriage or family planning toward the fag end of your fertility period. The need to build career, fierce competition for jobs and survival, stability in life, self-sufficiency, and ability to bear the burden of a family are some of the reasons that have resulted in increased age at the first child. Most of the reasons discussed above may be justified in terms of the requirement for a proper and confident parenthood. Nevertheless, in certain cases the harm can be more than good, if delaying takes a toll on fertility, resulting in loss of fertility in either of the partners. Therefore, planning of family as soon as possible or immediately after achieving a certain level of societal or job security is the key to utilizing good fertility that may no longer remain so after a few years. Fertility is naturally kept under a check in advanced age in order to prevent the transmission of faulty gametes to the coming generations that can result in the generation of individuals with certain disorders (Harris et al. 2011). Among other reasons for decline in fertility with time are decreased frequency of sex, erectile problems, and other health problems that are more likely with advancing age (Harris et al. 2011). The incidence of a number of disorders has been reported to be high in the children of old parents. It has been suggested that the paternal age should ideally not be above 35 and maternal age above 30 for the best results. With delayed age at the first child, we not only increase the risk of losing fertility but also increase the risk of giving birth to a child with congenital deformities.

Conclusion

Every individual undergoes loss of fertility with age. Fertility is at its best from puberty till the age of 30 years. It is believed that biological aging starts at the age of 30 years and fertility is no exception. The loss of fertility may be more accelerated in some individuals due to numerous genetic, environmental, circumstantial, and other unknown factors and, if unchecked, may result in infertility. General precautionary measures to look after fertility are inexpensive and can be fun at times. Hence, one must pay heed to the above suggestions in order to strive their fertility to the best and remain capable of procreation as and when required. The rewards can be extraordinary in the susceptible cases where accelerated loss of fertility could have rendered one incapable of procreation. Therefore, in your best interest and that of the species, always upkeep your fertility.

Acknowledgments I would like to thank the Council of Scientific and Industrial Research (CSIR), Govt. of India, for financial help under the network project PROGRAM (BSC0101).

References

Agarwal A, Deepinder F, Sharma RK, Ranga G, Li J (2008) Effect of cell phone usage on semen analysis in men attending infertility clinic: an observational study. Fertil Steril 89(1):124–128

Alvarenga TA, Hirotsu C, Mazaro-Costa R, Tufik S, Andersen ML (2015) Impairment of male reproductive function after sleep deprivation. Fertil Steril 103(5):1355–1362

Bahadur G, Almossawi O, Zaid RZ, Ilahibuccus A, Al-Habib A, Muneer A, Okolo S (2016) Semen characteristics in consecutive ejaculates with short abstinence in subfertile males. Reprod Biomed Online 32(3):323–328

Cannon E (2013) Emma Cannon's total fertility: how to understand, optimize and preserve your fertility. Pan Macmillan, London

Durairajanayagam D, Agarwal A, Ong C (2015) Causes, effects and molecular mechanisms of testicular heat stress. Reprod Biomed Online 30(1):14–27

Esch T, Stefano GB (2005) The neurobiology of love. Neuroendocrinol Lett 26(3):175–192

Fakhri Y, Khoei HH, Sistanian F, Moloudizargari M, Adel M, Keramati H, Bay A, Avazpour M, Zandsalimi Y, Moradi B, Amirhajeloo LR (2016) Effects of electromagnetic wave from mobile phones on human sperm motility and viability: a systematic review and meta-analysis. Int J Med Res Health Sci 5(6):172–182

Gamble KL, Resuehr D, Johnson C (2013) Shift work and circadian dysregulation of reproduction. Front Endocrinol 4:92

Harris ID, Fronczak C, Roth L, Meacham RB (2011) Fertility and the aging male. Rev Urol 13(4):e184

Imler PB, Willbanks D 2010. The essential guide to getting pregnant. http://americanpregnancy.org/getting-pregnant-ebook/p7M7O0q1c71703C/gettingpregnant.pdf. Accessed 18 Oct 2016

Michie S (2002) Causes and management of stress at work. Occup Environ Med 59(1):67–72

Nordkap L, Jensen TK, Hansen ÅM, Lassen TH, Bang AK, Joensen UN, Jensen MB, Skakkebæk NE, Jørgensen N (2016) Psychological stress and testicular function: a cross-sectional study of 1,215 Danish men. Fertil Steril 105(1):174–187

Sharma R, Harlev A, Agarwal A, Esteves SC (2016) Cigarette smoking and semen quality: a new meta-analysis examining the effect of the 2010 World Health Organization laboratory methods for the examination of human semen. Euro Urol 70(4):635–645

Sunanda P, Panda B, Dash C, Padhy RN, Routray P (2014) Effect of age and abstinence on semen quality: a retrospective study in a teaching hospital. Asian Pac J Reprod 3(2):134–141

Vaamonde D, Agarwal A, du Plessis SS, Algar-Santacruz C, Kruger TF (2016) Impact of physical activity and exercise on male reproductive potential: semen alterations. In: Vaamonde D, du Plessis SS, Agarwal A (eds) Exercise and human reproduction. Springer, New York, pp 101–124

Wilton LJ, Temple-Smith PD, Baker HG, de Kretser DM (1988) Human male infertility caused by degeneration and death of sperm in the epididymis. Fertil Steril 49(6):1052–1058

Specific and Generalized Treatments of Male Infertility

24

Sandeep Kumar Bansal and Rajender Singh

Abstract
Approximately, 10–15% of all couples have fertility problems, undergo fertility assessment, and seek treatment. Infertility is a common and complex disorder attributed to a number of etiological factors. Due to the complexity of this disorder, its diagnosis and treatment are not straightforward. There is no standardized drug available for the treatment of idiopathic infertility. Generally, medicinal therapy is recommended on the basis of actual or probable cause of infertility. Antiestrogen therapy is the most common treatment for idiopathic infertility. Besides this, vitamins and antioxidants are also prescribed as dietary supplements to improve the semen quality. However, assisted reproductive techniques can be used when medicinal therapy fails to restore fertility or initiate pregnancy. In this chapter, we have discussed specific and generalized therapies for the management of male infertility.

Keywords
Infertility treatment • Hormones and gonadotropins in male infertility Antiestrogens in male infertility • Antioxidants in male infertility

Key Points

- Due to highly complex nature of the disorder, there is no standardized form of male infertility treatment, which can be prescribed to most of the patients.
- Poor understanding of the etiology of male infertility is the prominent reason behind inadequate therapeutic measures available for treating it.

S.K. Bansal • R. Singh (✉)
Male Reproductive Biology Laboratory, CSIR-Central Drug Research Institute, Lucknow, Uttar Pradesh, India
e-mail: rajender_singh@cdri.res.in

R. Singh, K. Singh (eds.), *Male Infertility: Understanding, Causes and Treatment*, DOI 10.1007/978-981-10-4017-7_24

- Specific treatment upon identification of the reason or generalized/empirical treatment in idiopathic cases is recommended to begin with.
- In many cases, vitamins and antioxidants are also prescribed as dietary supplements to improve the semen quality.
- Failure of the above treatments makes a case for assisted reproduction, and the patient may be advised to go for an appropriate method of ART.

24.1 Introduction

Approximately, 10–15% of all couples have fertility problems, undergo fertility assessment, and seek treatment. Infertility is a common and complex disorder attributed to a number of etiological factors. The list of etiological factors includes genetic (chromosomal abnormalities, classical and microdeletions), epigenetic (DNA methylation), environmental (exposure to hazardous chemicals), lifestyle, and nutritional (malnutrition) aspects (Oliva et al. 2001; Rajender et al. 2011; Bansal et al. 2016). Due to the complexity of this disorder, its diagnosis and treatment are not straightforward. Identification of an etiological factor would facilitate directed therapy with significant chances of success. The remaining cases are often prescribed empirical therapies, which are usually prescribed based on the theoretical concepts. Assisted reproductive techniques (ARTs) are recommended after the failure of initial treatment (Cocuzza and Agarwal 2007) and need a number of considerations.

In the case of male factor infertility, the main goal of management is to diagnose the causes of infertility and to provide appropriate medications to achieve improvements in semen parameters. After exploring all etiological factors, the cause of seminal abnormalities in 25% remains unknown (Greenberg et al. 1978). There is no standardized drug available for treatment of infertility of idiopathic infertility. A variety of non-specific medical treatments has been recommended to treat these patients. Some of these treatments have been effective in improvement in semen parameters, but none of them ensures improvements in pregnancy rates. Moreover, the efficacies of medical treatments are doubtful due to lesser number of studies being conducted on such therapies, inappropriate study designs, lack of the placebo/controls, and problems in patient's follow-up. Normally, empiric treatments last for more than 3–6 months to cover one spermatogenic cycle. In this chapter, we have discussed the current medical treatments available for male infertility.

24.2 Hormonal Treatments

Hypogonadotropic hypogonadism (HGH) or secondary hypogonadism is a clinical syndrome of undeveloped gonads due to either inadequate or absent hypothalamic GnRH secretion or less or no pituitary gonadotropin secretion. Pulsatile secretion of gonadotropin-releasing hormone (GnRH) by hypothalamic neurons is crucial for initiating the release of pituitary gonadotropins, secretion of sex steroids, pubertal development, and gametogenesis. HGH may be congenital (Kallmann syndrome,

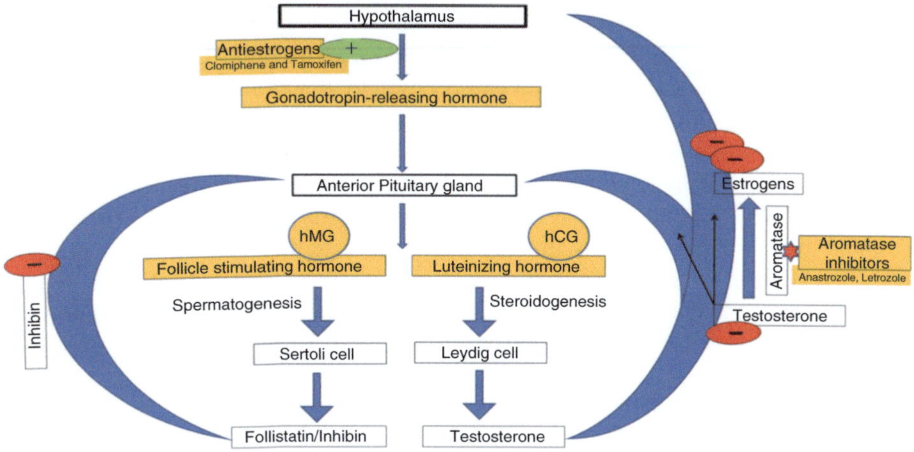

Fig. 24.1 Therapeutics and their targets for male infertility treatment

Prader-Willi syndrome), acquired (pituitary tumors, steroid abuse, panhypopituitarism, pituitary trauma, and testosterone replacement therapy), or functional (functional gonadotropin deficiency due to chronic systemic disease, malnutrition, acute illness, obesity, hyperprolactinemia). In Kallmann syndrome, this defect occurs at the level of hypothalamic GnRH secretion due to the malformation of the midline cranial structures (Cunningham and Lipshultz 1986). In HGH patients, testosterone therapy is given to the adult men to induce and maintain the secondary sexual characteristics and sexual function, but it does not restore fertility. When fertility is desired, gonadotropin therapy is given to induce spermatogenesis in HGH males (Ho and Tan 2013). Treatment protocols of gonadotropin therapy vary with the patient. In patients with acquired HGH, administration of exogenous GnRH or gonadotropins can restore normal spermatogenesis (Fig. 24.1).

24.2.1 Gonadotropin-Releasing Hormone (GnRH) Therapy

Gonadotropin-releasing hormone (GnRH) is secreted by hypothalamic neurons in a pulsatile manner and is transported to the anterior pituitary gland, which in turn secretes follicle-stimulating hormone (FSH) and luteinizing hormone (LH). FSH and LH control gonadal gametogenesis and steroidogenesis, respectively, in both sexes (Fink 1988). GnRH therapy has been used in the treatment of different reproductive endocrinopathies (Kiesel et al. 2002). Since GnRH has been discovered, many GnRH-I analogs have been made and studied broadly (Conn and Crowley 1994). Exogenous GnRH administration can increase the pituitary's production of FSH and LH and could potentially increase spermatogenesis. Badenoch et al. (1988) examined prolonged GnRH treatment in idiopathic OAT (oligoasthenoteratozoospermia) patients, but no effect was observed on either semen parameters or circulating gonadotropins.

Because GnRH is secreted in a pulsatile manner, a pulsatile GnRH therapy has also been tried using the portable mini-pumps. Pulsatile GnRH therapy has been effective in gonadotropin deficiency caused by hypothalamic or pituitary diseases (Mortimer et al. 1974; Crowley et al. 1985), but not in patients with loss of pituitary gonadotropin function (Wang et al. 1989). Moreover, GnRH therapy has been effective in restoring fertilization capacity in men undergoing the treatment of testicular tumors. In two different studies, men undergoing cisplatin and radiation therapy restored the fertility completely from germ cell damage when GnRH treatment was provided (Kreuser et al. 1990; Brennemann et al. 1994). However, the use of this therapy is restricted by the pituitary malfunction, formation of anti-GnRH antibodies (Lindner et al. 1981), the cumbersome wearing of the pulsatile pump, and high cost of the therapy.

24.2.2 Gonadotropins

The anterior pituitary gland produces and secretes two gonadotropins (FSH and LH), which stimulate spermatogenesis and steroidogenesis, respectively. hCG and hMG are also gonadotropins but are exogenous in nature. hCG is secreted by the chorionic cells of the placenta. It is analogous to LH and can stimulate the secretion of testosterone from the Leydig cells. hMG is extracted from the urine of postmenopausal women and has both FSH and LH activity. Generally, pituitary insufficiency is treated by hCG or hMG or urine FSH or recombinant human FSH (r-hFSH) alone or in combinations (Fig. 24.1). Treatment with gonadotropins has been very effective in the management of hypogonadotropic hypogonadism (HGH) (phenotypically hypogonadotropic oligozoospermia/azoospermia). Human chorionic gonadotropin (hCG), which contains LH-like activity, and human menopausal gonadotropin (hMG), which contains both FSH and LH activity, are used for replacement therapy in these patients. Normally, hCG, at the dose of 1500–3000 IU, is subcutaneously administered three times per week. However, in cases of congenital HGH, after 3 months of hCG therapy, FSH is administered intramuscularly at the dose of 37.5–75 IU three times/week. Semen parameters and testosterone levels are measured during the treatment. Normally, spermatozoa appear in ejaculate in 6–9 months, but can take much longer time. Once sperm concentration reaches to the satisfactory level, FSH administration can be stopped, and spermatogenesis may be maintained with hCG alone.

The significance of this therapy has been controversial in the treatment of normogonadotropic oligozoospermia (Siddiq and Sigman 2002). Moreover, the effects of this therapy on pregnancy rates/outcomes have been contradictory. Two randomized controlled trials have reported no improvement in pregnancy rates with either purified hMG (Matorras et al. 1997) or r-hFSH therapy (Kamischke et al. 1998), while one has shown positive outcomes after a post hoc analysis in a selected subpopulation (Matorras et al. 1997). The use of this therapy is limited by its expensiveness and the lack of studies showing its significance on pregnancy outcomes. Moreover, this therapy could not be prescribed to the men without demonstrable hormonal abnormalities.

In some patients, who do not respond to hCG/FSH combination therapy, GnRH therapy can be given. GnRH is administered intravenously or subcutaneously in a pulsatile fashion with a portable infusion pump. Pulsatile GnRH therapy depends on how well the anterior pituitary responds to exogenous GnRH. Pulsatile GnRH therapy is effective in gonadotropin deficiency caused by hypothalamic diseases (Mortimer et al. 1974), but not in the loss of pituitary gonadotropin function (Wang et al. 1989). GnRH therapy is also very effective in restoring fertility in men undergoing treatment for testicular tumors (Kreuser et al. 1990; Brennemann et al. 1994).

24.3 Inhibitors of Hormone Synthesis/Action

24.3.1 Antiestrogens

Antiestrogen therapy is a most common treatment for idiopathic infertility. Though estrogen is a female hormone, many studies have shown its role in male reproduction (Hess et al. 1997). Estrogen receptors are expressed on male germ cells, suggesting the importance of estrogens in spermatogenesis (Zondek 1934; Dorrington et al. 1978; Nitta et al. 1993; Carreau and Hess 2010). This hormone negatively regulates gonadotropin secretion (Finkelstein et al. 1991) and maintains the sexual behavior in adult males (Lauber et al. 1997). Antiestrogens work by blocking the estrogen and testosterone receptors in the hypothalamus, which increases the GnRH secretion, which in turn stimulates the secretion of FSH and LH from the anterior pituitary. The two commonly used antiestrogens are clomiphene and tamoxifen.

Clomiphene is a nonsteroidal drug, which has a structure similar to diethylstilbestrol (Fig. 24.1). Generally, clomiphene citrate is prescribed at a dose of 25 mg/day (doses range from 12.5 to 400 mg/day). The significance of clomiphene treatment on sperm count and pregnancy rates has been contradictory. Many randomized controlled studies on clomiphene citrate failed to show its efficacy over placebo (Foss et al. 1973; Paulson et al. 1977; Rönnberg 1980; Sokol et al. 1988). Only two studies have shown its positive effects on sperm count as well as on pregnancy rates (Wang et al. 1983; Check et al. 1988). Side effects of clomiphene treatment are mild and include headache, weight gain, nausea, change in libido, dizziness, allergic dermatitis, and gynecomastia. Moreover, regular monitoring of FSH, LH, and testosterone levels and frequent semen analysis are required in patients undergoing the clomiphene therapy because increased testosterone levels could negatively affect spermatogenesis (Gilbaugh and Lipshultz 1994).

Tamoxifen is also an antiestrogen and is commonly used for idiopathic male infertility treatment (Fig. 24.1). Tamoxifen citrate is prescribed at a dose of 10–30 mg orally per day. Recently, one study on infertile oligozoospermic men with different FSH levels revealed that tamoxifen citrate significantly increased the sperm count and concentration in men having lower FSH levels in comparison to those having higher FSH levels (Kadioglu 2009). Though uncontrolled studies have reported that tamoxifen citrate treatment increased sperm concentration/counts and pregnancy rates (Vermeulen and Comhaire 1978; Bartsch and Scheiber 1981; Buvat et al. 1983), yet many controlled studies using tamoxifen citrate (at the dose of 10–20 mg/day) did not find such an association (Willis et al. 1977; AinMelk et al. 1987; Krause et al. 1992). Side effects of tamoxifen treatment are milder than clomiphene citrate because of its weaker estrogenic properties.

Antiestrogens are comparatively inexpensive and safe oral drugs for the treatment of idiopathic male infertility. However, the efficacy of this treatment is doubtful. Therefore, prolonged courses of this therapy should not be recommended.

24.3.2 Aromatase Inhibitors

In the testis, the Leydig and Sertoli cells have high aromatase activity (Inkster et al. 1995). Aromatase is an enzyme that converts circulating testosterone into estrogen in fat cells. Therefore, obese men might have an excessive conversion of testosterone into estrogen. Theoretically, changes in the ratios of estrogen and testosterone systemically or within the testes could manipulate pituitary levels of LH and FSH and impair sperm production (Kulin and Reiter 1972; Veldhuis et al. 1985). Aromatase inhibitors suppress the conversion of testosterone to estrogen and increase spermatogenesis (Ciaccio et al. 1978).

Aromatase inhibitors are expensive pharmaceutical agents that fall into two categories: steroidal (testolactone) and nonsteroidal (letrozole, anastrozole, and exemestane). Anastrozole are the fourth generation of aromatase inhibitors. They are highly potent as well as specific for the aromatase enzyme (Fig. 24.1). These drugs are safe and well tolerated. These drugs can be prescribed to men with idiopathic oligozoospermia with abnormal testosterone/estrogen ratio. During the treatment, patients are followed at regular intervals for serum testosterone, estrogen levels, and seminal parameters. Some studies have shown very impressive results with this treatment (Pavlovich et al. 2001; Raman and Schlegel 2002). Treatment with the aromatase inhibitor (testolactone at the dose of 50–100 mg twice daily) in infertile men with a low serum testosterone-to-estradiol ratio significantly increased sperm count and motility as well as corrected the hormonal abnormality (Pavlovich et al. 2001; Raman and Schlegel 2002). Similar changes were also observed when patients were treated with the more selective aromatase inhibitor, anastrozole, at the dose of 1 mg/day (Raman and Schlegel 2002). However, more numbers of placebo-controlled, randomized trials are required to assess the efficacy of aromatase inhibitors in idiopathic male infertility.

24.3.3 Hyperprolactinemia

Hyperprolactinemia is a condition of elevated serum prolactins, which results in HGH and infertility. Prolactin is a 198-amino acid protein (23kDa), which is secreted by lactotroph cells of the anterior pituitary gland. Normally, prolactin is present in both men and women in a small amount in their blood. Its main function is to enhance breast development in women during pregnancy and to induce lactation after a baby is born. In men, prolactin regulates sperm production by controlling the secretion of GnRH. Normal fasting values of prolactin in men are less than 25 ng/mL. In hyperprolactinemia, elevated levels of prolactin inhibit the hypothalamic secretions of GnRH.

Hyperprolactinemia may occur due to pituitary tumors (micro- or macroadenomas), stress, hypothyroidism, medical illness, medications such as antidepressants and antihypertensives, and idiopathic factors. Pituitary micro- or macroadenomas are the most common causes of hyperprolactinemia. Generally, prolactin-secreting pituitary adenomas result in lowering of the gonadotropin and testosterone levels and elevation in prolactin levels. In macroadenomas, prolactin levels are high to greater than 250 ng/mL, while in microadenomas, the levels remain between 100 and 250 ng/mL.

In patients with hyperprolactinemia, pituitary MRI with gadolinium contrast is recommended to rule out a pituitary tumor. Prolactin levels are repeatedly checked many times in a day as prolactin levels vary throughout the day and with physical activity. In most of the patients with hyperprolactinemia or pituitary adenomas (especially microadenomas), medical therapy is the first line of treatment, but in macroadenomas where the condition is more serious, surgery may be recommended. In our body, prolactin levels are regulated by other hormones, called prolactin-inhibiting factors (PIFs), such as dopamine. Initially, in hyperprolactinemia, bromocriptine, a strong dopamine D_2 receptor agonist, is prescribed with doses ranging from 2.5 to 7.5 mg/day. Bromocriptine has been shown to significantly reduce the serum prolactin levels in oligozoospermic men with hyperprolactinemia and to increase the sperm count to a level sufficient for pregnancy initiation (Chuang and Howards 1998). In cases in which bromocriptine is not very effective and not well tolerated, a new long-lasting drug cabergoline is prescribed. Cabergoline shows fewer side effects and requires less frequent dosing than bromocriptine. Cabergoline is given at the dose of 1.0 mg/week. When prolactin levels get in normal range, the dose can be reduced to 0.5 mg/week (Verhelst et al. 1999).

24.4 Steroids and Antioxidants

24.4.1 Steroids for Anti-sperm Antibodies

Immunologic infertility is referred to as a condition in which anti-sperm antibodies (ASAs) are produced by the body as a response against sperm proteins. ASA may be present in serum and/or in seminal plasma or on the sperm surface. Normally, a man does not develop antibodies against his own spermatozoa because genital tract is a closed tube and is separated from the immune system. When blood cells and sperm come in contact, a male can produce antibodies against his own sperm. The presence of ASAs in the body fluids can block sperm-egg interactions via immobilizing and/or agglutinating the spermatozoa. They can also block the implantation and/or the development of embryo (Haas 1986; Koide et al. 2000). In all infertile couples with ASAs, IgG and IgA anti-sperm antibodies were found either on spermatozoa or in cervical mucus (Kremer et al. 1978).

ASAs have been found in a large number of infertile men and have been shown to compromise the male fertility (Rumke and Hellinga 1959). The most common causes of ASA include genital tract infections, surgical treatments such as testicular biopsy and vasectomy, testicular trauma, and testicular torsions (Broderick et al. 1989; Koide et al. 2000; Arap et al. 2007). In males, genital tract infections can weaken the blood-testis barrier (BTB), leading to the leakage of sperm and influx of immunologically competent cells. According to an estimate, 50–80% of men having undergone for vasectomy have circulating anti-sperm antibodies (Haas 1987). ASAs are one of the major causes of obstructive azoospermia associated with infertility after surgical treatments (Alexander and Anderson 1979; Linnet 1983; Mandelbaum et al. 1987). Moreover, ASAs are present in 80% of men having unilateral ductal obstruction (Hendry et al. 1986).

Generally, infertile men with ASAs are treated with oral corticoids to suppress the antibody production. However, no double-blinded, randomized trial has been done to confirm its efficacy till date. Prednisolone, a synthetic form of corticosteroid hormone, is the first line of the medical therapy. In a study, two men treated with 96 mg methylprednisolone per day for 7 days resulted in a slight decrease of the sperm-agglutination titer; however, no pregnancy was achieved (Kremer et al. 1978). In severe sperm autoimmunity, intracytoplasmic sperm injection (ICSI) may be a treatment of choice (Check et al. 2000). ICSI has shown no significant differences in clinical pregnancy rates (19% vs 12%) between ASA-positive and ASA-negative patient groups (Clarke et al. 1997). Recently, meta-analysis also revealed that semen ASAs are not related to the pregnancy rates after ICSI or IVF, indicating that both ART techniques can be used in infertile couples with semen ASAs (Zini et al. 2011).

24.4.2 Vitamins and Antioxidants

Elevated levels of ROS have been identified as an independent cause of male infertility (reviewed in Agarwal et al. 2006). In an estimate, increased levels of ROS in semen have been detected in 25–40% of infertile male patients (De Lamirande and Gagnon 1995; Padron et al. 1997). ROS can be beneficial or damaging depending upon the type and concentration of the ROS as well as length and location of the exposure to ROS (Agarwal and Saleh 2002). An excess amount of ROS can modify cell functions and increase cell death (Agarwal and Saleh 2002). Although ROS level in spermatozoa is controlled and maintained by the antioxidants present in seminal plasma, yet insufficient check on ROS could lead to oxidative stress, which in turn could be harmful to spermatozoa (Agarwal and Anandh Prabakaran 2005). Sperms are very susceptible to ROS because their plasma membrane has a large amount of polyunsaturated fatty acids (Alvarez and Storey 1995).

In most cases, damage induced by the ROS can be repaired. Seminal plasma has two different types of antioxidants to reduce the ROS level: enzymatic and nonenzymatic antioxidants. Enzymatic antioxidants include superoxide dismutase (SOD), catalase, and glutathione peroxidase (GPX), and nonenzymatic antioxidants are vitamin E, vitamin C, glutathione, pyruvate, and carnitine (Agarwal et al. 2004). In the treatment of infertility, antioxidants are often prescribed to idiopathic infertile men as a supplement to reduce the ROS level/oxidative stress. In a randomized, double-blinded controlled trial, asthenozoospermic patients were supplemented with oral vitamin E (300 mg/day). This treatment significantly decreased the malondialdehyde (MDA, a marker for lipid peroxidation) concentration and improved sperm motility (Suleiman et al. 1996). In another study, vitamin E and selenium supplementation significantly decreased the MDA concentration and improved sperm motility (Keskes-Ammar et al. 2003).

Vitamin C is a potent chain-breaking antioxidant and contributes up to 65% antioxidant capacity of the seminal plasma. Vitamin C concentration is ten times higher in seminal plasma than that in the blood plasma (Lewis et al. 1997). Fraga et al. (1991) reported that repletion of dietary vitamin C for 28 days (from 5 to 250 mg/day) doubled the vitamin C level in seminal plasma and reduced the 8-hydroxy-2′-deoxyguanosine (8-OHDG, a marker of oxidative stress) by 36%. This study indicated that dietary supplementation could be used to protect spermatozoa from endogenous oxidative damage. Some of the vitamins and their sources have been discussed in detail in Chap. 20.

24.5 Other Treatments

24.5.1 Genital Tract Infections

Genital tract infections account for about 15% of male infertility cases (Pellati et al. 2008). A number of microorganisms are involved in such infections. Some of them are *Streptococcus faecalis*, *Escherichia coli*, *Chlamydia trachomatis* (sexual transmission), *Ureaplasma urealyticum*, *Mycoplasma genitalium*, and *Mycoplasma hominis* (genital mycoplasma). According to one study, an overnight co-incubation of *M. hominis* with human spermatozoa showed small but statistically significant differences in sperm motility, morphology, and fertilization potential (Rose and Scott 1994). Moreover, *U. urealyticum* has been found to be associated with the generation of reactive oxygen species, even in the absence of leucocytospermia, and *M. genitalium* has been found to be attached to human spermatozoa (Taylor-Robinson 2002). Among viruses causing the infections in the genital tract are *herpesviruses (HSV)*, *human papilloma viruses (HPV)*, and *human immunodeficiency viruses (HIV)*. The contribution of these infections to infertility has been discussed in detail in Chap. 12.

Once an infection of genital tract is identified, antibiotic therapy is given. In culture-negative patients, anti-inflammatory therapy can be prescribed. According to the presence of the microorganism, the following antibiotics can be prescribed: for *C. trachomatis* infection, azithromycin 1 g single dose orally or doxycycline 100 mg orally twice daily for 7 days can be given. For *N. gonorrhoeae* infection, ceftriaxone (125 mg intramuscularly single dose) or fluoroquinolones (ciprofloxacin 500 mg, ofloxacin 400 mg, levofloxacin 250 mg/day) can be prescribed. For *Mycoplasma spp.*, macrolides (erythromycin/roxithromycin) are usually given. These drugs are usually prescribed for 2–3 weeks, depending on the severity of the infections (Haidl and Schill 1991).

24.5.2 Disorders of Ejaculation

Ejaculatory dysfunctions in males include premature ejaculation (PE), delayed ejaculation (DE), anejaculation (AE), and retrograde ejaculation (RE). Except for PE, all other ejaculatory dysfunctions interfere with the delivery of sperms to the female genital tract and are important etiological factors for male subfertility. While PE and DE are common causes of sexual dissatisfaction in men and their partners, these disorders are not associated with male infertility (Barazani et al. 2012). On the other hand, men with RE and AE are not able to deliver sperm to the female genital tract and are subfertile. RE is referred to as a condition in which ejaculates flow abnormally backward and toward the bladder. RE is a common ejaculatory dysfunction but contributes to only 0.3–2% of male infertility (Vernon et al. 1988; Yavetz et al. 1994). The diagnosis of RE is made by the post-ejaculate urine test. In patients with low-volume ejaculates (<1.0 mL semen), the presence of sperm (>10–15/hpf) in urine indicates the etiology of RE. On the other hand, in patients with AE, the absence of sperms in the urine indicates the failure of emission.

Initially, pharmacologic therapy is recommended to the patients with RE. This therapy is only successful in patients who do not have bladder neck abnormalities (which are caused by the surgery done earlier for the treatment of other problems of genital tract, such as prostate surgery) and the problem of anejaculation. In treatment, alpha-adrenergic agonists such as ephedrine sulfate (25–50 mg q.i.d), pseudoephedrine (60 mg q.i.d), and imipramine (25 mg b.i.d) are prescribed. Moreover, medical therapy for ejaculatory dysfunction has to be synchronized with female's ovulatory cycles. This therapy is more effective if given at least 7–10 days before the ejaculation is planned. If medical therapy fails to recover the normal ejaculation, ART techniques can be used to achieve the pregnancy. In such situations, spermatozoa can be retrieved from the post-ejaculatory urine (Shangold et al. 1990); however, urine may damage the sperm by its acidity, contamination, and change in osmolarity (Crich and Jequier 1978).

24.5.3 Miscellaneous Treatment Regimens

Other non-hormonal treatments have also been used for the treatment of idiopathic male infertility. One of these treatments included L-carnitine, which is present in epididymal secretions. Approximately, 50% of total carnitine in human seminal plasma is found as acetyl-carnitine, which plays a major role in energy metabolism and sperm membrane stabilization. L-Carnitine is given as a nutritional supplement and is available over the counter. Carnitine also possesses antioxidant capacity that protects spermatozoa from oxidative stress/damage (Agarwal and Said 2004). However, studies have shown no direct association between semen L-carnitine levels and fertility (Soufir et al. 1984). Uncontrolled studies have revealed improvements in semen parameters, but not in fertility rate (Costa et al. 1994; Vitali et al. 1995). Two randomized controlled trials using carnitine and acetyl-L-carnitine for idiopathic male infertility (Lenzi et al. 2003, 2004) reported statistically significant improvements in seminal parameters, but neither in carnitine levels in semen nor in pregnancy rates (Lenzi et al. 2003: 8%; Lenzi et al. 2004: 13%). There are little evidences on the effectiveness of carnitine treatment; therefore, more number of studies are required for validation of its efficacy.

Tamoxifen treatment has been prescribed either alone or in combination with kallikrein/testosterone. Tamoxifen has been effective in oligozoospermia while kallikrein in asthenozoospermia. In this context, the combination of these two should be useful in the treatment of oligoasthenozoospermia. Three studies using more than 84 oligoasthenozoospermic patients showed an increment in sperm count (Höbarth et al. 1990; Maier and Hienert 1990) and motility (Maier and Hienert 1990). However, these studies did not follow up the patients for pregnancy outcomes. Tamoxifen in combination with testosterone has been reported to be effective in men with idiopathic oligoasthenoteratozoospermia (OAT). Recently, one study using a combination of tamoxifen citrate and testosterone undecanoate has shown improvements in total sperm count, functional sperm count, and motility in men with OAT (Adamopoulos et al. 2003). Interestingly, they have also reported good pregnancy rates (Adamopoulos et al. 2003).

Conclusions

Selection of the therapy, whether specific or empirical, depends on the fertility status of infertile men. Once the etiology of the disorder is diagnosed, the treatment is provided accordingly. Generally, medicinal therapy is recommended for the treatment of idiopathic male infertility. However, ARTs can be used when medicinal therapy fails to restore fertility or initiate pregnancy. Empirical (nonspecific) therapy can be provided to patients when no specific etiology is identified. There are some side effects of these medicinal therapies; therefore, proper caution and regular checkups are required during the course of treatment. Nevertheless, prior to treatment, infertile couples should be informed about the inconsistency of therapy outcomes and low conception rates. If semen parameters do not improve significantly or a pregnancy is not achieved after at least two spermatogenic cycles, it is an indication to proceed with ART. Although few studies are available (even fewer studies have proper study designs) on the effectiveness of these therapies, more number of studies having better study designs are required.

In a large number of patients, the etiology remains unknown and treatment a challenge. Unfortunately, the number of such patients with male infertility is very high. Therefore, development of other therapeutic strategies is much needed. Inadequate treatment regimens for male infertility are in part due to poor understanding of the molecular cues to spermatogenesis and sperm fertility. Further research needs to focus on the identification of new molecular players critical to the process of spermatogenesis. Identification of new molecular targets would open new avenues for drug development. Empirical therapies lack support by appropriately designed studies; nevertheless, it is not a bad idea to try these therapies in the cases where specific and targeted therapies are either not possible or fail to yield results.

Acknowledgment The authors would like to thank the Indian Council of Medical Research (Ref. No. 5/10/FR/5/2012-RHN) and the Council of Scientific and Industrial Research (CSIR), Government of India, for financial help. Sandeep Kumar Bansal would like to acknowledge the fellowship from Department of Biotechnology (ref. no. Dbt-jrf/2012-13/125).

References

Adamopoulos DA, Pappa A, Billa E, Nicopoulou S, Koukkou E, Michopoulos J (2003) Effectiveness of combined tamoxifen citrate and testosterone undecanoate treatment in men with idiopathic oligozoospermia. Fertil Steril 80(4):914–920

Agarwal A, Anandh Prabakaran S (2005) Oxidative stress and antioxidants in male infertility: a difficult balance. Int J Reprod BioMed 3(1):1–8

Agarwal A, Said TM (2004) Carnitines and male infertility. Reprod Biomed Online 8(4):376–384

Agarwal A, Saleh RA (2002) Role of oxidants in male infertility: rationale, significance, and treatment. Urol Clin N Am 29(4):817–827

Agarwal A, Nallella KP, Allamaneni SS, Said TM (2004) Role of antioxidants in treatment of male infertility: an overview of the literature. Reprod Biomed Online 8(6):616–627

Agarwal A, Sharma RK, Nallella KP, Thomas AJ, Alvarez JG, Sikka SC (2006) Reactive oxygen species as an independent marker of male factor infertility. Fertil Steril 86(4):878–885

AinMelk Y, Belisle S, Carmel M, Jean-Pierre T (1987) Tamoxifen citrate therapy in male infertility. Fertil Steril 48(1):113–117

Alexander NJ, Anderson DJ (1979) Vasectomy: consequences of autoimmunity to sperm antigens. Fertil Steril 32(3):253–260

Alvarez JG, Storey BT (1995) Differential incorporation of fatty acids into and peroxidative loss of fatty acids from phospholipids of human spermatozoa. Mol Reprod Dev 42(3):334–346

Arap MA, Vicentini FC, Cocuzza M, Hallak J, Athayde K, Lucon AM, Arap S, Srougi M (2007) Late hormonal levels, semen parameters, and presence of antisperm antibodies in patients treated for testicular torsion. J Androl 28(4):528–532

Badenoch DF, Waxman J, Boorman L, Sidhu B, Moore HD, Holt WV, Blandy JP (1988) Administration of a gonadotropin releasing hormone analogue in oligozoospermic infertile males. Actaendocrinologica 117(2):265–267

Bansal SK, Jaiswal D, Gupta N, Singh K, Dada R, Sankhwar SN, Gupta G, Rajender S (2016) Gr/gr deletions on Y-chromosome correlate with male infertility: an original study, meta-analyses, and trial sequential analyses. Sci Rep 6:19798

Barazani Y, Stahl PJ, Nagler HM, Stember DS (2012) Management of ejaculatory disorders in infertile men. Asian J Androl 14(4):525

Bartsch G, Scheiber K (1981) Tamoxifen treatment in oligozoospermia. Eur Urol 7(5):283–287

Brennemann W, Brensing KA, Leipner N, Boldt I, Klingmüller D (1994) Attempted protection of spermatogenesis from irradiation in patients with seminoma by D-Tryptophan-6 luteinizing hormone releasing hormone. Clin Investig 72(11):838–842

Broderick GA, Tom R, McClure RD (1989) Immunological status of patients before and after vasovasostomy as determined by the immunobead antisperm antibody test. J Urol 142(3):752–755

Buvat J, Ardaens K, Lemaire A, Gauthier A, Gasnault JP, Buvat-Herbaut M (1983) Increased sperm count in 25 cases of idiopathic normogonadotropic oligospermia following treatment with tamoxifen. Fertil Steril 39(5):700–703

Carreau S, Hess RA (2010) Oestrogens and spermatogenesis. Philos Trans R Soc Lond B: Biol Sci 365(1546):1517–1535
Check JH, Chase JS, Nowroozi K, Wu CH, Adelson HG (1988) Empiricaltherapy of the male with clomiphene in couples with unexplained infertility. Int J Fertil 34(2):120–122
Check ML, Check JH, Katsoff D, Summers-Chase D (2000) ICSI as an effective therapy for male factor with antisperm antibodies. Arch Androl 45(3):125–130
Chuang AT, Howards SS (1998) Male infertility: evaluation and nonsurgical therapy. Urol Clin N Am 25(4):703–713
Ciaccio LA, Joseph AA, Kincl FA (1978) Direct inhibition of testicular function in rats by estriol and progesterone. J Steroid Biochem 9(12):1257–1259
Clarke GN, Bourne H, Baker HG (1997) Intracytoplasmic sperm injection for treating infertility associated with sperm autoimmunity. Fertil Steril 68(1):112–117
Cocuzza M, Agarwal A (2007) Nonsurgical treatment of male infertility: specific and empiric therapy. Biologics 1(3):259–269
Conn PM, Crowley WF Jr (1994) Gonadotropin-releasing hormone and its analogs. Annu Rev Med 45(1):391–405
Costa M, Canale D, Filicori M, D'lddio S, Lenzi A (1994) L-carnitine in idiopathic asthenozoospermia: a multicenter study. Andrologia 26(3):155–159
Crich JP, Jequier AM (1978) Infertility in men with retrograde ejaculation: the action of urine on sperm motility, and a simple method for achieving antegrade ejaculation. Fertil Steril 30(5):572–576
Crowley WF Jr, Filicori M, Spratt DI, Santoro NF (1985) The physiology of gonadotropin-releasing hormone (GnRH) secretion in men and women. Recent Prog Horm Res 41:473
Cunningham GR, Lipshultz LI (1986) Diseases of the testes and male sex organs. In: Basic clinical endocrinology. Wiley, New York, pp 263–278
De Lamirande E, Gagnon C (1995) Impact of reactive oxygen species on spermatozoa: a balancing act between beneficial and detrimental effects. Hum Reprod 10(suppl 1):15–21
Dorrington JH, Fritz IB, Armstrong DT (1978) Control of testicular estrogen synthesis. Biol Reprod 18(1):55–64
Fink G (1988) Gonadotropin secretion and its control. Physiol Reprod 1:1349–1377
Finkelstein JS, Whitcomb RW, O'dea LSL, Longcope C, Schoenfeld DA, Crowley WF Jr (1991) Sex steroid control of gonadotropin secretion in the human male. I. Effects of testosterone administration in normal and gonadotropin-releasing hormone-deficient men. J Clin Endocrinol Metab 73(3):609–620
Foss GL, Tindall VR, Birkett JP (1973) The treatment of subfertile men with clomiphene citrate. J Reprod Fertil 32(1):167–170
Fraga CG, Motchnik PA, Shigenaga MK, Helbock HJ, Jacob RA, Ames BN (1991) Ascorbic acid protects against endogenous oxidative DNA damage in human sperm. Proc Natl Acad Sci 88(24):11003–11006
Gilbaugh JH 3rd, Lipshultz LI (1994) Nonsurgical treatment of male infertility. An update. Urol Clin North Am 21(3):531–548
Greenberg SH, Lipshultz LI, Wein AJ (1978) Experience with 425 subfertile male patients. J Urol 119(4):507–510
Haas GG Jr (1986) The inhibitory effect of sperm-associated immunoglobulins on cervical mucus penetration. Fertil Steril 46(2):334
Haas GG Jr (1987) Antibody-mediated causes of male infertility. Urol Clin North Am 14(3):539–550
Haidl G, Schill WB (1991) Guidelines for drug treatment of male infertility. Drugs 41(1):60–68
Hendry WF, Treehuba K, Hughes L, Stedronska J, Parslow JM, Wass JA, Besser GM (1986) Cyclic prednisolone therapy for male infertility associated with autoantibodies to spermatozoa. Fertil Steril 45(2):249–254
Hess RA, Bunick D, Lee KH, Bahr J, Taylor JA, Korach KS, Lubahn DB (1997) A role for oestrogens in the male reproductive system. Nature 390(6659):509–512
Ho CCK, Tan HM (2013) Treatment of the hypogonadal infertile male—a review. Sex Med Rev 1(1):42–49

Höbarth K, Lunglmayr G, Kratzik C (1990) Effect of tamoxifen and kallikrein on sperm parameters including hypoosmotic swelling test in subfertile males. A retrospective analysis. Andrologia 22(6):513–517
Inkster S, Yue W, Brodie A (1995) Human testicular aromatase: immunocytochemical and biochemical studies. J Clin Endocrinol Metab 80(6):1941–1947
Kadioglu TC (2009) Oral tamoxifen citrate treatment is more effective in normogonadotropic patients who have follicle-stimulating hormone levels within the lower half of normal. Int Urol Nephrol 41(4):773–776
Kamischke A, Behre HM, Bergmann M, Simoni M, Schäfer T, Nieschlag E (1998) Recombinant human follicle stimulating hormone for treatment of male idiopathic infertility: a randomized, double-blind, placebo-controlled, clinical trial. Hum Reprod 13(3):596–603
Keskes-Ammar L, Feki-Chakroun N, Rebai T, Sahnoun Z, Ghozzi H, Hammami S, Zghal K, Fki H, Damak J, Bahloul A (2003) Sperm oxidative stress and the effect of an oral vitamin E and selenium supplement on semen quality in infertile men. Arch Androl 49(2):83–94
Kiesel LA, Rody A, Greb RR, Szilagyi A (2002) Clinical use of GnRH analogues. Clin Endocrinol 56(6):677–687
Koide SS, Wang L, Kamada M (2000) Antisperm antibodies associated with infertility: properties and encoding genes of target antigens. Proc Soc Exp Biol Med 224(3):123–132
Krause W, Holland-Moritz H, Schramm P (1992) Treatment of idiopathic oligozoospermia with tamoxifen—a randomized controlled study. Int J Androl 15(1):14–18
Kremer J, Jager S, Kuiken J (1978) Treatment of infertility caused by antisperm antibodies. Int J Fertil 23(4):270–276
Kreuser ED, Hetzel WD, Hautmann R, Pfeiffer EF (1990) Reproductive toxicity with and without LHRHA administration during adjuvant chemotherapy in patients with germ cell tumors. Horm Metab Res 22(09):494–498
Kulin HE, Reiter EO (1972) Gonadotropin suppression by low dose estrogen in men: evidence for differential effects upon FSH and LH. J Clin Endocrinol Metab 35(6):836–839
Lauber ME, Sarasin A, Lichtensteiger W (1997) Sex differences and androgen-dependent regulation of aromatase (CYP19) mRNA expression in the developing and adult rat brain. J Steroid Biochem Mol Biol 61(3):359–364
Lenzi A, Lombardo F, Sgrò P, Salacone P, Caponecchia L, Dondero F, Gandini L (2003) Use of carnitine therapy in selected cases of male factor infertility: a double-blind crossover trial. Fertil Steril 79(2):292–300
Lenzi A, Sgro P, Salacone P, Paoli D, Gilio B, Lombardo F, Santulli M, Agarwal A, Gandini L (2004) A placebo-controlled double-blind randomized trial of the use of combined l-carnitine and l-acetyl-carnitine treatment in men with asthenozoospermia. Fertil Steril 81(6):1578–1584
Lewis SE, Sterling ESL, Young IS, Thompson W (1997) Comparison of individual antioxidants of sperm and seminal plasma in fertile and infertile men. Fertil Steril 67(1):142–147
Lindner J, McNeil LW, Marney S, Conway M, Rivier J, Vale W, Rabin D (1981) Characterization of human anti-luteinizing hormone-releasing hormone (LRH) antibodies in the serum of a patient with isolated gonadotropin deficiency treated with synthetic LRH. J Clin Endocrinol Metab 52(2):267–270
Linnet L (1983) Clinical immunology of vasectomy and vasovasostomy. Urology 22(2):101–114
Maier U, Hienert G (1990) Tamoxifen and kallikrein in therapy of oligoasthenozoospermia: results of a randomized study. Eur Urol 17(3):223–225
Mandelbaum SL, Diamond MP, DeCherney AH (1987) The impact of antisperm antibodies on human infertility. J Urol 138(1):1–8
Matorras R, Perez C, Corcostegui B, Pijoan JI, Ramon O, Delgado P, Rodriguez-Escudero FJ (1997) Treatment of the male with follicle-stimulating hormone in intrauterine insemination with husband's spermatozoa: a randomized study. Hum Reprod 12(1):24–28
Mortimer CH, McNeilly AS, Fisher RA, Murray MAF, Besser GM (1974) Gonadotrophin-releasing hormone therapy in hypogonadal males with hypothalamic or pituitary dysfunction. Br Med J 4(5945):617–621

Nitta H, Bunick D, Hess RA, Janulis L, Newton SC, Millette CF, Osawa Y, Shizuta Y, Toda K, Bahr JM (1993) Germ cells of the mouse testis express P450 aromatase. Endocrinology 132(3):1396–1401
Oliva A, Spira A, Multigner L (2001) Contribution of environmental factors to the risk of male infertility. Hum Reprod 16(8):1768–1776
Padron OF, Brackett NL, Sharma RK, Lynne CM, Thomas AJ, Agarwal A (1997) Seminal reactive oxygen species and sperm motility and morphology in men with spinal cord injury. Fertil Steril 67(6):1115–1120
Paulson DF, Hammond CB, de Vere White R, Wiebe RH (1977) Clomiphene citrate: pharmacologic treatment of hypofertile male. Urology 9(4):419–421
Pavlovich CP, King P, Goldstein M, Schlegel PN (2001) Evidence of a treatable endocrinopathy in infertile men. J Urol 165(3):837–841
Pellati D, Mylonakis I, Bertoloni G, Fiore C, Andrisani A, Ambrosini G, Armanini D (2008) Genital tract infections and infertility. Eur J Obstet Gynecol Reprod Biol 140(1):3–11
Rajender S, Avery K, Agarwal A (2011) Epigenetics, spermatogenesis and male infertility. Mutat Res 727(3):62–71
Raman JD, Schlegel PN (2002) Aromatase inhibitors for male infertility. J Urol 167(2):624–629
Rönnberg L (1980) The effect of clomiphene treatment on different sperm parameters in men with idiopathic oligozoospermia. Andrologia 12(3):261–265
Rose BI, Scott B (1994) Sperm motility, morphology, hyperactivation, and ionophore-induced acrosome reactions after overnight incubation with mycoplasmas. Fertil Steril 61(2):341–348
Rumke PH, Hellinga G (1959) Autoantibodies against spermatozoa in sterile men. Am J Clin Pathol 32(4):357–363
Shangold GA, Cantor B, Schreiber JR (1990) Treatment of infertility due to retrograde ejaculation: a simple, cost-effective method. Fertil Steril 54(1):175–177
Siddiq FM, Sigman M (2002) A new look at the medical management of infertility. Urol Clin N Am 29(4):949–963
Sokol RZ, Steiner BS, Bustillo M, Petersen G, Swerdloff RS (1988) A controlled comparison of the efficacy of clomiphene citrate in male infertility. Fertil Steril 49(5):865–870
Soufir JC, Ducot B, Marson J, Jouannet P, Feneux D, Soumah A, Spira A (1984) Levels of seminal free L (-) carnitine in fertile and infertile men. Int J Androl 7(3):188–197
Suleiman SA, Ali ME, Zaki ZM, el-Malik EM, Nasr MA (1996) Lipid peroxidation and human sperm motility: protective role of vitamin E. J Androl 17(5):530–537
Taylor-Robinson D (2002) Mycoplasma genitalium-an up-date. Int J STD AIDS 13(3):145–151
Veldhuis JD, Sowers JR, Rogol AD, Klein FA, Miller N, Dufau ML (1985) Pathophysiology of male hypogonadism associated with endogenous hyperestrogenism: evidence for dual defects in the gonadal axis. N Engl J Med 312(21):1371–1375
Verhelst J, Abs R, Maiter D, van den Bruel A, Vandeweghe M, Velkeniers B, Mockel J, Lamberigts G, Petrossians P, Coremans P, Mahler C (1999) Cabergoline in the treatment of hyperprolactinemia: a study in 455 patients. J Clin Endocrinol Metab 84(7):2518–2522
Vermeulen A, Comhaire F (1978) Hormonal effects of an antiestrogen, tamoxifen, in normal and oligospermic men. Fertil Steril 29(3):320–327
Vernon M, Wilson E, Muse K, Estes S, Curry T (1988) Successful pregnancies from men with retrograde ejaculation with the use of washed sperm and gamete intrafallopian tube transfer (GIFT). Fertil Steril 50(5):822–824
Vitali G, Parente R, Melotti C (1995) Carnitine supplementation in human idiopathic asthenospermia: clinical results. Drugs Exp Clin Res 21(4):157–159
Wang C, Chan CW, Wong KK, Yeung KK (1983) Comparison of the effectiveness of placebo, clomiphene citrate, mesterolone, pentoxifylline, and testosterone rebound therapy for the treatment of idiopathic oligospermia. Fertil Steril 40(3):358–365
Wang C, Tso SC, Todd D (1989) Hypogonadotropic hypogonadism in severe β-thalassemia: effect of chelation and pulsatile gonadotropin-releasing hormone therapy. J Clin Endocrinol Metab 68(3):511–516

Willis KJ, London DR, Bevis MA, Butt WR, Lynch SS, Holder G (1977) Hormonal effects of tamoxifen in oligospermic men. J Endocrinol 73(1):171–178

Yavetz H, Yogev L, Hauser R, Lessing JB, Paz G, Homonnai ZT (1994) Retrograde ejaculation. Hum Reprod 9(3):381–386

Zini A, Fahmy N, Belzile E, Ciampi A, Al-Hathal N, Kotb A (2011) Antisperm antibodies are not associated with pregnancy rates after IVF and ICSI: systematic review and meta-analysis. Hum Reprod 26(6):1288–1295

Zondek B (1934) Mass excretion of oestrogenic hormone in the urine of the stallion. Nature 133(3354):209–210

Oncofertility: Fertility Preservation in Cancer

25

Sameer Trivedi

Abstract

The number of cancer patients in young age has increased in the recent years, which is coupled with late marriages and family planning. This makes an important place for a new and emerging field of cryo-preservation of gametes for cancer patients. A significant proportion of oncological treatments utilize chemotherapy or radiation exposure, both of which are detrimental to spermatogenesis. In post-cancer treatment period, fertility may resume in some but not all patients. Therefore, these patients need to be counselled about the methods of fertility preservation before commencing anti-cancer therapy. The present chapter brings a comprehensive overview of the fertility preservation options for cancer patients and the techniques used in this process.

Keywords

Oncofertility • Cancer and infertility • Cancer and fertility preservation • Sperm cryopreservation

Key Points

- The concept of developing an interdisciplinary specialty addressing fertility concerns of cancer patients was put forward in 2006 by Dr. Teresa K. Woodruff with the formation of Oncofertility Consortium.
- With 10% of cancer patients being younger than age 40, a substantial number of cancer patients are likely to have fertility preservation issues as a key component of their treatment plan.
- As the number of cancer survivors increases, the ability to have children assumes a vital role for patients and their families.

S. Trivedi
Department of Urology and Division of Andrology, Institute of Medical Sciences, Banaras Hindu University, Varanasi 221005, Uttar Pradesh, India
e-mail: drsameertrivedi@gmail.com

R. Singh, K. Singh (eds.), *Male Infertility: Understanding, Causes and Treatment*, DOI 10.1007/978-981-10-4017-7_25

- Majority of the cancer drugs act by interfering with the process of cell division (mitosis) and inducing apoptosis and are detrimental to spermatogenesis.
- Spermatogonia are particularly sensitive to the effects of irradiation and even a dose as low as 4–6 Gy can cause permanent damage to the germ cells.
- Lazzaro Spallanzani in 1776 first reported that the motility of human spermatozoa could be preserved after freezing and thawing.

25.1 Introduction

The demographic profile of oncology patients has undergone a sea change over the last few decades, and an increasing number of men are getting diagnosed with cancers of diverse organs in childhood or at an early age. Additionally, the improvements in diagnostic and treatment modalities and the resultant increase in survival have resulted in the emergence of a new pool of young cancer survivors for whom quality of life issues in general and fertility preservation in particular have been paramount. Parenthood is the dream of every couple, and cancer patients have traditionally been deprived of this gift as the conventional modalities of cancer treatment like chemotherapy and radiation therapy tend to have a deleterious effect on the reproductive functions of both men and women. Oncofertility is an evolving discipline at the intersection of oncology and reproductive medicine. The aim of this chapter is to review the available fertility preservation options for young males with cancers and sneak a glimpse into the advancements taking place in the field of fertility preservation.

Reproductive process and the underlying physiology have intrigued mankind for thousands of years. Hippocrates and Galen described human conception as occurring from two "seeds" and propagated different philosophies of male and female genitalia and of conception. The research and studies over the subsequent centuries and a better understanding of the reproductive physiology demystified the process of conception and paved the way for introduction of amazing modalities of treatment like in vitro fertilization (IVF), intracytoplasmic sperm injection (ICSI) and cryopreservation of gametes and embryos. Although the technique of sperm banking or cryopreservation of sperm to enable fertility in future has been around for many decades, the concept of oncofertility is relatively new and seeks to provide a holistic approach towards fertility preservation of cancer survivors.

25.2 Oncofertility: A New Approach to Fertility Preservation

The domain of oncofertility acts as a link between reproductive medicine and oncology with the objective of enabling fertility in future for cancer survivors. The concept of developing an interdisciplinary specialty addressing fertility concerns of cancer patients was put forward in 2006 by Dr. Teresa K. Woodruff with the formation of Oncofertility Consortium (Woodruff 2007). It is an interdisciplinary network of doctors, researchers and scientists which deals with fertility issues of young cancer patients. Conventional modalities of cancer treatment like chemotherapy, radiotherapy and surgery can have a deleterious effect on fertility, and oncofertility aims to expand fertility preservation options in these patients. The scope of oncofertility includes (Woodruff 2007) efforts to develop novel fertility preservation options for cancer patients; (Nandakumar 2001) collaboration among diverse clinical specialties to assimilate reproductive science, social and family counselling and fertility management after cancer treatment; and (Dikshit et al. 2012) expansion of awareness about oncofertility.

The incidence of cancer in India is rising significantly with an increasing number of cases being diagnosed at an earlier age. Over one million new cases of cancer are being diagnosed every year in India, and the estimated number of people living with cancer in India is approximately eight million (Nandakumar 1990; Dikshit et al. 2012). With 10% of cancer patients being younger than age 40, a substantial number of cancer patients are likely to have fertility preservation issues as a key component of their treatment plan (National Cancer Registry Programme 2001; Saranath and Khanna 2014). As the number of cancer survivors increases, the ability to have children assumes a vital role for patients and their families. Knowledge about fertility preservation and potential for ability to have a family in the future offers these patients a ray of hope even before the treatment for cancer has commenced. The domain of oncofertility encompasses a wide range of issues besides techniques of cryopreservation of tissues. It seeks to develop the contemporary understanding and research for a number of issues to improve clinical practice and enable better training of health-care providers in reproductive medicine. These issues include:

- Advancing the basic knowledge of gametogenesis and using this to improve fertility preservation techniques
- A better understanding of the mechanisms of gonadotoxic effects of chemotherapeutic drugs
- Improving the techniques of cryopreservation, storage, thawing and growing of gonadal tissue
- Enabling better communication between cancer patients and health-care providers
- Addressing ethical and legal issues regarding the use of fertility preservation techniques in cancer patients
- Providing counselling to cancer survivors and their families about issues like family planning, contraception, donor insemination, surrogacy and adoption

25.3 Impact of Cancer Treatments on Fertility

Chemotherapy—Cytotoxic chemotherapeutic agents form the mainstay of treatment for majority of cancers in the body. Majority of these drugs act by interfering with the process of cell division (mitosis) and inducing apoptosis (Wallace et al. 2005). These drugs act mainly on the rapidly dividing cells, which include normal cells of the body, and this action is responsible for the toxic side-effects of chemotherapeutic agents. The process of spermatogenesis involves a population of rapidly dividing cells, and it makes them particularly susceptible to the cytotoxic actions of chemotherapy drugs which can cause interstitial fibrosis and hyalinization in the testicular tissue. Among all the chemotherapy drugs, alkylating agents like cyclophosphamide and cis-platinum have the highest gonadotoxicity with the maximum risk for prolonged azoospermia. The gonadotoxic effects of chemotherapy depend on the age of the patient, type of drug used and the dosage administered.

Radiation therapy—The role of radiation therapy to treat cancers is based on use of high-energy rays. Radiation can cause primary testicular damage if administered directly to the testis or by scatter radiation if adjacent tissues are irradiated. The harmful effects of radiotherapy on fertility are usually due to the damage to the germinal epithelium. The gonadotoxic effects of radiation may be transient or permanent and depend on the dose, amount of scatter, site of radiation in relation to the testis, fractionation and patient age. Spermatogonia are particularly sensitive to the effects of irradiation, and even a dose as low as 4–6 Gy can cause permanent damage to the germ cells (Wallace et al. 2005). Secondary testicular failure can be a consequence of radiation to the brain where radiation-induced injury to the pituitary gland can hamper the production of luteinizing hormone and follicle-stimulating hormone leading to impaired spermatogenesis.

Surgery—Surgical procedures involving genitourinary organs like removal of both testes, prostate, urinary bladder and organs involved in sperm transport can lead to impaired fertility. Certain surgical procedures in the pelvis or retroperitoneum like retroperitoneal lymph node dissection can damage the pelvic nerves and lead to an ejaculation or retrograde ejaculation.

25.4 Role of Oncofertility

Over the last few decades, improvements in technology and introduction of novel techniques have opened up new avenues for fertility preservation in men with cancer. With the significant progress made in the field of oncofertility, many of these patients can hope to realize their dream of having progeny despite undergoing cancer therapy.

Oncofertility can play a role at different stages during the cancer management of a patient.

1. Diagnosis stage: The patients and their families are going through a tempestuous period at the time of diagnosis and are overwhelmed by the enormity of the disease and the treatment options. The reassurance that they have options to preserve their fertility and counselling regarding fertility preservation options can provide solace to the patients and help them in making decisions with far-reaching implications. The patients need to be informed about the likely impact of cancer treatments on their future fertility potential and assisted in realizing their fertility aspirations.
2. Treatment stage: The issues concerning fertility preservation are difficult to address once the treatment for cancer has started, and the role of oncofertility is

limited in this scenario. Nevertheless, counselling about available options and providing psychological support can alleviate the concerns of patients and their families about fertility issues to a certain extent.

3. Recuperation phase: Post-treatment fertility outcomes have been extensively studied. Besides the type of cancer treatment, pretreatment fertility status plays an important role in predicting future sperm recovery. Oncofertility team can play an important role in counselling, prognosticating and guiding patients realistically about their fertility prospects.

25.5 Fertility Preservation Options for Men with Cancer

The established fertility preservation options for men scheduled to undergo potentially gonadotoxic cancer therapies include sperm cryopreservation, testicular sperm extraction and the use of gonadal shielding to protect the gonads during radiation therapy. In addition, in prepubertal young males, there are new technologies to biopsy or remove portions of the testes, which can be frozen and then transplanted back into the testes, and mature human sperm can be created (Fig. 25.1). Although both of these treatment options are still experimental, a number of important studies are underway to make these options available for routine clinical use.

- Experimental technique—still under trial.

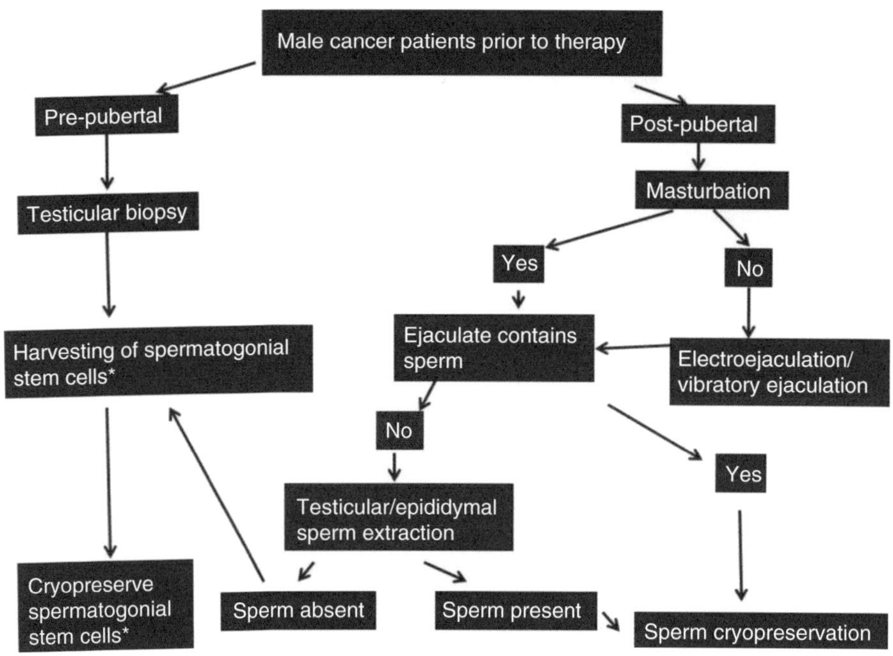

Fig. 25.1 Algorithm for sperm cryopreservation in male cancer patients

25.6 Genesis of Cryopreservation

Cryopreservation is a method of preserving cells, tissues, organs or any other biological material by cooling to very low temperatures (Pegg 2007) to prevent damage caused by unchecked biochemical processes. Cryopreservation techniques aim to reach freezing temperatures without formation of ice crystals during freezing. Lazzaro Spallanzani in 1776 first reported that the motility of human spermatozoa could be preserved after freezing and thawing (Spallanzani and Bonnet 1780). Sperm preservation by freezing had been attempted for many centuries by researchers with disappointing results. Freezing leads to rapid ice crystal formation from water both inside and outside of the cells causing damage to critical cell metabolic processes and secondary injury due to dehydration and an increase in concentration of solutes as progressively more ice is formed. The understanding of the mechanism of freezing injury to cells led to introduction of techniques utilizing controlled or slow cooling to obtain maximum survival on thawing of the living cells. It was realized that a slow- and controlled-rate cooling process which enables tissues to equilibrate to optimal physical parameters in a cryoprotectant prior to controlled cooling is vital to prevent freezing injury. Cryoprotectants are substances that prevent cells from freezing injury during cryopreservation. Discovery of cryoprotection was serendipitous in an accidental laboratory event in the United Kingdom when Christopher Polge unintentionally added glycerol to his experimental material while looking for a suitable cryoprotectant (Polge et al. 1949). This path-breaking discovery of cryoprotectants opened up the avenues for cryopreservation of sperm and embryos, which led to the advent of revolutionary in vitro fertilization techniques like intracytoplasmic sperm injection. The successful introduction of cryopreservation techniques for freezing of sperm laid down the foundation for the concept of oncofertility enabling cancer patients to cryopreserve sperm for subsequent in vitro fertilization.

Although most reproductive health-care providers would agree that sperm cryopreservation options should routinely be offered to all patients at the risk for impaired fertility during forthcoming cancer therapy, this is not yet reflected in the current practice patterns (Kliesch et al. 1997; Achille et al. 2006). A study showed that only 27% of men diagnosed with cancer chose semen cryopreservation, and paucity of awareness was the most common reason for failure of sperm banking in this study. In a survey of American physicians, only 10% affirmed suggesting routine sperm banking (Schover et al. 2002). The main cause for underutilization of semen cryopreservation is the lack of physicians' awareness regarding the necessity for fertility preservation and the efficacy of this modality (Joint Council for Clinical Oncology 1998). Additionally, oncologists may be unaware of recent advancements in assisted reproductive technology like intracytoplasmic sperm injection (ICSI) and get influenced by the suboptimal semen parameters and consider cryopreservation an exercise in vain (Lee et al. 2006). This aspect was highlighted in a survey in which 74% of oncologists were not cognisant of developments in assisted reproductive techniques (ART) (Bonetti et al. 2009). Failure to offer cryopreservation to cancer patients deprives them of the only possible reproductive option available. All men with fertility potential encountering gonadotoxic therapy should consider sperm cryopreservation before the therapy damages their spermatogenic process (Bonetti et al. 2009). The responsibility of education and awareness of these patients about sperm cryopreservation rest with physicians and oncologists, and the use of educational leaflets and multimedia tools can be an effective method to achieve this objective.

25.7 Process of Sperm Cryopreservation

The whole process of sperm cryopreservation consists of a number of steps and is achieved in the following six stages.

25.7.1 Sperm Collection

Sperm collection is performed in the usual way in a sterile, non-spermatotoxic and wide-mouthed container, preferably at the laboratory or the storage facility, with masturbation being the preferred method. An abstinence period of 3–5 days is optimal, and the use of lubricants should be avoided. In patients with obstructive or nonobstructive azoospermia, several techniques of direct sperm retrieval from testis or epididymis can be employed. These include testicular sperm aspiration (TESA), testicular sperm extraction (TESE) as well as by microscopic epididymal sperm aspiration (MESA), percutaneous epididymal sperm aspiration (PESA), micro-TESE and testicular fine needle aspiration mapping.

25.7.2 Sperm Preparation

Sperm preparation prior to cryopreservation involves removal of seminal plasma and enhancement of collected sample by density gradient centrifugation or swim-up techniques and washing. Cryopreserved sperm isolated after swim-up technique has been demonstrated to have better linear and forward progressive velocity, better capacitation abilities, higher fraction of intact acrosomes and superior results in sperm penetration assays as compared to untreated cryopreserved sperm (Russell and Rogers 1987; Esteves et al. 2000). Another sperm preparation technique which has been used to select high-quality sperm for cryopreservation is magnetic-activated cell sorting (MACS), which utilizes annexin microbeads to immunolabel and remove apoptotic spermatozoa (Said et al. 2008). Annexin is a phospholipid-binding protein which binds avidly to phosphatidylserine which is an indicator of apoptosis when present on the outer aspect of the cell membrane as during diminished cell membrane integrity. Various studies have demonstrated that non-apoptotic sperm selected from MACS shows higher motility, higher cryopreservation survival ratios, higher levels of intact mitochondria and an overall significantly better fertilization potential (Grunewald et al. 2001; Said et al. 2005; Grunewald et al. 2006).

25.7.3 Medium Preparation

Preparation of sperm cryoprotective buffering media is a vital component of sperm cryopreservation and has a direct bearing on the successful freeze-thaw of spermatozoa. The most commonly used cryoprotectant for sperm cryopreservation is glycerol, which promotes cell dehydration, decreases the harmful effect of ice crystal formation and limits the toxic effects of solute build-up (Mack and Zaneveld 1987; Mortimer 2004). Although all cryomedia require the cryoprotectant glycerol, glycerol itself can be damaging to human spermatozoa. It is necessary to keep the final concentrations of glycerol in the sperm mixture below 7.5% and to minimize the contact of sperm with glycerol by starting the cooling/freezing process immediately and by immediate washing after thawing. Various constituents called extenders have been added to the sperm suspension besides glycerol in order to improve the survival of cryopreserved sperm. A number of studies have demonstrated improved recovery of sperm on thawing when these substances are added to glycerol cryoprotectant suspension. These compounds include egg yolk, milk powder and serum proteins, among others. Their mechanism of action is not clear, but they are believed to interact with membrane proteins and phospholipids, protect from fluctuations in pH and prevent cold shock during freezing (Bergeron and Manjunath 2006). However, their optimal concentration has not been standardized, and the efficacy remains questionable, precluding their routine utilization in sperm cryopreservation media. The addition of cryoprotectants should be in a controlled manner to reduce osmotic stress, but with care to avoid prolonged contact with glycerol (Watson 1979). The precise addition time is not clearly specified, but most laboratories currently prefer to keep it within 10 min (Royere et al. 1996), although few modified techniques utilize a quicker addition (Gao et al. 1995).

25.7.4 Packaging

Packaging is an important component of sperm cryopreservation, and optimal storage containers should have certain fundamental features in order to ensure a secure and durable stowage. The ideal sperm packaging systems should be simple to manage in terms of storage, handling, packing and cataloguing; should be ergonomic and prevent wastage; should enable consistent cooling and augment heat exchange by offering a greater surface area to volume ratio; and should establish a leak-proof seal capable of maintaining integrity in freezing temperatures. Use of conventional plastic vials for cryopreservation while commonly practised is far from ideal. The heat exchange in these vials is uneven with higher cooling at the centre than at the periphery. Additionally, the sealing mechanism in these plastic vials is prone to leaks when stored in liquid nitrogen dewars, which are used commonly in cryopreservation facilities, and carries a probable explosion hazard due to entry of liquid nitrogen into the vials. Probably the best packaging currently available is in the form of straws which ensure a homogeneous cooling, effective sealing by soldering at both ends and easier filling by means of a sterile nozzle which also avoids contamination. A high-security vitrification ionomeric resin straw (Cryo Bio System, France) ostensibly provides higher tensile strength and extra safety at ultra-low temperatures.

25.7.5 Freezing

The process of cooling of sperm during cryopreservation varies widely at different laboratories, and there is no standardized protocol. Ideal cooling rates for sperm are believed to range from 1 to 10 °C per minute (Medeiros et al. 2002). Although both slow- and fast-freezing protocols have been described, it is important to keep in mind the fact that very rapid cooling can damage sperm by intracellular ice formation, while too tardy cooling rates can subject the sperm to disproportionate osmotic forces and solute build-up (Henry et al. 1993). The two most commonly used methods for freezing of sperm are controlled-rate freezing (slow freezing) and static vapour cooling method (rapid freezing). The slow and controlled freezing, also called as Cleveland Clinic Foundation (CCF) method, comprises of measured and sequential addition of freezing media, subsequent storage at −20 °C for 8 min followed by storage in nitrogen vapours at −96 °C for 2 h and finally immersion in liquid nitrogen at −196 °C (Kobayashi et al. 2001; Nallella et al. 2004). Few studies have shown that controlled freezing using programmed and computerized approaches by means of automated freezers gives better quality sperm after thawing and curbs damage of low-quality sperm (Ragni et al. 1990), but use of such automated freezers is constrained by prohibitive cost, requirement for specialized equipment and need for increased amount of liquid nitrogen (Paras et al. 2008). The rapid freezing technique, also referred to as the Irvine Scientific (IS) method, is a quicker cryopreservation method, which enables rapid freezing of sperm. In this technique, the full amount of freezing medium is added in one batch, and the sample is then immersed in liquid nitrogen (Kobayashi et al. 2001; Nallella et al. 2004). Although it is generally accepted that controlled and gradual freezing with resultant acclimatization of sperm shields sperm from cryodamage, many reports have indicated that rapid freeze technique gives better sperm motility and survival after thawing as compared to the slow-freeze technique (Hallak et al. 2000; Nallella et al. 2004). However, a number of studies have failed to provide conclusive evidence of advantages of one technique over the other. Paras et al. (2008) failed to demonstrate any significant difference in sperm survival or motility rates in comparison of controlled freezing technique to vapour freeze technique. McLaughlin et al. (1990) in a similar comparative study showed that controlled-rate freezing technique resulted in a higher survival of motile sperm, but the proportion of viable sperm and comparative velocity were identical after the two freezing techniques.

Vitrification is a relatively new cryopreservation technique in which sperm is centrifuged to remove the plasma components and then resuspended in a sucrose solution before being directly dipped into the liquid nitrogen to fast freeze with a cooling rate approximating 50,000 K/min or more (Nawroth et al. 2002). The vitrified sperm can then be stored either in liquid nitrogen or in an ultra-cold deep freeze at

−86 °C. The benefits of this technique include non-requirement of any specialized equipment, low cost, simplicity and speed of performance. Additionally, vitrification is claimed to provide a significantly improved motility, and a higher number of viable sperm are available after thawing. The potential downside of vitrification process is the need for relatively higher concentrations (30–50%) of cryoprotectants, which can lead to a significant deterioration of sperm motility. The current focus of studies regarding vitrification is on developing validated protocols and finding out the right balance of cryoprotectants to optimize the outcomes of cryopreserved sperm.

25.7.6 Storage

The storage of cryopreserved sperm is conventionally done in the liquid nitrogen medium at −196 °C because of its relative inertness, being in liquid state at this temperature and the ease of storage at low pressures. However, cryopreserved sperm can be kept in storage at higher temperatures for short durations as is practised by some donor sperm banking facilities for storage during transit at −79 to −80° C in dry ice. It must be kept in mind that prolonged storage at these temperatures is likely to lead to poorer quality sperm after thawing (Ackerman 1967; Behrman and Ackerman 1969). Human sperm are most appropriate for cryosurvival because they are one of the smallest cells in the body and have the highest ratio of surface area to volume among all cells of the body. The most common method of storage of cryopreserved sperm is in dewars which are specialized metal vacuum flasks for storing liquid nitrogen. The walls of dewars are made of two or more layers with a vacuum in between to provide optimal thermal insulation. The capacity of medium-sized dewars usually ranges between 2000 and 8000 straws depending on the internal arrangement and the volume of the vessel. The dewars are designed in a manner to take care of the gas which forms as the cryogenic liquid boils gradually. The suitability of dewars in cryostorage is based on their simplistic design, durability, relative safety, low maintenance and provision of consistent temperatures. For facilities which require mass storage or storage for very long periods, automated vapour storage systems are required. The functioning of these automated vapour storage systems is much more sophisticated than dewars and relies on constant supervision of parameters like temperature and liquid levels. The critical component of automated storage systems pertains to sensing of fluid levels and autofilling of liquids. Malfunctioning of the fluid-level sensors can result in overfilling or under filling, both of which can pose serious threats to the samples and/or handlers (Tomlinson 2005). Additionally, vapour phase storage systems are considerably more expensive and require much higher amounts of liquid nitrogen as compared to dewars. The decision to use dewars or automated vapour storage systems should be individually taken by the concerned facility depending on a variety of factors like the number of samples to be stored, the type of packaging used, the kind of labelling and inventory required and the available floor space. Centres which need to store large number of samples for very long periods with limited storage space and where vials are used for cryopreservation should preferably be using automated vapour storage systems.

25.8 Sperm Function After Cryopreservation

The effects of cryopreservation on sperm parameters have been extensively studied, and researchers have attempted to analyse the impact of both the duration of cryopreservation and the freeze-thaw process itself on sperm characteristics like motility, viability, DNA stability and acrosomal integrity. It is vital to remember that baseline sperm parameters prior to freezing play an important role in determining the effects of cryopreservation on sperm characteristics. Cryopreservation prior to starting anticancer therapy in cancer patients may be handicapped by the suboptimal sperm features that are present in certain cancers and can be attributed to the disease process itself. For instance, as many as 30–60% patients with testicular malignancies may have impaired sperm parameters at baseline (Petersen et al. 1998). Likewise, patients with other cancers like Hodgkin's disease can also have an impairment of spermatogenesis prior to the initiation of therapy (Naysmith et al. 1998). However, cancer stage does not appear to have any bearing on sperm parameters. Impaired spermatogenesis at baseline should not preclude sperm cryopreservation because ART techniques like IVF/ICSI may be successful with even a solitary good quality sperm. In a study on cryopreserved sperm acquired from men with cancers, the outcome of IVF procedures was 60% in terms of fertilization rate and 40% in terms of pregnancy rate (Khalifa et al. 1992). In another study in which cryopreserved sperm from ten cancer patients were used, the pregnancy rate per cycle of ICSI was 36%, which is comparable to the usual ICSI protocols (Opsahl et al. 1996; Hallak et al. 1998; Naysmith et al. 1998). The deterioration in sperm parameters like motility and viability following cryopreservation is comparable in cancer patients and normal individuals. A study demonstrated a mean recovery rate after thawing of cryopreserved sperm to be around 30% in healthy men as well as in cancer patients (Bonetti et al. 2009). Poor post-thaw sperm parameters in cancer patients are mainly ascribed to the suboptimal semen quality prior to cryopreservation (Williams et al. 2009). The postulated reasons for impaired sperm parameters in cancer patients include defects in germ cells, a probable history of cryptorchidism, local hormonal changes due to the malignant lesion, anti-sperm antibody formation secondary to autoimmune disturbances, generalized endocrine imbalance, paraneoplastic effects and the strains and anxieties of the disease itself (Hallak et al. 1999; Williams et al. 2009). This impairment of spermatogenesis needs to be considered while formulating sperm cryopreservation strategies in order to counteract the decline in post-thaw semen quality.

Studies have tried to determine the reasons responsible for deterioration in sperm parameters following cryopreservation. Freeze-thaw process significantly diminishes sperm viability, and many studies have demonstrated that the decrease in number of viable sperm after cryopreservation is approximately 50%. Studies have failed to reveal any correlation between the duration of storage and sperm parameters after cryopreservation (Edelstein et al. 2008); likewise, no correlation was observed between the age of patient and sperm quality after thawing (Hourvitz et al. 2008). With successful IUI reported using sperm stored for as long as 28 years, it seems the freeze-thaw procedure itself leads to impaired sperm characteristics rather than the period of cryopreservation. Extracellular ice crystal formation results in solute toxicity causing osmotic damage and altering sperm morphology. Likewise, cryoprotectants like glycerol used during cryopreservation cross the plasma membrane during cooling and are subsequently eliminated during thawing. This leads to osmotic imbalance causing bulging of the cells during freezing and shrinkage during the thawing process. This may damage the cell membrane leading to impaired motility (Hallak et al. 2000) and can also damage the intracellular structures like mitochondria. Freeze-thaw processes can lead to the formation of free radicals like reactive oxygen species, which can cause peroxidation of plasma membrane and again have a deleterious effect on sperm motility. Studies have focussed on the role of antioxidants in improving post-thaw sperm parameters. Pentoxifylline pretreatment of sperm samples has been demonstrated to improve sperm motility, acrosome reaction and fertilization potential by virtue of elimination of the reactive oxygen species and enhanced intracellular cAMP levels (Esteves et al. 1997; Schmidt et al. 2004).

25.9 Risks During Cryopreservation

The potential hazards related to cryopreservation can be substantial as the cryopreserved specimen probably is the only hope of parenthood for the patient. The risks associated with cryopreservation of sperm include not only the hazards of working with liquid nitrogen but also the danger of loss or mixing up of patient's samples, incomplete thawing, contaminated samples or the breach of storage process. Prevention of such risks must be an essential component of protocol at any sperm storage facility. Labels on the samples should be unequivocally clear and should be able to stay legible for prolonged periods. The key patient parameters on the labels should be sufficient to avoid any possibility of misidentification. Premature thawing should be prevented by thorough vigilance during removal or substitution of samples and by using measures like freezer alarm systems to avoid the possibility of equipment failure. Correctly suited packing, vapour phase storage and screening for discernible pathogens like HIV and hepatitis B viruses can minimize the hazards of specimen breach and cross-contamination.

25.10 Looking Towards the Future

In prepubertal young boys where spermarche has not commenced, sperm collection for cryopreservation by conventional techniques is not possible. In this subset of patients, research is underway to remove the testicular tissue and harvest spermatogonial stem cells for cryopreservation and later transplant these spermatogonial stem cells subsequently to produce normal mature human sperm. Currently, these techniques are in experimental phase but hold promise for assisted reproductive techniques in the future. In a rodent model, Brinster et al. were able to perform a successful spermatogonial stem cell transplantation leading to the restoration of spermatogenic process (Brinster and Avarbock 1994). Another experimental technique currently being studied relates to the role of testicular tissue allografting. In a study on mice by Ohta et al., donor testicular tissue was harvested from the cloned donor mice and subsequently transplanted into testes of the recipient mice (Ohta and Wakayama 2005). The outcome analysis at 3 months was promising with the transplanted donor testicular tissue being taken up by the seminiferous tubules of some recipient mice and generating spermatogenesis. While these novel concepts look promising, certain serious issues need to be addressed before these can be adapted in clinical settings. The biggest concern is regarding the possibility of reinsertion of cancer cells through the cryopreserved and transplanted testicular tissue. Since most of the childhood malignancies have the ability to infiltrate testicular tissue, the threat of reintroduction of malignant cells is very real (Jahnukainen et al. 2001). While cell separation techniques can theoretically surmount this problem, it is critical to remember that even a tiny proportion of cancer cells can cause recurrent cancer (Fujita et al. 2005). Harvesting of testicular tissue from young boys has attendant hazards associated with it. The removal of testicular tissue from these young boys can seriously hamper the ability of the native testis to regain its function after the cancer therapy is complete.

Conclusion

Oncological therapeutic modalities have far-reaching implications in terms of gonadotoxicity and resultant impairment of fertility. It is imperative on part of the treating physicians to counsel the patients and their families about available fertility preservation options before commencing the anticancer therapy. Oncofertility is an emerging discipline that addresses these issues in a holistic manner. Oncofertility combines the two diverse domains of cancer therapy and reproductive science with the goal of broadening the fertility preservation options of cancer survivors. Cryopreservation is an advanced technique for storage of sperm at ultra-low temperatures. Cryopreservation offers a chance of paternity to these patients by successfully preserving and banking the sperm for long periods. Centuries of research and technological advancements have led to improvements in sperm cryopreservation such as choosing correct cryoprotectants, finding their optimal dose, improvements in freeze-thaw techniques and development of correct protocols. The cryopreserved sperm can be used for either IUI or IVF with or without ICSI as per the clinical indications. Further advancements in the field of cryopreservation are likely to be based on research in topics like cryoprotectant-free preservation by means of sperm vitrification and further improving the freezing and thawing procedures. Sperm can be harvested for cryopreservation by conventional methods in post-pubertal males. For prepubertal males, there are no scientifically assured methods for cryopreservation of sperm, but ongoing research, particularly in animal models, may make it possible to harvest and store spermatogonial stem cells for maturation and xenografting in future. Advent of novel chemotherapeutic agents which are more target specific and less toxic along with advancements in the field of assisted reproductive techniques is likely to have better fertility-related outcomes in cancer survivors.

References

Achille MA, Rosberger Z, Robitaille R, Lebel S, Gouin JP, Bultz BD, Chan PT (2006) Facilitators and obstacles to sperm banking in young men receiving gonadotoxic chemotherapy for cancer: the perspective of survivors and health care professionals. Hum Reprod 21(12):3206–3216

Ackerman DR (1967) Damage to human spermatozoa during storage at warming temperatures. Int J Fertil 13(3):220–225

Behrman SJ, Ackerman DR (1969) Freeze preservation of human sperm. Am J Obstet Gynecol 103(5):654–664

Bergeron A, Manjunath P (2006) New insights towards understanding the mechanisms of sperm protection by egg yolk and milk. Mol Reprod Dev 73(10):1338–1344

Bonetti T, Pasqualotto FF, Queiroz P, Iaconelli A Jr, Borges E Jr (2009) Sperm banking for male cancer patients: social and semen profiles. Int Braz J Urol 35(2):190–198

Brinster RL, Avarbock MR (1994) Germline transmission of donor haplotype following spermatogonial transplantation. Proc Natl Acad Sci U S A 91(24):11303–11307

Dikshit R, Gupta PC, Ramasundarahettige C, Gajalakshmi V, Aleksandrowicz L, Badwe R, Kumar R, Roy S, Suraweera W, Bray F, Mallath M (2012) Cancer mortality in India: a nationally representative survey. Lancet 379(9828):1807–1816

Edelstein A, Yavetz H, Kleiman SE, Hauser R, Botchan A, Paz G, Yogev L (2008) Effect of long-term storage on deoxyribonucleic acid damage and motility of sperm bank donor specimens. Fertil Steril 90(4):1327–1330

Esteves SC, Sharma RK, Thomas AJ Jr, Agarwal A (1997) Effect of in vitro incubation on spontaneous acrosome reaction in fresh and cryopreserved human spermatozoa. Int J Fertil Womens Med 43(5):235–242

Esteves SC, Sharma RK, Thomas AJ, Agarwal A (2000) Improvement in motion characteristics and acrosome status in cryopreserved human spermatozoa by swim-up processing before freezing. Hum Reprod 15(10):2173–2179

Fujita K, Ohta H, Tsujimura A, Takao T, Miyagawa Y, Takada S, Matsumiya K, Wakayama T, Okuyama A (2005) Transplantation of spermatogonial stem cells isolated from leukemic mice restores fertility without inducing leukemia. J Clin Invest 115(7):1855–1861

Gao DY, Liu J, Liu C, McGann LE, Watson PF, Kleinhans FW, Mazur P, Critser ES, Critser JK (1995) Andrology: prevention of osmotic injury to human spermatozoa during addition and removal of glycerol. Hum Reprod 10(5):1109–1122

Grunewald S, Paasch U, Glander HJ (2001) Enrichment of non–apoptotic human spermatozoa after cryopreservation by immunomagnetic cell sorting. Cell Tissue Bank 2(3):127–133

Grunewald S, Paasch U, Said TM, Rasch M, Agarwal A, Glander HJ (2006) Magnetic-activated cell sorting before cryopreservation preserves mitochondrial integrity in human spermatozoa. Cell Tissue Bank 7(2):99–104

Hallak J, Hendin BN, Thomas AJ, Agarwal A (1998) Investigation of fertilizing capacity of cryopreserved spermatozoa from patients with cancer. J Urol 159(4):1217–1220

Hallak J, Kolettis PN, Sekhon VS, Thomas AJ, Agarwal A (1999) Cryopreservation of sperm from patients with leukemia. Cancer 85(9):1973–1978

Hallak J, Sharma RK, Wellstead C, Agarwal A (2000) Cryopreservation of human spermatozoa: comparison of TEST-yolk buffer and glycerol. Int J Fertil Womens Med 45(1):38–42

Henry MA, Noiles EE, Gao D, Mazur P, Critser JK (1993) Cryopreservation of human spermatozoa. IV. The effects of cooling rate and warming rate on the maintenance of motility, plasma membrane integrity, and mitochondrial function. Fertil Steril 60(5):911–918

Hourvitz A, Goldschlag DE, Davis OK, Gosden LV, Palermo GD, Rosenwaks Z (2008) Intracytoplasmic sperm injection (ICSI) using cryopreserved sperm from men with malignant neoplasm yields high pregnancy rates. Fertil Steril 90(3):557–563

Jahnukainen K, Hou M, Petersen C, Setchell B, Söder O (2001) Intratesticular transplantation of testicular cells from leukemic rats causes transmission of leukemia. Cancer Res 61(2):706–710

Joint Council for Clinical Oncology (1998) Management of gonadal toxicity resulting from the treatment of adult cancer. Royal College of Physicians, London

Khalifa E, Oehninger S, Acosta AA, Morshedi M, Veeck L, Bryzyski RG, Muasher SJ (1992) Successful fertilization and pregnancy outcome in in-vitro fertilization using cryopreserved/thawed spermatozoa from patients with malignant diseases. Hum Reprod 7(1):105–108

Kliesch S, Bergmann M, Hertle L, Nieschlag E, Behre HM (1997) Semen parameters and testicular pathology in men with testicular cancer and contralateral carcinoma in situ or bilateral testicular malignancies. Hum Reprod 12(12):2830–2835

Kliesch S, Kamischke A, Cooper TG, Nieschlag E (2010) Cryopreservation of human spermatozoa. In: Nieschlag E, Behre HM, Nieschlag S (eds) Andrology. Springer, Berlin, pp 505–520

Kobayashi H, Ranganathan P, Mahran AM, Sharma RK, Thomas AJ, Agarwal A (2001) Comparison of two cryopreservation protocols for freezing human spermatozoa. Fertil Steril 76(3):S229–S230

Lee SJ, Schover LR, Partridge AH, Patrizio P, Wallace WH, Hagerty K, Beck LN, Brennan LV, Oktay K (2006) American Society of Clinical Oncology recommendations on fertility preservation in cancer patients. J Clin Oncol 24(18):2917–2931

Mack SR, Zaneveld LJ (1987) Acrosomal enzymes and ultrastructure of unfrozen and cryotreated human spermatozoa. Gamete Res 18(4):375–383

McLaughlin EA, Ford WCL, Hull MGR (1990) A comparison of the freezing of human semen in the uncirculated vapour above liquid nitrogen and in a commercial semi-programmable freezer. Hum Reprod 5(6):724–728

Medeiros CMO, Forell F, Oliveira ATD, Rodrigues JL (2002) Current status of sperm cryopreservation: why isn't it better? Theriogenology 57(1):327–344

Mortimer D (2004) Current and future concepts and practices in human sperm cryobanking. Reprod Biomed Online 9(2):134–151

Nallella KP, Sharma RK, Allamaneni SS, Aziz N, Agarwal A (2004) Cryopreservation of human spermatozoa: comparison of two cryopreservation methods and three cryoprotectants. Fertil Steril 82(4):913–918

Nandakumar A (1990) National cancer registry programme. Indian Council of Medical Research, Consolidated report of the population based cancer registries, New Delhi, India, 96

Nandakumar A (2001) National Cancer Registry Programme (2001) Consolidated report of the population based cancer registries1990–1996. Indian Council of Medical Research, New Delhi

Nawroth F, Isachenko V, Dessole S, Rahimi G, Farina M, Vargiu N, Mallmann P, Dattena M, Capobianco G, Peters D, Orth I (2002) Vitrification of human spermatozoa without cryoprotectants. CryoLetters 23(2):93–102

Naysmith TE, Blake DA, Harvey VJ, Johnson NP (1998) Do men undergoing sterilizing cancer treatments have a fertile future? Hum Reprod 13(11):3250–3255

Ohta H, Wakayama T (2005) Generation of normal progeny by intracytoplasmic sperm injection following grafting of testicular tissue from cloned mice that died postnatally. Biol Reprod 73(3):390–395

Opsahl MS, Fugger EF, Sherins RJ, Schulman JD (1996) Preservation of reproductive function before therapy for cancer: new options involving sperm and ovary cryopreservation. Cancer J Sci Am 3(4):189–191

Paras L, Freisinger J, Esterbauer B, Schmeller N, Szlauer R, Jungwirth A (2008) Cryopreservation technique: comparison of test yolk buffer versus SpermCryo and vapour versus computerised freezing. Andrologia 40(1):18–22

Pegg DE (2007) Principles of cryopreservation. In: Clifton NJ (ed) Methods in molecular biology 368:39–57

Petersen PM, Giwercman A, Skakkebaek NE, Rørth M (1998) Gonadal function in men with testicular cancer. Semin Oncol 25(2):224–233

Polge C, Smith AU, Parkes AS (1949) Revival of spermatozoa after vitrification and dehydration at low temperatures. Nature 164(4172):666

Ragni G, Caccamo AM, Dalla Serra A, Guercilena S (1990) Computerized slow-staged freezing of semen from men with testicular tumors or Hodgkin's disease preserves sperm better than standard vapor freezing. Fertil Steril 53(6):1072–1075

Royere D, Barthelemy C, Hamamah S, Lansac J (1996) Cryopreservation of spermatozoa: a 1996 review. Hum Reprod Update 2(6):553–559

Russell LD, Rogers B (1987) Improvement in the quality and fertilization potential of a human sperm population using the rise technique. J Androl 8(1):25–33

Said TM, Grunewald S, Paasch U, Rasch M, Agarwal A, Glander HJ (2005) Effects of magnetic-activated cell sorting on sperm motility and cryosurvival rates. Fertil Steril 83(5):1442–1446

Said TM, Agarwal A, Zborowski M, Grunewald S, Glander HJ, Paasch U (2008) ANDROLOGY LAB CORNER*: utility of magnetic cell separation as a molecular sperm preparation technique. J Androl 29(2):134–142

Saranath D, Khanna A (2014) Current status of cancer burden: global and Indian scenario. Biomed Res J 1(1):1–5

Schmidt KLT, Larsen E, Bangsbøll S, Meinertz H, Carlsen E, Andersen AN (2004) Assisted reproduction in male cancer survivors: fertility treatment and outcome in 67 couples. Hum Reprod 19(12):2806–2810

Schover LR, Brey K, Lichtin A, Lipshultz LI, Jeha S (2002) Knowledge and experience regarding cancer, infertility, and sperm banking in younger male survivors. J Clin Oncol 20(7):1880–1889

Spallanzani L, Bonnet C (1780) Dissertazioni di fisica animale, e vegetabaile (vol. 2). Societa tipografica, Modena

Tomlinson M (2005) Managing risk associated with cryopreservation. Hum Reprod 20(7):1751–1756

Wallace WHB, Anderson RA, Irvine DS (2005) Fertility preservation for young patients with cancer: who is at risk and what can be offered? Lancet Oncol 6(4):209–218

Watson PF (1979) The preservation of semen in mammals. Oxf Rev Reprod Biol 1:283–350

Williams DH, Karpman E, Sander JC, Spiess PE, Pisters LL, Lipshultz LI (2009) Pretreatment semen parameters in men with cancer. J Urol 181(2):736–740

Woodruff DTK (2007) The emergence of a new interdiscipline: oncofertility. In: Woodruff TK (ed) Oncofertility fertility preservation for cancer survivors. Springer, New York, pp 3–11

Assisted Reproductive Technologies in Infertility Treatment: Opportunities and Challenges

26

Pawan K. Dubey, Anima Tripathi, and Akhtar Ali

Abstract

A tremendous rise in the fertility clinics providing ART services is seen worldwide with the birth of first IVF baby (Louise Joy Brown) in 1978. ART comprises various types of medical treatments designed to assist in achieving pregnancy. IVF and other ART-associated technologies of fertilization (ICSI, IUI, PZD, SUZI, MESA, and PESA) offer an opportunity to become parent even in severe cases of infertility. These technologies have allowed millions of individuals to fulfill their parenting wish. A positive attitude combined with an appropriate treatment can help most of the infertile couples experience the joy of parenthood. This chapter provides a thorough overview of the assisted reproductive technologies with opportunities for patients and challenges for clinical professionals or researchers.

Keywords

Assisted reproductive technologies (ARTs) • In vitro fertilization (IVF) Intracytoplasmic sperm injection (ICSI) • Surrogacy • Infertility treatment

P.K. Dubey (✉) • A. Ali
Centre for Genetic Disorders, Institute of Science, Banaras Hindu University, Varanasi, Uttar Pradesh, India
e-mail: pawandubey1981@gmail.com

A. Tripathi
Department of Pharmaceutics, IIT-BHU, Varanasi 221005, India

R. Singh, K. Singh (eds.), *Male Infertility: Understanding, Causes and Treatment*, DOI 10.1007/978-981-10-4017-7_26

Key Points

- In 1978, for the first time, manipulation of the gametes was done under in vitro conditions by conventional IVF methods that resulted in the successful birth of Louise Joy Brown.
- In a number of cases, failure of other treatments leaves the patients with the option of ART as the last hope.
- ARTs in the form of IUI, IVF, and ICSI are used very commonly because of a variety of reasons.
- Other variants of ARTs, such as SUZI, ZIFT, GIFT, and PZD, are good alternative techniques to routine ARTs.
- ARTs have revolutionized the field of infertility treatment as theoretically even men with one or few sperm can have a hope to father of a child.

26.1 Introduction

Infertility can be considered as inability of a female individual to conceive pregnancy for the full term. Infertility occurs mainly because of two factors, the male factor and the female factor. One third of both the male and the female factors are responsible for infertility, and the remaining one third is because of unexplained infertility. Despite the progresses in the field of reproductive biology, the etiology of infertility is still unknown, and about 50% of the cases are termed as "idiopathic." The diagnosis and treatment of infertility may involve targeted or empirical therapies depending upon the nature of infertility, depth of investigations, and success in identifying the underlying cause. Unfortunately, a large number of individuals who are suffering from the infertility do not get benefit from the traditional medications or treatments; therefore, they need to move for the next line of therapy, i.e., assisted reproduction.

For a number of infertile couples, leaving the few exceptions, assisted reproductive technologies (ARTs) are the only effective treatments that allow conception even in severe infertility cases, including azoospermia. These technologies include in vitro fertilization (IVF), intracytoplasmic sperm injection (ICSI), intrauterine insemination (IUI), percutaneous epididymal sperm aspiration (PESA), microsurgical epididymal sperm aspiration (MESA), testicular sperm extraction (TESE), partial zona dissection (PZD), and subzonal sperm injection (SUZI). Earlier, infertile men were dependent on sperm donor insemination or adoption, but at present, even in more severe infertility cases, IVF and other ART fertilization technologies (ICSI, IUI, PZD, SUZI) provide them an opportunity to become parent. This chapter provides an overview of the available ARTs for infertile individuals with focus on their suitability, advantages, and disadvantages.

26.2 ARTs Are a Boon

Assisted reproductive technologies acted as a boon for millions of people worldwide by providing them the opportunity to become the parent of their biological children that would rather not been possible ever. According to the European Society of

Table 26.1 ARTs used in treatment of infertility

Cause of infertility	ART	Outcome
Ejaculatory disorders (oligo-, azoo-, and zoospermia)	In vitro fertilization (IVF)	Pregnancy
Repeated fertilization failure by natural method or IVF	Intracytoplasmic sperm injection (ICSI)	Pregnancy
Repeated embryo transfer failure		
Asthenozoospermia (progressive motility), teratozoospermia, oligozoospermia	Partial zona dissection (PZD) and subzonal sperm injection (SUZI)	Pregnancy

Table 26.2 Different parameters of male fertility considered for the diagnosis of infertility

Parameters	Clinical examination of male infertility
History	Time and duration of infertility, any previous pregnancy, medical details of female partner, intercourse frequency and timing, any existing and past disease, alcohol, smoking, and drug consumption
Examination	Details of treatment for testicular maldescent, size of testis, vas deferens diameter and blockage, epididymis diameter and blockage, hydrocele/varicocele, semen analysis by CASA system
Investigation	Endocrine profile (T3, T4, TSH, FSH, LH, testosterone)

Human Reproduction and Embryology (ESHRE), three million babies have born with the help of ARTs in the last 30 years which suggest that ARTs can be used to treat any form of infertility either related to women and men, lesbians and gays, or transgender couples. The arrival of these fascinating technologies in the form of ARTs has changed the opinion about the reproductive world and has generated new hopeful possibilities for infertile couples to have their own baby. The following ARTs may be considered for the treatment of infertility-related problems (Table 26.1).

Before undertaking an ART procedure, a number of investigations must be completed in order to know the symptoms, cause, and type of infertility that would help them choose the best possible therapy (Table 26.2). Traditionally, the gynecologist or reproductive endocrinologist starts observing and evaluating the infertility-related problems mainly with the female partner in comparison with a little analysis of the male partner. Analysis of male factors which are also equally responsible for infertility must be done by a urologist with specialization in male infertility. In fact, it is very necessary to diagnose and identify the real problem in order to suggest the best-suited ARTs to cure infertility-related problems. Hence, infertile couple should opt for a complete clinical checkup by a physician specialized either in male or female infertility, respectively.

26.3 Sperm Recovery Techniques

The preference of sperm retrieval technique and its success rate is based on the type of male infertility either obstructive or non-obstructive. Some of the important preoperative tools for diagnosis are clinical history, physical examination, and endocrine assessment like measurement of follicle-stimulating hormone (FSH) and testosterone levels in patient serum. In conditions like vasectomy, failed vasectomy reversal, primary testicular failure, or congenital obstruction of the sperm ducts where there is no spermatozoa in the patient's semen, different types of technologies are available to retrieve sperm for ART purposes. This includes electroejaculation/vibratory stimulation, percutaneous epididymal sperm aspiration (PESA), microsurgical epididymal sperm aspiration (MESA), and testicular sperm extraction (TESE). For azoospermic patient where there is no obstruction, PESA and MESA are extremely useful to retrieve sperm for ART purposes. In contrast, MESA is a more advanced technology in which a number of sperms can be retrieved with less epididymal damage (Kim and Lipshultz 1997). In TESE, an open surgical procedure is adopted to retrieve the sperm. The procedure is performed in those cases where the process of epididymal sperm retrieval fails. However, it provides similar pregnancy rates with micro-assisted fertilization technologies (Ghazzawi et al. 1998). Considering the number of sperm retrieval methods which plays an important role to achieve pregnancy, each infertile male partner must be carefully examined to determine the best sperm retrieval method because it would affect the overall outcome of ART. The most common and useful methods for sperm retrieval are summarized in Table 26.3 with their approximate costs summarized in Table 26.4.

Table. 26.3 Common and useful methods for sperm retrieval

Technique	Common name	Indications
Percutaneous epididymal sperm aspiration	PESA	Obstructive azoospermia
Microsurgical epididymal sperm aspiration	MESA	Obstructive azoospermia
Testicular sperm aspiration	TESA	Failed PESA, epididymal agenesis, non-obstructive azoospermia
Microsurgical testicular sperm extraction	Micro-TESE	Non-obstructive azoospermia

Table 26.4 The approximate cost of different ARTs

ART	Approximate cost/ cycle (USD)	Success rate
IUI	$120–$400	5–30%, depending on the age of woman
IVF	$10,000–$14,000	34% successful pregnancies per cycle
ICSI	$12,000–$16,000	31% success rate
ZIFT/GIFT	$12,000–$20,000	39–45% success rate
PGD	$2500–$5000	Success rates are 90% for testing for medical conditions and close to 100% for sex selection

26.4 Intrauterine and Donor Insemination

Intrauterine insemination (IUI) is the simplest form of assisted reproductive technology where, at the time of egg ovulation, washed ejaculated sperm is placed in the uterus, beyond the cervix. This technology is used to treat infertility problem in males related to low sperm count or low motility, antisperm antibodies, and erectile dysfunction. In this technology, if the husband's sperm is used, it is considered as AI (artificial insemination), and if the donor sperm is used, it is considered as DI (donor insemination). Donor insemination is an alternative form of AI that offers an effective advantage to those couples who fail to conceive despite repeated clinical therapies. In case of the absence of persisting female factor infertility, IUI technology is extremely successful (70%). However, donor insemination is unacceptable at social level, and it may be considered an illegal practice by some societies. Moreover, where adoption is not desired and the male wants to have his own genetic offspring, assisted reproductive technologies may prove to be a good option.

In recent years, IUI along with superovulation technology has become a famous method for the treatment of male-related infertility. Insemination using a high number of motile and morphologically normal sperms after superovulation by gonadotropins has a theoretical advantage and maximum chance of successful pregnancy. The success rate of IUI varies widely and is closely related to female age and reproductive potential as, for example, the IUI success rate is higher for younger women. The IUI is very beneficial in case of male infertility; however, if the ovaries are stimulated with drugs to increase the number of eggs obtained every month, we can further enhance the overall pregnancy rate, for example, if the IUI is performed with superovulation in comparison with the IUI alone, a fourfold increase in the pregnancy rate will be observed (Kemmann et al. 1987). Furthermore, a study conducted by Serhal et al. (1988) showed that pregnancy rate/cycle is significantly greater for the combination of IUI and gonadotropin superovulation (26.4%) as compared to IUI (2.7%) or superovulation alone (6.1%). Further, a review by Dodson and Haney (1991) showed that the fecundity rate with IUI and superovulation in male factor infertility is 8.7% as compared to 17% for unexplained infertility. It appears that combination of IUI and superovulation may offer some limited benefits to infertile men but may improve the success rate of IUI in infertility due to female factors.

26.4.1 IUI Procedure

The IUI can be conducted mainly using three procedures, viz., natural, clomiphene, or gonadotropin stimulation. Natural cycle is recommended when treatment is with donor sperm or infertility is secondary to difficulties with intercourse. Cycles of fertility drugs such as clomiphene (Clomid) or gonadotrophins (Gonal-F, Puregon, and Menopur) are generally prescribed if there is a case of unexplained or mild male factor infertility. In case of IUI, lower doses of drugs are used than in IVF with an

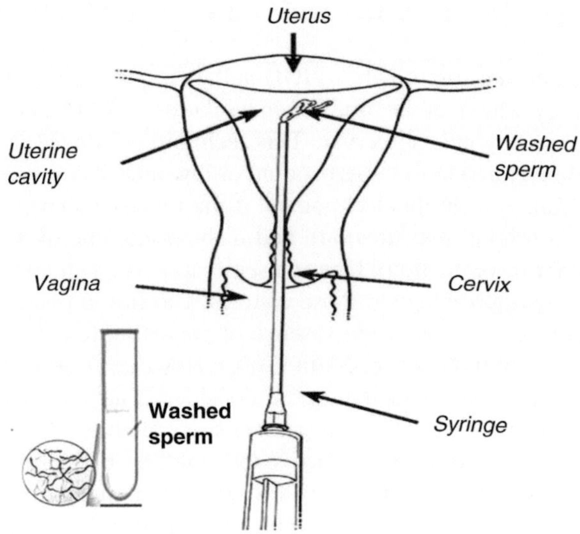

Fig. 26.1 Diagrammatic representation of IUI procedure

aim to increase the incidences of successful fertilization by stimulating the production of more than one follicle, for example, production of two or three follicles.

After induction of ovulation by injection of HCG or detection of natural ovulation by urine, freshly prepared sperm from male partner or donor is prepared as per the established procedure and drawn into a syringe with small amount of culture medium. In insemination procedure, a fine plastic catheter is used to transfer the processed sperm through female partner's cervix into the uterus (Fig. 26.1). Following IUI, there is no need to take time off or limit work, but the patient is advised to visit an embryologist and IVF clinic for post-insemination checkup to assure the pregnancy. The IUI procedure is more advantageous because of its less invasive nature and is better tolerated as compared to IVF. However, the major disadvantage of IUI is that its success rate is low and there is a high chance of occurrence of multiple pregnancies as compared to the IVF.

26.5 In Vitro Fertilization (IVF)

Among the various ART treatments, IVF is the most popular and invasive technology. It has been seen that in general, women who are trying for the pregnancy or live birth adopt other methods first and finally move on to the IVF when those methods become unsuccessful. In contrast to artificial insemination, fertilization in IVF gets done outside of the woman's body in which eggs (retrieved from the woman trying to get pregnant or from an egg donor) are fertilized with the sperm (from a male partner or donor-derived sperm) in a petri dish. In 1978, for the first time, manipulation of the gametes has been done under in vitro conditions by conventional IVF methods that resulted in the successful birth of Louise Joy Brown. Since the delivery of the first IVF baby some 38 years ago, the technology has spread worldwide and is still in high practice because of its consistent results. Here, we summarize the basic procedures of IVF.

26.5.1 IVF Procedure

26.5.1.1 Step 1: Controlled Ovarian Hyperstimulation (COH)

For ovarian hyperstimulation, GnRH agonist (Lupron) protocol is used to suppress the secretion of gonadotropin hormone to avoid premature ovulation. The next stage is the multiple follicular recruitment by the use of gonadotropin injections daily once the suppression of gonadotropic hormone is achieved to optimum level. The follicular development is monitored by the use of the technologies like ultrasound imaging and hormone assessments. Physician using ultrasound examinations and blood testing can determine whether the follicles are ready for egg retrieval or not. Generally, 8–14 days of stimulation are required. The hCG administration is given for final maturation of the egg when the follicles are ready and reach an appropriate size. Egg retrieval is scheduled 34–36 h after hCG injection.

26.5.1.2 Step 2: Egg Retrieval

Egg retrieval is usually performed by transvaginal ultrasound aspiration, a minor surgical procedure, which is performed for egg retrieval process. To retrieve eggs from the patients, clinicians generally administer some pain medications. During egg retrieval process to identify the follicles, an ultrasound probe is inserted into the vagina, and a needle is guided through the vagina into the follicles. Thereafter, to locate all the available eggs, the follicular fluid is scanned by the embryologist. The cumulus-oocyte complexes (COCs) consisting of the first polar body (PB) (Fig. 26.2) are placed in a special media and cultured in a CO_2 incubator until insemination. Laparoscopy technique is also used to retrieve the eggs using a small telescope placed in the umbilicus. For more information on laparoscopy, consult with an ART center or a specialized doctor.

26.5.1.3 Step 3: Fertilization and Embryo Culture

After assessment of maturity and quality, the retrieved eggs are placed in an IVF culture medium. For fertilization, simultaneously sperms are processed from a male's partner or donor-derived semen. Alternatively, if the male's partner is azoospermic or having any kind of obstruction, sperm can be obtained from the testicle, epididymis, or vas deferens using sperm retrieval technology as described above. For fertilization purposes, approximately 50,000–100,000 motile sperms are mixed with the meiotically competent eggs under in vitro condition. After 16–18 h of co-incubation of the egg and

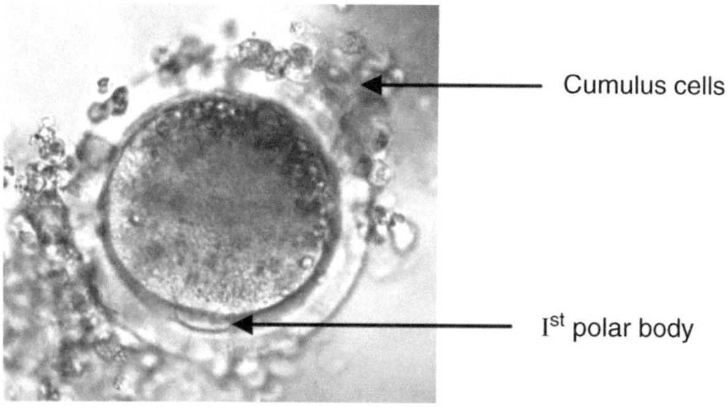

Fig. 26.2 Cumulus-enclosed mature oocyte showing first polar body extrusion

sperm, fertilization is assessed by visualization of two pronucleus formation. Once the fertilization is confirmed, the presumptive zygotes (fertilized eggs) are cultured into a specially formulated culture medium that supports the growth and development of embryos. After 24 h of postfertilization, the fertilized eggs are divided into two to four cell embryos (Fig. 26.3). For transfer purpose, the embryos are grown till the blastocyst stage which have higher potential for implantation.

26.5.1.4 Step 4: Embryo Transfer

In general, 4–8 cell stage embryos (day 3 of postfertilization) are used to transfer for implantation purpose. Although, the later stages like 8–16, morula or blastocyst stage (Fig. 26.4) of embryo can be transferred into female partner or surrogate mother to get pregnancy. However, before transfer, clinicians must perform the pre-implantation diagnosis to examine the embryo for any fragmentation or diseases. In practice, transferable embryos should be free from any kind of diseases and must be classified into grades 1–4 on the basis of several parameters where grade 1 represents the best-quality embryos to maximize the chance of successful pregnancy.

26.5.2 Assisted Hatching (AH)

In general, after transfer of embryo in the uterus, embryo must expand and rupture the zona pellucida (ZP) allowing to implant. In some cases, embryo does not hatch out from ZP and as a result implantation does not occur. In such cases, assisted hatching (AH) is used to overcome this problem. AH is a technology which is used to create a hole in the ZP with the help of an instrument prior to embryo transfer which facilitates hatching in utero. However, it has been seen that AH does not improve the rates of live birth though it is useful for aged women or couples to get maximum chance of pregnancy.

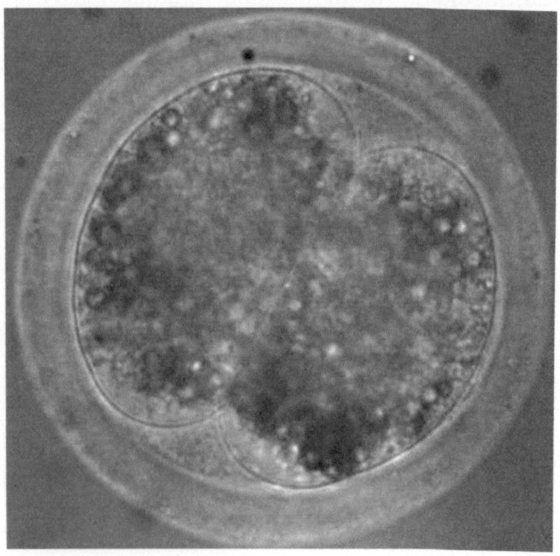

Fig. 26.3 Cell stage embryo

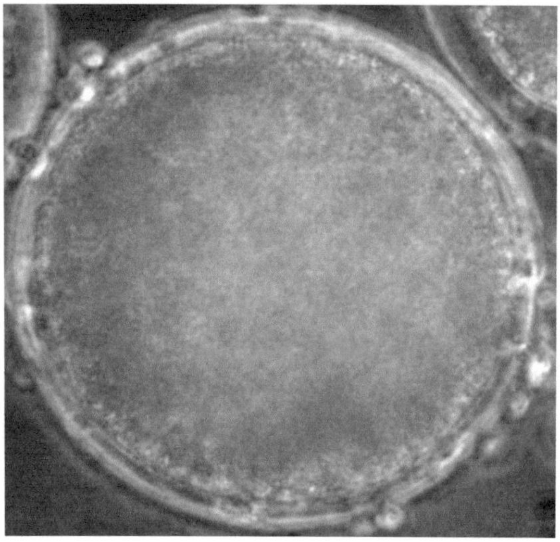

Fig. 26.4 Blastocyst stage of embryo

26.5.3 Preimplantation Genetic Diagnosis (PGD)

PGD is a technology which is used to diagnose inherited diseases by screening of preimplantation embryo. In this procedure, one or two cells called blastomere are retrieved from the presumptive zygote and diagnosed the probability for different genetic disease using different molecular approaches. After diagnosis, embryos that free from any type of diseases are selected for transfer in the uterus. However, for conducting these procedures, a specialized clinician and equipment are needed. PGD is helpful for couples who are carriers of some genetic diseases, and these couples must perform embryo screening to reduce the risk of having an affected child. There are some other methods like chorionic villus sampling (CVS) and amniocentesis which can be used for diagnosis of genetic diseases during gestational period.

26.5.4 Cryopreservation

Cryopreservation is a technology in which any type of cells, tissues, or body organs can be preserved at very low temperature (−196 °C) in a natural state for future use. In ART, cryopreservation technology can be used to store or preserve extra embryos or oocytes for future use. The most significance of this technology is that ART clinicians may use the preserved oocyte or embryo for the fertilization or embryo transfer purposes rather than initiating a new IVF cycle in infertile patient. Moreover, live births that have been reported using frozen embryos showed the importance of this technology. However, there are some risks like chromosomal aberrations or biochemical level associated with frozen embryo. Therefore, it is advisable for infertile couples and clinicians both that before using cryopreserved embryos, they must ensure that embryos are healthy and free from any type of aberrations at cellular and

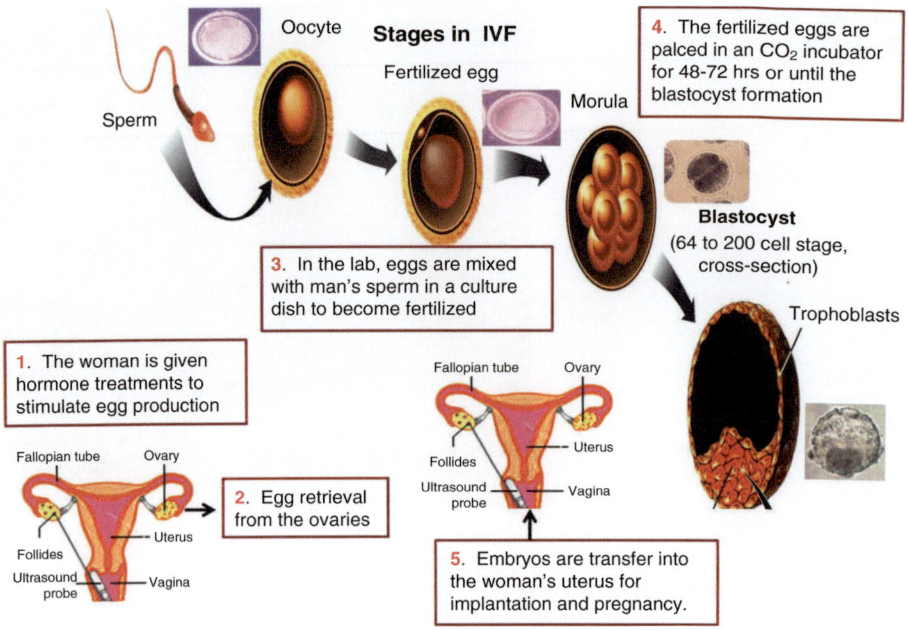

Fig. 26.5 Diagrammatic representation of stages used in IVF procedure

molecular level. In general, in the field of ART, cryopreservation plays an important role because it is less expensive, time saving, and an invasive procedure (Fig. 26.5).

26.5.5 Advantages and Disadvantages of IVF

- This technology may help infertile couple to get a baby of their own.
- IVF can be performed with less number of motile sperms (50,000–100,000/oocyte) compared to natural (2–6 million/oocyte) fertilization.
- It has higher success rate as compared to IUI and other ARTs.
- Preimplantation genetic diagnosis can be done for identification and prevention of genetic abnormalities.
- The procedure is relatively safe and has been utilized for a long time to produce a baby via egg or sperm donors.
- Sperm as well as embryos which are unused can be cryopreserved and may be utilized for stem cell research that would help cure various kinds of degenerative diseases in the future.

26.5.6 Disadvantages of IVF

- The main drawback of IVF is multiple births, i.e., delivery of more than one baby. To get higher success rate, clinics and doctors generally transfer more than one embryo that can result undesired multiple births.

- IVF may cause ovarian hyperstimulation syndrome due to heavy use of hormones and drugs during the procedure. IVF can lead to ectopic pregnancy in which implantation of embryo occurs outside the uterus.
- The success rate of IVF depends on the age of the female, the quality of eggs, the quality of sperm, the quality of the uterus, etc. IVF can cause some abdominal pain due to the use of some minor surgical procedure besides the use of drugs and hormones.
- IVF technique is little costly and it may not be affordable for some.

Failure of IVF could be devastating for any infertile couple. If it happens, what would be offered to the patient? In recent years, many advanced technologies have been developed in the field of ART, some of which promise to provide positive results even in severe infertility cases, such as severe oligospermia, asthenospermia, and teratospermia. In case of IVF failure, infertile couple can choose other options such as gamete intrafallopian transfer (GIFT), zygote intrafallopian transfer (ZIFT), and micro-assisted fertilization.

26.6 Gamete Intrafallopian Transfer (GIFT) and Zygote Intrafallopian Transfer (ZIFT)

Gamete intrafallopian tube transfer (GIFT) and zygote intrafallopian transfer (ZIFT) are variants of IVF which are used in the case of female infertility or some other infertility treatments that have been unsuccessful. From the last few years, both of the technologies have increased attraction of the clinicians because these technologies reasonably increased the clinical pregnancy rates in comparison with IVF. It is predicted that higher clinical pregnancy rate is due to the in vivo environment of the fallopian tube. As like IVF, ZIFT and GIFT technologies begin with ovarian stimulation and egg retrieval. In ZIFT, retrieved eggs are fertilized outside of the body, and the resulting zygote(s) is directly transferred into the woman's fallopian tube by the help of laparoscopic surgery. In contrast to ZIFT, in GIFT, the processed eggs and sperms are both transferred directly into the woman's fallopian tube. Principally, GIFT is more useful for the type of unexplained infertility. However, the disadvantage of GIFT is we can't confirm whether transferred eggs become fertilized or not if the pregnancy is not achieved. Therefore, ZIFT is better than GIFT, and mostly ART clinicians preferred ZIFT over GIFT for the treatment of unexplained infertility. Moreover, some studies have been randomized comparing the ZIFT and IVF and found no advantage of ZIFT over IVF for the treatment of male-related infertility (Tournaye et al. 1992a, b). Furthermore, Tournaye and associates conducted a comparative study between IVF, GIFT, and ZIFT and stated that take-home baby rates are 13.5%, 7%, and 20%, respectively (Tournay et al. 1991). Remarkably, IVF is still the first choice of infertile couple when compared to other ARTs due to the ease of access, availability, cost, and success rate. However, as per the report, the technology like ZIFT also may be useful for treating infertile couple with little more success rate.

26.7 Micro-assisted Fertilization

Though IVF, IUI, GIFT, and ZIFT are very useful for infertility treatment in the field of ART with variable success rate, however, still in many severe or unexplained infertility cases, these technologies are not able to treat the patient. Therefore, there is a need to treat such cases with more advanced technologies. When there is severe deficiency of sperm number and/or limited ability of sperm to fertilize during IVF, GIFT, and ZIFT, adjunctive micromanipulation technologies such as intracytoplasmic sperm injection (ICSI) may be useful in providing a reasonable chance of pregnancy.

26.7.1 Micromanipulation Technique

Micromanipulation is an advanced technology which is used to manipulate the gametes (egg and sperm) under in vitro condition for different purposes. Micromanipulation technology have revolutionized the field of ART where maximum pregnancy rate can be achieved. In this technology, single oocyte and sperm can manipulated as per the need with the help of a holding and injection pipette equipped with an inverted microscope. First, processed oocyte is immobilized by the holding pipette and then injected a single sperm or even any other chemicals into cytoplasm of oocyte with the help of injection pipette. This technology can be divided into zonal, subzonal, and intracytoplasmic procedure (Fig. 26.6). In zonal procedure, a tiny hole is created in zona pellucida, an acellular layer surrounding the egg by the help of laser-guided beam.

Basically, this procedure has been broadly termed as "zona drilling" which is successfully adopted for treating male patient-related infertility. This method is also called partial zona dissection (PZD). Subzonal procedure of micromanipulation technology directly facilitate sperm-egg interaction are known as subzonal insertion of sperm (SUZI). In SUZI, sperm is directly placed into the perivitelline space of egg for the fertilization purposes. The third and most invasive form of microsurgical

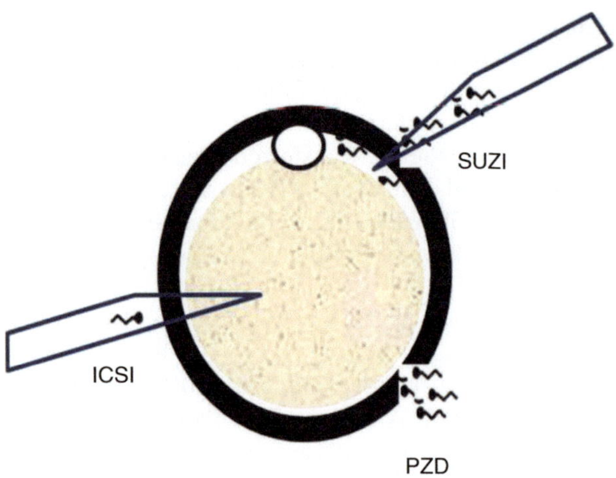

Fig. 26.6 Diagrammatic representation of different micromanipulation technologies which are used to fertilize meiotically competent egg

fertilization is the microinjection of a single sperm into the cytoplasm of oocyte, referred to as intracytoplasmic sperm injection (ICSI).

In the field of male factor infertility where sperm production is nil or zero sperm count, another micromanipulation technology known as round spermatid nucleus injection (ROSNI) can be used. In ROSNI, round spermatid is directly extracted from male testicles and after removing the nucleus injected into the female partner's eggs. However, this process has yet to give live birth and has to be clinically validated, though clinicians believe that it will eventually become a successful technology that will allow men, who previously had no hope, to be a father of a biological child.

26.7.2 Intracytoplasmic Sperm Injection (ICSI)

One of the major leading technologies for the treatment of male factor infertility is ICSI where a single sperm is injected into the cytoplasm of an egg. ICSI is performed in case of low sperm count or when there is no sperm present in the ejaculate, in case of abnormally shaped sperm, low sperm motility, as well as when the IVF has been previously unsuccessful. ICSI has become the ART of choice for male infertility and is much more effective therapy than other assisted fertilization technologies. ICSI is carried out using automated instrument called as micromanipulator, which is equipped by a holding and injection pipette. In ICSI, first a single healthy and motile sperm and then meiotically competent egg are immobilized by the help of injection and holding pipette, respectively. After insuring that everything is right, then single sperm is injected into the cytoplasm of the egg by the help of injection pipette (Fig. 26.7). However, this technology has the possibility of transmitting genetic defects of spermatogenesis or other genetic defects to a future offspring.

In a study, fertilization rate of 55% for ICSI versus 17% for SUZI has been reported (Van Steirteghem et al. 1993). ICSI has become the most quickly adopted technology for those couples who are unable to conceive from conventional IVF. Further, a study conducted by Palermo and associates showed 69% fertilization rate and a 38% ongoing pregnancy rate using ICSI (Palermo et al. 1995). The use of ICSI may prevent such complete failures; however, fertilization failure may still occur even when ICSI is used. Therefore, taking into consideration the added expenses and the potential

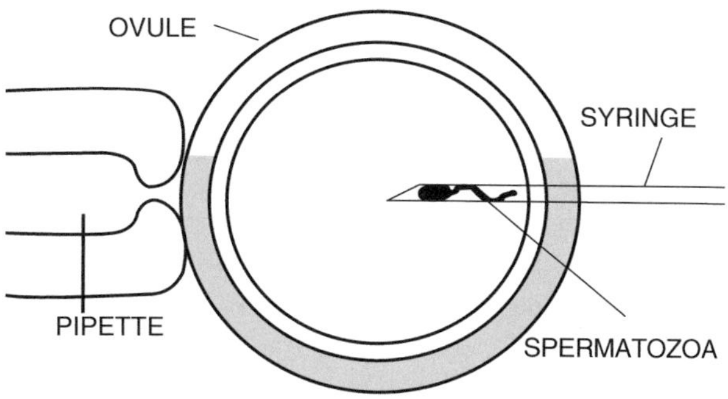

Fig. 26.7 Intracytoplasmic sperm injection

risks of the procedure, it is still debatable whether ICSI should be exclusively used for all the patients in place of the conventional IVF method.

In addition, concerns regarding disruption of chromosomal or cytoskeletal elements or fertilization consequences with genetically abnormal sperm remain to be a matter of further discussion and research. Approximately, half of the 7% of infertile men that harbor major sex chromosome abnormality accounts for a mosaic Klinefelter condition. The incidence of sex chromosome abnormality rises from 2% in men with normal sperm concentration to 20% in those with azoospermia (Baker et al. 1993). Despite these data and an apparently high risk for chromosome abnormalities in ICSI fetuses, Bonduelle and colleagues found the risk of chromosomal abnormalities to be approximately 1%, similar to the general newborn population (Bonduelle et al. 1996). There is a growing concern of germ line mutation that may result in heritable defects because ICSI allows fertilization by sperm, which under natural conditions is incapable of ZP penetration and oocyte-sperm fusion. The inheritance of susceptibility to infertility is another concern that remains unrecognized until late in the next generation. Once sexing of the spermatozoa becomes routinely available, prevention of sex-linked diseases may be prevented by selecting the healthier gender. Thus, during the ICSI procedure careful evaluation, genetic consultation with the couples, as well as follow-up of the pregnancies, is necessary.

Five important steps in the ICSI procedure involve:

1. The sperm sample is either surgically removed from the testes or epididymis or taken from male partner's semen.
2. Eggs are collected by surgical method from hormonally induced ovarian follicles. Single motile sperm is injected carefully from male partner into meiotically matured egg of the female partner by using a tiny hollow needle.
3. The fertilized egg is observed for growth and development after injection.
4. Once the normal growth is seen, the presumptive embryo is delivered into the female uterus where it has a chance to implant and grow.

26.7.3 Advantage and Disadvantages of ICSI

Although ICSI has become an established ART procedure, however, many concerns have been raised on the resultant embryos and children over its potential detrimental effects. First and foremost, ICSI is a procedure, where the sperm cells are directly introduced into an egg which effectively eliminates male infertility. One of the major concerns is that ICSI bypasses the natural selection of sperm for fertilization and so the sperm having defects that would have been prevented from fertilizing oocytes may do so with the aid of ICSI that ultimately passes the defects on to the next generation. Apart from facilitating the transmission of genetic defects, the sperm injection process may inevitably cause physical damage to the oocyte that finally interferes with subsequent embryo development. The ICSI-generated embryos have been found to less likely attain the blastocyst stage in vitro and have a greater chance of developing fragments in comparison with embryos from the conventional IVF method. Despite the observed and potential detrimental effects on the embryos at the genetic and cellular levels, ICSI has not been associated with increased incidence of birth defects (Van Steirteghem et al. 2002; Hansen et al. 2002). Public concerns over the safety of ART were raised again by a report showing a higher incidence of birth defects in IVF babies (Hansen et al. 2002). The limited information available suggests that ICSI children may have a small delay in mental development although it is unknown if this mental impairment is caused by the ICSI procedure or by factors inherent to the patients who require ICSI in the first place (Bowen et al. 1998). Obviously, further studies on the long-term effects of ICSI on the offspring involving multiple centers with well-controlled study designs are needed to minimize confounding variables, such as operator/technical variations and population variations. Until conclusive data become available, patients should be counseled carefully before ICSI is offered as an ART treatment.

26.8 Surrogacy

In severe cases of infertility, the infertile couple may choose surrogacy. In surrogacy, the infertile couple does the legal contract with fertile women in which fertile woman becomes pregnant and gives birth to a child. If the surrogates used their own egg for the fertilization purposes, then the condition is referred to as "genetic surrogate." On the other hand, if embryos are generated using another woman's eggs and then implanted into the surrogate, condition is referred to as "gestational surrogate" and has no genetic tie with the child. From the last decade, it has been seen that hiring a surrogate becomes a business to earn money worldwide. In the United States alone, for surrogacy, ART clinic can charge $40,000–$100,000, including the surrogate fee, insemination or IVF costs, and costs related to medical care, transportation, and legal services. Due to the high cost, recently it has been seen that some of the couples started to hire women in the developing countries. In Indian subcontinent, hiring a surrogate costs from $5000 to $12,000, and the surrogate gets paid $3000–$6000.

26.9 Regulation of ART

In the United States alone, it is estimated that ART is a $3–5-billion-dollar industry. Considering the modern lifestyle and increasing infertility rate, ART clinics are rapidly expanding including egg brokers, sperm banks, and surrogacy services worldwide. As many ethical, social, and critical issues are associated with the ARTs, it is necessary that ART clinics must be regulated by the government agencies; otherwise, it would be at risk. There also must be some international law which can regulate individuals, couples, and ART clinics to find the quality services whether in their own country or another.

26.10 Risks in ART

Beyond the fact that ARTs offer the possibility for infertile couples to attain pregnancy, however, it poses potential risk in health issue for the mother as well as the infant. As in majority of ART procedures, multiple embryos are transferred, and there is a risk of multiple gestational pregnancy and multiple births. The risk of multiple births at maternal interface includes high rates of cesarean deliveries, maternal hemorrhage, pregnancy-related high blood pressure, and gestational diabetes. At fetal interface, it includes prematurity, low birth weight, infant death, elevated risk for birth defects, and developmental disability. Further, even singleton infants conceived with ART have a higher risk for low birth weight compared with singleton infants conceived with normal procedures. If during ART a maximum of two embryos are transferred rather than multiple embryos, the risk of high-order multiple births can be restricted. For patients who are seeking ART, twin pregnancy can be treated as necessary but manageable complication of infertility treatment. Double embryo transfer for the patient undergoing ART may be an option provided their health is good and they are in proper condition to conceive a twin pregnancy for 34 weeks and wish to have more than one child. ART programs should not be penalized for providing patients the option of double embryo transfer, by not counting twin births when reporting IVF "success." Nevertheless, IVF and ICSI technologies have revolutionized the treatment of male infertility with new hope of having their own genetic offspring as well as disease-free newborns.

Conclusion

ART has become one of the widely accepted and most desirable technologies since last one decade. ART is a rising hope to millions of couples facing the problem of infertility. In the coming years, advancing technology is likely to exacerbate ethical, legal, and social concerns associated with ART. Further, due to the rapidly evolving nature of the ART, legislation is often unable to keep pace and address all of the ethical and legal issues that are constantly emerging in the field. It is therefore incumbent upon physicians to continuously monitor these issues and ensure that ART technologies are offered and delivered in a manner that balances patient care with social and moral responsibility. Furthermore, medical professionals should be keenly aware of their professional as well as social and ethical responsibilities in the pursuit of technical advancement. Of course, the latest advances in ART have not only enhanced the possibility of pregnancy but have also made today's women conceive in situations which would not have been possible decades ago.

Acknowledgment Authors are thankful to Banaras Hindu University for providing necessary platform to write this chapter and also to Springer for providing opportunity to review the status and outcome of ART.

References

Baker HWG, Liu DY, Bourne H, Lopata A (1993) Assisted fertilization: diagnosis of sperm defects in selecting patients for assisted fertilization. Hum Reprod 8(11):1779–1780

Bonduelle M, Legein J, Buysse A, Van Assche E, Wisanto A, Devroey P, Van Steirteghem AC, Liebaers I (1996) Prospective follow-up study of 423 children born after intracytoplasmic sperm injection. Hum Reprod 11(7):1558–1564

Bowen JR, Gibson FL, Leslie GI, Saunders DM (1998) Medical and developmental outcome at 1 year for children conceived by intracytoplasmic sperm injection. Lancet 351(9115): 1529–1534

Dodson WC, Haney AF (1991) Controlled ovarian hyperstimulation and intrauterine insemination for treatment of infertility. Fertil Steril 55(3):457–467

Ghazzawi IM, Sarraf MG, Taher MR, Khalifa FA (1998) Comparison of the fertilizing capability of spermatozoa from ejaculates, epididymal aspirates and testicular biopsies using intracytoplasmic sperm injection. Hum Reprod 13(2):348–352

Hansen M, Kurinczuk JJ, Bower C, Webb S (2002) The risk of major birth defects after intracytoplasmic sperm injection and in vitro fertilization. N Engl J Med 346(10):725–730

Kemmann E, Bohrer M, Shelden R, Fiasconaro G, Beardsley L (1987) Active ovulation management increases the monthly probability of pregnancy occurrence in ovulatory women who receive intrauterine insemination. Fertil Steril 48(6):916–920

Kim ED, Lipshultz LI (1997) Male subfertility: diagnostic and therapeutic advances. Br J Urol 80(4):633–641

Palermo GD, Cohen J, Alikani M, Adler A, Rosenwaks Z (1995) Intracytoplasmic sperm injection: a novel treatment for all forms of male factor infertility. Fertil Steril 63(6):1231–1240

Serhal PF, Katz M, Little V, Woronowski H (1988) Unexplained infertility—the value of Pergonal superovulation combined with intrauterine insemination. Fertil Steril 49(4):602–606

Tournay H, Camus M, Khan I, Staessen C, Van Steirteghem AC, Devroey P (1991) In-vitro fertilization, gamete-or zygote intra-fallopian transfer for the treatment of male infertility. Hum Reprod 6(2):263–266

Tournaye H, Devroey P, Camus M, Staessen C, Bollen N, Smitz J, Van Steirteghem AC (1992a) Comparison of in-vitro fertilization in male and tubal infertility: a 3 year survey. Hum Reprod 7(2):218–222

Tournaye H, Devroey P, Camus M, Valkenburg M, Bollen N, Van Steirteghem AC (1992b) Zygote intrafallopian transfer or in vitro fertilization and embryo transfer for the treatment of male-factor infertility: a prospective randomized trial. Fertil Steril 58(2):344–350

Van Steirteghem A, Liu J, Nagy Z, Joris H, Tournaye H, Liebaers I, Devroey P (1993) Assisted fertilization: use of assisted fertilization. Hum Reprod 8(11):1784–1785

Van Steirteghem A, Bonduelle M, Liebaers I, Devroey P (2002) Children born after assisted reproductive technology. Am J Perinatol 19(02):059–066

MIX
Papier aus verantwortungsvollen Quellen
Paper from responsible sources
FSC® C105338

If you have any concerns about our products,
you can contact us on
ProductSafety@springernature.com

In case Publisher is established outside the EU,
the EU authorized representative is:
Springer Nature Customer Service Center GmbH
Europaplatz 3, 69115 Heidelberg, Germany

Printed by Libri Plureos GmbH
in Hamburg, Germany